Progress in Speech Synthesis

Springer
New York
Berlin
Heidelberg
Barcelona
Budapest
Hong Kong
London
Milan
Paris
Santa Clara
Singapore
Tokyo

Jan P.H. van Santen Richard W. Sproat
Joseph P. Olive Julia Hirschberg
Editors

Progress in Speech Synthesis

With 158 Illustrations

Springer

Jan P.H. van Santen
Bell Laboratories
Room 2D-452
600 Mountain Avenue
Murray Hill, NJ 07974-0636 USA

Richard W. Sproat
Bell Laboratories
Room 2D-451
600 Mountain Avenue
Murray Hill, NJ 07974-0636 USA

Joseph P. Olive
Bell Laboratories
Room 2D-447
600 Mountain Avenue
Murray Hill, NJ 07974-0636 USA

Julia Hirschberg
AT&T Research
Room 2C-409
600 Mountain Avenue
Murray Hill, NJ 07974-0636 USA

Library of Congress Cataloging-in-Publication Data
Progress in speech synthesis/Jan P.H. van Santen . . . [et al.], editors.
p. cm.
Includes bibliographical references and index.
ISBN 0-387-94701-9 (hardcover: alk. paper)
1. Speech synthesis. I. Santen, Jan P.H. van.
TK7882.S65S6785 1996
006.5′4–dc20 96-10596

Printed on acid-free paper.

Production managed by Natalie Johnson; manufacturing supervised by Jacqui Ashri.
Camera-ready copy prepared using LaTeX.
Printed and bound by Edwards Brothers, Inc., Ann Arbor, MI.
Printed in the United States of America.

9 8 7 6 5 4 3 2 1

ISBN 0-387-94701-9 Springer-Verlag New York Berlin Heidelberg SPIN 10524632

Preface

Text-to-speech synthesis involves the computation of a speech signal from input text. Accomplishing this requires a system that consists of an astonishing range of components, from abstract linguistic analysis of discourse structure to speech coding.

Several implications flow from this fact. First, text-to-speech synthesis is inherently multidisciplinary, as is reflected by the authors of this book, whose backgrounds include engineering, multiple areas of linguistics, computer science, mathematical psychology, and acoustics. Second, progress in these research areas is uneven because the problems faced vary so widely in difficulty. For example, producing flawless pitch accent assignment for complex, multi-paragraph textual input is extremely difficult, whereas producing decent segmental durations given that all else has been computed correctly is not very difficult. Third, the only way to summarize research in all areas relevant for TTS is in the form of a multi-authored book—no single person, or even small group of persons, has a sufficiently broad scope.

The most important goal of this book is, of course, to provide an overview of these research areas by having invited key players in each area to contribute to this book. This will give the reader a complete picture of what the challenges and solutions are and in what directions researchers are moving.

But an important second goal is to allow the reader to judge what all this work adds up to—the scientific sophistication may be impressive, but does it really produce good synthetic speech? The book attempts to answer this question in two ways. First, we have asked authors to include results from subjective evaluations in their chapter whenever possible. There is also a special section on perception and evaluation in the book. Second, the book contains a CD-ROM disk with samples of several of the synthesizers discussed. Both the successes and the failures are of interest—the latter in particular because it is unlikely that samples are included demonstrating major flaws in a system. We thank Christian Benoit for suggesting this idea.

A brief note on the history of this volume: In 1990, the First ESCA Workshop on Text-to-Speech Synthesis was organized in Autrans, France. The organizers of this workshop, Gerard Bailly and Christian Benoit, felt that there was a need for a book containing longer verions of papers from the workshop proceedings, resulting in *Talking Machines*. In 1994, the editors of the current volume organized the Second ESCA/IEEE/AAAI Workshop on Text-to-Speech Synthesis and likewise decided that a book was necessary that would present the work reported in the proceedings in a more complete, updated, and polished form. To ensure the highest possible quality of the chapters for the current volume, we asked members of the scientific committee of the workshop to select workshop papers for possible inclusion. The editors added their selections, too. We then invited those authors whose work received an unambiguous endorsement from this process to contribute to the book.

Finally, we want to thank the many people who have contributed: the scientific committee members who helped with the selection process; fourteen anonymous reviewers; Bernd Moebius for work on stubborn figures; Alice Greenwood for editorial assistance; Mike Tanenblatt and Juergen Schroeter for processing the speech and video files; David Yarowsky for providing the index; Thomas von Foerster and Kenneth Dreyhaupt at Springer-Verlag for expediting the process; Cathy Hopkins for administrative assistance; and Bell Laboratories for its encouragement of this work.

Jan P. H. van Santen
Richard W. Sproat
Joseph P. Olive
Julia Hirschberg

Murray Hill, New Jersey
October 1995

Contents

Contributors

Masanobu Abe, Speech and Acoustic Laboratory, NTT Human Interface Laboratories, 1-2356 Take Yokosuka-Shi, Kanagawa 238-03, Japan (ave@nttpsch.hil.ntt.jp)

Ali Adjoudani, Institut de la Communication Parle'e, INPG/ENSERG–Université Stendhal, BP 25X, 38040 Grenoble Cedex 9, France (adjouani@icp.grenet.fr)

Christophe R. d'Alessandro, LIMSI-CNRS, PO BOX 133, Orsay, F-91403, France (cda@limsi.fr)

Plínio Almeida Barbosa, IEL/Unicamp, CP 6045, 13081-970 Campinas-SP, Brazil (plinio@iel.unicamp.br)

Benjamin Ao, First Byte, 19840 Pioneer Ave, Torrance, CA 90503, USA (bao@firstbyte.davd.com)

Véronique Aubergé, Institut de la Communication Parleé, UA CNRS 368–INPG/Université Stendhal, 46, av. Félix Viallet, 38031 Grenoble Cedex, France (auberge@icp.grenet.fr)

Gérard Bailly, Institut de la Communication Parleé, UA CNRS 368–INPG/Université Stendhal, 46, av. Félix Viallet, 38031 Grenoble Cedex, France (bailly@icp.grenet.fr)

Frédéric Beaugendre, Hearing and speech department, Institute for Perception Research (IPO), PO Box 513, 5600 MB Eindhoven, The Netherlands (beaugend@prl.philips.nl)

Mary E. Beckman, Ohio State University, Department of Linguistics, 222 Oxley Hall, 1712 Neil Ave., Columbus, OH 43210, USA (mbeckman@ling.ohio-state.edu)

Karim Belhoula, Lehrstuhl für Allgemeine Elektrotechnik und Akustik, Ruhr-Universität Bochum, Universitätsstr. 150, Bochum, 44801, Germany (belhoula@aea.ruhr-uni-bochum.de)

Christian Benoît, Institut de la Communication Parleé, INPG/ENSERG–Université Stendhal, BP 25X, 38040 Grenoble Cedex 9, France (benoit@icp.grenet.fr)

Corine A. Bickley, Research Laboratory of Electronics, Massachusetts Institute of Technology, 77 Massachusetts Avenue, Cambridge, MA 02139, USA, and Sensimetrics Corp., 26 Landsdowne St., Cambridge, MA 02139, USA (bickley@speech.mit.edu)

Alan W. Black, ATR Interpreting Telecommunications Laboratories, 2-2 Hikaridai, Seika-cho, Soraku-gun, Kyoto 619-02, Japan (awb@itl.atr.co.jp)

Antal P. J. van den Bosch, MATRIKS, Maastricht University, P.O. Box 616, 6200 MD Maastricht, The Netherlands (antal@cs.rulimburg.nl)

Nick Campbell, ATR Interpreting Telecommunications Laboratories, 2-2 Hikaridai, Seika-cho, Soraku-gun, Kyoto 619-02, Japan (nick@itl.atr.co.jp)

Michael M. Cohen, Program in Experimental Psychology, 433 Clark Kerr Hall, University of California–Santa Cruz, Santa Cruz, CA 95064 USA (mcohen@fuzzy.ucsc.edu)

John S. Coleman, Oxford University Phonetics Laboratory, 41 Wellington Square, Oxford, OX1 2JF, UK (John.Coleman@Phonetics.Oxford.ac.UK)

René Collier, Institute for Perception Research, P.O. Box 513, 5600 MB Eindhoven, The Netherlands (collier@natlab.research.philips.nl)

Alistair D. Conkie, Centre for Speech Technology Research, University of Edinburgh, 80 South Bridge, Edinburgh EH1 1HN, Scotland (adc@cstr.ed.ac.uk)

Peter-Arno J. M. Coppen, Department of Language and Speech, University of Nijmegen, P.O. Box 9103, 6500 HD Nijmegen, The Netherlands (coppen@let.kun.nl)

Walter M. P. Daelemans, Computational Linguistics, Tilburg University, PO Box 90153, 5000 LE Tilburg, The Netherlands (Walter.Daelemans@kub.nl)

Arthur Dirksen, Institute for Perception Research, P.O. Box 513, 5600 MB Eindhoven, The Netherlands (adirksen@prl.philips.nl)

José Gregorio Escalada-Sardina, Speech Technology Services, Telefonica I+D, Emilio Vargas, 6, 28043 Madrid, Spain (goyo@craso.tid.es)

Giuliano Ferri, New Markets Department–Advanced Technologies and Applications, IBM Semea, Piazzale G. Pastore, 6, 00144 Rome, Italy (ferrig@vnet.ibm.com)

Marcus K. D. Filipsson, Dept. of Linguistics and Phonetics, University of Lund, Helgonabacken 12, S-223 62, Lund, Sweden (Marcus.Filipsson@ling.lu.se)

Stefan Frisch, Department of Linguistics, Northwestern University, 2016 Sheridan Road, Evanston, IL 60202, USA (s-frisch@nwu.edu)

Bertrand Le Goff, Institut de la Communication Parlee, Universite Stendhal, 1180, avenue Centrale, 38400 Saint Martin d'Heres, France (legoff@icp.grenet.fr)

Nina Gronnum, Institute of General and Applied Linguistics, University of Copenhagen, 80 Njalsgade, DK-2300 Copenhagen, Denmark (ng@cphling.dk)

Thierry Guiard-Marigny, Institut de la Communication Parlee, Université Stendhal, 1180, avenue Centrale, 38400 Saint Martin d'Heres, France (guiard@icp.grenet.fr)

Carlos H. M. Gussenhoven, Department of English, University of Nijmegen, P.O. Box 9103, 6500 HD Nijmegen, The Netherlands (u260003@vm.uci.kun.nl)

Wolfgang J. Hess, Institut für Kommunikationsforschung und Phonetik, Universtät Bonn, Poppelsdorfer Allee 47, 53115 Bonn, Germany (wgh@ikp.uni-bonn.de)

Norio Higuchi, Department 2, ATR Interpreting Telecommunications Research Laboratories, 2-2, Hikaridai, Seika-cho, Soraku-gun, Kyoto, 619-02, Japan (higuchi@itl.atr.co.jp)

Toshio Hirai, Department 2, ATR Interpreting Telecommunications Research Laboratories, 2-2, Hikaridai, Seika-cho, Soraku-gun, Kyoto, 619-02, Japan (thirai@itl.atr.co.jp)

Julia Hirschberg, Human-Computer Interface Research Department, AT&T Research, 600 Mountain Avenue, Murray Hill, NJ 07974 USA (julia@research.att.com)

Florian Höfer, Institut für Kommunikationsforschung und Phonetik, Universtät Bonn, Poppelsdorfer Allee 47, 53115 Bonn, Germany (fho@ikp.uni-bonn.de)

Merle A. Horne, Dept. of Linguistics and Phonetics, University of Lund, Helgonabacken 12, S-223 62, Lund, Sweden (Merle.Horne@ling.lu.se)

Stephen Isard, Centre for Speech Technology Research, University of Edinburgh, 80 South Bridge, Edinburgh EH1 1HN, Scotland (S.Isard@ed.ac.uk)

Naoto Iwahashi, Research Center, SONY, 6-7-35 Kitashinagawa, Shinagawa-ku, Tokyo, 141, Japan (naoto@av.crl.sony.co.jp)

Ute Jekosch, Lehrstuehl fuer Allgemeine Elektroakustik and Akustik, Ruhr-Universität Bochum, D-44780 Bochum, Germany (jekosch@aea.ruhr-uni-bochum.de)

Dan Kahn, MCC-1C253R, Bellcore, 445 South Street, Morristown, NJ 07960, USA (dk@thumper.bellcore.com)

Klaus J. Kohler, Institut fur Phonetik und digitale Sprachverarbeitung, Universität Kiel, D-24098 Kiel, Germany (kk@ipds.uni-kiel.de)

Regina Krüger, Philips Kommunikations Industrie AG, Postfach 3538, Nürnberg, 90327, Germany (krueger@pki-nbg.philips.de)

Marianne Kugler, Philips Kommunikations Industrie AG, Postfach 3538, Nürnberg, 90327, Germany (kugler@pki-nbg.philips.de)

Andrej Ljolje, Linguistics Research Department, AT&T Research, 600 Mountain Avenue, Murray Hill, NJ 07974, USA (alj@research.att.com)

John Local, Department of Language and Linguistic Science, University of York, Heslington, York, Y01 5DD, UK (lang4@york.ac.uk)

Marian J. Macchi, MCC-1C221R, Bellcore, 445 South Street, Morristown, NJ 07960, USA (mjm@thumper.bellcore.com)

Domenic W. Massaro, Program in Experimental Psychology, 433 Clark Kerr Hall, University of California–Santa Cruz, Santa Cruz, CA 95064 USA (massaro@fuzzy.ucsc.edu)

Erwin C. Marsi, Department of Language and Speech, University of Nijmegen, P.O. Box 9103, 6500 HD Nijmegen, The Netherlands (marsi@let.kun.nl)

Piet Mertens, Department of Linguistics–Centre for Computational Linguistics, K.U. Leuven, Blijde-Inkomststraat 21, Leuven, B-3000, Belgium (Piet.Mertens@arts.kuleuven.ac.be)

Bernd Möbius, Linguistics Research Department, Bell Laboratories, 600 Mountain Avenue, Murray Hill, NJ 07974, USA (bmo@bell-labs.com)

Christine H. Nakatani, Aiken Computation Laboratory, Harvard University, 33 Oxford Street, Cambridge, MA 02138, USA (chn@das.harvard.edu)

Sieb G. Nooteboom, Research Institute for Language and Speech, Utrecht University, Trans 10, 3512 JK Utrecht, The Netherlands (sieb.nooteboom@let.ruu.nl)

Richard Ogden, Department of Language and Linguistic Science, University of York, Heslington, York, Y01 5DD, UK (rao1@york.ac.uk)

Joseph P. Olive, Linguistics Research Department, Bell Laboratories, 600 Mountain Avenue, Murray Hill, NJ 07974, USA (jpo@bell-labs.com)

Luís C. Oliveira, INESC/IST, Rua Alves Redol 9, 1000 Lisboa, Portugal (lco@inesc.pt)

Joseph S. Perkell, Research Laboratories of Electronics, Massachusetts Institute of Technology, Building 36, Room 591, 50 Vassar Street, Cambridge, MA 02139, USA (perkell@speech.mit.edu)

Janet B. Pierrehumbert, Department of Linguistics, Northwestern University, 2016 Sheridan Road, Evanston, IL 60202, USA (jbp@nwu.edu)

Piero Pierucci, New Markets Department–Advanced Technologies and Applications, IBM Semea, Piazzale G. Pastore, 6, 00144 Rome, Italy (mc7784@mclink.it)

Jan Roelof de Pijper, Institute for Perception Research (IPO), P.O. Box 513, 5600 MB Eindhoven, The Netherlands (pyper@prl.philips.nl)

David B. Pisoni, Speech Research Laboratory, Indiana University, Department of Psychology, Bloomington, IN, 47405, USA (pisoni@indiana.edu)

Louis C. W. Pols, Institute of Phonetic Sciences, University of Amsterdam, Herengracht 338, 1016 CG Amsterdam, The Netherlands (pols@fon.let.uva.nl)

Thomas Portele, Institut für Kommunikationsforschung und Phonetik, Universität Bonn, Poppelsdorfer Allee 47, 53115 Bonn, Germany (tpo@ikp.uni-bonn.de)

Gaël Richard, Center for Computer Aids for Industrial Productivity (CAIP), Rutgers University, P.O. Box 1390, Piscataway, NJ 08855, USA (gael@caip.rutgers.edu)

Toni C. M. Rietveld, Department of Language and Speech, University of Nijmegen, P.O. Box 9103, 6500 HD Nijmegen, The Netherlands (rietveld@let.kun.nl)

Miguel Ángel Rodríguez-Crespo, Speech Technology Services, Telefonica I+D, Emilio Vargas, 6, 28043 Madrid, Spain (miguel@craso.tid.es)

Hans-Wilhelm Rühl, Philips Kommunikations Industrie AG, Postfach 3538, Nürnberg, 90327, Germany (hwr@pki-nbg.philips.de)

Yoshinori Sagisaka, Department 1, ATR Interpreting Telecommunications Research Laboratories, 2-2 Hikaridai Seika-cho, Soraku-gun, Kyoto, 619-02, Japan (sagisaka@itl.atr.co.jp)

Jan P. H. van Santen, Linguistics Research Department, Bell Laboratories, 600 Mountain Avenue, Murray Hill, NJ 07974, USA (jphvs@bell-labs.com)

Donatella Sanzone, New Markets Department–Advanced Technologies and Applications, IBM Semea, Piazzale G. Pastore, 6, 00144 Rome, Italy (dsanzone@selfin.it)

Juergen Schroeter, Linguistics Research Department, AT&T Research, 600 Mountain Avenue, Murray Hill, NJ 07974, USA (jsh@bell-labs.com)

Luis Monzón Serrano, Speech Technology Services, Telefonica I+D, Emilio Vargas, 6, 28043 Madrid, Spain (luis@craso.tid.es)

Chilin Shih, Linguistics Research Department, Bell Laboratories, 600 Mountain Avenue, Murray Hill, NJ 07974, USA (cls@bell-labs.com)

Richard W. Sproat, Linguistics Research Department, Bell Laboratories, 600 Mountain Avenue, Murray Hill, NJ 07974, USA (rws@bell-labs.com)

Kenneth N. Stevens, Electrical Engineering and Computer Science Department, and Research Laboratory of Electronics, Massachusetts Institute of Technology, 77 Massachusetts Avenue, Cambridge, MA 02139, USA, and Sensimetrics Corp., 26 Landsdowne St., Cambridge, MA 02139, USA (stevens@speech.mit.edu)

David T. Talkin, Entropic Research Laboratory, Suite 202, 600 Pennsylvania Ave. S.E., Washington, D.C. 20003, USA (dt@entropic.com)

Pilar Sanz Velasco, Speech Technology Services, Telefonica I+D, Emilio Vargas, 6, 28043 Madrid, Spain (pilar@craso.tid.es)

Colin W. Wightman, Electrical Engineering Department, New Mexico Institute of Mining and Technology, Campus Station, Socorro, NM 87801, USA (cww@ee.nmt.edu)

Reiner F. Wilhelms-Tricarico, Research Laboratories of Electronics, Massachusetts Institute of Technology, Building 36, Room 565, 50 Vassar Street, Cambridge, MA 02139, USA (reiner@speech.mit.edu)

David R. Williams, Sensimetrics Corp., 26 Landsdowne St., Cambridge, MA 02139, USA (williams@sens.com)

David Yarowsky, Department of Computer and Information Science, University of Pennsylvania, 200 S. 33rd St., Philadelphia, PA 19104-6389, USA (yarowsky@unagi.cis.upenn.edu)

Section I

Signal Processing and Source Modeling

1

Section Introduction. Recent Approaches to Modeling the Glottal Source for TTS

Dan Kahn
Marian J. Macchi

1.1 Modeling the Glottal Source: Introduction

One of the most important discoveries in speech science was the finding that voiced speech, which accounts for the major part of most utterances, could be fairly well modeled by a pulsive *source* function serving as excitation for an all-pole *transfer* function. The source function represents airflow through the glottis, and the transfer function captures the filtering effect of the vocal tract.

The all-pole transfer function can be implemented as the concatenation of four or five second-order IIR filters, or resonators, corresponding to the formants, or resonances, typically seen in speech spectrograms. With proper choice of coefficients, a single $2n$th-order IIR filter can be made equivalent to n second-order resonators, and if the order of such a filter is chosen to be p, somewhat greater than $2n$, and if its coefficients are chosen by a best-match criterion to real speech (as in linear predictive coding, LPC [Atal and Hanauer]), then some of the poles will typically be found to reside on the real axis, that is, to provide some overall spectral tilt rather than additional resonances. It is sometimes even observed that the best-match criterion assigns several pole-pairs (resonances) in a way as to model a *zero* (antiresonance) in the spectrum.

These nonformant manifestations in the transfer function are generally seen as being due to the approximate nature of the source plus all-pole-filter model. In fact, it was long common for purposes of text-to-speech (TTS) synthesis and digital transmission of speech to represent the glottal source as a single impulse per pitch period (i.e., as spectrally flat) and rely on some of the poles in the pth-order IIR filter to capture the spectral content of the glottal pulse.

In the case of digital speech coding, in which speech is simply analyzed, then resynthesized on a frame-by-frame basis, this may be a reasonable approach in principle, because if p is made high enough, the resulting filter can hide a multitude of glottal-modeling sins. However in TTS synthesis, even if one is starting from recorded speech units, the situation is more difficult because of the need to implement changes in pitch and duration. The high-order spectral model that represents well the actual speech that was analyzed may be inappropriate for prosodically altered speech. For example, the poles may model the actual harmonics, rather than just formants; so when the fundamental frequency (F_0) is changed, it becomes inappropriate to continue having poles at those old harmonic positions.

Also, we know that the glottal source function varies with prosody. But in systems of the type under consideration, since source influences on the spectrum are incorporated into the vocal-tract function, these influences will not change as prosody is varied. As an alternative to high-order modeling, the glottal source may be derived from inverse filtering and used as the excitation signal in synthesis. However, the resultant source may be inappropriate for prosodically altered speech if prosodic alterations are effected by truncating or extending the pitch period in a simple way.

Further, even if glottal source parameters are extracted from the inverse filtered signal according to some theoretically motivated model of glottal behavior, these parameters represent well the actual speech that was analyzed, but may be inappropriate for prosodically altered speech.

In addition, the simple model of the preceding paragraphs views speech as either fully periodic (to be represented as outlined above) or having no periodic structure (can be modeled as filtered noise). However, we know that this model is oversimplified. For example, there are speech sounds like voiced fricatives that have both periodic voicing and frication noise from the supraglottal constriction. We also know that even vowels often have appreciable amounts of glottally induced noise superimposed on the basic periodic waveform.

For these reasons, then, seekers of high-quality TTS have rejected the simple noise-free, single-pulse-per-pitch-period model of the glottal source.

1.2 Alternatives to Monopulse Excitation

Although the high-order modeling discussed above can represent speech well for transmission, it is not an efficient approach, due to the need to transmit the many coefficients. Workers in speech coding have found an approach that uses either *multi-pulse* or *LPC-residual* excitation [Atal and Remde, Caspers and Atal] and a relatively small p to produce better speech quality at transmission rates of interest.

Multipulse excitation is found through a perceptual and bit-rate optimization technique and does not typically correspond to observed glottal functions. The LPC residual[1] does perhaps somewhat better in this regard, but it too provides far from accurate representations of glottal waveforms, due to the approximate nature of LPC's all-pole assumption. Nevertheless, improved speech quality in

[1]Given a segment of speech and a pth-order LPC analysis thereof, the LPC residual is the source function that, when passed through the pth-order filter, will exactly reproduce the speech signal. Practical coders use a simplified representation of the residual.

TTS is obtainable through use of the multipulse or LPC-residual approach, with prosodic modifications achieved through simple modifications of the multipulse or residual functions [Macchi et al]. Another recent alternative approach to improving the speech quality has been the use of a time-domain technique such as PSOLA [Moulines and Charpentier].

However, many workers in speech synthesis believe that we will never achieve human-like synthesis quality until we take seriously the task of accurately modeling the glottal source. Through actual physical airflow and pressure measurements, much is known about the the typical shape of the glottal waveform [Klatt and Klatt]. The chapters in this section make serious attempts to model the glottal source and thereby achieve more natural synthetic speech. In addition, several of the chapters argue that parameters of the glottal source vary with linguistic structure and therefore must be controlled by rule in order to achieve fully natural-sounding speech in a TTS system.

1.3 A Guide to the Chapters

As a first approximation, the glottis can be regarded as a source of periodic excitation, but we know that there is often an appreciable noise component and other aperiodicities in the glottal signal. The chapters in this section all relate to this general observation; three of the chapters attempt to explicitly account for a noise component in their glottal models, and the fourth studies the "glottal stop" phenomenon, whereby the expected periodic pulse stream is slowed and stopped.

In chapter 4 by Richard and D'Alessandro, a quite complex analysis of the glottal source is proposed. The core of their approach is a decomposition of an approximation of the source signal into periodic and aperiodic components. The approximation is obtained through inverse filtering and the decomposition through an iterative Fourier/cepstral technique. The authors argue for the importance of maintaining the temporal coherence of the aperiodic component, which they achieve through a bandpass analysis with formant/bandwidth estimation and preservation of the temporal envelope in each passband.

Chapter 3 by Oliveira also views the glottis as a source of both periodic and noise excitation, but differs from the previous chapter in elaborating an existing model for these components rather than proposing a new one. Oliveira seeks to analyze a corpus of natural speech in order to determine the parameters of the model of the glottal excitation function proposed by Rosenberg (1971) with modifications by Klatt and Klatt (1990). The key parameters of this model can be expressed as: the *open quotient* (the fraction of a pitch period for which the glottis is open), the *aspiration ratio* (the fraction of total amplitude due to noise), and a parameter characterizing a low-pass filter that serves to smooth the transition between open and closed phases of the glottis. Once these parameters have been extracted from the natural-speech database, Oliveira presents histograms of their values and, more interestingly, demonstrates a correlation between their values and segment duration, which is directly applicable to parameter-value control in synthesizers incorporating a realistic glottal source model.

Like the previous chapters, the one by Crespo et al. takes seriously the observation that even those sections of speech normally viewed as periodic often have considerable noise components. But rather than attempt to explicitly model the glottal source function, Crespo et al. argue for use of a sinusoidal analysis in modeling the speech signal. Their model is a simplified version of models proposed by Quatieri and McCauley (see the several references in chapter 5) in which harmonic-only sinusoids are assumed up to a cut-off frequency, above which non-harmonic sinusoids come into play. Although there is no explicit glottal waveform generator, and the source amplitudes are subsumed into the system (vocal-tract) amplitudes, the model does derive independent phase contributions for source and system, facilitating the natural modification of prosodic features in synthesis.

Chapter 2 by Pierrehumbert and Frisch differs from the others in not presenting a new speech model or studying model parameters in a general way. Rather, Pierrehumbert and Frisch investigate in detail a particular source-based phenomenon, the glottal stop. The chapter shows that the occurrence and degree of "glottalization" is dependent on both segmental (phoneme-based) and suprasegmental (prosodic) factors. It also demonstrates that the same Klatt glottal model that we saw in chapter 3 can be used to represent the glottalization observations for purposes of synthesis. Chapter 2 also shows that, contrary to the traditional view that segmental features are cued acoustically by spectral properties, and suprasegmental features are cued by F_0, what is considered to be a *segmental* effect (the glottal stop) can be cued acoustically solely by F_0.

1.4 Summary

The "classical" model of voiced speech—a simple pulse generator and all-pole filter—is an excellent first approximation but has well-known limitations. Each of the chapters in this section relies on a more complex model, in particular giving up the simplistic simulation of the glottal source in order to achieve more natural synthetic speech in which prosody-induced changes are required, as in text-to-speech synthesis. Further, the implication of these chapters is that TTS systems will need to incorporate rules to control the glottal source in order to approach natural-sounding speech quality.

References

[Klatt and Klatt] D. H. Klatt and L. C. Klatt. Analysis, synthesis and perception of voice quality. *J. Acoust. Soc. Amer.* 87:820–857, 1990.

[Atal and Hanauer] B. S. Atal and S. L. Hanauer. Speech analysis and synthesis by linear prediction of the speech wave. *J. Acoust. Soc. Amer.* 50:637–655, 1971.

[Atal and Remde] B. S. Atal and J. R. Remde. A new model of LPC excitation for producing natural-sounding speech at low bit rates. *Proceedings International*

Conference on Acoustics, Speech, and Signal Processing. ICASSP-82, 614–617, 1982.

[Caspers and Atal] B. E. Caspers and B. S. Atal. Changing pitch and duration in LPC synthesized speech using multipulse excitation. *J. Acoust. Soc. Amer.* suppl. 1, 73:S5, 1983.

[Macchi et al] M. J. Macchi, M J. Altom, D. Kahn, S. Singhal, and M. Spiegel. Intelligibility as a function of speech coding method for template-based speech synthesis, Eurospeech. In *Proceedings Eurospeech '93*, 893–896, 1993.

[Moulines and Charpentier] E. Moulines and F. Charpentier. Pitch-synchronous waveform processing techniques for text-to-speech synthesis using diphones. *Speech Communication* 9:453–467, 1990.

[Rosenberg] A. Rosenberg. Effect of glottal pulse shape on the quality of natural vowels. *J. Acoust. Soc. Amer.* 49:583–590, 1971.

2

Synthesizing Allophonic Glottalization

Janet B. Pierrehumbert
Stefan Frisch

ABSTRACT This chapter presents a method for synthesizing allophonic glottalization. The method is motivated by empirical studies of the phonological context for glottalization and of its acoustic consequences. A baseline study of production explored glottalization in two situations: (1) vowel-vowel hiatus across a word boundary, and (2) voiceless stops before sonorants. The study showed that allophonic glottalization depends on the segmental context, syllabic position, and phrasal prosody. Successful synthesis of contextually appropriate glottalization requires an architecture with a running window over a fully parsed phonological structure, or its effective equivalent. The signal coding used was based on the source model and cascade formant synthesis presented by [Kla87]. Synthesis of glottalization can be achieved by lowering the fundamental frequency (F_0), keeping all other factors in formant synthesis constant. Thus, any synthesis procedure that has the ability to directly control F_0 will be able to reproduce glottalization in a similar manner. For fully natural, theoretically correct synthesis, additional control parameters are needed to control the length of the glottal pulse and for spectral tilt.

2.1 Introduction

A traditional view of speech distinguished between speech segments and suprasegmentals (e.g., intonation, stress, and phrasing); suprasegmentals were considered to be manifested in f_0, duration, and amplitude whereas segments were responsible for spectral characteristics and voicing. A large body of experimental work has demonstrated the existence both of suprasegmental influences on the spectrum and of segmental influences on f_0, duration, and amplitude; nonetheless, this viewpoint is still implicitly reflected in the architecture of many speech synthesis systems. One consequence is that the interaction of suprasegmental and segmental factors is not accurately modeled, with a resulting penalty in the naturalness of the result. For synthetic speech to sound completely natural, it will be necessary to model the ramifications of all aspects of phonological structure for all phonetic parameters.

This chapter presents a case study of this interaction. The segment we examine is the consonant produced by adducting the vocal folds. Following common practice we will transcribe it as the glottal stop, /ʔ/, despite the fact that a full stop, defined by absence of voicing or silence, is found only in the most extreme realizations. More typically, there is merely a disturbance in the voice quality. /ʔ/ is of particular interest because it is executed using the larynx, and thus presents an example of a segmental effect on the voice source. In general, voice source allophony has been underinvestigated compared to other types of allophony, but there is increasing recognition of its importance.

/ʔ/ arises in two different ways in English. As shown by the pronunciation of words such as "night-rate" and "Danton," voiceless stops can be glottalized in the context of sonorant consonants; syllable final /t/ in particular is very reliably glottalized and often loses its oral articulation altogether. In addition, /ʔ/ is often supplied as a syllable onset to words beginning in a vowel, especially when the words are utterance initial or following a vowel, as in the phrase "Calgary ʔairfield."

The two prosodic factors we examine are syllable structure and phrasal stress. The relevance of syllable structure is demonstrated by the fact that /t/ is affricated rather than glottalized in "nitrate," where it is in the onset. Extending the results of [PT92], we show that phrasal stress affects the reliability and extent of glottalization. We also demonstrate an interaction between syllable structure and stress: Stress has more effect on the onset of the syllable than on the coda.

This chapter presents both experimental results establishing the contexts for glottalization, and results on synthesis of glottalization. Glottalization was synthesized using a formant synthesizer (a modification of the *formsy* program from Entropic Research Laboratory) driven by a voice source parameterized according to [Kla87]. The results argue for using a physically perspicuous coding of speech, as Klatt and others have proposed. They also have implications for the control structure in speech synthesis, as discussed further below.

2.2 Experimental Data

In order to determine the factors controlling the existence and extent of glottalization of voiceless stops in English, an experimental study was conducted. A more extensive description of the methods and results can be found in [Pie95a, Pie95b]. The study examined both the glottalization of the voiceless stops /t/ and /p/ and the glottalization found on vowel initial words. The voiceless stops were put in varying segmental contexts by constructing assorted compounds in which the first word ended in a stop and the next began with /m/, /r/, /w/, /l/, /st/, /f/, or a stop. Some examples of the compounds used (with the target region indicated with an underscore) are: "print maker," "pep rally," "foot wear," "sip lid," "chop sticks," "Flint stones," and "paint brush." Vocalic contexts were created by placing V-initial words after words ending in the syllable /ri/, for example, "Calgary airfield" or "cherry armrests." All target words or compounds had initial stress. The compounds were recorded in continuous speech by embedding them in discourse segments that ended in open-ended lists, as in [1], the first sound sample on the CDROM of this study (see Appendix).

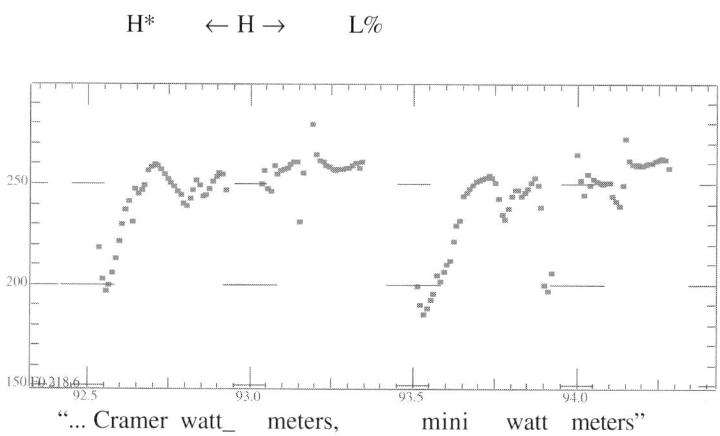

FIGURE 2.1. Example of the H*HL% pitch contour.

[1] We offer the latest looks from head to toe. We have designer knits, high-fashion foot wear, creative accessories ....

Because of the extended contexts, subjects were not aware of which aspect of their speech was under investigation. This particular discourse context was selected because it tends to elicit the intonation pattern transcribed as H*HL% in the ToBI transcription system [PBH94], which has a sustained f_0 value toward the middle of the pitch range in the area of interest. This was done to eliminate extraneous glottalization associated with low f_0. One example of this pattern is shown in figure 2.1. The phrasal stress in the target region was manipulated by controlling the novelty of the information in the list items. For example, [1] has a nuclear accent on the word "footwear," whereas in [2], the nuclear accent on "high-fashion" causes "footwear" to be deaccented (second sound sample on the CDROM).

[2] This store has every kind of footwear for the whole family. It has athletic footwear, high-fashion foot wear, everyday foot wear...

High-quality recordings of two speakers were made. (One of these speakers also participated in the synthesis experiment presented in section 2.3.) Because glottalization is produced by adducting the vocal folds, its primary acoustic hallmarks are irregular and long pitch periods (with f_0 going as low as 40 Hz for both male and female speakers), a sharp closure of the vocal folds, which boosts the upper spectral frequencies, and to some extent a reduction of formant bandwidths (See [Ana84, Gob89, Lad93, PT92]). These acoustic features were examined using Entropic's *waves+* speech software. The presence of irregular pitch periods, indicating glottalization, was easily seen on a speech waveform (see figure 2.3). In questionable cases, spectrograms were used to examine the upper frequencies and formant bandwidths for additional evidence of glottalization. Evaluation of the extent of glottalization was done subjectively, as an objective measure of glottalization is difficult to define. There are two major obstacles to providing an exact measure for glottalization. First, the boost in high frequencies and the reduced bandwidths of the formants are not always found, and second, the long and irregular pitch periods indicative of glottalization can result from other causes [PT92, Pie95a]. Thus, data on the degree of glottalization was analyzed by pairwise comparisons of minimal pairs involving stress, and the effects of stress were checked for significance using a sign test, as discussed in [Pie95a].

There were three major findings. First, contrary to previous reports, glottalization of /t/ was found before sonorant consonants only, not before fricatives or stops. Glottalization of /p/ occurred only before nasals. Second, the /ʔ/ inserted as an onset to vowel initial words showed much more glottalization under nuclear stress than in postnuclear position. Finally, the degree of glottalization of coda stops was not sensitive to phrasal stress, thus presenting a contrast to the behavior of /ʔ/ in onset position.

These results are very similar to the pattern of glottalization for German [Koh94]. For example, the German /ʔ/ can be used in place of syllable final plosives before sonorant consonants. Also, German frequently uses glottalization in a vowel-vowel hiatus between words. Kohler also reported that word initial glottalization can replace supraglottal coda consonants such as /t/ and /k/ from the preceding word. As in English, word initial stressed vowels are glottalized more frequently and to a greater degree than unstressed ones. Thus we see a pattern that is very close to that of English. German appears to be more aggressive in its use of word initial /ʔ/ before vowels in all contexts, and this possible difference between glottalization in English and German warrants further study.

2.3 Synthesis Experiments

Modeling glottalization in synthesis requires modeling both the context in which it occurs and its acoustic consequences. The experimental results presented above show that the relevant contextual factors are the segmental context, even across a word boundary; the phrasal prosody; and the position in the syllable. The remainder of this section explains, in detail, our model of the acoustic consequences of glottalization. Experimentation with the model suggests that there is a single necessary and sufficient acoustic factor: an abrupt reduction in f_0 to below normal speaking levels. Two other acoustic correlates of glottalization are a reduced amplitude and a flattened source spectrum [Gob89]. However, these correlates are not sufficient, in and of themselves, to signal glottalization. They should be used in a theoretically correct model of glottalization. We also found that glottalization can be synthesized equally well for male and female voices using the same methods.

2.3.1 *Materials*

To collect examples of glottalized and nonglottalized speech for use in analysis and synthesis, we made new recordings of a male speaker (SF) and a female speaker (ST). New data were obtained rather than using data from the previous experiment described above in order to study and resynthesize near minimal pairs involving allophonic glottalization. The new sentences were constructed so that the target regions in the minimal pairs would have similar phonetic and prosodic contexts. We used two segmental contexts of glottalization, vowel-vowel hiatus and /t/ before sonorant. An example of glottalization in a vowel-vowel hiatus contrasting with a lack of glottalization is shown in the third sound sample, [3].

[3] The crane couldn't lift the heavy oak.
The water buffalo couldn't lift the heavy yoke.

Similar sentences were also constructed for the use of /t/ before a sonorant. The sentences were read in a randomized order to avoid contrastive stress. The talkers were instructed to speak at a normal to above average rate. The complete set of near-minimal pairs (shown in [4]) was designed to give examples of glottalization in various vocalic contexts for ST (fourth sound sample), and for SF (fifth sound sample).

[4]	twenty years	twenty ears
	steady yawning	steady awning
	heavy yoke	heavy oak
	flue maker	flute maker
	sea liner	seat liner
	spa removal	spot removal

Figures 2.2 and 2.3 display the acoustic contrast between "heavy yoke" and "heavy oak," which was to be reproduced. As is evident from the spectrograms 2.2a and 2.2b, which show the sonorant regions of "heavy yoke" and "heavy oak" from /v/ to /k/, the formant transitions are very similar. Arrows indicate the region

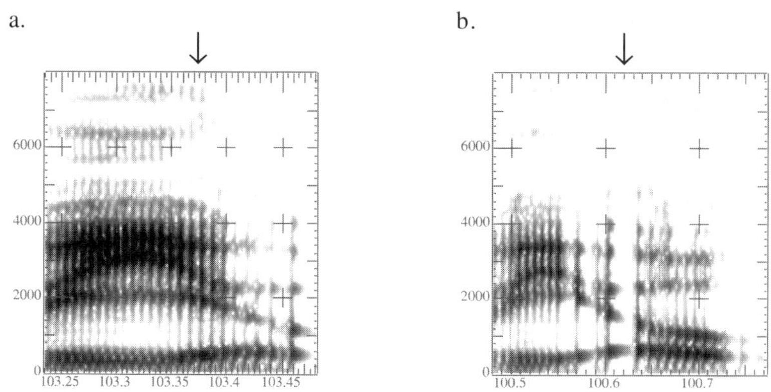

FIGURE 2.2. Spectrograms of the regions from /v/ to /k/ in "heavy yoke" and "heavy oak."

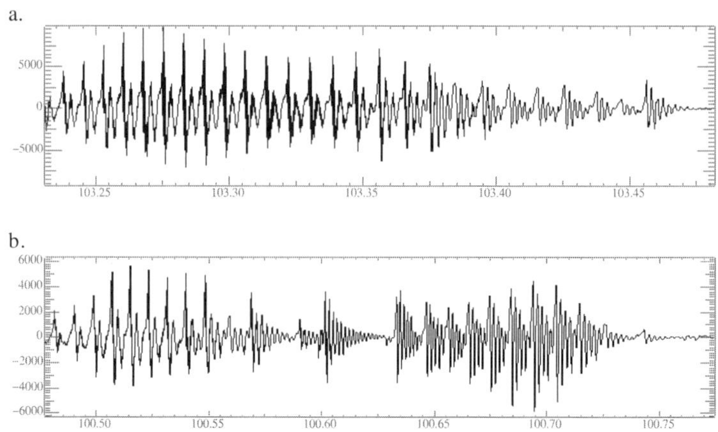

FIGURE 2.3. Waveforms of "heavy yoke" and "heavy oak" from /v/ to /k/.

in which a difference in the excitation pattern is superimposed on the formant transitions. The contrasting waveforms in this region are shown in figure 2.3. The long irregular pitch periods at the onset of "oak," which are characteristic of glottalization, are conspicuous in figure 2.3b.

As is apparent from figure 2.2, a true minimal pair for source characteristics is not possible. The /y/ target is more fully achieved in "heavy yoke," as /y/ is present as both an offglide and the onset to the following syllable. Because of coarticulation with the onset, the second formant for /o/ is higher than in the "heavy oak" case. In addition, the formant transitions are steeper for "heavy oak" than for "heavy yoke", because the /y/ is present only as an offglide in "heavy oak." Similarly, if the (optional) coronal articulation is carried out in compounds such as "seat liner," it results in formant transitions that are absent from "sea liner." Nevertheless, each pair of cases is similar enough to permit a demonstration of the effects of source characteristics alone. A fully natural text-to-speech system must model both the source effects and the coarticulation effects (see section 2.5 for additional discussion).

In our data, the new recordings as well as those from the previous experiment show that the disturbance of f_0, which is characteristic of glottalization, was as long as 0.14 seconds for both SF and ST. The disturbance covered 2 to 3 phonemes. f_0 dropped as low as 27 Hz for SF and 43 Hz for ST.

To determine which acoustic factors are necessary for perceiving /ʔ/, we used these recorded minimal pairs as data for synthesis. In particular, we took original recordings with no /ʔ/, such as "heavy yoke" and doctored the source characteristics to produce "heavy ʔoak." In addition, we resynthesized original recordings with /ʔ/ for comparison. The remainder of this section details our synthesis model and procedure.

2.3.2 Model

We modeled the shape of the glottal pulse (or equivalently, the shape of the source spectrum) using the complete excitation parameterization reviewed in [Kla87]. The synthesis contained in the demonstration was created by manipulating only the source parameters available in this model.

Synthesis was done with Entropic's *waves+* speech software, using the *Formsy* utility and a custom excitation function based on the polynomial flow model [Ros71], supplemented by a low-pass filter to control spectral tilt, as proposed in [Kla87]. *Formsy* is a formant based cascade synthesizer, which has smoothly varying parameters that are updated at 10 ms intervals. Our synthesis used four formants obtained with the *formant* utility, and a baseline f_0 contour obtained with *get_f0*. All formant and fundamental frequency values were hand checked for errors. In addition, we used two constant high-pole formants to correct the spectral balance and improve the synthesis quality.

We modeled the glottal flow (U_g) during the open part of the glottal period based on the equation $U_g = aT^2 - bT^3$. The constants a and b are determined by the time of peak excitation (T_e), and the maximum glottal flow (U_{gmax}) by the following equations. Since T_e is the moment of glottal closure, it represents the length of the open period of the glottal pulse.

$$a = \frac{27U_{gmax}}{4T_e^2} \qquad b = \frac{a}{T_e} \tag{2.1}$$

The wave produced by these equations is then smoothed by a low-pass filter, which produces a more realistic glottal flow wave. The resulting waveform is then combined with a noise component, which can be used to enhance a breathy voice quality, or to add an aspiration source, as in the synthesis of /h/ in "heavy yoke." In our synthesis, a smaller bandwidth and larger noise component was used for ST, to more accurately reflect the female voice quality [Kar90]. A relatively high bandwidth and little noise was used for SF, who has a pressed voice. Since glottal allophones involve adduction, which reduces losses to the subglottal system, they are very pressed and less breathy than the ordinary voice quality for the speaker. Thus, we used a high bandwidth for the low-pass filter and no breathiness noise during glottalization.

2.3.3 Method

To produce glottalization, alterations to the source parameters were made over a 0.1 s interval. We began with source parameters based on the study of different voice qualities in [Gob89], and our own observations of the recorded tokens with /ʔ/. We then used trial and error to match the parameters to each talker's voice quality. In our first experiment, we altered three source characteristics simultaneously. We lowered fundamental frequency by increasing the length of the closed period of the glottal cycle. Amplitude was lowered by reducing the value of U_{gmax}, which has the consequence of reducing the effective excitation, U_g' [Ana84]. The source

TABLE 2.1. Synthesis parameters for SF's "heavy /ʔ/oak" and "heavy yoke."

Frame	1	2	3	4	5	6	7	8	9	10	11	12
f_0 (Hz)	131	75	74	72	69	35	35	69	72	74	75	124
amp (arbitrary units)	2	1.8	1.5	1.2	0.9	0.9	0.9	0.9	1.5	2.1	2.7	3.3
bandwidth (kHz)	9	9	12	15	16	16	16	16	15	12	9	9
T_e (ms)	5	5	4.5	4	3.5	2.5	2.5	3.5	4	4.5	5	5
f_0 (Hz)	131	129	128	127	127	127	131	132	127	127	126	124
amp (arbitrary units)	2	1.9	1.7	1.7	1.7	2	2.6	2.8	2.7	3.1	3.1	3.3
bandwidth (kHz)	9	9	9	9	9	9	9	9	9	9	9	9
T_e (ms)	5	5	5	5	5	5	5	5	5	5	5	5

spectrum was flattened by reducing the length of the open period of the glottal pulse (T_e) and by increasing the bandwidth of the low-pass filter in the Klatt excitation model.

The first section of table 2.1 gives the values used in synthesizing "heavy oak" from "heavy yoke" for SF. The second section of the table gives the values used to resynthesize "heavy yoke" without modification. Each frame represents 0.01 s. Frames 1 and 12 are unaltered frames based on the original recording. Frames 2 to 11 covered the target range for glottalization, which was aligned to overlap 0.03s (three frames) with the following vowel. Thus, the most glottalized portion is in the transition and syllable onset, and some of the effect of glottalization is realized on the preceding vowel. Figure 2.4 is a graphic representation of the changes in parameter values shown in table 2.1. The values of f_1 and f_2 (which were not altered) are displayed at the top of figure 2.4, to show how the source parameter variation is aligned with the speech.

The same f_0 contour, pulse length, and spectral tilt bandwidth were used to synthesize all six examples for SF. The target amplitude varied in each case. The amplitude contour was computed by taking one-fourth of the average amplitude of the endpoints as the target amplitude (in rms) for the frames that had the lowest f_0. Thus, the minimum amplitude was used over four frames in the middle of the target area. The remaining values were filled in by linear transitions (in rms) to each endpoint.

Similar parameters were used to resynthesize "heavy oak" from "heavy yoke" for ST (table 2.2). ST had a higher f_0 overall, 1.5 to 2 times that of SF, and a higher minimum f_0 was observed in the glottalized examples of ST's speech. Thus, a minimum f_0 of 50 Hz was used in place of SF's 35 Hz. Also, we used a more gradual transition, with an additional intermediate step, to avoid too abrupt a change in f_0. A shorter pulse length was used for the female voice to accommodate the higher f_0.

Once again, the same f_0 contour was used to synthesize all examples for ST. The same pulse length and tilt bandwidth were used, but the amplitude varied and was computed separately for each example as it was for SF.

Figure 2.5 shows an example of the source signal used to synthesize "heavy yoke" and the doctored "heavy /ʔ/oak" using all of the acoustic correlates of glottalization. Figure 2.5a shows the reconstruction of the glottal flow for resynthesis

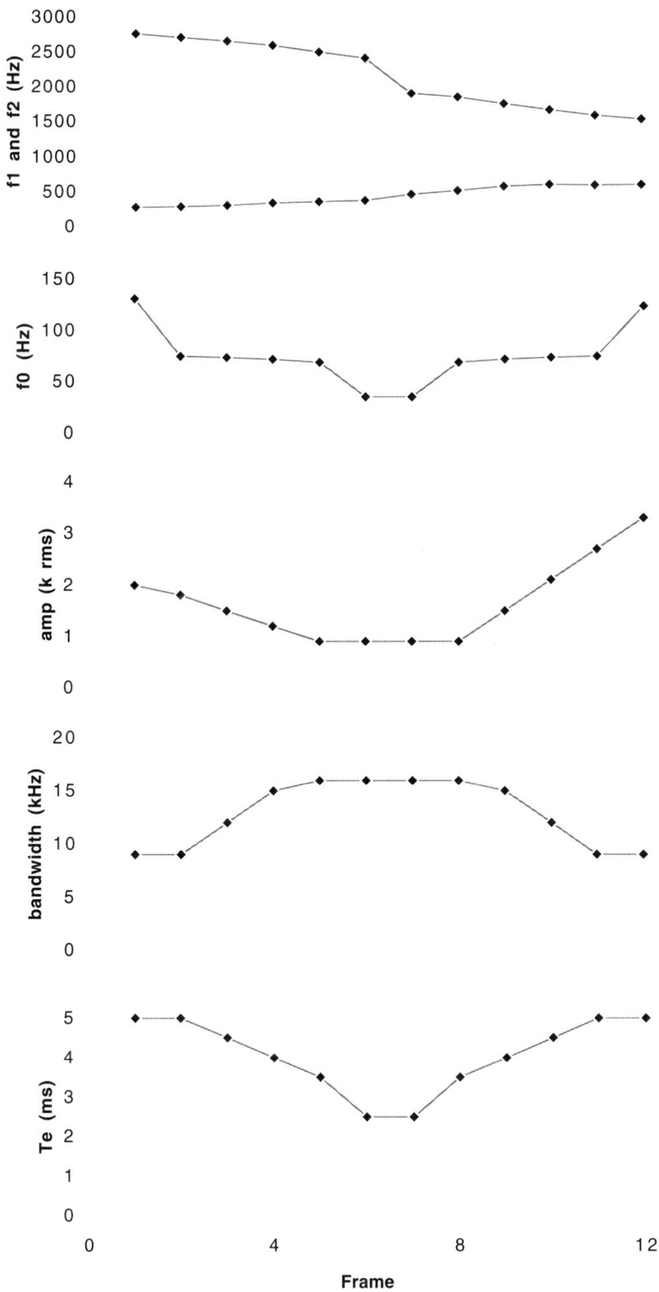

FIGURE 2.4. Contours of formants, f_0, amplitude, pulse length, and tilt bandwidth for SF.

TABLE 2.2. Synthesis parameters for ST's "heavy /ʔ/oak" and "heavy yoke."

Frame	1	2	3	4	5	6	7	8	9	10	11	12
f_0 (Hz)	184	150	100	98	95	50	50	95	98	100	150	200
amp (arbitrary units)	0.8	0.7	0.6	0.4	0.3	0.2	0.2	0.5	0.6	0.9	1.0	1.2
bandwidth (kHz)	5	5	11	15	16	16	16	16	15	11	5	5
T_e (ms)	3.5	3.5	3	2.5	2	1	1	2	2.5	3	3.5	3.5
f_0 (Hz)	184	178	181	182	184	192	196	196	201	201	203	200
amp (arbitrary units)	0.8	0.5	0.5	0.4	0.4	0.5	0.9	1.0	1.0	1.1	1.2	1.2
bandwidth (kHz)	5	5	5	5	5	5	5	5	5	5	5	5
T_e (ms)	3.5	3.5	3.5	3.5	3.5	3.5	3.5	3.5	3.5	3.5	3.5	3.5

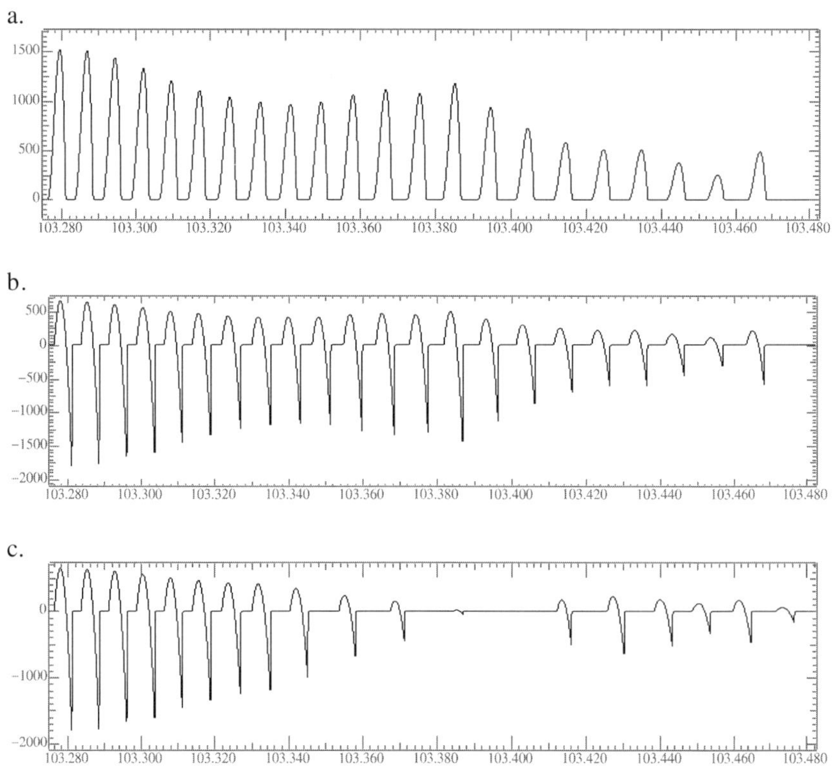

FIGURE 2.5. From top to bottom: glottal flow (U_g) and excitation (U_g') for original "heavy yoke" and excitation for doctored "heavy yoke" to produce /ʔ/. Horizontal axis denotes time lines).

without modification, using the parameters in the bottom of table 2.1. Figure 2.5b shows the flow derivative, which is the effective excitation [Ana84]. Figure 2.5c shows the modification to the flow derivative, which gives the impression of a glottal stop, using the parameters in the top half of table 2.1 and shown also in figure 2.4.

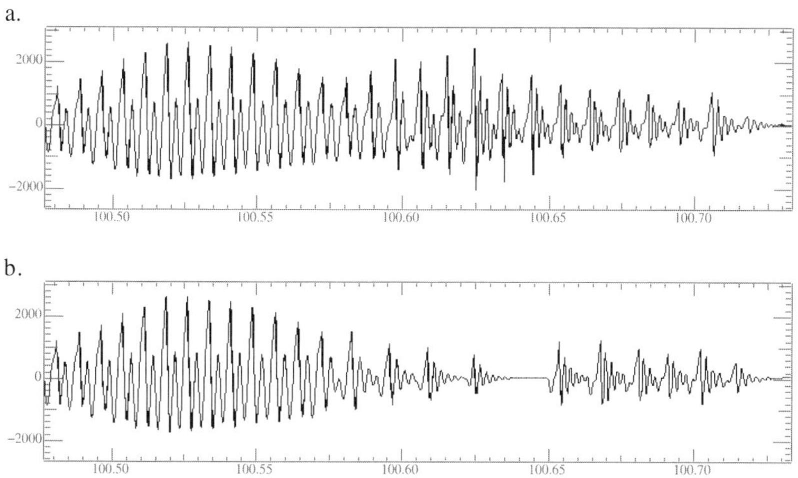

FIGURE 2.6. Resynthesized original "heavy yoke" and doctored "heavy yoke" with /ʔ/ added.

To resynthesize the original recordings that had a glottal stop such as "heavy oak," we used the original amplitude and f_0 contours derived by computer analysis. In addition, we added appropriate source spectrum characteristics by reducing the pulse length (T_e) and raising the bandwidth of the smoothing filter. These additions flattened the source spectrum in the same manner as the doctored examples. All of the synthesized examples for each speaker are arranged in ordered triplets on the demonstration, as shown in [5], the sixth sound sample, for ST, and the seventh sound sample for SF. The first example is the resynthesized example with no glottal stop. The second example uses the same formants as the first, but has had the source characteristics doctored in the region of the word boundary to produce /ʔ/. The third example is the original example with the glottal stop, resynthesized for comparison. Figure 2.6 shows the waveforms for resynthesized "heavy yoke" and doctored "heavy yoke" in the region from /v/ to /k/ (cf. figure 2.3).

[5]	twenty years	twenty years (doc)	twenty ears
	steady yawning	steady yawning (doc)	steady awning
	heavy yoke	heavy yoke (doc)	heavy oak
	flue maker	flue maker (doc)	flute maker
	sea liner	sea liner (doc)	seat liner
	spa removal	spa removal (doc)	spot removal

We listened to the resynthesized utterances that originally contained a glottal stop and also to the utterances in which a glottal stop had been introduced by doctoring the source signal to simulate glottalization. Informal listening indicated that the doctored utterances sounded like they contained glottal stops. That is, the "source doctoring" was sufficient to convey the perception of glottalization.

2.4 Contribution of Individual Source Parameters

While glottalization, in theory, should be modeled using a combination of lowered f_0, reduced amplitude, and flattened source spectrum [Gob89], our experiments in synthesizing /ʔ/ suggest that lowering the f_0 is sufficient, as we will describe.

To assess the relative contribution of each of the source parameters to the perception of glottalization, we synthesized selected examples for both ST and SF with each of the source parameters altered individually. That is, we produced four versions of the doctored source signals, and used these to produce doctored utterances: one in which only the source amplitude was reduced, one in which only the source spectrum was flattened (via both a raised filter bandwidth and shortened pulse length), one in which only the f_0 was lowered, and one with all three alterations. The resulting excitations are illustrated in figure 2.7.

Informal listening showed that the versions in which only the f_0 was lowered sounded as though they contained glottal stops, whereas the versions with the doctored amplitude and spectral tilt did not. That is, the dominant cue for /ʔ/ is sharply lowered f_0. The reduced f_0 alone was sufficient to cue /ʔ/, although the quality was judged to be more natural in the doctored version in which all three parameters were altered.

The demonstration material presents selected examples for both ST and SF with each of the source characteristics altered individually (eighth and ninth sound samples). These examples are recorded in quadruplets shown in [6]. The first example has only the amplitude reduced, by reducing the value of U_{gmax} in the excitation, as was done previously. The second example has only the source spectrum flattened, by shortening the pulse length and increasing the bandwidth of the low-pass filter. In this case, the amplitude of the glottal wave also had to be reduced to compensate for the increased excitation caused by a shorter glottal pulse that would otherwise result from these modifications. This correction was carried out by hand so that the amplitude contour of the result matched that of the original as closely as possible. The third example has only the f_0 lowered. The final example has all three alterations for comparison.

[6] heavy yoke (amp) heavy yoke (tilt) heavy yoke (f_0) heavy yoke (all)
flue maker (amp) flue maker (tilt) flue maker (f_0) flue maker (all)

2.5 Discussion

Glottalization is an important allophonic variation in English. In order to synthesize natural-sounding speech in, for example, a text-to-speech (TTS) system, the results presented in this chapter will have to be incorporated into the TTS system.

It is well known that speech segments exert some influence on f_0. Our results on coda /t/ before sonorants show that this segment can be realized solely as a disturbance in the source, with the f_0 disturbance being by far the most important effect. This seriously undermines the traditional distinction between "segmental"

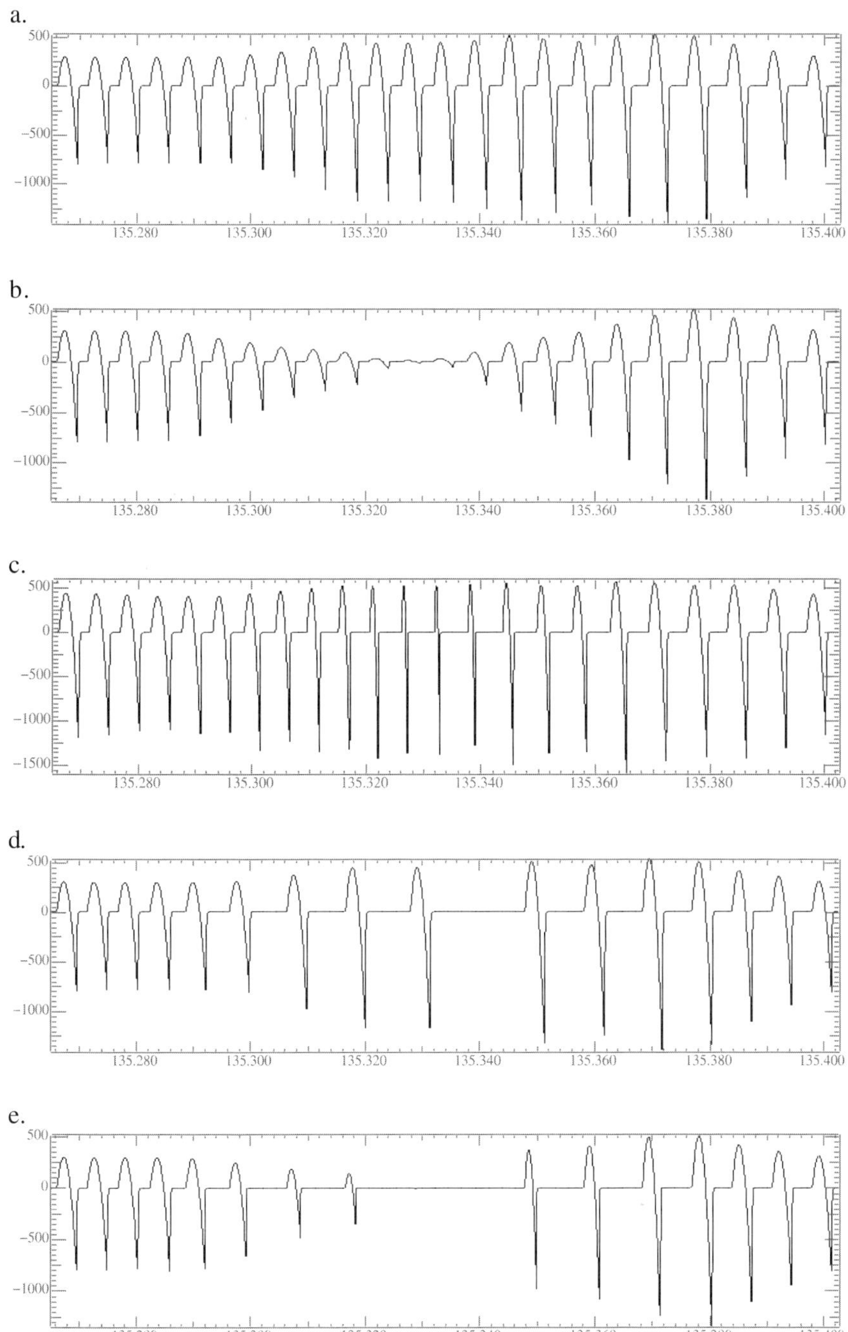

FIGURE 2.7. Excitation (U'_g) for "heavy yoke" with individual source characteristics altered. (a) is the excitation for resynthesis without modification; (b) has reduced amplitude; (c) has a flattened spectrum; (d) has lowered f_0; (e) has all alterations made (note that there is a minute pitch pulse at 135.324).

and "suprasegmental" acoustic properties of speech. The architecture of speech synthesis systems should support mapping of any phonological parameter onto any phonetic parameter.

Additional consequences for system architecture arise from the fact that the system must be able to identify the contexts in which glottalization is found and model its effects. Recall from section 2.2 that allophonic /ʔ/ is found in coda /t/ before sonorant sequences, even across word boundaries, and in a vowel-vowel hiatus. Further, the prosodic context influences the strength of the phenomenon. In a faithful TTS implementation of the glottalization rule, the phonological structure will have to be parsed up to the intonation phrase, and the rule will have to have access to both the immediate segmental context and the prosodic strength of the surrounding syllables. This information is available, for example, if the system is implemented with a running window on fully parsed phonological structures, as in [PB88]. In a running window model, elements (phonemes or tones) are implemented one after another in temporal order, mimicking the production of speech in time. The realization of any particular element depends on what it is, what its prosodic position is, and what appears in its near environment. Although the hierarchical structure renders properties of large units more "visible" at a greater distance than properties of small units, the model does not support generalized nonlocal computations. For example, the model asserts that the pronunciation of phrase-medial /t/ cannot depend on what phoneme ends the last word in the phrase.

By contrast, in a head-based implementation model [Col92, Loc92, DC94], phonological structures with no explicit temporal orientation are phonetically evaluated based on hierarchical organization. Heads are implemented before nonheads. The result is that codas are implemented after nuclei, and onsets are implemented after rhymes. Temporal order is derived in the course of the implementation; for example, the implementation of the onset is timed to occur before the maximal expression of the nucleus.

Consider glottalization of /t/ before a sonorant, as in the case of "seat liner." The /t/ is a coda consonant, and thus will be implemented as a dependent of the nucleus of its syllable, /i/, before the onset of that syllable is implemented. "Seat" is the prosodic head of the compound. Hence, under the head-based implementation, the entire syllable "seat" will be processed before "liner," which contains the following sonorant in a weak syllable; the environment is crucial in triggering glottalization. Notice that no matter what the relative prosodic positions of the syllables involved, the two syllables will be processed independently. The only way to implement the glottalization rule would be at the level of the prosodic unit including both words in the compound, where the relative strength of both syllables is available for processing. Any rule of this kind, which applies between adjacent tokens of prosodic units at one hierarchical level, must be implemented at the next level up, no matter how local the rule may be.

Head-wrapping implementation leads to several specific problems in implementing glottalization and other allophonic processes like it. First, it is cumbersome to determine the rule target from the prosodic level at which the rule must be implemented. The rule targets the coda of one syllable and the onset of the next syllable, regardless of their places within the prosodic structure. Thus, there are many possible paths through the prosodic hierarchy from the level at which glottalization applies to the possible target segments, and these paths are all equivalent in determining where glottalization can apply. Although the relative prominence of the syllables influences the likelihood and the degree of glottalization, it does not influence the segmental context to which the rule may apply.

Second, the head-based approach introduces unnecessary nonmonotonicity into the computational system. Consider the head-based implementation in the case where /t/ is realized only as a disturbance in voice quality, as in our synthesis of "seat liner" from "sea liner." In this case, there are no formant transitions from the coronal articulation. A head-based implementation would compute coronal transitions when determining the rhyme of "seat" (this would be necessary in order to supply the transitions in more generic medial positions, as in "seat assignment" or "seat sales"). These transitions would have to be removed at the level containing the entire compound, where the glottalization rule is implemented. As a result, the computation is nonmonotonic, and the head-based implementation is less constrained than a running window implementation, which can compute glottalization monotonically by bringing to bear all relevant contextual factors at the same time.

Third, because the head-based implementation of glottalization must rely on complex paths downward through the prosodic structure to identify the target phoneme and the relevant context, it appears that there is no principled way to exclude rules of segmental allophony depending on arbitrary nonlocal segmental relations. That is, adjacency that cross-cuts the hierarchical structure simply has no special status under this approach. For example, a possible rule would be to glottalize /t/ in the onset of one syllable based on the sonority of the coda of the following syllable. There is no mechanism that forces the kind of locality we find in the glottalization rule. Once again, the running window implementation is a more constrained system for phonetic implementation.

Similar difficulties for the head-wrapping approach emerge when considering examples of glottalization or full /ʔ/ insertion in a vowel-vowel hiatus. Recall that phrasal stress on a vowel initial word greatly increases the likelihood and extent of glottalization, both in English and in German [Koh94, Pie95a]. In cases where a full /ʔ/ is achieved, a great deal of the accompanying glottalization is realized as creaky voice in the *preceding vowel*, as shown in figure 2.8. Because the preceding vowel is in the rhyme of the preceding syllable, it will be processed separately from the following syllable, and the same problems arise from the interaction of the prosodic conditioning with the local context.

Any post hoc repair designed to model the glottalization rule would have to involve examining the segmental context as well as the surrounding prosodic context in both directions, and would thus be equivalent to the results obtained using a running implementation window, as discussed above. In summary, any workable treatment of allophonic glottalization in the head-based implementation is more cumbersome and less constrained than the running window implementation.

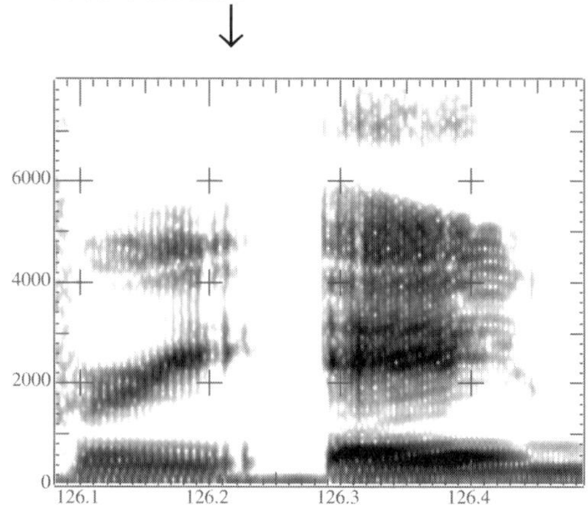

FIGURE 2.8. Spectrogram of "savory ʔeggrolls." The arrow indicates the effects of the word initial /ʔ/ on the /i/ of "savory."

Our results also have implications for the signal coding used in TTS systems. To properly model the effects of the glottalization rule, a TTS system would need direct control of f_0 as an independent parameter over a great range. In our experiments, we used the Klatt model, which takes f_0 directly as a synthesis parameter. Similarly, any articulatory-based synthesis that has direct control of the excitation function can capture the effects we have presented. Modeling /ʔ/ using waveform concatenation of diphones, as in the pitch synchronous overlap add (PSOLA) system [MC90], is more difficult. Test cases in which f_0 was reduced by 0.7 of its original value in a PSOLA model caused signal degradation [BYB94]. In this study recognition dropped from near perfect to 65% to 75%, with the lowest values occurring with resynthesis of speakers with very high f_0. It is unclear whether the degradation was due to a signal that was unacceptable in any context or conceivably to a successful but contextually inappropriate example of glottalization. Further research on ths issue is needed.

2.6 Summary

This chapter shows that glottalization is contingent on both segmental and prosodic factors. Successful synthesis of glottalization can be achieved by manipulation of the source characteristics alone in a Klatt model of the glottal wave. The resulting synthesis clearly shows that the f_0 contour is the predominant cue for glottal allophones. Other factors, such as reduced amplitude and flattened source spectrum, are at most contrast enhancing. These factors can be ignored in favor of computational simplicity, or can be included if a theoretically correct and possibly more natural model of speech is the desired result.

In addition, we argue that incorporation of glottalization into a TTS system requires simultaneous access to phonemic segments and prosodic structure, as proposed in [PB88]. In order to properly model glottalization, a speech synthesizer must have the ability to make extreme modifications to f_0 without loss of naturalness.

Acknowledgments: We would like to thank Michael Broe, Marian Macchi, and David Talkin for their comments on an earlier draft of this paper. Thanks also to David Talkin and the Entropic Research Laboratory, Inc. for providing the necessary resources to implement formant synthesis based on the Rosenberg/Klatt model. This work was supported by NSF Grant No. BNS-9022484.

References

[Ana84] T. Ananthapdmanabha. Acoustic analysis of voice source dynamics. *Speech Transmission Laboratory* 2–3:1–24, 1984.

[BYB94] H. T. Bunnell, D. Yarrington, and K. E. Barner. Pitch control in diphone synthesis. In *Proceedings of the Second ESCA/IEEE Workshop on Speech Synthesis*, 127–130, 1994.

[Col92] J. Coleman. The phonetic interpretation of headed phonological structures containing overlapping constituents. *Phonology* 9(1): 1–44, 1994.

[DC94] A. Dirksen and J. Coleman. All-prosodic synthesis architecture. In *Proceedings of the Second ESCA/IEEE Workshop on Speech Synthesis*, 232–235, 1994.

[Gob89] C. Gobl. A preliminary study of acoustic voice quality correlates. *Speech Transmission Laboratory* 4:9–27, 1989.

[Kar90] I. Karlsson. Voice source dynamics for female speakers. In *Proceedings of ICSLP '90*, 69–72, 1990.

[Kla87] D. H. Klatt. Text-to-speech conversion. *J. Acoust. Soc. Amer.* 82(3): 737–793, 1988.

[Koh94] K. J. Kohler. Glottal stops and glottalization in German. *Phonetica* 51: 38–51, 1994.

[Lad93] P. Ladefoged. *A Course in Phonetics*, 3rd edition. Harcourt Brace Jovanovich, Fort Worth, 1993.

[Loc92] J. Local. Modelling assimilation in a non-segmental rule-free phonology. In *Papers in Laboratory Phonology II*, G. Docherty and D. Ladd, eds. Cambridge University Press, Cambridge, 190–223, 1992.

[MC90] E. Moulines and F. Charpentier. Pitch synchronous waveform processing techniques for text to speech synthesis using diphones. *Speech Comm.* 9:453–467, 1990.

[PB88] J. Pierrehumbert and M. Beckman. *Japanese Tone Structure*. MIT Press, Cambridge, Mass., 1988.

[PBH94] J. Pitrelli, M. Beckman, and J. Hirschberg. Evaluation of prosodic transcription labelling reliability in the ToBI framework. In *Proceedings of ICSLP '94*, 18–22, 1994.

[Pie95a] J. Pierrehumbert. Prosodic effects on glottal allophones. In *Vocal Fold Physiology* 8, O. Fujimura, ed. Singular Publishing Group, San Diego, 39–60, 1995.

[Pie95b] J. Pierrehumbert. Knowledge of variation. In *Papers from the 30th Regional Meeting of the Chicago Linguistic Society*. University of Chicago, Chicago, 1995.

[PT92] J. Pierrehumbert and D. Talkin. Lenition of /h/ and glottal stop. In *Papers in Laboratory Phonology II*, G. Docherty and D. Ladd, eds. Cambridge University Press, Cambridge, 90–116, 1992.

[Ros71] A. E. Rosenberg. Effect of glottal pulse shape on the quality of natural vowels. *J. Acoust. Soc. Amer.*, 49:583–590, 1971.

Appendix: Audio Demos

The CDROM contains nine demonstrations.

3

Text-to-Speech Synthesis with Dynamic Control of Source Parameters

Luís C. Oliveira

ABSTRACT This chapter describes the study of some characteristics of source-parameter dynamics to derive a preliminary set of rules that were integrated in text-to-speech (TTS) systems. An automated procedure estimated the source parameters of 534 seconds of voiced speech from a set of 300 English sentences spoken by a single female speaker. The results showed that there is a strong correlation between the values of the source parameter in the vowel midpoint and the vowel duration. The same parameters tend to decrease on vowel onsets and to increase on vowels offsets. This seems to indicate a prosodic nature of these parameters requiring special treatment in concatenative-based TTS systems that use source modification techniques, such as pitch synchronous overlap add (PSOLA) and multipulse.

3.1 Introduction

The increasing demand for higher quality and naturalness of the synthetic speech generated by text-to-speech systems has emphasized the need for a more realistic modeling of the dynamics of the vocal tract excitation. The synthesis of convincing female and child voices and of voice qualities such as breathy, press, or loud requires variations at the source level.

In this work we will focus on the integration in text-to-speech systems of a simplified, noninteractive model for the glottal flow and the control of its parameters from rules based on the results of an automatic analysis of natural speech.

3.2 Source Model

3.2.1 Periodic Excitation

Several formulations have been suggested for the voice source model. In this work we will use a polynomial model based on the model suggested by [Ros71]. This source model was previously used in the Klattalk system [Kla87] and in the Bell Laboratories TTS system [TR90]. The model can be derived by imposing three restrictions to a generic third-order polynomial [Oli93]:

$$u_g(t) = \begin{cases} \frac{E_e}{T_e^2}\left(T_e t^2 - t^3\right) & \text{if } 0 \le t < T_e \\ 0 & \text{if } T_e \le t < T_0, \end{cases}$$

where T_e is the duration of the open glottis phase. This parameter is often related with the fundamental period (T_0) in a ratio referred as the *open quotient* ($O_q = T_e/T_0$).

The radiation characteristic at lips (a pole at the origin) is usually incorporated in the source model to avoid an additional differentiation in the synthesis process. For this reason we will use the derivative of the glottal flow:

$$u'_g(t) = \begin{cases} \frac{E_e}{T_e^2}\left(2T_e t - 3t^2\right) & \text{if } 0 < t \le T_e \\ 0 & \text{if } T_e < t < T_0. \end{cases} \quad (3.1)$$

In this model the location of the maximum flow (T_p) is:

$$u'_g(T_p) = 0 \Leftrightarrow T_p = \frac{2}{3}T_e.$$

This fixed location of T_p is the major drawback of the model: It cannot model changes in the skewness of the glottal pulse. The maximum flow value is proportional to the duration of the open phase:

$$u_{gmax} = u_g(\frac{2}{3}T_e) \Leftrightarrow u_{gmax} = \frac{4}{27}E_e T_e.$$

Equation (3.1) expresses the derivative glottal flow with an abrupt glottal closure at T_e: $u'_g(T_e) = -E_e$. To synthesize less tense voices, the closing region of $u'_g(t)$ is usually modeled by a decaying exponential with a time constant dependent on the vocal folds closing speed. As suggested in [Kla87], the effect can be achieved by filtering the abrupt closure model with a first-order low-pass filter:

$$u'_v(n) = (1 - a_{st})u'_g(n) + a_{st}u'_v(n-1).$$

This approach has an easier frequency domain representation than using separate models for the open and closed glottis regions. The low-pass filter coefficient, a_{st}, can be related to the closing time constant T_a by matching the impulse response of the filter:

$$a_{st} = e^{-1/(T_a F_s)}$$

where F_s is the sampling frequency.

3.2.2 Nonperiodic Excitation

The turbulence generated at the glottis is an important characteristic of breathy and whispery voice qualities. The correct modeling of this phenomenon requires a more detailed modeling of the vocal apparatus: glottal opening area, impedance

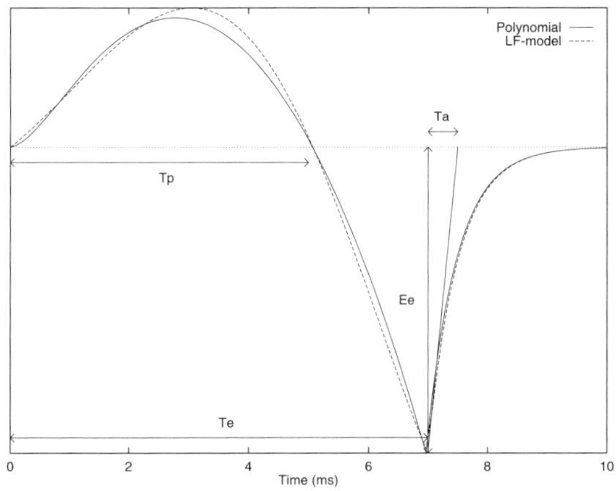

FIGURE 3.1. Polynomial and LF models for the derivative of the glottal flow.

of the vocal tract at the glottis, etc. [SS92]. In the noninteractive model that we are using, this information is not available and this effect can be incorporated only in a minimal form. As suggested in [TR90] and in [KK90], the derivative of the turbulent flow was modeled with amplitude-modulated flat-spectrum noise.

3.2.3 LF-Model Equivalence

The polynomial model of equation (3.1) can be related to the well-known four-parameter LF-model [FLL85]:

$$u_g'(t) = \begin{cases} E_0 e^{\alpha t}\sin(\omega_g t) & \text{if } 0 < t \le T_e \\ -\frac{E_e}{\varepsilon T_a}\left[e^{-\varepsilon(t-T_e)} - e^{-\varepsilon(T_0-T_e)}\right] & \text{if } T_e < t < T_0 \end{cases}$$

where

$$\omega_g = \frac{\pi}{T_p}$$

and

$$E_e = E_0 e^{\alpha T_e}\sin(\pi\frac{T_e}{T_p}).$$

This model is uniquely defined by the parameters T_p, T_e, T_a and E_e. Figure 3.1 shows the polynomial model (three parameters excluding the turbulence: T_e, T_a, E_e) and the four-parameter LF-model for the derivative of the glottal flow. As shown earlier, the polynomial model has one less parameter due to the fixed maximum flow location (T_p) inside the open glottis region (T_e):

$$T_p = \frac{2}{3}T_e \Leftrightarrow r_k = \frac{T_e - T_p}{T_p} = \frac{1}{2}.$$

3.3 Analysis Procedure

In order to integrate the model in a text-to-speech system some insight was needed on the dynamics of the source parameters in natural speech.

The correct determination of the source parameters requires a good estimate of the glottal waveform. One approximation of this signal can be computed by inverse filtering of the speech waveform, using an estimate of the vocal tract transfer function. The most important requirement for this procedure is to have the speech signal recorded without phase distortion. This can be accomplished by using a high-quality microphone and pre-amplifier in a silent environment and a very high sampling rate, avoiding the main source for phase distortion: the anti-aliasing analog filters. The filtering required to down-sample the signal to more workable sampling rates can be performed with a digital linear-phase filter.

The next requirement for the correct estimation of the glottal waveform is the determination of the vocal tract transfer function. The ideal linear model involves the estimation of the resonances due to formants as well as antiresonances (nasalization, etc.). In this work we have adopted a non-ideal model by using only the resonances estimated by pitch-synchronous linear prediction analysis. This method, described in [TR90], uses an accurate epoch finder [Tal89] to locate the linear prediction analysis window synchronously with the glottal activity. Although the linear prediction analysis tends to flatten the overall spectral slope of the inverse filtered signal as a result of the least-squares error minimization, the error will partially be undone during LPC synthesis using a direct filter with parameters computed by the same method. It must be noted, however, that the computed values for the source parameters cannot be generalized for other vocal tract models. Figure 3.2 shows an example of the inverse-filtered signal computed with this method.

The source model parameters were estimated from the inverse-filtered signal using the frequency-based method described in [Oli93]. This method assumes the frequency representation for the voice-source model:

$$\left|U'_{vh}\left(e^{j\omega}\right)\right| = \underbrace{A_v \left|\tilde{U}'_g\left(e^{j\omega}\right)\right| \left|\frac{1-a_{st}}{1-a_{st}e^{j\omega}}\right|}_{\text{periodic}} + \underbrace{A_h \left|U'_h\left(e^{j\omega}\right)\right|}_{\text{random}} \tag{3.2}$$

where $\tilde{U}'_g\left(e^{j\omega}\right)$ is the Fourier transform of the discrete-time version of the polynomial model, $U'_h\left(e^{j\omega}\right)$ models the turbulence noise at the glottis, and A_v and A_h are the amplitudes of the periodic and random components, respectively.

Having an equation for the frequency representation of the source model, we can fit the equation to the signal spectrum and estimate the four source parameters: duration of the open phase, spectral tilt, voicing amplitude, and aspiration amplitude.

Figure 3.3 depicts the several steps of the inverse-filtered signal analysis procedure.

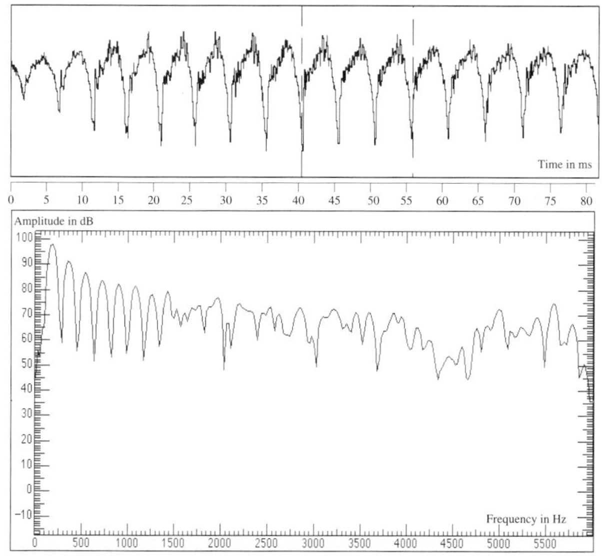

FIGURE 3.2. Inverse-filtered signal and the magnitude of the short-time Fourier transform thereof (Hanning window).

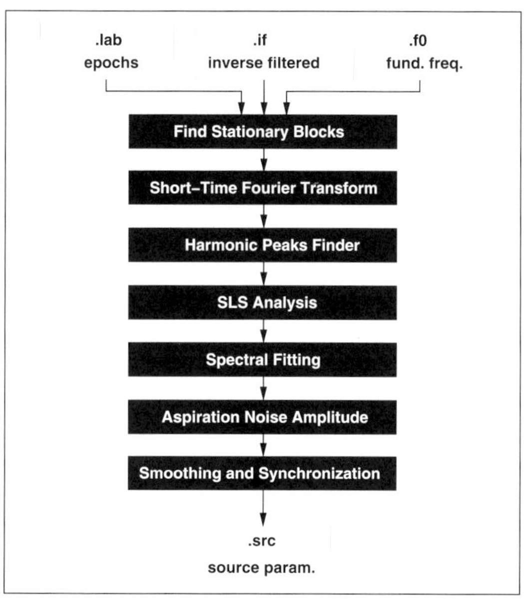

FIGURE 3.3. Block diagram of the analysis procedure.

Harmonic Peak Picking

The analysis procedure starts by locating all the local maxima in the spectrum of the inversed-filtered signal. Then, the peaks in the vicinity of multiples of the fundamental frequency are selected as harmonics. The selection ends when the distance from the frequency of the peak to the nearest harmonic frequency exceeds a predefined value. The harmonic spectrum cut-off frequency, F_{hc}, is defined as the frequency of the last harmonic peak plus half the fundamental frequency.

Removal of the Window Spectrum: SLS Analysis

The numeric computation of the spectrum of a slowly varying quasi-periodic waveform requires the usage of some window to truncate the signal. The resulting representation, the short-time spectrum, is the convolution of the quasi-periodic signal spectrum with the spectrum of the window. The shape of the lobes observed in the low-frequency region of the magnitude of the short-time Fourier transform of figure 3.2 is the result of this convolution. The Hamming window minimizes the interference between adjacent harmonics, due to its low leakage (the spectral envelope decreases with $1/\omega^3$).

The problem of finding the real amplitude of the harmonic pulses from the short-time spectrum was addressed in the context of the harmonic modeling of voiced speech in [AT83]. The method was later extended to the unvoiced regions and was named stationary least squares (SLS) analysis [MA89].

The analysis assumes a sinusoidal representation for the signal to be estimated:

$$\hat{s}(t) = \sum_{k=-L}^{L} a_k e^{j\omega_k t}$$

with $\omega_{-k} = -\omega_k$ and $a_{-k} = a_k^*$. In the harmonic region of the spectrum the frequencies of the exponentials are located at multiples of the fundamental frequency, ω_0: $\omega_k = k\omega_0$.

The complex amplitudes, a_k, are estimated by minimizing the weighted least squares criterion:

$$\int_{-\infty}^{+\infty} w^2(t) \left|s(t) - \hat{s}(t)\right|^2 dt.$$

Non-Linear Fitting of the Spectral Envelope

Having the Fourier coefficients, a_k, determined by the SLS analysis, we can now estimate the parameters of the periodic model: A_v, N_{op} and a_{st}.

By using the N_{hp} harmonic frequencies, ω_k, and amplitudes, a_k, a non-linear fitting can be performed by using the Levenberg-Marquardt method to minimize the sum:

$$\chi^2(A_v, N_{op}, a_{st}) = \sum_{k=1}^{N_{hp}} \left(|a_k| - A_v \frac{|1 - a_{st}|}{2\pi} \frac{\left|U_g'\left(e^{j\omega_k}\right)\right|}{\left|1 - a_{st} e^{j\omega_k}\right|} \right)^2$$

where $N_{op} = T_e F_s$ is the duration of the open glottis phase in the samples. Because the frequency effect of the duration of the open phase is related to the glottal resonance frequency (F_g) located in the vicinity of the fundamental frequency, the method is first applied with the first few harmonics (usually three). This gives a good estimate of the open phase duration. Next, using all the harmonic peaks, the spectral tilt and the voiced amplitude are estimated together.

As discussed earlier, the inverse filtered signal is only an approximation of the glottal waveform, due to the simplified model. In regions where this model is not valid, the fitting method can diverge or converge to invalid values for the parameters (e.g., $O_q > 1$). When this happens the solution is discarded and the glottal pulse is marked as nonanalyzable.

Aspiration Noise Amplitude

At this point of the analysis procedure, all the parameters of the periodic component have been determined. It is now necessary to know the amplitude of the the random component, A_h. Figure 3.2 shows the time and frequency representation of the inverse-filtered signal for voiced speech. The low-pass characteristic of the glottal waveform makes the random component predominant in the higher frequencies. This suggests the determination of aspiration amplitude by the average difference between the short-time spectrum of the inverse-filtered signal and the model for the periodic component, in the random region of the spectrum ($F > F_{hc}$).

Stationary Blocks of Fundamental Periods

Until now we have assumed the stationarity of the inverse-filtered signal. In general this assumption is false, but in short segments the signal can have a quasi-stationary behavior.

Because the method requires more than two glottal cycles to be able to locate the harmonic peaks on the short-time spectrum, it is necessary to avoid abrupt changes of the signal inside the analysis window. To prevent this, the inverse-filtered signal is scanned to group clusters of glottal cycles with slow-varying differences in durations. Each cluster is then divided in overlapping analysis blocks containing from three to five cycles. This ensures the significance of the results but disables the analysis in the transitional regions where the inverse-filtered signal cannot be considered quasi-stationary. In these regions the inverse-filtered signal is not a good estimate of the glottal waveform because the LPC analysis also requires the stationarity of the input signal.

3.4 Analysis Results

Having a method to estimate the parameters of the source model, we next analyzed speech material in order to devise some basic rules to control the parameter trajectories.

3.4.1 Speech Material

A set of 300 utterances was selected from the recordings of the female speaker whose voice was used in the acoustic inventory of the Bell Laboratories TTS systems [Oli90]. The material was chosen from different recording sessions with the intention of representing the speaker's normal phonation style. There was no attempt in selecting particular voice qualities, namely breathy or laryngeal, although this speaker's voice can be considered more breathy than average. The speech signal had a bandwidth of 6 kHz and a fourteenth-order pitch-synchronous LPC analysis was performed.

The previously described pitch-synchronous analysis of this material was successful for 22799 glottal pulses. The spectral fitting process either did not converge or converged to invalid values for 4571 pulses (17% of the cases). The estimated model parameters N_{op}, a_{st}, A_v, and A_h were converted to more meaningful formats:

open quotient: $$O_q = \frac{N_{op}}{F_s T_0}$$

spectral tilt frequency: $$F_a = \frac{1}{T_a} = -F_s \ln(a_{st})$$

aspiration ratio: $$R_h = \frac{A_h}{A_v + A_h}$$

Figure 3.4 shows the histograms of the parameters.

3.4.2 Vowels

To integrate the control of the source parameters in a text-to-speech system we need to relate its values with the corresponding phonetic segments. The study started by considering only the vowels. Because a vowel usually includes several glottal pulses, the parameters were averaged using a Hanning weighting function aligned with the vowel. This resulted in a set of parameters for each of the 3276 vowels in the selected utterances.

The average values in the middle of the vowels were correlated with the fundamental frequency observed in the same location. This resulted in a correlation coefficient that was weak for the open quotient ($r_{O_q} = 0.40$) and very weak for F_a and R_h ($r_{Fa} = 0.15$ and $r_{R_h} = 0.16$). Previous work reported a similar result for the open quotient [Kar85, KK90].

The midsegment average values of the source parameters were also correlated with the phonetic segment duration. However, in this case a difficulty arises due to the distribution of the observations being dominated by the short segments. To avoid this effect, the average values for the parameters were computed for each range of durations. A strong negative correlation was found between the duration and the average values of the open quotient and of the aspiration ratio ($r_{O_q} = -0.978$ and $r_{R_h} = -0.927$, $p < 0.001$). A less strong correlation was found for the spectral tilt frequency ($r_{F_a} = -0.881$, $p < 0.01$). Figure 3.5 shows

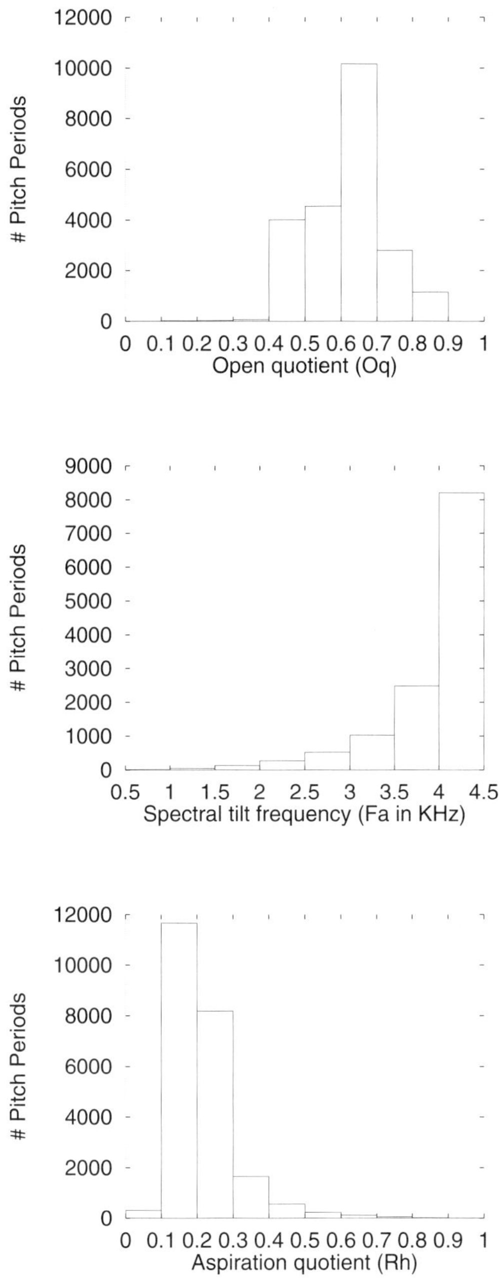

FIGURE 3.4. Histograms of the source parameters for 22799 glottal pulses.

the average midsegment values of the source parameters plotted against vowel duration.

3.4.3 *Vowel Boundaries*

After the analysis of the behavior of the source parameters in the middle of the vowel, we studied the influence of the adjacent voiceless segments on the parameter trajectories inside the vowel.

For that purpose, the unvoiced phonetic segments were characterized by the values of the source parameters in the closest boundary of the adjacent vowel. When the voiceless segment had two adjacent vowels, the left and right boundary values were characterized separately. To minimize the effect of analysis errors due to irregularities of the glottal wave, a weighted average of the parameters of the glottal pulses close to the transition was used. A similar procedure was applied to the values of the source parameters in sentence-final vowels.

Table 3.1 shows the average values of the parameter variations between the boundary and the midpoint of the vowel. The values presented must be used only as an indication because the effect of the vowel duration was not taken into account. It shows, for instance, in the transition from a voiceless consonant to a vowel, the usual presence of an interval of breathy voicing (high O_q and R_h, low F_a) [GC88]. It also shows the raising of the open quotient from the midpoint of sentence-final vowels to the end of the utterance.

3.5 Conclusions

In this chapter we have presented the results of the study of some of the characteristics of the source-parameter dynamics in an English-speech database. A similar study is being conducted for the Portuguese language, and the preliminary results seem to lead to very similar conclusions.

The results for the parameter variation at phonetic segment boundaries could indicate its segmental nature and that they could be incorporated in the units ta-

TABLE 3.1. Average source parameter variation from the midpoint to the vowel boundary.

Transition	ΔO_q	ΔF_a	ΔR_h
vowel to voiceless fricative	+0.15	+356	+0.12
voiceless fricative to vowel	−0.11	+990	−0.10
vowel to obstruent	+0.11	−654	+0.11
obstruent to vowel	+0.09	−927	+0.12
vowel to nasal	+0.08	+901	+0.06
nasal to vowel	−0.06	−1034	−0.05
sentence-final vowel	+0.20	+194	+0.26

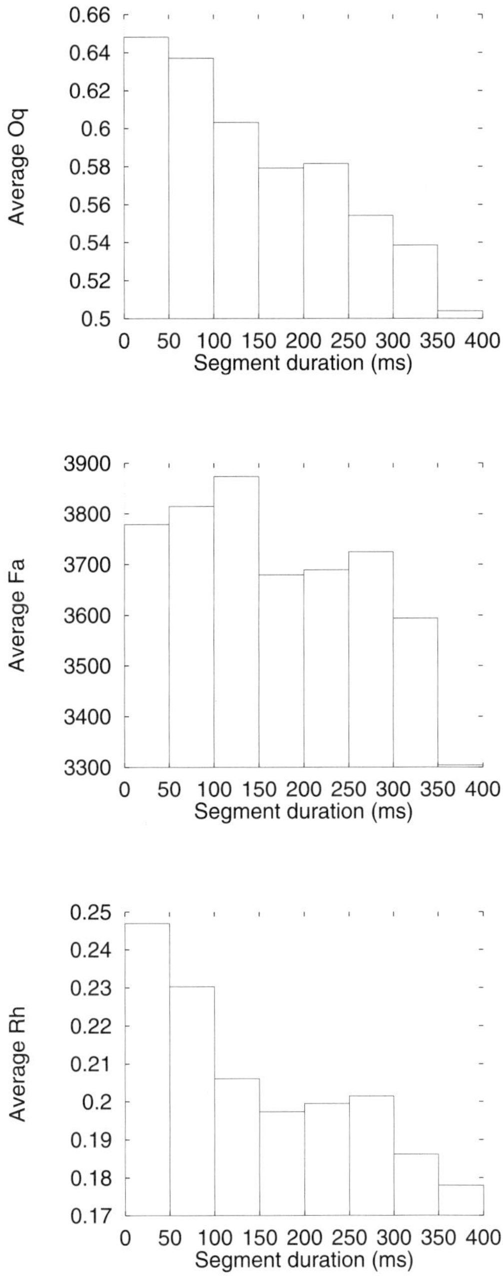

FIGURE 3.5. Average values of source parameters computed at the midpoint of the 3276 vowels as a function of the vowel duration.

ble of a concatenative-based system. The midsegment parameter values and their variation with the segment duration show, on the other hand, a prosodic characteristic. For this reason we have started with a rule-based approach to control the source parameter trajectories. The rules based on the current study were integrated in the English version of the Bell Laboratories text-to-speech synthesizer, which is concatenative based, and is being integrated in the DIXI text-to-speech system for Portuguese using the Klatt formant synthesizer. Although no formal tests were performed yet, informal listening showed some improvements in the naturalness, especially on segments with longer durations.

The prosodic characteristic of these effects is easily integrated in systems with a parametric model of the source. In concatenative-based systems that store the source signal in the unit table, this effect can be accomplished by using multiple representatives for different vowel durations.

References

[AT83] L. B. Almeida and J. M. Tribolet. Non-stationary spectral modeling of voiced speech. *Transactions on Acoustic Speech and Signal Proc.* ASSP-31(3):664–678, June 1983.

[GC88] C. Gobl and A. Chasaide. The effects of adjacent voiced/voiceless consonants on the vowel voice source: A cross language study. *Speech Transmission Laboratory – QPSR* Stockholm, Sweden, 2–3, 1988.

[FLL85] G. Fant, J. Liljencrants, and Q. Lin. A four parameter model of glottal flow. *Speech Transmission Laboratory – QPSR* Stockholm, Sweden, 4:1–13, 1985.

[Kar85] I. Karlsson. Glottal waveforms for normal female speakers. *Speech Transmission Laboratory – QPSR* Stockholm, Sweden, 31–36, 1985.

[KK90] D. H. Klatt and L. C. Klatt. Analysis, synthesis and perception of voice quality variations among female and male talkers. *J. Acoust. Soc. Amer.* 87(2):820–857, 1990.

[Kla87] D. H. Klatt. Review of text-to-speech conversion for English. *J. Acoust. Soc. Amer.* 82(3):737–793, 1987.

[MA89] J. Marques and L. Almeida. Sinusoidal modeling of voiced and unvoiced speech. In *Proceedings of the European Conference on Speech Communication and Technology*, September 1989.

[Oli90] J. P. Olive. A new algorithm for a concatenative speech synthesis system using an augmented acoustic inventory of speech sounds. In *ESCA Workshop on Speech Synthesis*, Autrans, France, 25–29, September 1990.

[Oli93] L. C. Oliveira. Estimation of source parameters by frequency analysis. In *Proceedings of the European Conference on Speech Communication and Technology*, Berlin, vol. 1, 99–102, September 1993.

[Ros71] A. E. Rosenberg. Effect of glottal pulse shape on the quality of natural vowels. *J. Acoust. Soc. Amer.* 49(2 (Part 2)):583–590, 1971.

[SS92] J. Schroeter and M. M. Sondhi. Speech coding based on physiological models of speech production. In *Advances in Speech Signal Processing*, S. Furui and M. Mohan Sondhi, eds., Marcel Dekker, Inc., New York, 231–268, 1992.

[Tal89] D. Talkin. Voice epoch determination with dynamic programming. *J. Acoust. Soc. Amer.* 85(S1):S149, 1989.

[TR90] D. Talkin and J. Rowley. Pitch-synchronous analysis and synthesis for TTS systems. In *ESCA Workshop on Speech Synthesis*, Autrans, France, 55–58, September 1990.

4

Modification of the Aperiodic Component of Speech Signals for Synthesis

Gaël Richard
Christophe R. d'Alessandro

ABSTRACT Modeling the excitation component of speech signals is a challenging problem for speech synthesis. Recently, several works have been devoted to periodic/aperiodic decomposition of the speech signal: a decomposition that permits a better characterization of the source. This chapter introduces a new analysis/synthesis algorithm for representing the aperiodic component of the excitation source in speech signals. This component is decomposed as a sum of random formant wave forms (FWF), which correspond to formant filters impulse responses. The time of arrivals of the FWF define the virtual excitation source. The signal is decomposed in subbands, and, according to the random modulation theory, each passband signal is represented as an envelope modulating an oscillating term. All parameters (formant filters and excitation sources) are estimated in the time domain. This new representation scheme gives a very good fusion of the aperiodic component with the quasi-periodic component of speech. The method proposed provides new relevant parameters for manipulating the voice quality features that are linked to noise. For example, it is possible to perform voice quality modifications such as time scaling, formant or modulation depth modifications of the aperiodic component, or modification of the periodic/aperiodic ratio.

4.1 Introduction

Modeling the excitation source of speech signal is a key problem for speech synthesis. The excitation source encompasses the glottal source (periodic flow and aspiration noise), frication noise (occurring at a constriction of the vocal tract), and burst releases (occurring after a sudden release of a closure in the vocal tract). Voice quality modification, voice conversion, and prosodic processing are highly dependent on our ability to analyze, model, and modify the excitation source. In the context of both rule-based and concatenation-based synthesis, it seems important to develop signal representation methods that are able to deal with natural excitation sources. Traditionally, the most important prosodic parameters are pitch, duration, and intensity. These parameters are intimately linked to voice quality.

Other important aspects of voice quality, which have not received as much attention in synthesis research, are the vocal effort and the phonatory quality. Ideally, a synthesizer should be able to simulate various types of phonatory styles, the extreme situations being whispered speech and shouting. To reach this goal, it appears necessary to deal with various types of excitation, including noise excitation, which can occur at the glottis or at various locations of the vocal tract.

The analysis and synthesis of the speech noise component has recently become a focus of interest for several reasons. On the one hand, it is well-known that this component is responsible for a part of the perceived voice quality (e.g., breathiness, creakiness, softness). There is a long history, especially in the fields of voice pathology and voice perception, of studies involving parameters such as jitter (random variation of the source periodicity), shimmer (random variation of the glottal flow amplitude), or diplophony. On the other hand, some new results [LSM93], [DL92] indicate that separate processing of the periodic and aperiodic components of speech signals may improve the quality of synthetic speech, in the framework of concatenative synthesis. Finally, for some methods, the two components obtained seem to be relevant from the acoustic point of view [Cha90, DAY95]: It is possible to associate the periodic and aperiodic components to physical components in the voice source. This is important in achieving realistic modifications of the source.

Different terminologies have been used by various authors (e.g., "harmonic + noise (H+N) model" in [LSM93], "multiband excitation (MBE) vocoder" in [GL88] and [DL92], "deterministic and stochastic components" in [SS90]). We prefer the terminology "periodic and aperiodic (PAP) components." The precise meaning attached to the terms "periodic" and "aperiodic" must be discussed in some detail. The acoustic model of speech production is a source/filter model, with an excitation source $e(t)$ and a filter $v(t)$. An exactly periodic vibration of vocal chords is not a reasonable assumption, even for sustained vowels, because of the complex nature of this phenomenon. More accurately, the excitation source can be decomposed into a quasi-periodic component $p(t)$ and an aperiodic component $a(t)$:

$$s(t) = e(t) * v(t) = [p(t) + a(t)] * v(t). \tag{4.1}$$

Along this line, the periodic component represents the regular vibratory pattern of the vocal chords, and the aperiodic component represents all the irregularities presented in both voiced and unvoiced sounds. It must be emphasized that the aperiodic component may represent signals of different natures. It is generally acknowledged that both modulation noise (i.e., noise due to the source aperiodicity, such as jitter, shimmer, and F_0 variations) and additive random noise are present in the aperiodic component. This additive random noise includes transients (e.g., bursts of plosives), steady noises (e.g., unvoiced fricatives), or time-modulated noises (e.g., noise in voiced fricatives or in breathy vowels). Ideally, a decomposition method should be able to separate these different sources of aperiodicity. However, having solely the speech signal $s(t)$, obtaining the two components $p(t)$ and $a(t)$ is not straightforward. For this study, a new PAP decomposition algorithm is presented, which yields an aperiodic component with a realistic acoustic meaning [AYD95], [DAY95].

The first step is therefore to achieve an acoustically relevant decomposition. The second step is to define a model for the aperiodic component. However, poor modeling of the aperiodic component could introduce a lack of perceptual fusion between the quasi-periodic component and the aperiodic component. Recent studies show that this perceptual separation is a consequence of the weak coherence between these two components in the time domain [Cha90], [Der91]. Therefore, methods for representing the aperiodic component are needed that provide an accurate control in both time and frequency. In this chapter, a new analysis/synthesis model for the aperiodic component is introduced. The synthesis method is based on previous work on the elementary wave form representation of speech [Ale90], and on speech noise synthesis using random formant impulse responses [Ric92]. Furthermore, this new coding scheme provides relevant parameters for manipulating the voice quality features that are linked to noise. Breathiness, creakiness, or roughness of a voice represent such features.

The chapter is organized as follows. The next section gives a detailed description of the speech signal decomposition algorithm. Section 4.3 introduces the random formant wave form model and describes the various steps of the analysis/synthesis algorithm. Section 4.4 presents some evaluation results. Section 4.5 demonstrates some of the voice modification abilities of the method. Finally, the results are discussed and some conclusions are suggested in the last section.

4.2 Speech Signal Decomposition

Even though there is a long history of research on aperiodicities in speech, particularly in the field of voice analysis research, most studies do not explicitly perform a separation of the two components (a periodic component and an aperiodic component), but rather measure a harmonic-to-noise ratio (HNR) to describe different types of voices (see, e.g., [Hil87, Kro93]).

On the contrary, in the field of speech and music synthesis, explicit PAP decomposition of signals has become a focus of interest, without paying much attention to the underlying acoustic or perceptual aspects. Various algorithms based on sinusoidal or harmonic models have been proposed. The MBE vocoder [GL88] is based on linear predictive coding (LPC). The LPC residual signal is coded in the frequency domain in terms of different frequency bands, which are labeled either "harmonic" or "noise" depending on their resemblance to ideal harmonic structures. Although this is an efficient coding scheme, it is difficult, if not impossible, to interpret the different frequency bands in terms of speech production. In the H+N model ([SM93]), a low-pass harmonic signal is subtracted from the original signal. The synthetic noise signal is obtained by modulation by an energy envelope function of LPC-filtered noise. Nevertheless, in this technique, there is no noise for frequencies below 2-3 kHz and no harmonics above. Although it might improve

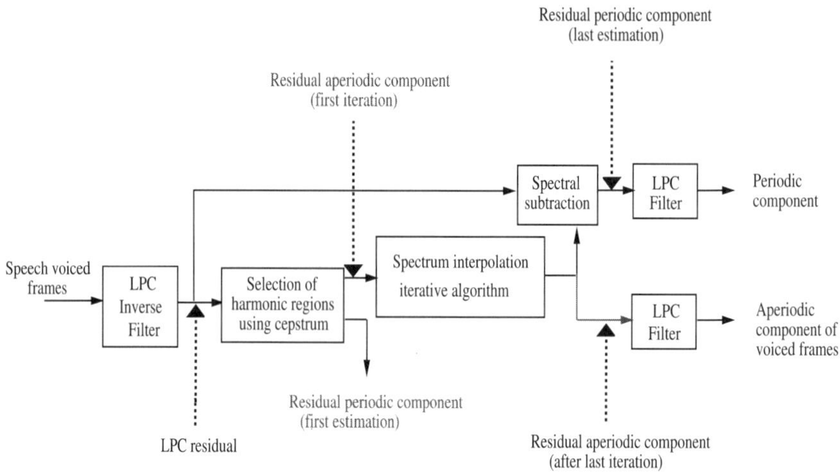

FIGURE 4.1. Schematized diagram of the PAP decomposition algorithm.

the quality of concatenative synthesis, this is not realistic from the acoustic point of view. Decomposition methods based on sinusoidal coding have also been proposed [SS90]. Some criteria for harmonicity are applied to the sinusoidal tracks to form the so-called deterministic component. The remaining frequency points are associated with the so-called stochastic component. Impressive decomposition is obtained for both speech and musical sounds. The problem with this approach is that a binary decision is taken for each frequency point: a region is either "deterministic" or "stochastic." This is not the case in actual speech, where noise is present even in harmonic regions.

A new algorithm for PAP decomposition has been proposed in [AYD95]. One of its aims is to obtain an aperiodic component that represents the real features of speech including voice quality features such as breathiness or roughness. This algorithm is applied only to voiced frames, the unvoiced frames being merged with the aperiodic component. Following the voiced-unvoiced decision, this algorithm may be decomposed into five main steps (see figure 4.1):

- An approximation of the excitation source signal is obtained by inverse filtering the speech signal (typically 8 kHz sampling rate, 10 LPC coefficients, 20 ms window size). The excitation source signal is then processed, on a frame-by-frame basis, using short-term Fourier analysis-synthesis (20 ms window size (200 pt), 5 ms overlap, 512 points fast Fourier transform (FFT)).

- For each frame, the ratio of periodic to aperiodic frequency points is measured in three steps, based on [Kro93]:

    1. The cepstrum of the original signal is computed.
    2. The region of the main peak (pitch) is isolated.

3. An inverse Fourier transform is then applied to this region. Ideally, the main peak is a Dirac distribution, and his inverse Fourier transform is a complex exponential. By considering only the real part of the spectrum, one obtains a sinusoid whose frequency is given by the location of the cepstrum main peak. The positive peaks of the sinusoid define the location of the harmonics, and all other positive values provide an estimation of the bandwidth of each harmonic. This last information is particularly important in practice as the cepstrum main peak is not an ideal Dirac distribution. The remaining part, the negative values of the sinusoid, provides an indication of frequency points where the valleys between harmonics are located. These frequency points will serve as a basis for a first approximation of the aperiodic component.

At this stage of processing, only a primary identification of the frequency points associated with the aperiodic component is formed and each frequency point is labeled as either periodic or aperiodic.

- The secondary estimation of the aperiodic component is then performed using an iterative algorithm based on Papoulis-Gershberg extrapolation algorithm ([Pap84]). Starting with the initial estimation, the signal is successively transformed from the frequency domain to the time domain and back to the frequency domain, imposing finite duration constraints in the time domain and the known noise samples in the frequency domain (frequency points in valleys between harmonics). After a few iterations (10 to 20), the obtained stochastic component possesses a continuous spectrum with extrapolated values in the harmonic regions (see figure 4.2). Thus, for each frequency point the complex values of both periodic and aperiodic components are available.

- The periodic component is then obtained by subtracting the complex spectrum of the aperiodic signal from the complex spectrum of the residual signal. The synthetic source components are obtained by inverse Fourier transform and overlap-add synthesis.

- Finally, the two components of the residual signal are filtered by the time-varying all-pole filter to obtain the final aperiodic and periodic components of the voiced frames. The complete aperiodic component is obtained by including the unvoiced frames of the original signal. The result of the PAP decomposition algorithm is depicted in figure 4.3.

This algorithm was tested using natural and synthetic signals [DAY95]. The results showed that the PAP decomposition algorithm is able to separate additive random noise and periodic voicing for a wide range of F_0 variation. Therefore, in normal natural speech, we feel justified in using the aperiodic component as an estimate of additive noise in the source, when jitter and shimmer are reasonably low. This is usually the case for speech synthesis databases. However, in the case

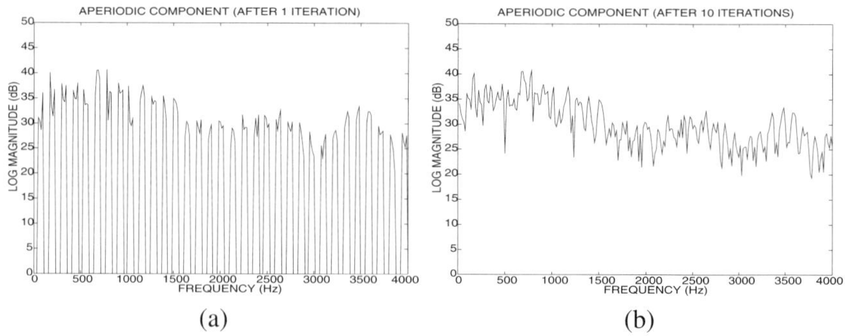

FIGURE 4.2. The effect of iterative noise reconstruction. The initial estimate of the aperiodic component has energy only between harmonic regions (a). After 10 iterations of the algorithm, a continuous spectrum is obtained (b).

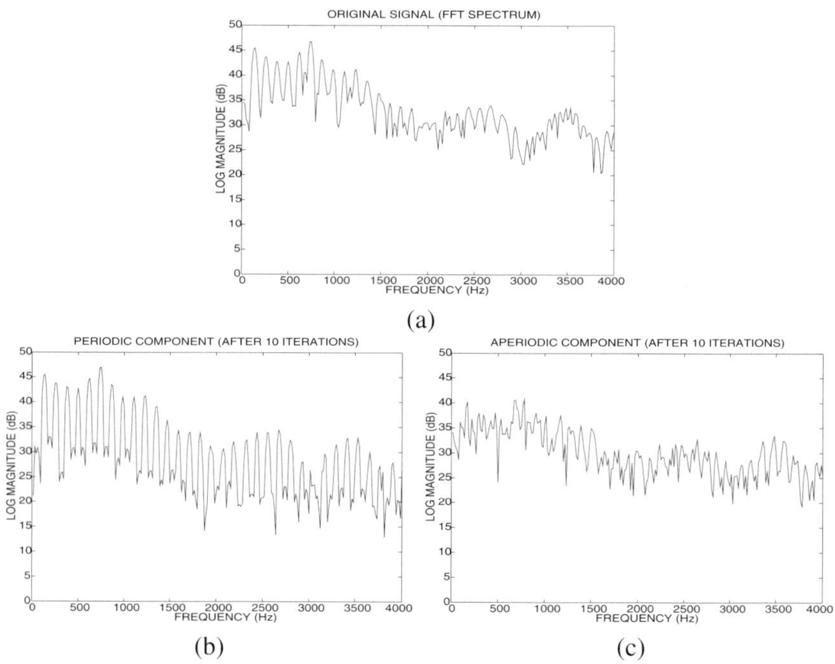

FIGURE 4.3. Result of the PAP decomposition algorithm: (a) displays the log magnitude spectrum of one frame of the original signal; (b) and (c) respectively represent the periodic and the aperiodic components obtained after decomposition.

of large jitter or shimmer values, additive random noise and modulation noise are merged in the aperiodic component. Although it is still possible to achieve separation of a periodic and an aperiodic component, it seems difficult in this case to separate the different physical components of the aperiodic component.

In the following discussion, we make the following assumptions:

1. The aperiodic and periodic components exist in speech signals. They are not artifacts due to a signal-processing method.
2. These components represent actual features of speech production that are linked to voice quality.
3. The components can be measured with accuracy using an appropriate PAP decomposition algorithm.

Sound example 1 of the audio demo (see Appendix) illustrates the results of the PAP decomposition algorithm on several sentences of natural speech.

4.3 Aperiodic Component Analysis and Synthesis

Several algorithms have been proposed for coding or synthesizing the aperiodic component of acoustic signals (see, for example, [Kla80, SS90, GL88, MQ92]). Most use a Gaussian excitation source and a slowly time-varying filter. It is clear that some time modulation of the aperiodic component is needed for speech because a white noise excitation source shaped by the spectral envelope of a slowly time-varying filter is not sufficiently precise in the time domain. As a matter of fact, it is acknowledged that it is important to take into account the temporal structure of noises if one wants to obtain a good perceptual fusion of the periodic and aperiodic components in the final reconstructed signal [CL91, Her91]. In formant synthesis, it is also common to modulate a noise source by the glottal flow [Kla80]. To trace this time modulation, [LSM93] proposed to time-modulate the high-pass noise excitation source by the time-domain envelope of the signal. However, this technique cannot be successfully applied to wideband noises and especially not for noises with significant energy in the lower part of the spectrum. This may be due to the fact that the rate of fluctuation of the envelope is of the same order of magnitude as the rate of fluctuation of the excitation noise signal. In other words, when some energy is present in the lower part of the spectrum, the maxima and minima of the time modulation (deduced from the envelope of the original noise signal) are almost never synchronized with the maxima and minima, respectively, of the white noise excitation signal. Thus, the precise time-domain control is lost and the resulting signal has a different modulation structure than the desired one.

Furthermore, for voice modification it may be important to control the spectral maxima (related to formants) and to get a description of the aperiodic component as a sum of well-localized spectro-temporal items.

For these reasons, we decided to develop an algorithm in the framework of source/filter decomposition, in which the filter is decomposed into several formant filters excited by separate random sources. Within a formant region, the passband noise signal is described as a random point process, which defines the random times of arrival of the formant filter impulse responses. The random point process is deduced from the maxima of the time-domain envelope.

The formant filters chosen are the Formant Wave Forms (FWF) introduced by [Rod80], which are close to second-order resonator impulse responses. A FWF is defined as a modulated sinusoid:

$$s(t) = \Lambda(t) \sin(2\pi f_c t + \phi) \tag{4.2}$$

where the FWF time domain envelope is given by:

$$\Lambda(t) = \begin{cases} 0 & \text{if } t \leq 0 \\ \frac{1}{2} A(1 - \cos(\beta t))e^{-\alpha t} & \text{if } 0 < t \leq \pi/\beta \\ Ae^{-\alpha t} & \text{if } t > \pi/\beta \end{cases} \tag{4.3}$$

and $\pi/\beta, f_c, \alpha/\pi, A, \phi$ are the excitation duration, the formant center frequency, the -3 dB bandwidth, the formant amplitude, and the FWF initial phase, respectively.

An iterative algorithm was designed for automatic extraction of both source and filter parameters. The random excitation source is a set of points along the time axis, and the filter parameters are FWF parameters.

According to random modulation theory, any passband stochastic signal $x(t)$ can be represented using the real envelope $r(t)$ and the instantaneous phase $2\pi f_m t + \Psi(t)$, where f_m is arbitrary:

$$x(t) = r(t) \cos[2\pi f_m \, t + \Psi(t)]. \tag{4.4}$$

A more detailed theoretical background may be found in [Ric94]. The basic ideas of the algorithm are:

- to define the excitation point process according to the envelope maxima locations;
- to compute the FWF envelope $\Lambda(t)$ using the envelope $r(t)$ between two successive minima;
- to estimate the FWF center frequency from the center of gravity of the instantaneous frequency of $x(t)$.

More precisely, the analysis/synthesis algorithm is the following (see figure 4.4):

1. Band-pass filtering of the signal $x(t)$ (e.g., 6 bands, for a sampling rate of 8 kHz).

2. For each band-pass signal $x_b(t)$:

   a. Computation and low-pass filtering of the real envelope, using the Hilbert transform, $\hat{x}_b(t)$ of $x_b(t)$:

   $$r(t) = \sqrt{x_b^2(t) + \hat{x}_b^2(t)} \tag{4.5}$$

   b. Definition of the excitation point process according to the real envelope maxima (see figure 4.5).

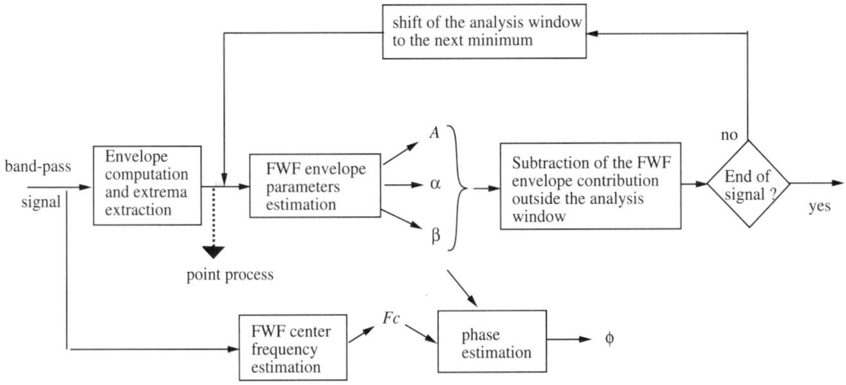

FIGURE 4.4. General diagram of the FWF estimation procedure for one band-pass signal. The same analysis must be applied to all band-pass signals (e.g., six bands for a sampling rate of 8 kHz) to obtain a complete description of the aperiodic component.

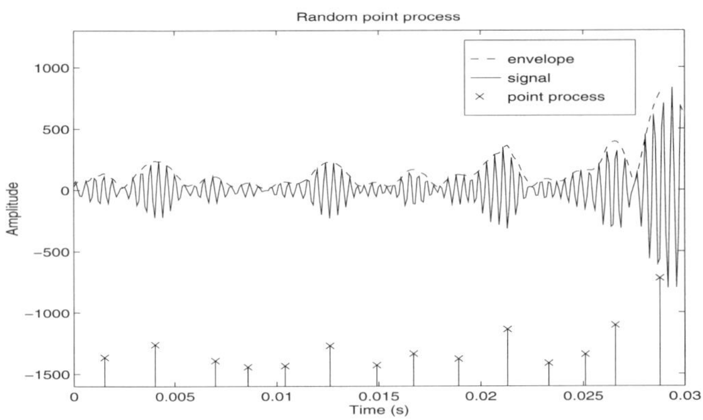

FIGURE 4.5. The point-process impulses are defined from the (time-domain) envelope maxima location.

c. Estimation of the FWF envelope parameters by fitting the FWF envelope and the real envelope between two successive minima of the envelope ($t \in [t_{m1}, t_{m2}]$). This gives α, β, and A.

d. Estimation of f_c as the optimal frequency f_m in equation (4.4). It is the weighted average of the instantaneous frequency $f_i(t)$:

$$f_c = \frac{\sum_{t \in [t_{m1}, t_{m2}]} \Lambda^2(t - t_{m1}) f_i(t - t_{m1})}{\sum_{t \in [t_{m1}, t_{m2}]} \Lambda^2(t - t_{m1})} \tag{4.6}$$

where the instantaneous frequency (time-derivative of the instantaneous phase) is given by:

$$f_i(t) = \frac{1}{2\pi} \times \frac{x_b(t)\hat{x_b}'(t) - x_b'(t)\hat{x_b}(t)}{r^2(t)} \tag{4.7}$$

e. The initial phase, ϕ, is set as a function of f_c and β in order to give a maximum at the exact place defined by the envelope maximum.

f. Subtraction of the FWF envelope contribution outside the analysis window (that is, for $t > t_{m2}$).

g. Iteration of steps c–f of the algorithm, until the end of the signal is reached.

3. FWF synthesis is performed using the estimated FWF parameters.

Sound example 2 of the audio demo illustrates the results of this algorithm on the aperiodic component of natural speech signals.

4.4 Evaluation

Perceptual tests (degrading category rating (DCR), see [CCI92]) were run to measure the quality obtained with the random FWF method compared to the LPC analysis/synthesis method. Ten subjects were asked to give an appreciation of the degradation of a synthetic signal (second of a pair) compared with a natural signal (first of the pair). Four conditions were tested:

Condition 1: Whispered speech (eight sentences of at least 1 s duration, 4 males/4 females).

Conditions 2,3 and 4: Normal speech (eight sentences of at least 1 s duration, 4 males/4 females). Periodic and aperiodic parts were separated. The aperiodic part was then modeled by either the LPC or random FWF model, scaled by a gain factor (1, 2, and 3 for tests 2, 3, and 4, respectively) before being added to the periodic component. The aim of this test was to measure the degree of fusion of the aperiodic and the periodic components and to test the robustness of this method when the aperiodic component is modified.

The results of the DCR test are given in figure 4.6. It is noticeable that both methods show similar results for condition 1 (whispered speech). This is not surprising, as the LPC model is excellent for this type of speech. The results for conditions 2 to 4 show a greater degradation for LPC than for random FWF. We think that these results are linked to the better time and frequency accuracy of our method: formants are well represented, and the time domain control gives a better perceptual fusion between the periodic and aperiodic components. In fact, the LPC analysis/synthesis method cannot trace the modulated structures that are present in the aperiodic component (see figure 4.7).

An informal listening test was also performed to compare the FWF representation to a simpler representation that takes into account the temporal structure of the noise. This simpler model (similar to the noise representation used in the

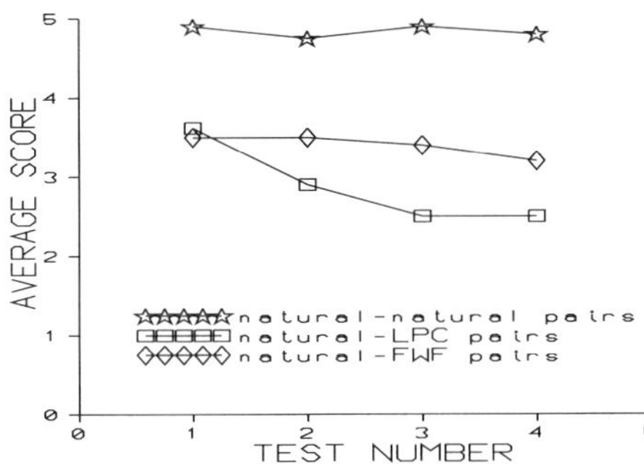

FIGURE 4.6. Perception test results. X-axis: condition number. Y-axis: average DCR score. A score of 5 corresponds to the answer "the signals within a pair are equals," and a score of 1 corresponds to the answer "the degradation in the second signal is very annoying." Stars denote pairs of identical signals. Diamonds denote pairs in which the second signal is reconstructed using the random FWF method. Squares denote pairs in which the second signal is reconstructed using LPC analysis/synthesis (from [Ric94]).

H+N model) consists of modulating (by an energy function) the signal obtained by filtering, with a normalized LPC filter, a white noise excitation source.

Our model seems to better represent the exact temporal structure of the noise and does not have the audible artifacts that can be seen in the other model (these artifacts are a consequence of a high amplitude of the excitation source (white noise) occurring at the same time as a high amplitude of the energy envelope function) (see figure 4.7). However, it seems that the two methods lead to results of comparable quality.

4.5 Speech Modifications

At the output of the analysis procedure, the aperiodic component is represented as a set of elementary wave forms well localized in the spectro-temporal domain. These wave forms are described by relevant acoustic parameters in the frequency domain (formant center frequencies, bandwidths, and amplitudes) as well as in the time domain (excitation times, instants of reference, initial phases) and thus provide various signal modification abilities.

In the context of realistic speech modifications, it is not sufficient to simply modify the speech signal. It is necessary to perform only those modifications that are possible in the speech production process. Although it is possible to separate a periodic and an aperiodic component, many voice-quality modifications affect both components. For example, the decay in intensity observed at the end of utterance results in changes in the glottal wave form, a higher spectral tilt, a lower periodic to aperiodic ratio, a lower aperiodic signal impulsiveness, etc. On the other hand, an increased vocal effort results in lower spectral tilt, a higher periodic to aperiodic

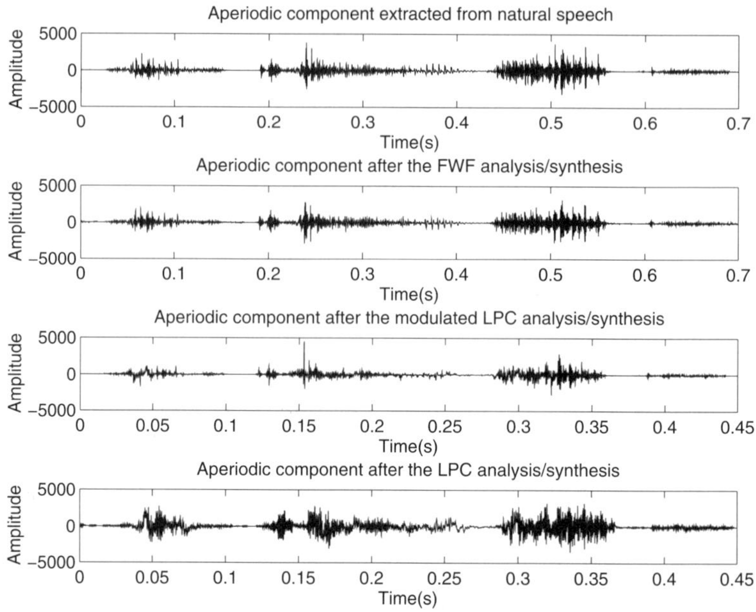

FIGURE 4.7. Time domain wave form of the (top curve) aperiodic component extracted from the speech signal with the PAP algorithm; (second curve) aperiodic component after the FWF analysis/synthesis method; (third curve) aperiodic component after the modulated LPC analysis/synthesis method; (bottom curve) aperiodic component after the LPC analysis/synthesis method.

ratio, a higher aperiodic signal impulsiveness, etc. Our current knowledge of these kinds of covariation of periodic and aperiodic parameters seems rather limited.

Due to the signal representation method proposed, several types of modification of the aperiodic components are straightforward. These modification capabilities are linked to the parameters that are available.

4.5.1 *Time Scaling*

Time scaling may be performed by simply modifying the reference instant (the time of generation) of each FWF. The results obtained are fairly good either for compression or dilation. However, for a large dilation coefficient, a more sophisticated procedure is needed, such as duplication in time of each wave form with lower amplitudes.

This type of time scaling results in a global dilation or compression of the signals without affecting the (possible) underlying periodicity of noise modulation.

4.5.2 *Spectral Modifications*

Format wave forms are defined by formant parameters. It is therefore easy to modify these parameters. Modifying formant center frequencies can be achieved simply by changing the corresponding parameter. It is also possible to change the formant spectral amplitudes. This allows us to change relevant parameters such as spectral tilt and noise amplitude in selected regions, and to shift the formants. These parameters are important for voice quality modification.

4.5.3 *Modification of the Aperiodic Component Impulsiveness*

For each FWF, it is also possible to control the individual time-domain envelope through the excitation time and bandwidth parameters. The time-domain envelope characterizes the modulation structure of the noise. Thus, it becomes possible to modify the overall depth of the time modulation of the stochastic component. This has an important consequence in the perceptual point of view as a deeper modulation gives a rougher voice with an impression of evident vocal effort, and a smoother modulation gives a softer and more whispery voice. Figure 4.8 illustrates the modification of the impulsiveness of a synthetic modulated signal produced by the FWF synthesizer.

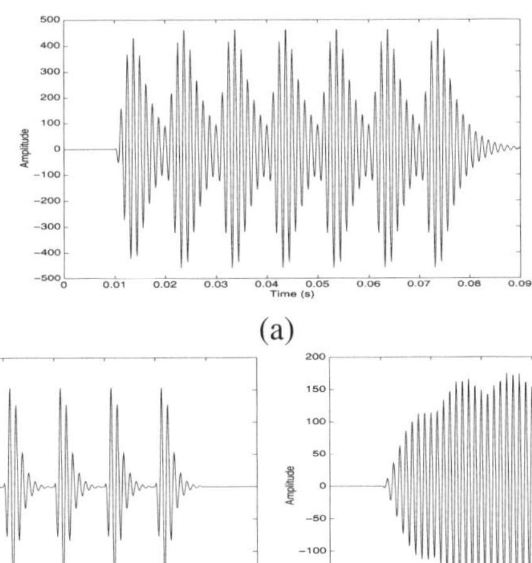

(a)

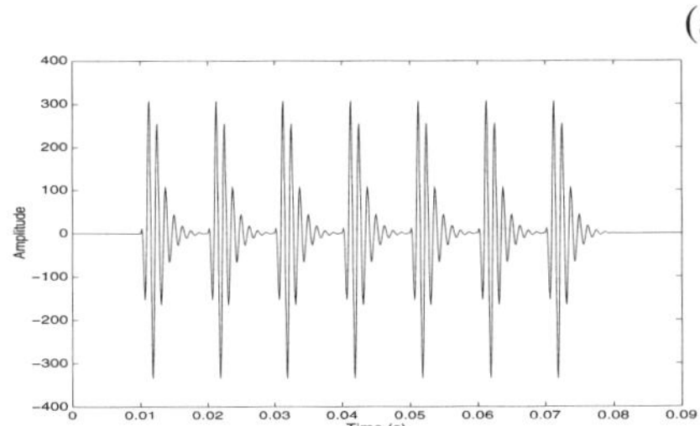

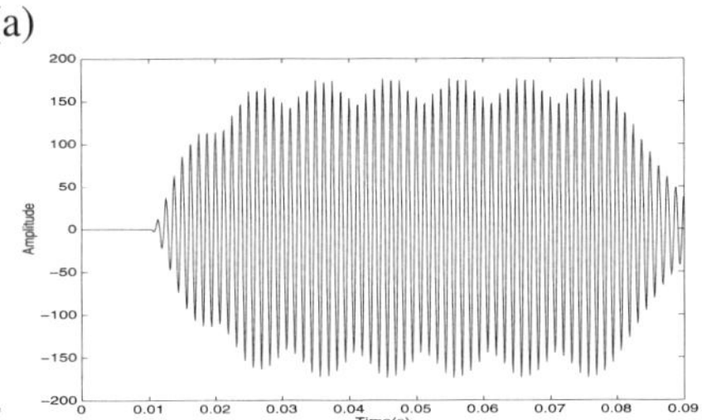

(b) (c)

FIGURE 4.8. Modification of the aperiodic component impulsiveness: curve (a) displays the original synthetic signal; curve (b) gives an example with a deeper modulation structure obtained by simultaneously increasing the bandwidth and decreasing the excitation time (or onset time); in contrast, curve (c) gives an example with a smoother modulation structure obtained by simultaneously decreasing the bandwidth and increasing the excitation time.

4.5.4 *Modification of the Periodic/Aperiodic Ratio*

The periodic/aperiodic ratio can be easily modified. This can be done either globally (for all the formants), or locally (for particular frequency regions). A joint modification of impulsiveness and periodic/aperiodic ratios makes it possible to change continuously from voiced to whispered speech.

Sound examples 3 to 6 of the audio demo illustrate these various speech signal modifications using the PAP decomposition algorithm and the FWF model for the aperiodic component.

4.6 Discussion and Conclusion

The PAP decomposition of the speech signal seems relevant for studying voice quality features and, in particular, breathiness, roughness, or whisperiness of a voice. Compared to sinusoidal-coding-based methods, our decomposition method has the advantage of a better modeling of the aperiodic component. Both periodic and aperiodic components are defined at each frequency point in the complex frequency domain. Therefore, there is no binary decision between harmonic or noise regions, but a variable amount of noise at each frequency. This better reflects the acoustic reality.

As for aperiodic component coding, a time-frequency elementary wave form decomposition was preferred to the widely used LPC synthesis scheme.

The algorithm presented in this chapter for the analysis and representation of the aperiodic component proved to be efficient for modeling this component, including the strongly modulated segments of it.

Although the synthesized noise quality and naturalness is better with our method than with a conventional LPC model, the complexity (both in terms of computation and data rate) is much higher. The aim of this study was to design a technique that is able to represent noise with accuracy and with various voice quality modification capabilities, but not to perform a parameter rate reduction. However, the complexity in terms of data rate does not seem to be excessive for practical synthesis. Typically, the number of FWF per second is of the order of 1000, which leads to a data rate of 5000 parameters per second. Furthermore, this data rate can be easily lowered by suppressing the low-energy FWF. Actually, more than 50 percent of FWF are nearly inaudible.

The drawbacks of this new analysis-synthesis method are of two types. The method is more expensive than other methods, especially in terms of computation. In addition, it depends heavily on the success of the PAP decomposition method. This decomposition method is acoustically relevant in the case of little random modulation of the voice source. If this is not the case, it is more difficult to assign a meaning to the two components, and therefore the speech modification quality degrades. Unfortunately, it is likely that this last drawback will be shared by all decomposition methods.

We think that this first attempt to modify the aperiodic component of the voice source brings new capabilities for voice quality modification. Therefore it opens new ways for modeling different voice qualities and different voice styles. It also offers new challenges because so little is currently known about the production and perception of the aperiodic component of speech signals.

Acknowledgments: We wish to thank Daniel J. Sinder for reading and commenting on the manuscript.

References

[Ale90] C. d'Alessandro. Time-frequency speech transformation based on an elementary wave form representation. *Speech Comm.* 9:419–431, 1990.

[AYD95] C. d'Alessandro, B. Yegnanarayana, and V. Darsinos. Decomposition of speech signals into deterministic and stochastic components. In *Proceedings of IEEE ICASSP'95,* Detroit, 760–764, 1995.

[DAY95] V. Darsinos C. d'Alessandro, and B. Yegnanarayana. Evaluation of a periodic/aperiodic speech decomposition algorithm. In *Proceedings of Eurospeech'95*, Madrid, Spain, 1995.

[Cha90] C. Chafe. Pulsed noise in self-sustained oscillations of musical instruments. In *Proceedings of IEEE ICASSP'90,* Albuquerque, 1157–1160, 1990.

[CCI92] CCITT. *Revised recommendation P.80 - "Methods for subjective determination of transmission quality."* SQEG, COM XII-118 E, International Telegraph and Telephone Consultative Committee (CCITT) (from Recommendation P.80, Blue Book, Volume V, 1989), 1992.

[CL91] D. G. Childers and C. K. Lee. Vocal quality factors: Analysis, synthesis and perception. *J. Acoust. Soc. Amer.* 9(5):2394–2410, 1991.

[DL92] T. Dutoit and H. Leich. Improving the TD-PSOLA text-to-speech synthesizer with a specially designed MBE re-synthesis of the segments database. In *Proceedings of EUSIPCO'92,* Brussels, Belgium, 343–346, 1992.

[GL88] D. Griffin D and J. S. Lim. Multiband excitation vocoder. *IEEE Trans. ASSP* ASSP-36(8):1223–1235, 1988.

[Her91] D. J. Hermes. Synthesis of breathy vowels : some research methods. *Speech Comm.* 10:497–502, 1991.

[Hil87] J. Hillenbrand. A methodological study of perturbation and additive noise in synthetically generated voice signals. *J. Speech and Hearing Res.* 30:448–461, 1987.

[Kla80] D. H. Klatt. Software for a cascade/parallel formant synthesizer. *J. Acoust. Soc. Amer.* 67(3):971–995, 1980.

[Kro93] G. de Krom. A cepstrum-based technique for determining a harmonics-to noise ratio in speech signals. *J. Speech and Hearing Res.* 36:254–266, 1993.

[LSM93] J. Laroche, Y. Stylianou, and E. Moulines. HNS: Speech modification based on a harmonic + noise model. In *Proceedings of IEEE ICASSP'93,* Minneapolis, 550–553, 1993.

[MQ92] R. J. McAulay and T. F. Quatieri. Low-rate speech coding based on the sinusoidal model. In *Advances in Speech Signal Processing,* S. Furui and M. M. Sondhi, eds., Marcel Dekker, New York, 165–208, 1992.

[Pap84] A. Papoulis. *Signal Analysis*. Mc Graw-Hill, New York, 1984.

[Ric92] G. Richard, C. d'Alessandro, and S. Grau. Unvoiced speech analysis and synthesis using Poissonian random formant-wave-functions. In *Proceedings of EUSIPCO'92,* Brussels, Belgium, 347–350, 1992.

[Ric94] G. Richard. *Modelisation de la composante stochastique de la parole.* PhD thesis, Universite de Paris-XI, Orsay, France, (in French), 1994.

[Rod80] X. Rodet. Time-domain formant-wave-function synthesis. In *Spoken Language Generation and Understanding,* J. C. Simon, ed., D. Reidel, Dordrecht, Netherlands, 1980. Also in *Comp. Music J.* 8(3):9–14, 1980.

[SS90] X. Serra and J. Smith. Spectral modeling synthesis: A sound analysis/synthesis system based on a deterministic plus stochastic decomposition. *Comp. Music J.* 14(4), 1990.

Appendix: Audio Demos

The audio demo contains sound examples, with explanations given.

5

On the Use of a Sinusoidal Model for Speech Synthesis in Text-to-Speech

Miguel Ángel Rodríguez Crespo
Pilar Sanz Velasco
Luis Monzón Serrano
José Gregorio Escalada Sardina

ABSTRACT We discuss the adequacy of a model that represents the speech signal as a sum of sine waves to the requirements of concatenative speech synthesis in text-to-speech (TTS). This model allows wide-range and high-quality prosodic modifications, and produces smooth transitions where the speech segments are joined. A preference test was carried out between speech synthesized using a pitch synchronous linear predictive coding (PS-LPC) synthesizer and the sinusoidal synthesizer.

5.1 Introduction

One of the most successful and widely used approaches to generate synthetic speech in TTS is by means of speech-segment concatenation. With this approach, it is necessary to have the capacity to modify the prosodic parameters (duration, fundamental frequency, amplitude) of the speech segments so that they match the synthetic prosody produced by the TTS. It is also necessary that the model used to encode, store, and synthesize the segments protect the speech signal from having discontinuities at the segment boundaries.

In Telefónica I+D, we have a TTS system for the Spanish language that uses speech-segment concatenation [Mac90]. This system has been shown to have very good quality [AGFLM94], although it still needs to be refined at the linguistic, the prosodic, and the acoustic levels.

Regarding this latter level, we are currently using a PS-LPC-based method in our TTS system for synthesizing the speech signal. This method gives very good results, as noted above, but we have found it has some disadvantages:

- The method is pitch synchronous. This means that before doing the LPC analysis it is necessary to get the epochs in the speech files that contain the speech segments in order to locate the points where the analysis windows are to be placed. Although there are good methods available to do this [TR90], some errors may arise.

- The frames obtained from the LPC analysis have to be classified as voiced or unvoiced using a binary decision. But real speech does not follow such a simple classification. As the treatment given in the synthesis to voiced and unvoiced frames is different (for example, in the duration modification or in the generation of the excitation signal), the voiced-unvoiced classification has to be corrected to avoid errors in the synthetic speech.

- In the PS-LPC method, each voiced frame represents a pitch period of the speech signal. When modifying the fundamental frequency and duration of the speech segments from their original values, it is found that this cannot be managed in a straightforward way, as fundamental frequency and duration modifications are closely related—the pitch period assigned to a voiced frame is also the duration of that frame. The only way to perform fundamental frequency and duration modifications is by inserting or deleting frames. So, the duration adjustment can be done only in steps equal to the duration of one frame (one pitch period). Therefore, this adjustment can be poor when the fundamental frequency is very low. Another problem to be considered is the selection of the places where a frame has to be inserted or deleted. This is not a trivial matter. The method used to do this is basically heuristic, trying to preserve the frames that represent the essential parts of the different sounds and the characteristics of the transitions between them.

A method for representing the speech signal as a sum of sine waves has been proposed [MQ86]. This method allows prosodic modifications in a direct and flexible way [QM92] and has built-in interpolation procedures that help smooth the synthetic speech. In addition, the sinusoidal model that we have explored as an alternative to the PS-LPC model does not show the disadvantages reported above for the following reasons:

- It uses a fixed frame rate: the analysis windows are located at equally spaced time intervals. Therefore, there is no need to obtain the epoch locations prior to the analysis.

- The sinusoidal model can manage mixed-excitation frames (frames that are partially voiced and partially unvoiced). The parameters that represent one frame of speech include a voicing probability that accounts for this fact and that is used during the synthesis.

- The fundamental frequency and duration modifications can be done independently and in a straightforward way. In particular, the duration adjustment can be done very accurately (with a precision of one speech sample) and without the need of inserting or deleting frames.

In the following sections we describe the sinusoidal model, its characteristics, the experiments carried out to test its adequacy to synthesize speech in TTS, and the results obtained.

5.2 Overview of the Sinusoidal Model

A detailed presentation of the sinusoidal model can be found in [MQ86], [QM92], and [FS91]. We give here a brief overview for convenience (see figure 5.1).

As usual, the sinusoidal model considers the speech signal as the result of passing a vocal cord excitation function $e(n)$ through a time-varying linear system $h(n)$ representing the characteristics of the vocal tract. For simplicity, it is assumed that this linear system includes the effects of the glottal pulse shape and the vocal tract impulse response.

Under the quasi-stationarity assumption, one frame of the excitation signal is represented as a sum of sine waves, avoiding the voiced-unvoiced decision :

$$e(n) = \sum_{k=1}^{L} a_k(n) \cdot \cos\left[(n - n_0) \cdot \omega_k\right] \tag{5.1}$$

where ω_k is the frequency of each sine wave and $a_k(n)$ is the amplitude associated with it. L is the number of sine waves in the speech bandwidth and n_0 represents the pitch pulse onset time. A pitch pulse occurs when all the sine waves add coherently (i.e., are in phase) [QM89]. The time $n = 0$ is the center of the analysis frame.

The time-varying vocal tract transfer function (the Fourier transform of h(n)) is represented by :

$$H(\omega, n) = M(\omega, n) \cdot \exp(j \cdot \Psi(\omega, n)) \tag{5.2}$$

where $M(\omega, n)$ and $\Psi(\omega, n)$ represent the amplitude and phase of the system transfer function.

As a result of passing the excitation function $e(n)$ through the time-varying linear system $h(n)$, the speech signal is represented as another sum of sine waves:

$$s(n) = \sum_{k=1}^{L} A_k(n) \cdot \cos(\Theta_k(n)) \tag{5.3}$$

where $A_k(n) = a_k(n) \cdot M_k(n)$ and $\Theta_k(n) = (n - n_0) \cdot \omega_k + \Psi_k(n)$. $M_k(n)$ and $\Psi_k(n)$ represent the amplitude and phase of the system function along the frequency track given by ω_k.

This separation of the excitation and system contributions to the signal amplitudes and phases allows us to treat them independently when making the prosodic transformations.

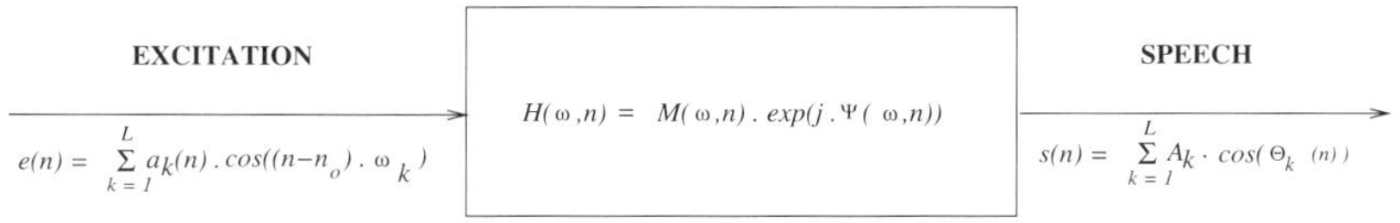

FIGURE 5.1. Sinusoidal model of speech production.

5.3 Sinusoidal Analysis

If we want to represent the speech signal using the general sinusoidal model outlined in the previous section, we need to take windows from the original speech signal and obtain the frequencies, amplitudes and phases used in equation (5.3).

This can be done by calculating a STFT using a pitch-adaptive Hamming window, with a width of at least 2.5 times the local average pitch period to reduce the distortion produced by the Hamming window.

The frequencies ω_k are the points in the frequency axis where there is a peak in the magnitude of the STFT. A_k $(n = 0)$ are the values of those peaks, and Θ_k $(n = 0)$ are the values of the phase of the STFT at the frequencies ω_k.

5.4 Sinusoidal Synthesis

In this section we deal with recovering the original speech signal using the general sinusoidal model.

Once we have a set of amplitudes, frequencies, and phases for each frame, we need to obtain the amplitude and phase functions, $A_k(n)$ and $\Theta_k(n)$ in order to obtain values of $s(n)$ for each value of the time index n between two analysis frames, using equation (5.3). If the number of samples between two consecutive analysis frames is N, this means we have to obtain values of $A_k(n)$ and $\Theta_k(n)$ from $n = 0$ (center of the first analysis frame) to $n = N - 1$ (one sample before the center of the second analysis frame), using an interpolation algorithm.

Before making any interpolation between amplitude or phase values associated with peaks of two consecutive frames, it is necessary to make a match between the peaks of one frame j and the following one $j + 1$. This match procedure is based on *connecting* one peak of the frame j to the peak with the nearest frequency of the frame $j + 1$. This frame-to-frame peak matching has to allow the *birth* and *death* of the spectral peaks in order to account for rapid changes in both the location and number of peaks between consecutive frames (as can happen in voiced-unvoiced transitions).

As a result of the frequency matching algorithm, each amplitude and phase of frame j is associated with a corresponding amplitude and phase of frame $j+1$ (an association can be a real *connection*, a *birth*, or a *death*).

A linear interpolation is used for the amplitudes, and a cubic polynomial with phase unwrapping is used for interpolating the phases. A detailed description of the frame-to-frame peak matching and the interpolation procedures can be found in [MQ86].

5.5 Simplification of the General Model

The general sinusoidal model as introduced has some drawbacks for use in TTS synthesis:

- The general sinusoidal model has too many parameters to be considered. Each frame of speech needs to be represented by a set of frequencies, each with its associated amplitude and phase. As a lot of memory would be needed to code and store the speech segment inventory of a TTS system, a reduction in the required number of parameters is sought.
- The fundamental frequency is not an explicit parameter of the general sinusoidal model. We need to find a sinusoidal representation of the signal that has the fundamental frequency as a parameter, so we can easily perform fundamental frequency modifications when synthesizing speech.

With these goals in mind, the general sinusoidal model has been simplified as described in the following subsections.

5.5.1 Harmonic Sinusoidal Model

The harmonic sinusoidal model is obtained by changing the general sinusoidal representation of the speech signal to another for which all of the frequencies are harmonically related (the sine wave frequencies are multiples of a fundamental frequency ω_0).

A method that estimates a *pitch* value has been described in [QM90], in which all the details of the method can be found.

The method uses a mean squared error (MSE) criterion to fit a harmonic set of sine waves to the set of sine waves obtained using the sinusoidal analysis presented in section 5.3. This pitch estimation method is inherently unambiguous, uses pitch-adaptative resolution, uses small-signal suppression to provide enhanced discrimination, and uses amplitude compression to eliminate the effects of pitch-formant interaction.

This method also gives a quality measure of the fit as a voicing probability. This voicing probability is used to define a voicing cut-off frequency that divides the frequency band into a voiced region and an unvoiced region (below and above the cut-off frequency).

As a result of using the harmonic sinusoidal model, the frequencies of the sine-wave components of the speech signal can be written as:

$$\omega_k = \begin{cases} k \cdot \omega_0 & \text{if } k \cdot \omega_0 \leq \omega_c(P_v) \\ (k - k^*) \cdot \omega_u + k^* \cdot \omega_0 & \text{otherwise} \end{cases} \tag{5.4}$$

where $k = 1, 2, \ldots$ (until ω_k reaches the bandwidth limit)

ω_0 is the voiced pitch value

$\omega_c(P_v)$ is the voicing cut-off frequency defined by the voicing

probability P_v
ω_u is the unvoiced region pitch value (a fixed value of 100 Hz is adequate)
k^* is the largest k for which $k \cdot \omega_0 \leq \omega_c(P_v)$

A *pitch* value is obtained for all the frames, regardless of the voicing characteristics of the frame. So ω_0 will be referred to as the *pitch*, although in the frames that are not voiced this terminology is not meaningful in the usual sense.

Some other harmonic models use a different technique (nonsinusoidal) for synthesizing the unvoiced regions in the spectrum [RG93], but this additional complexity is not necessary, as noted in [FS91].

We have to point out that the sinusoidal pitch extractor needs to know the values of an average pitch contour of the speech files to be processed. This contour can be previously obtained using some other common method of pitch estimation.

5.5.2 *System Amplitudes and Phases*

In section 5.2 it was noted that if the system and amplitude contributions to the amplitudes and phases in equation (5.3) are separated, then they can be manipulated independently to adequately perform the prosodic transformations.

In this subsection we present how the system amplitudes and phases can be coded and recovered.

Individual coding of the system amplitudes and phases for each frequency ω_k in a frame is not practical for two reasons:

- When we want to change the fundamental frequency from the original value, we need to know the system amplitudes and phases at frequencies that are different from the original set of frequencies ω_k.
- For normal pitch values encountered in adult speech, there are a lot of system amplitudes and phases to be coded.

System Amplitudes

The approach taken is to obtain an amplitude envelope that can be efficiently coded and that can be sampled at the appropriate points in the frequency axis.

As a first approximation, the amplitude envelope can be taken as the linear interpolation between the amplitudes of successive peaks obtained in the general sinusoidal model (the peaks in the magnitude of the STFT). The problem with this simple envelope estimator is that some spurious peaks (due to the side lobes of the window spectrum) could seriously distort the envelope estimation.

These problems can be avoided using the technique proposed in [Paul81], which was used in the development of the spectral envelope estimation vocoder (SEEVOC).

The SEEVOC algorithm assumes that a local average pitch value, $\overline{\omega}_0$, is known for every speech frame. The peak with the largest amplitude in the interval $[\overline{\omega}_0/2, 3\,\overline{\omega}_0/2]$ is searched. Once the frequency of this peak is found at frequency ω_1, the search is repeated in the interval $[\omega_1 + \overline{\omega}_0/2, \omega_1 + 3\,\overline{\omega}_0/2]$, and a new peak is found at frequency ω_2. The process is continued until the limit of the speech bandwidth is reached. If no peak is found in a frequency interval, then the largest of the two STFT magnitude values calculated at the interval end points is taken and placed at the interval center as a virtual peak.

The SEEVOC envelope is then obtained by applying linear interpolation to the amplitudes of the peaks selected using the SEEVOC peak-picking algorithm.

A cepstral representation can be used to provide a convenient parametric model for coding the SEEVOC envelope. Using this representation, the system amplitude at the center of a particular frame ($n = 0$), for a frequency ω_k, is given by the expression:

$$log(M_k(n = 0)) = c_0 + 2 \cdot \sum_{m=1}^{NCEP} c_m \cdot cos(m \cdot \omega_k) \tag{5.5}$$

where the c_m are the cepstral coefficients that are obtained by calculating the inverse FFT of the logarithm of the SEEVOC envelope. *NCEP* is the number of cepstral coefficients (a value of *NCEP* = 32 has been used).

The SEEVOC envelope allows the use of a high number of cepstral coefficients to adequately model the spectral zeros, avoiding the influence of the fine structure of the spectrum.

System Phases

Once we have obtained a cepstral representation for the system amplitude envelope, the system phase can be calculated at any frequency ω_k, if the system function is assumed to be minimum phase. The system phase at the center of a particular frame ($n = 0$) is then given by:

$$\Psi_k(n = 0) = -2 \cdot \sum_{m=1}^{NCEP} c_m \cdot sin(m \cdot \omega_k) \tag{5.6}$$

5.5.3 *Excitation Amplitudes and Phases*

The SEEVOC envelope was obtained as the linear interpolation of a selected set of peaks in a frame. As this envelope passes through the values of the STFT magnitude at the location of those peaks, the excitation contributions a_k ($n = 0$) to the amplitudes A_k ($n = 0$) in equation (5.3) can be set to 1.0 for all the peaks in a frame without loss of generality. Therefore, $A_k(n)$ will be equal to $M_k(n)$ and the only amplitude we have to take into account is the system amplitude.

The excitation phases for the center of a frame ($n = 0$) are given by $-n_0\omega_k$. Therefore, the calculation of the excitation contributions to the phases Θ_k ($n = 0$) relies on the calculation of the onset time n_0.

The onset time can be calculated using a mean squared error criterion [FS91]. The problem with this approach is that any small error in the calculation of n_0

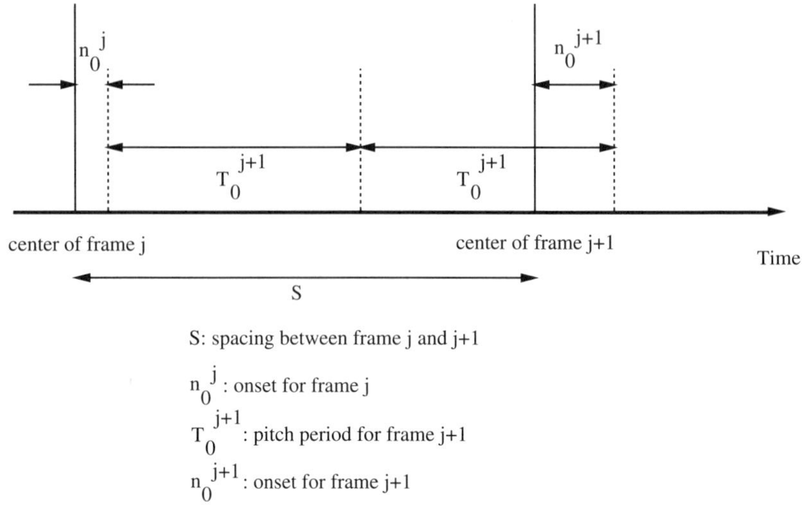

FIGURE 5.2. Onset time calculation.

can introduce large phase errors, especially in high frequencies. This makes the synthetic speech *rough*.

There is an alternative and more adequate way of calculating the onset time during the speech synthesis [FS91]. Because the function of the onset time is to bring the sine waves into phase at the times corresponding to the sequence of pitch periods, the onset time for a frame is obtained from the value of the onset time in the previous frame (but referred to the center of the present frame), and accumulating pitch periods of the present frame until the point nearest the center of the present frame is reached.

$$n_0^{j+1}(l) = n_0^j + T_0^{j+1} \cdot l \qquad l = 0, 1, 2, \ldots \tag{5.7}$$

Figure 5.2 illustrates how the synthetic onset times are determined.

A sine wave that is in the unvoiced region of the spectrum does not have to be phase locked with the voiced sine waves. The voicing-dependent model developed for estimating the speech signal phases from the excitation and system components [MQ91] includes a random component for the unvoiced frequencies to avoid periodicity during unvoiced speech.

5.6 Parameters of the Simplified Sinusoidal Model

As a result of the simplifications done to the general sinusoidal model, the set of parameters that a sinusoidal frame contains is:

- The set of cepstral coefficients.

- The pitch ω_0.
- The voicing probability.
- The spacing (number of speech samples) between a frame and the preceding one.

The original pitch values are preserved for the analysis-resynthesis experiments, but they are changed when modifying the fundamental frequency. It is not necessary to specify the spacing between frames in every frame for analysis-resynthesis, as it is constant due to the fixed frame rate used in the analysis, but it is necessary when doing nonuniform duration modifications (different frames could then have different spacing between frames).

The onset times are obtained during the synthesis using the pitch and the spacing between frames.

The values A_k $(n = 0)$ and Θ_k $(n = 0)$ that are needed to make the sinusoidal synthesis using the general sinusoidal model can be obtained from the system and excitation contributions, as explained in the previous section.

The minimum-phase assumption was introduced to derive an estimate of the system phase. Although this assumption removes the precise details of the temporal structure of the speech signal, the resulting resynthetic speech is of good quality, but not equivalent to that obtained using the phases measured in the spectrum [MQ91, FS91].

Some preliminary experiments showed that the quality of the speech resynthesized using the simplified sinusoidal model was good enough for use in a TTS system. Therefore, we decided to maintain the minimum phase assumption because it allows a relatively small number of parameters to code the speech signal and produces a resynthetic speech whose quality was considered better than the one produced by other minimum phase models. This is probably due to some advantages of the simplified sinusoidal model, such as the ability to manage partially voiced and partially unvoiced frames or its powerful built-in interpolation procedures [MQ86].

5.7 Fundamental Frequency and Duration Modifications

As stated in section 5.1, the sinusoidal model allows independent and straightforward fundamental frequency and duration modifications. Only two parameters in a sinusoidal frame need to be changed to perform those modifications:

- the pitch ω_0, to modify the fundamental frequency
- the spacing between the frame and the preceding one, to modify the duration.

The modifications in the pitch and the spacing affect the calculation of the system and excitation contributions to A_k $(n = 0)$ and Θ_k $(n = 0)$ in the ways that are explained in the following subsections.

5.7.1 System Contributions

Only the fundamental frequency changes affect the system contributions to the amplitude and phase. The only calculation that is necessary is to obtain the values of the new frequencies ω_k using equation (5.4) of the harmonic sinusoidal model with the new value of the pitch ω_0. Once the frequencies ω_k are known, equations (5.5) and (5.6) are used to calculate the values M_k $(n = 0)$ and Ψ_k $(n = 0)$ for every ω_k.

5.7.2 Excitation Contributions

The excitation contribution to the amplitude is assumed to be 1.0, as explained before.

The excitation phases $-n_0\omega_k$ are obtained for every frequency ω_k. The values of ω_k are derived from the pitch of the frame, using equation (5.4). The value of n_0 is obtained from the value of n_0 in the preceding frame and the values of the pitch and the spacing (affected by the duration modification) in the present frame, as illustrated in figure 5.3.

The phases for the frequencies ω_k that are in the unvoiced region of the spectrum, as defined in equation (5.4), are considered random.

5.8 Analysis and Resynthesis Experiments

Some experiments have been done to test the ability of the sinusoidal model to modify the prosody of male and female speech.

The sentences were digitized using an 8 kHz sampling frequency. An average fundamental frequency contour was generated prior to sinusoidal analysis, as it is needed for obtaining the SEEVOC envelope and performing the sinusoidal pitch and voicing detection. The sinusoidal analysis was done using Hamming windows of length 3 times the local average pitch period. The frame step was kept fixed at 5 msec. A 1024-point STFT was calculated for each frame, and the frequencies corresponding to peaks in the magnitude spectrum were picked and refined using quadratic interpolation. The SEEVOC algorithm was used to estimate the system amplitude envelope, and the sinusoidal pitch and voicing probability were obtained. Then a set of 32 cepstral coefficients was calculated from the SEEVOC envelope.

The speech was resynthesized with the harmonic version of the sinusoidal model, using frame-to-frame peak matching, linear interpolation for the amplitudes, and cubic polynomial interpolation with phase unwrapping for the phases, and allowing the original pitch and duration (spacing between adjacent analysis frames) to be modified. Wide-range prosodic transformations were done, halving and doubling the original pitch and duration values (an example of these can be seen in figures 5.4 and 5.5). The transformations were applied to all phonemes, regardless of their nature. In spite of the extreme and unnatural transformations that were done, the resulting synthetic speech did not have severe artifacts.

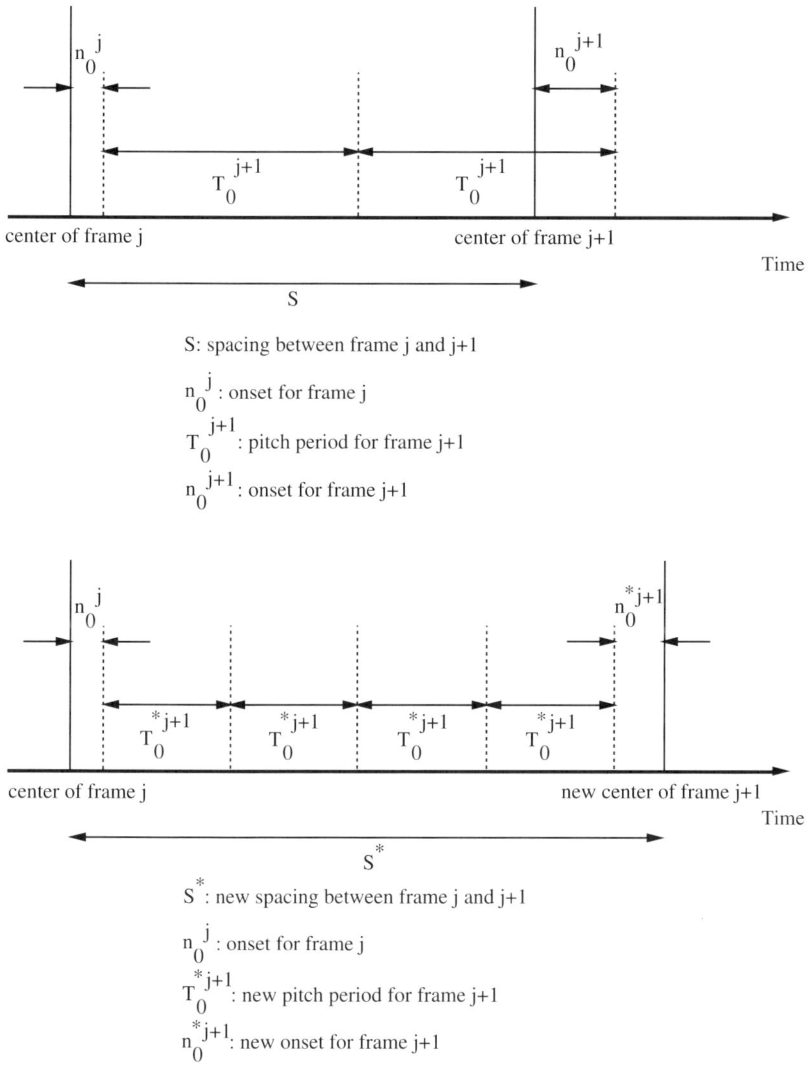

FIGURE 5.3. Influence of fundamental frequency and duration modifications on the onset calculation.

Some sentences were selected and synthesized from speech segments using the sinusoidal model (SINU) described above and the PS-LPC model of our TTS system. The PS-LPC model of our TTS system uses 16 reflection coefficients. The analysis windows are centered at the epoch locations during voiced speech and spaced 5 msec during unvoiced speech.

The speech segments used to synthesize the sentences were taken from the segment dictionaries for one female voice and one male voice available in our TTS system, and the same boundaries between allophones were used. The synthesis was

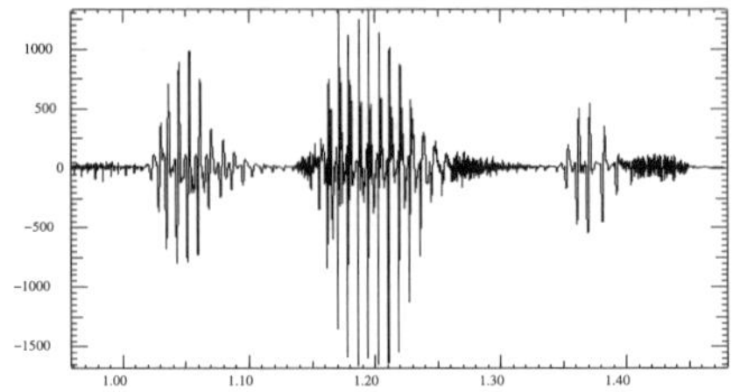

FIGURE 5.4. Waveform resynthesized without modifications.

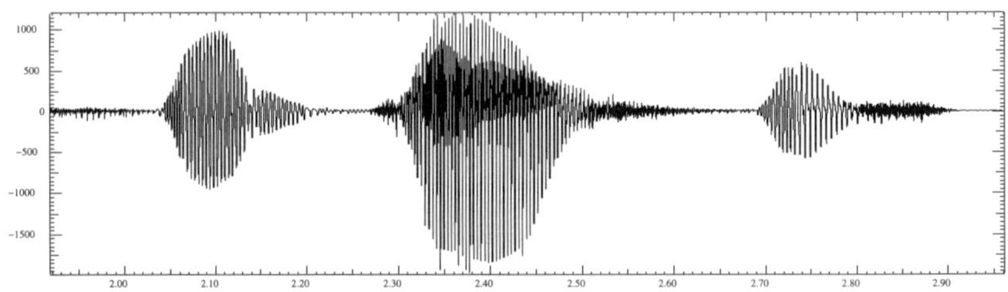

FIGURE 5.5. Waveform resynthesized doubling fundamental frequency and duration.

TABLE 5.1. Male speech.

Natural prosody		Synthetic prosody	
PS-LPC	SINU	PS-LPC	SINU
13.3% (2/15)	86.7% (13/15)	6.7% (1/15)	93.3% (14/15)

done with natural prosody (fundamental frequency and duration) extracted from the sentence uttered by the same speakers used to build the segment dictionaries, and also with the synthetic prosody produced by the TTS. Fifteen people were asked to listen to a pair of synthetic sentences with the same prosody, one using the PS-LPC model and the other using the sinusoidal model, and to select one of them. The result of the test, which is summarized in tables 5.1 and 5.2, shows that the sinusoidal synthetic speech was clearly preferred, especially in the case of male speech and synthetic prosody.

These results encouraged us to integrate the sinusoidal model in our TTS system. A first integration has recently been achieved, and a first complete speech segment sinusoidal inventory has been created, following the structure presented in [Oli90]. The sinusoidal synthetic speech produced by the TTS has been consid-

TABLE 5.2. Female speech.

Natural prosody		Synthetic prosody	
PS-LPC	SINU	PS-LPC	SINU
40% (6/15)	60% (9/15)	33.3% (5/15)	66.7% (10/15)

ered in informal listenings as clearly better than the PS-LPC synthetic speech. The sinusoidal synthesis has been shown to be less sensitive to amplitude normalization problems in the speech segment inventory.

It is interesting to point out that the identity of the speaker whose voice was used to create the speech segment inventory seems to be more easily recognized using the sinusoidal model.

5.9 Conclusions

A sinusoidal speech model has been proposed that has many desirable properties for use in TTS, overcoming some disadvantages found in the PS-LPC model. The sinusoidal model is able to perform wide-range and high-quality prosodic transformations. The duration can be modified without being influenced by durations (pitch periods) of the original frames, and without the need to insert or delete frames (although frame insertion or deletions can still be done if needed). Mixed-excitation frames (partially voiced and partially unvoiced) are taken into account, avoiding the common errors that arise during binary voiced-unvoiced decisions (although the voicing nature of the frames can still be corrected). A fixed frame rate is used, so there is no need to obtain the epochs prior to the analysis. A comparison has been made between concatenative synthetic speech generated using the PS-LPC and the sinusoidal model. This comparison has shown that the sinusoidal synthetic speech is clearly preferred, although this preference is not so clear in the case of female speech. This can be due to the use of a sampling frequency value (8 kHz) that is too low to adequately reflect the characteristics of female speech. Further work has to be done to clarify this point.

Once the sinusoidal synthesizer has been integrated in our TTS system, a more formal evaluation has to be done to assess the quality of the sinusoidal synthetic speech and to obtain clearer evidence of its advantages over the PS-LPC model. It would also be very interesting to compare the performance of the sinusoidal model to PSOLA-based models used in other high-quality TTS systems.

Regarding memory requirements, the PS-LPC speech segment inventory (including about 800 units) needs 260 Kbytes of memory. The same inventory for the sinusoidal synthesizer needs 400 Kbytes. This size is judged to be reasonable, but it could easily be reduced at the expense of a small voice quality penalty.

The major drawback of the sinusoidal synthesizer is its computational requirements. Up to now, only the most obvious optimizations have been done to the algorithm, and the computational load is about 10 times the computational load of the PS-LPC synthesizer for a low-pitched male voice (the computational load of the sinusoidal synthesizer increases as the pitch gets lower).

Although further work has to be done to reduce the computational load, the whole TTS system with the sinusoidal synthesizer runs in real time on a Sun Sparcstation 10 with no additional hardware (no digital signal processing (DSP) board).

References

[AGFLM94] L. Aguilar, J. M. Garrido, J. M. Fernández, J. Llisterri, A. Macarrón, L. Monzón, and M. A. and Rodríguez. Evaluation of a Spanish text-to-speech system. In *Proceedings of the Second ESCA/IEEE Workshop on Speech Synthesis*, New Paltz, 207–211, 1994.

[FS91] S. Furui and M. M. Sondhi, eds. *Advances in Speech Signal Processing*. Marcel Dekker, New York, 165–208, 1991.

[Mac90] A. Macarrón-Larumbe. Design and generation of the acoustic database of a text-to-speech synthesizer for Spanish. In *Proceedings of ESCA Workshop on Speech Synthesis*, Autrans, 31–34, 1990.

[MQ86] R. J. McAulay and T. F. Quatieri. Speech analysis-synthesis based on a sinusoidal representation. *IEEE Trans. Acoust., Speech and Signal Proc.* ASSP-34:744–754, 1986.

[MQ91] R. J. McAulay and T. F. Quatieri. Sine-wave phase coding at low data rates. In *Proceedings of IEEE Int. Conf. Acoust., Speech and Signal Proc.* Toronto, 577–580, 1991.

[Oli90] J. P. Olive. A new algorithm for a concatenative speech synthesis algorithm using an augmented acoustic inventory of speech sounds. In *Proceedings of ESCA Workshop on Speech Synthesis*, Autrans, 25–29, 1990.

[Paul81] D. B. Paul. The spectral envelope estimation vocoder. In *Proceedings of IEEE Trans. Acoust., Speech and Sign. Proc.* ASSP-29:786–794, 1981.

[QM89] T. F. Quatieri and R. J. McAulay. Phase coherence in speech reconstruction for enhancement and coding applications. In *Proceedings of IEEE Int. Conf., Acoust., Speech and Signal Proc.* Glasgow, 207–252, 1989.

[QM90] T. F. Quatieri and R. J. McAulay. Pitch estimation and voicing detection based on a sinusoidal model. In *Proceedings of IEEE Trans. Acoust., Speech and Signal Proc.*, Albuquerque, 249–252, 1990.

[QM92] T. F. Quatieri and R. J. McAulay. Shape invariant time-scale and pitch modification of speech. In *Proceedings of IEEE Trans. Acoust., Speech and Signal Proc.* 40:497–510, 1992.

[RG93] E. Rodríguez-Banga and C. García-Mateo. New frequency domain prosodic modification techniques. In *Proceedings of ESCA Eurospeech '93*, Berlin, 987–990, 1993.

[Tal90] D. Talkin and J. Rowley Pitch synchronous analysis and synthesis for TTS. In *Proceedings of ESCA workshop on speech synthesis*, Autrans, 55–58, 1990.

Appendix: Audio Demos

The CDROM contains two demonstrations of our system.

Section II

Linguistic Analysis

6

Section Introduction. The Analysis of Text in *Text*-to-Speech Synthesis

Richard W. Sproat

The word "text" comprises half of the contentful portion of the phrase "text-to-speech," yet a count of the papers presented at recent workshops and colloquia on synthesis reveals that text-analysis has not received anything like half the attention of the synthesis community. A recent electronic survey, which queried synthesis researchers on what they felt were the most important obstacles to high-quality synthesis for the foreseeable future, asked hardly any questions specifically on text-analysis. Well-known compendia of research on speech synthesis, including [AHK87, HP93], do, of course, cover topics related to text-analysis, such as text "normalization," word pronunciation, and phrasing. Indeed, both of the texts cited show a very nice proportion of chapters devoted to text-analysis issues, but this does not reflect the situation in the literature at large. Naturally, there are important applications of speech synthesis technology in which text-analysis per se is not an issue: one presumes, for example, that *message-to-speech* systems fall into this category because in message-to-speech systems the machine "knows" the structure of the linguistic message it is composing and does not need to derive that structure from text. But probably most research synthesis systems, and certainly most commercial synthesis systems, have been aimed at full text-to-speech capabilities. Seen in that light it is puzzling that there has been so much attention to such matters as the correct modeling of the source and the vocal tract transfer function, or the computation of duration and intonation contours, and so little on how one might do a better job of extracting from text the relevant linguistic parameters for these computations. One supposes that, as with all such matters, those topics that are regarded as important in the field have much more to do with the areas of expertise of the workers in it, combined with the influence of a few people regarded generally as being key principals, than with any extrinsic plan to cover all of the terrain equally thoroughly.

Work on text-analysis for speech synthesis has traditionally been organized around the various subproblems that one faces when attempting to convert a textual representation of language into a linguistic representation appropriate for synthesis. Such subproblems include "normalization" issues, such as date and number expansion, and abbreviation expansion; morphological analysis, primarily for the purpose of word pronunciation; and syntactic analysis, primarily for the purpose of phrasing and accent assignment. There are various tried and true methods for handling these kinds of problems, with systems based on hand-built rules being the dominant approach. Good, solid systems can be constructed using such methods; chapter 10 by Ferri, Pierucci, and Sanzone discusses work that falls into this category.

Naturally this more traditional work builds upon rule-based methods for linguistic analysis that have been developed in the computational linguistics literature. One of the trends that has defined computational linguistics from the latter half of the 1980s to the present has been the move away from hand-constructed, rule-based methods, and the return to statistical corpus-based methods. In recent years, there have been several interesting applications of such techniques to text-analysis for synthesis, and the chapters by Yarowsky (12) and Daelemans and van den Bosch (7) are nice instances of this line of research.

Of course, just as important as the methods used for converting from text into linguistic representations (from which synthesis parameters are subsequently computed) are the nature of the linguistic representations themselves. Most work on linguistic analysis for synthesis has assumed string-based representations of the kind that were used in phonology a quarter of a century ago [CH68] and were first applied in synthesis in the MITalk system. Representations in phonology have evolved far beyond that stage, however, and recently workers, particularly those with a background in *declarative phonology*, have argued that the hierarchical structures used by current phonologists, along with declarative rather than procedural rule-based approaches to expressing linguistic generalizations, are useful in synthesis. The main argument is that such representations allow for a much more adequate characterization of various phonological phenomena, especially phenomena such as assimilation, lenition, and deletion, which might best be characterized as phonetic implementation processes. The chapters by Dirksen and Coleman (8) and Local and Ogden (9) do not deal with text-analysis per se, but do deal with the issue of what linguistic representation one should synthesize from, with both chapters arguing strongly for a declarative hierarchical approach.

As might be expected, research into text-analysis for speech synthesis has tended to focus on those areas that have seemed more tractable: one can do a very good job of pronouncing most words—and at least an acceptable job of phrasing, accentuation, and intonation assignment—by morphological and (partial) syntactic analysis, largely ignoring, or only modeling in a simple manner, the effects of paragraph- or narrative-level discourse information. Yet this kind information certainly *is* relevant for producing plausible output, as has been argued in various previous works and as Nakatani further shows in chapter 11. The results of such research could in principle be applied today in message-to-speech systems, but full application in text-to-speech must await more powerful methods in text-analysis than what we are currently able to provide.

References

[AHK87] J. Allen, M. S. Hunnicutt, and D. Klatt. *From Text to Speech: the MITalk System.* Cambridge University Press, Cambridge, 1987.

[CH68] N. Chomsky and M. Halle. *The Sound Pattern of English.* Harper and Row, New York, 1968.

[HP93] V. van Heuven and L. Pols. *Analysis and Synthesis of Speech: Strategic Research Towards High-Quality Text-to-Speech Generation.* Mouton de Gruyter, Berlin, 1993.

7

Language-Independent Data-Oriented Grapheme-to-Phoneme Conversion

Walter M. P. Daelemans
Antal P. J. van den Bosch

ABSTRACT We describe an approach to grapheme-to-phoneme conversion that is both *language-independent* and *data-oriented*. Given a set of examples (spelling words with their associated phonetic representation) in a language, a grapheme-to-phoneme conversion system is automatically produced for that language that takes as its input the spelling of words and produces as its output the phonetic transcription according to the rules implicit in the training data. We describe the design of the system and compare its performance to knowledge-based and alternative data-oriented approaches.

7.1 Introduction

Grapheme-to-phoneme conversion is an essential module in any text-to-speech system. It can be described as a function mapping the spelling form of words to a string of phonetic symbols representing the pronunciation of the word. The largest part of research on this process focuses on developing systems that implement various levels of language-specific linguistic knowledge (especially morphological and phonotactic knowledge, but often also syntactic knowledge). It is generally assumed that this is essential to solving the task. MITalk [AHK87] is a classic example of such a knowledge-based approach for English; for Dutch, Morpa-cum-Morphon [HH93, NH93] can be considered state-of-the-art. A clearly disadvantageous consequence of the knowledge-based strategy is the fact that it requires a large amount of handcrafting of linguistic rules and data during development. Furthermore, language-specificity of a grapheme-to-phoneme model tends to be incompatible with reusability of the developed implementation. That is, for each language a specific set of rules and principles has to be found in order to successfully run the model.

In this chapter we describe an implemented grapheme-to-phoneme conversion architecture and explore to what extent it allows data-oriented induction of a grapheme-to-phoneme mapping on the basis of examples, thereby alleviating the expensive linguistic engineering phase. Input to our system is a set of spelling

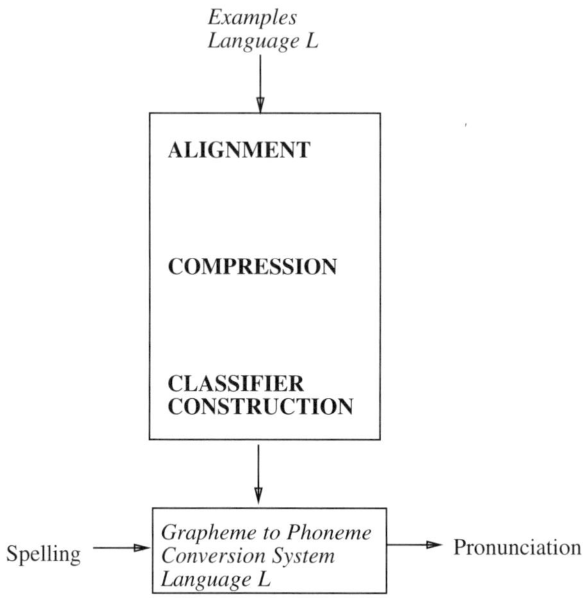

FIGURE 7.1. Overview of the architecture.

words with their associated pronunciations in a target phonemic or phonetic alphabet (the training data). Spelling and pronunciation do not have to be aligned. The phonetic transcription can be taken from machine-readable or scanned dictionaries, or from automatic phoneme recognition. The words may represent text in context (when effects transgressing word boundaries have to be modeled) or isolated words. Output of the system is a grapheme-to-phoneme conversion system that takes as its input the spelling of words and produces as its output the phonetic or phonemic transcription according to the rules implicit in the training data (see figure 7.1 for an overview of the architecture).

The approach has a number of desirable properties:

1. It is *data-oriented*. The output system is constructed automatically from the training data, thereby effectively removing some of the knowledge acquisition bottlenecks. Knowledge-based solutions to the problem need considerable handcrafting of phonological and morphological data structures, analysis, and synthesis programs.

2. It is in principle *language-independent and reusable*. Versions of the system for French, Dutch, and English have been constructed automatically using the same architecture on different sets of training data. In linguistic approaches, the handcrafting has to be redone for each new language. Languages with other writing systems may exist, however, for which the approach is less well suited.

3. It achieves a *high accuracy*. Output of the Dutch version has been extensively compared to the results of a state-of-the-art "hand-crafted" linguistic system. The data-oriented solution proved to be significantly more accurate in predicting phonetic transcriptions of previously unseen words.

7.2 Design of the System

The system consists of the following modules:

1. *Automatic alignment*: Strings of spelling symbols and strings of phonetic symbols have to be made of equal length in order to be processed by the other modules.
2. *Automatic training set compression*: Part of the training data is represented in a compact way using tree structures and information gain.
3. *Automatic classifier construction*: Using the compacted training data, a classifier is constructed that extrapolates from its memory structures to new, unseen input spelling strings.

We will discuss these stages in turn. Compression and classifier construction are achieved at the same time in our current implementation and will be discussed together.

7.2.1 Alignment

The spelling and the phonetic trancription of a word often differ in length. The grapheme-to-phoneme conversion module described in the next section demands, however, that the two representations be of equal length, so that each individual graphemic symbol can be mapped to a single phonetic symbol. The problem is to align the two representations in such a way that graphemes or strings of graphemes are consistently associated with the same phonetic symbols. This is not a trivial task: If this alignment has to be done by hand, it is extremely labor-intensive. Consider the example alignments of the words "rusty" and "rookie":

graphemes	r	u	s	t	y	r	oo	k	ie
	|	|	|	|	|	|	|	|	|
phonemes	r	ʌ	s	t	ɪ	r	u	k	i

Whereas the alignment of "rusty" is very straightforward (i.e., five one-to-one grapheme-to-phoneme mappings), the alignment of "rookie" involves the knowledge that the "oo" cluster maps to the /u/-phoneme, and that the "ie" cluster maps to the /i/-phoneme. No other partitioning of the spelling string is allowed, or at least is intuitively correct. Our automatic alignment algorithm attempts to make the length of a word's spelling string equal to the length of its transcription by adding *null* phonemes to the transcriptions. Nulls have to be inserted in the transcription at those points in the word where a grapheme cluster maps to one phoneme. In the example of the word "rookie," this would be done as follows ("–" depicts phonemic nulls):

graphemes	r	o	o	k	i	e
	|	|	|	|	|	|
phonemes	r	u	–	k	i	–

Note that it is arbitrary whether one uses the /ru-ki-/ alignment or the /r-uk-i/ alignment (i.e., whether one chooses to map the last or the first of a grapheme cluster to a phonetic null), as long as it is done consistently. Alignments such as /ruki - -/ or /- - ruki/ should never be generated because they imply highly improbable mappings. An alignment should always follow the principles that (i) its grapheme-to-phoneme mappings are viable (i.e., that they can be motivated intuitively or linguistically); (ii) the combination of all mappings within the alignment has a maximal probability; and (ii) the mapping is consistent with alignments of other, similar words.

The first part of the algorithm automatically captures the probabilities of all possible phoneme-grapheme mappings in an association matrix. For each letter in the spelling string, the association with the phoneme that occurs at the same position in the *unaligned* transcription is increased; furthermore, if a spelling string is longer than its transcription, phonemes that precede the letter position are also counted, as they may be associated with the target letter as well. In other words, the algorithm determines, for each letter in the graphemic string, all phonemes in the transcription to which the letter *may* map. Furthermore, these phonemes are weighted differently. As shown in the diagram below, the algorithm shifts the unaligned phonemic word from the left-aligned position to the right-aligned position. Taking the k as an example, the algorithm adds a score of 8 to the association between k and /i/, a score of 4 to the association between k and /k/, and a score of 2 to the association between k and /u/. Note that shifting is repeated at most three times; at the third shift, which may occur with longer words, associations are increased by a score of 1. Other values for these weights result in slightly (but not significantly) worse results. The idea behind this weighting is the simple assumption that left-aligned phonemic transcriptions *generally* contain more correct grapheme-to-phoneme mappings than right-aligned transcriptions.

graphemes	r	o	o	k	i	e	association
	|	|	|	|	|	|	score
phonemes	r	u	k	i	_	_	8
phonemes, 1 right shift	_	r	u	k	i	_	4
phonemes, 2 right shifts	_	_	r	u	k	i	2

Although a lot of noise is added to the association matrix by including associations that are less probable (e.g., in our example, the mapping between k and /i/), the use of this shifting association "window" ensures that the most probable associated phoneme is always captured in this window. More important, it turns out that the "intuitive" or "linguistically appropriate" mappings receive the highest scores in the end. When all words of the database are processed this way, the association scores in the association matrix are converted into association probabilities.

The second part of the alignment algorithm generates for each pair of unaligned spelling and phoneme strings all possible (combinations of) insertions of null phonemes in the transcription. For each hypothesised string, a total association probability is computed by multiplying the scores of all individual letter-phoneme association probabilities of each hypothesized alignment. The hypothesis with the highest total association probability is then taken as output of the algorithm.

The resulting alignment, albeit very consistent, is not always identical to the intuitive alignment applied by human coders. To test its efficacy, we compared classification accuracy of the complete system when using a hand-aligned training set as opposed to the automatically aligned training set. The results indicate that there is no significant difference in classification accuracy: the alignments result in equally accurate systems. The resulting IG-Tree (see following section) is on average about 3 percent larger with the automatically generated alignment, however.

7.2.2 *IG-Trees: Compression and Classifier Construction*

Rules and Patterns

The way grapheme-to-phoneme conversion works in our system can be seen as optimized, generalized lexical lookup. Reasoning is based on *analogy* in the overall correspondence of the writing system and the pronunciation of a certain language. Words that are spelled similarly are pronounced similarly. The system automatically learns to find those parts of words on which similarity matching can safely be performed. This is perhaps best illustrated in the example of the word <behave>. An "analogical" model, which operates with a certain similarity metric and which has already encountered (and learned) roughly similar words such as <shave>, <beehive>, and <have>, and perhaps even <behave> itself, will certainly have a number of clues as to how <behave> is pronounced. However, in this example, problems may arise with the <a> of <behave>. If the similarity matcher of the analogical model decides to retrieve the pronunciation of the word <have> as the pronunciation of <-have> in <behave>, the incorrect pronunciation /bihæv/ would result. Our system does not take such overgeneralization risks. The model is extremely sensitive to context in the sense that it will have stored the knowledge that <have> is not enough context to be certain of the pronunciation of the <a>. Instead, the system will look for more contextual information. In a current implementation of the system trained on English words, the system decides to take /e^{j}/ as output only when it finds the subword chunk <ehave> in the input word. Note that this system has encountered the word <behave> during training, but that for the case of the pronunciation of the <a> it was not necessary to store the complete word, as there were no other words with the subword chunk <ehave> with a different pronunciation of the <a>. In sum, the system stores single letter–phoneme correspondences with a minimal context that is sufficient to be certain that the mapping is unambiguous (in the training material).

Each of these subword–phoneme correspondences can be seen as a context-sensitive rewrite rule, which rewrites a letter in context to a phoneme. As the context may be of any width, many of these rewrite rules are much more specific than would be a typical rule in a rule-based grapheme-to-phoneme module; many even contain whole words. Although one would be tempted to categorize such an approach as rule-based, it could equally well be regarded as a lexical approach. Some of the rules are so specific that it would make more sense to call them lexical patterns. The rules are a compressed version of the text-to-speech corpus it is trained on. After training, they contain in a compressed format complete knowledge of the pronunciation of all words of the learning material (lossless compression), except for homographs such as <read> (pronounced as /rɛd/ or /rid/), of which only one pronunciation is stored. This is certainly a shortcoming for languages such as Russian, in which pronunciation depends on stress placement, which in its turn depends on lexical properties that cannot be deduced from the form of the word (which is the only information the current implementation of our system uses). We will return to this problem in our conclusion. See chapter 12 by Yarowsky for an approach to homograph disambiguation using decision lists, which could be integrated with our approach.

To illustrate the appearance of automatically extracted rules, table 7.1 lists some examples of the model extracted from English data and French data.

TABLE 7.1. Examples of automatically extracted subword–phoneme correspondences, with their associated phonemes, from English and French data. Example words containing the rules are given. Dots represent unused context positions; underscores represent word boundaries.

English											Phoneme	Example Word
.	.	.	.	.	v	o	.	.	.	.	v	**v**oucher
.	.	.	e	s	i	d	e	n	.	.	ə	pres**i**dent
.	.	.	.	w	o	_	.	.	.	.	u	tw**o**
.	.	.	_	h	a	v	e	_	.	.	æ	h**a**ve
French											Phoneme	Example Word
.	.	.	.	.	ç	.	.	.	.	.	s	fran**ç**ais
.	.	_	_	b	e	a	u	x	.	.	o	b**e**aux
.	.	.	.	v	i	n	_	.	.	.	E	v**i**n
.	.	.	.	n	c	o	.	.	.	.	k	fran**c**ophone

From table 7.1, it can clearly be seen that some correspondences express very general pronunciation knowledge, whereas others are used to disambiguate between only a few words (e.g., for <esiden> /ə / discriminates <president> from <reside>). There is no clear distinction between rules and lexical patterns; they are regarded as extremes in a continuum.

Compressing Knowledge into an IG-Tree

The system does not actually store a large list of context-sensitive rewrite rules and lexical patterns. It compresses the information contained in these rules even more by storing them in a decision tree. Each rule is represented as a path in this tree. A path consists of a starting node, which represents the target letter that is to be mapped to a phoneme; the consecutive nodes represent the consecutive context letters. The order in which these letters are attached to the path is governed by computing their overall relative importance in disambiguating the mapping. This is done using information gain, a computational metric based on information theory (hence the name IG-tree). A description of this metric is given in [DGD94]. Computation of the information gain of context positions renders a result that is constant for all corpora used. Trivially, the focus letter itself is the most important "context" letter. The farther the context position is removed from the focus letter, the less important that position is for disambiguation, on the average. Furthermore, there is an as yet unexplained difference between right and left context: Right context positions are computed to be slightly more important than their respective left context positions. In practice, this leads to an ordering in which the first character on the right is the first context expansion, that is, the first node down the tree. Then follows the first character on the left, then the second character on the right, then the second character on the left, and this alternating pattern simply repeats. To visualize the way in which knowledge is organized in the decision tree, figure 7.2 displays the part of the tree in which the pronunciation of the <a> in the word <behave> is stored.

With the <a>-node as a starting point, the node labeled with the first character on the right, <v>, is the second node accessed in the path. Then, the <h>-node, the first character to the left of the <a>, is taken. At that point, the only possible extensions stored in the tree are <have>, <havi> (from <having>) and <havo> (from <havoc>); the pronunciation at that point is still ambiguous. Then, the <e>-node is accessed, which leaves open the extensions < have> (underscores depict word boundaries), <ehave>, and <shave>. As mentioned earlier, at the next step, the model retrieves the unambiguous phonemic mapping /e^j/, when the final <e> node is reached.

It can be seen that the depth of a path reflects in a certain sense the ambiguity of the mapping it represents. End nodes near the top of the decision tree typically belong to highly regular pronunciations. For example, the French model contains at the top layer of the tree the end node <ç>, as this special character always maps to /s/ regardless of the context. An example of an extremely ambiguous mapping is that of the first <o> of <photograph>, /o/, which has the competitor word

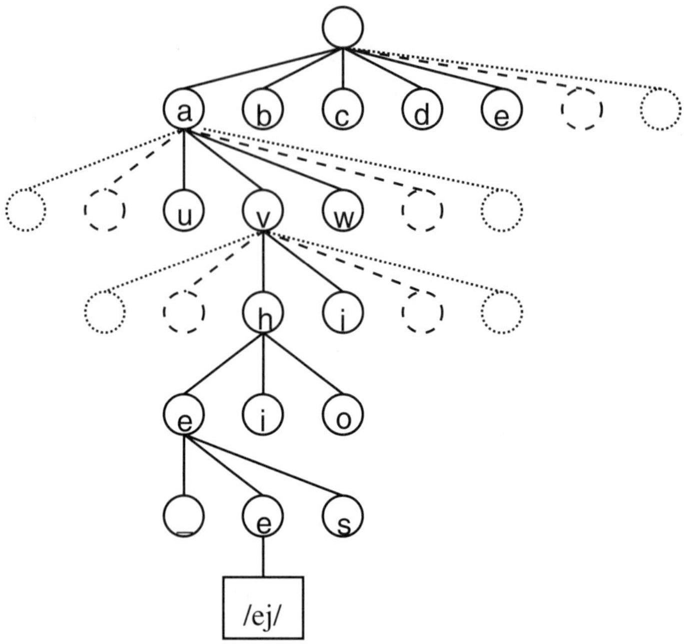

FIGURE 7.2. Retrieval of the pronunciation of < *a* >, /e^j/, of the word < *behave* >. The path represents the minimally disambiguating context < *ehave* >.

<photography>, in which the <o> maps to /ə/. In this case, a context to the right of width 8 is needed for disambiguation.

Best Guess Strategy

In our approach, all spelling-to-phonology knowledge contained within the learning material is stored lossless, with the exception of homographs, of which only one pronunciation is kept. The rule-based aspect of the decision tree, however, enables the model also to generalize to new cases. To retrieve the pronunciation of a word that was not in the learning material, each letter of the new word is taken as a starting point of tree search. The search then traverses the tree, up to the point at which either (1) the search successfully meets an end node, or (2) the search fails as the specific context of the new word was not encountered in the learning material and consequently was not stored as a path in the tree. In the first case, the phonemic label of the end node is simply taken as the phonemic mapping of the new word's letter. In the second case, the exact matching strategy is taken over by a *best-guess* strategy.

In present implementations of the system, the best-guess strategy is implemented in a straightforward way. When building a path in the tree, the construction algorithm constantly has to check whether an unambiguous phonemic mapping has been reached. At each node, the algorithm searches in the learning material for all phonemic mappings of the path at that point of extension. In cases for which there is more than one possible phonemic mapping, the algorithm computes what is the most *probable* mapping at that point. Computation is based on occurrences: the most frequent mapping in the learning material is preferred (in case of ties, a random choice is made). This extra information is stored with each nonending node. When a search fails, the system returns the most probable phonemic mapping stored in the node at which the search fails.

7.3 Related Approaches

The information gain metric used to select context features while building the IG-tree is used in a similar way in C4.5 decision tree learning [Qui93]. The main difference with C4.5's approach to decision-tree learning is the fact that our model computes the ordering only once for the complete tree, whereas in C4.5 the ordering is computed at every node. Another difference is that in the IG-tree, nodes are created until all training set ambiguity is resolved (there is no pruning of the tree).

In earlier versions of this system [DB93, BD93], a different approach was taken to handle strings that were not found in the IG-tree. A similar effect to defaults at leaf nodes was achieved by combining the compression and IG-tree building with a form of similarity-based reasoning (based on the k-nearest neighbor decision rule, see, e.g., [DK82]. During training, a memory base is incrementally built consisting of *exemplars*. In our domain of grapheme-to-phoneme conversion, an exemplar consists of a string of graphemes (one focus grapheme surrounded by context graphemes) with one or more associated phonemes and their distributions (as there may exist more phonemic mappings for one graphemic string). During testing, a test pattern (a graphemic string) is matched against all stored exemplars. If the test pattern is in memory, the category with the highest frequency associated with it is used as output. If it is not in memory, all stored exemplars are sorted according to the similarity of their graphemic string pattern to the test pattern. The (most frequent) phonemic mapping of the highest ranking exemplar is then predicted as the category of the test pattern. Daelemans and van den Bosch [DB92] extended the basic nearest-neighbor algorithm by introducing information gain as a means to assign different weights to different grapheme positions when computing the similarity between training and test patterns (instead of using a distance metric based on overlap of patterns).

In the combination of the IG-tree and the information gain–aided nearest neighbor algorithm, the IG-tree classification algorithm stops when a graphemic string is not found in the tree. Instead of using the information on the most probable phoneme at the nonending leaf node, the nearest-neighbor algorithm takes over and searches in its exemplar base for the best-matching exemplar to the graphemic pattern under consideration. The phoneme associated with the best-matching exemplar is then taken as "the best guess."

We experimented both with the nearest-neighbor approach independent from the IG-tree algorithm, and with the combination of the two. We found that the current approach (select most probable category at ambiguous leaf nodes), being computationally simpler and conceptually more elegant, achieves the same generalization accuracy. We therefore adopted this simpler approach. The current approach is also as accurate as using the nearest-neighbor technique independently.

Earlier work on the application of nearest-neighbor approaches (memory-based reasoning, [SW86, Sta87]) to the phonemization problem using the NetTalk data (MBRTalk) showed a better performance than NetTalk [SR87] itself; however, at the cost of an expensive, domain-dependent computational measure of dissimilarity that seems to be computationally feasible only when working on a massive parallel computer such as the Connection Machine. Another analogy-based system (or rather a hybrid combination of case-based reasoning and relaxation in a localist interactive activation network) is PRO [Leh87]. However, the reported performance of this system is not very convincing, nor is the need for a combination of connectionist and case-based techniques apparent. Dietterich and Bakiri [DB91] systematically compared the performance of ID3 (a predecessor of C4.5 [Qui93]) and BP on the NetTalk data. Their conclusion is that BP consistently outperforms ID3 because the former captures statistical information that the latter does not. However, they demonstrate that ID3 can be extended to capture this statistical information. Dietterich and Bakiri suggest that there is still substantial room for improvement in learning methods for text-to-speech mapping, and it is indeed the case that our approach significantly outperforms BP.

The application of compression techniques such as our IG-tree to the phonemization problem has not yet been reported on as such in the literature. Golding and Rosenbloom [GR91] studied the interaction of rule-based reasoning and case-based reasoning in the task of pronouncing surnames. They claimed that such a hybrid approach is preferable, in which the output of the rules is used *unless* a compelling analogy exists in the case base. If a compelling analogy is found, it overrides the rule output. In this approach, the (hand-crafted) rules are interpreted as implementing the defaults, and the cases the *pockets of exceptions*. Our IG-tree method works along a different dimension: both default mappings (rules) and pockets of exceptions are represented in the IG-tree in a uniform way.

7.4 Evaluation

In this section, we compare the accuracy of our approach to the knowledge-based approach and to alternative data-oriented approaches for Dutch grapheme-to-phoneme conversion. More detailed information about the comparison can be found in Van den Bosch and Daelemans [BD93].

7.4.1 Connectionism

In our approach, explicit use is made of analogical reasoning. Back-propagation learning in feed-forward connectionist networks (BP), too, uses similarity (or analogy), but more implicitly. An input pattern activates an output pattern which is similar to the activation pattern of those items that are similar to the new item. Complexity is added by the fact that an intermediate hidden layer of units "redefines" similarity by extracting features from the activation patterns of the input layer.

Automatic learning of grapheme-to-phoneme conversion of English (NetTalk, [SR87]) has been acclaimed as a success story for BP. The approach was replicated for Dutch in NetSpraak [WH90]. It is therefore appropriate to compare our alternative data-oriented approach to BP.

The performance scores on randomly selected, unseen test words (generalization accuracy) show a best score for the IG-tree approach. Similar results were obtained for different training and test sets.

Model	Generalization Accuracy on Phonemes
BP	91.3
IG-tree	95.1

7.4.2 *The Linguistic Knowledge-Based Approach*

The traditional linguistic knowledge-based approach of grapheme-to-phoneme conversion has produced various examples of combined rule-based and lexicon-based models. The developers of all of these models shared the assumption that the presence of linguistic (phonotactic, morphological) knowledge is essential for a grapheme-to-phoneme model to perform at a reasonably high level.

In Morpa-cum-Morphon [HH93, NH93], a state-of-the-art system for Dutch, grapheme-to-phoneme conversion is done in two steps. First, Morpa decomposes a word into a list of morphemes. These morphemes are looked up in a lexicon. Each morpheme is associated with its category and a phonemic transcription. The phonemic transcriptions of the consecutive morphemes are concatenated to form an underlying phonemic representation of the word. Morphon then applies a number of phonological rules to this underlying representation, deriving the surface pronunciation of the word. The system is the result of a five-year research effort sponsored by the Dutch government and industry, and is generally acclaimed to be the best system available.

We applied the IG-tree method to the same test data used to evaluate the Morpa-cum-Morphon system to make a comparison. Again, we see that the IG-tree scores significantly higher.[1]

---

[1]Note that the performance of Morpa-cum-Morphon on the benchmark has been boosted to 88.7% by incorporating a data-oriented probabilistic morphological analysis component [Hee93].

Model	Generalization Accuracy on Words
IG-tree	89.5
Morpa-cum-Morphon	85.3

7.5 Conclusion

The most surprising result of our research is that an extremely simple method (based on compressing a training set) yields the best accuracy results (judged by measuring *generalization* accuracy), suggesting that previous knowledge-based approaches as well as more computationally expensive learning approaches to at least some aspects of the problem were overkill.

The system described, the IG-tree, constructs a grapheme-to-phoneme conversion module on the basis of an unaligned corpus of spelling words and their phonetic representations. The approach is data-oriented (eliminates linguistic engineering), language-independent (reusable for different dialects or languages), and accurate (when compared to knowledge-based and alternative data-oriented methods).

Current limitations are the absence of word stress computation (but see [DGD94] for a compatible data-oriented approach to this problem) and sentence accent computation (for which syntactic, semantic/pragmatic and discourse information are required). The present word pronunciation modules output by our system can be combined with existing approaches to this problem, however. This raises a question concerning modularity in the design of data-oriented (learning) approaches to grapheme-to-phoneme conversion. Should stress assignment be integrated with grapheme-to-phoneme conversion or should two separate systems be trained to learn these two aspects of the problem, one using as input the output of the other? This is a subject for further research.

Finally, as we mentioned earlier, our approach fails miserably in all cases where different pronunciations correspond to the same spelling string (if two pronunciations differ in their spelling, if only in a single letter, there is no problem, but in that case, useful generalizations may be missed). For languages such as Russian this is clearly an important shortcoming. However, our approach is not limited to using only spelling information. Additional features (e.g., lexical category) can be added to the spelling features as input, and these would then be used in the construction of the IG-tree without any change to the present system. In that case, the added features would be used in generalization as well.

Acknowledgments: We thank the two anonymous referees for useful comments on an earlier version of this chapter.

REFERENCES

[AHK87] J. Allen, S. Hunnicut, and D. H. Klatt. *From Text to Speech: The MITalk System.* Cambridge University Press, Cambridge, 1987.

[BD93] A. van den Bosch and W. Daelemans, Data-oriented methods for grapheme-to-phoneme conversion. In *Proceedings of the Sixth Conference of the European Chapter of the ACL*, 45–53, 1993.

[DB91] T. G. Dietterich and G. Bakiri. Error-correcting output codes: A general method for improving multiclass inductive learning programs. In *Proceedings AAAI-91*, Menlo Park, CA, 572–577, 1991.

[DB92] W. Daelemans and A. van den Bosch. Generalization performance of back-propagation learning on a syllabification task. In *Proceedings of the 3rd Twente Workshop on Language Technology*, M. Drossaers and A. Nijholt, eds. Universiteit Twente, Enschede, 27–37, 1992.

[DB93] W. Daelemans and A. van den Bosch. TABTALK: Reusability in data-oriented grapheme-to-phoneme conversion. In *Proceedings of Eurospeech*, Berlin, 1459–1466, 1993.

[DGD94] W. Daelemans, S. Gillis, and G. Durieux. The acquisition of stress, a data-oriented approach. *Computational Linguistics* 20(3):421–451, 1994.

[DK82] P. A. Devijver and J. Kittler. *Pattern Recognition: A Statistical Approach.* Prentice-Hall, London, 1982

[GR91] A. R. Golding and P. S. Rosenbloom. Improving rule-based systems through case-based reasoning. In *Proceedings AAAI-91*, Menlo Park, CA, 22–27, 1991.

[HH93] J. Heemskerk and V. J. van Heuven. MORPA, a lexicon-based MORphological PArser. In *Analysis and Synthesis of Speech: Strategic Research Towards High-Quality Text-to-Speech Generation*, V. J. van Heuven and L. C. W. Pols, eds. Mouton de Gruyter, Berlin, 1993.

[Hee93] J. Heemskerk. A probabilistic context-free grammar for disambiguation in morphological parsing. In *Proceedings EACL-93*, Utrecht, 1993.

[Leh87] W. Lehnert. Case-based problem solving with a large knowledge base of learned cases. In *Proceedings AAAI-87*, Seattle, WA, 1987.

[NH93] A. Nunn and V. J. van Heuven. MORPHON, lexicon-based text-to-phoneme conversion and phonological rules. In *Analysis and Synthesis of Speech: Strategic Research Towards High-Quality Text-to-Speech Generation*, V. J. van Heuven and L. C. W. Pols, eds. Mouton de Gruyter, Berlin, 1993.

[Qui93] J.R. Quinlan, *C4.5: Programs for Machine Learning*. Morgan Kaufmann, San Francisco, CA, 1993.

[SR87] T. J. Sejnowski and C. R. Rosenberg. Parallel networks that learn to pronounce English text. *Complex Systems* 1:145–168, 1987.

[Sta87] C. W. Stanfill. Memory-based reasoning applied to English pronunciation. In *Proceedings AAAI-87*, Seattle, WA, 577–581, 1987.

[SW86] C. W. Stanfill and D. Waltz. Toward memory-based reasoning. *Communications of the ACM*, 29(12):1213–1228, 1986.

[WH90] A. Weijters and G. Hoppenbrouwers. NetSpraak: Een neuraal netwerk voor grafeem-foneem-omzetting. *Tabu* 20(1):1–25, 1990.

8

All-Prosodic Speech Synthesis

Arthur Dirksen
John S. Coleman

ABSTRACT We present a speech synthesis architecture, IPOX, which allows the integration of various aspects of prosodic structure at different structural levels. This is achieved by using a hierarchical, metrical representation of the input string in analysis as well as phonetic interpretation. The output of the latter step consists of parameters for the Klatt synthesizer. The architecture is based primarily on YorkTalk [Col92, Col94, Loc92], but differs in that it uses a rule compiler [Dir93], which allows a clean separation of linguistic statements and computational execution as well as a more concise statement of various kinds of generalizations.

8.1 Introduction

A major problem in speech synthesis is the integration of various aspects of prosodic structure at different structural levels. We present an architecture in which this problem is addressed in a linguistically sophisticated manner. Our system, IPOX, is based on the idea that it is possible to generate connected, rhythmically appropriate speech from a hierarchically structured representation, a prosodic tree. This metrical representation is assigned by parsing an input string using declarative, constraint-based grammars with a standard parsing algorithm. Each node in the metrical representation is then assigned a temporal domain within which its phonetic exponents are evaluated. This evaluation is done in a top-down fashion, allowing lower-level prosodic constituents to modify the exponents of higher-level nodes. The phonetic exponents of adjacent nodes in the metrical tree are allowed to overlap with one another. Also, the order in which constituents are evaluated depends on the prosodic configuration in which they appear. Within the syllable, heads are evaluated before nonheads, allowing metrically weak constituents such as onset and coda to adapt to their strong sister constituents (rime and nucleus, respectively) with which they overlap. Across syllables, the order of interpretation is left-to-right, so that each syllable is "glued" to the previous one. After all phonetic exponents have been evaluated, a parameter file for the Klatt formant synthesizer is generated.

The architecture of IPOX is rather similar to that of YorkTalk [Col92, Col94, Loc92], on which it is based, but is different in a number of respects:

- YorkTalk representations are implemented as arbitrary Prolog terms. In IPOX, metrical structure is made explicit in the representation [DQ93]. This has made it possible to define general algorithms to process these structures in various ways, whereas in YorkTalk such algorithms are spelled out on a case-by-case basis.

- The YorkTalk morphological and phonotactic parser is a Prolog DCG (definite clause grammar). IPOX, on the other hand, uses a rule compiler, which forces the developer to keep linguistic rules separate from the control logic, which is fixed.

- IPOX includes a facility to state feature co-occurrence restrictions separately from the phrase structure rules [Dir93].

More generally, IPOX aims to further formalize and extend the YorkTalk architecture, making it more flexible and easier to adapt to different languages, which is one of our long-term goals. Also, IPOX integrates all functions, including synthesis and sound output, in a single executable file (which runs under Windows on a PC with a standard 16-bit sound card), using graphics to display analysis trees, temporal structure, phonetic interpretation, and audio output waveforms. However, the system is still under development. Currently, there is no interface between morphosyntactic structure and phrase-level prosodic structure, although grammars for each of these modules have been developed separately. As a consequence, speech output temporarily suffers from certain linguistic limitations. We call this architecture "all-prosodic" because the phonological approach is based on metrical theory, making extensive use of distinctive features at nonterminal nodes, and the approach to phonetic interpretation is based on phonetic parameters computed in parallel rather than a sequence of concatenative units.

In this chapter, we discuss the various components of IPOX, illustrated with examples from British English that have been generated using rule sets adapted from (an earlier version of) YorkTalk.[1] First, in section 8.2, we present the basic architecture of IPOX, illustrated with the analysis and generation of isolated monosyllables. Section 8.3 discusses the use of overlap and compression in the generation of polysyllabic utterances and demonstrates how various kinds of vowel reduction can be obtained by varying the speech rhythm. Section 8.4 discusses the generation of connected speech, illustrated with a detailed consideration of the sentence "*This is an adaptable system*."

[1]All examples discussed in this chapter are provided on the CD-ROM (see Appendix) in the form of audio files generated by IPOX.

8.2 Architecture

8.2.1 Analysis

The analysis component of IPOX uses declarative, constraint-based phrase structure grammars to analyze input text and assign a metrical-prosodic representation. A declarative grammar is one in which the rules define the well-formedness of grammatical representations without specifying procedures for constructing such representations. In this section, we discuss how the internal structure of English syllables is defined in terms of such a grammar.

Basic syllable structure is assigned by the following two—presumably universal—rules:

```
syl --> (onset / rime).
rime --> (nucleus \ coda).
```

In these rules, the slash is used to encode which of two nodes is the prosodic head: the forward slash (/) indicates the pattern *weak-strong*, the backward slash (\) indicates *strong-weak*. Further (often language-specific) rules are used to define the internal structure of onset, nucleus, and coda (including the possibility that onset or coda may remain empty).

One of the properties that must be specified by the syllable grammar is syllable weight, which is an important factor in the distribution of stressed versus unstressed syllables in quantity-sensitive stress systems such as that of English. For example, the following rule forces heavy syllables (and light syllables that precede a light syllable) to be the head of a foot:

```
foot --> (syl \ syl:[-heavy]).
```

As is well known, syllable weight is usually determined only by the internal structure of the rime, disregarding the onset. For English, we assume that a syllable is heavy if the nucleus or the coda branches.[2] Thus, we might annotate the above rules for syllable and rime as follows (capital letters are used to indicate shared variables within a single rule):

```
syl:[heavy=A] --> (onset / rime:[heavy=A]).
rime:[heavy=A] --> (nucleus:[branching=A] \ coda:[branching=A]).
```

This works if we also write rules that assign the feature specification **[+branching]** to branching nuclei and codas,[3] while leaving nonbranching nuclei and codas unspecified for this feature. For example:

```
coda:[+branching] --> (cons \ cons).
coda --> cons.
```

[2]This particular formulation hinges on our assumption of maximal ambisyllabicity (subsection 8.3.1).

[3]Note that the name of a feature does not by itself imply any interpretation whatsoever and serves only a mnemonic purpose.

If we set up the grammar this way, light syllables remain unspecified for the feature *heavy*, and are accepted anywhere. Heavy syllables, on the other hand are specified as **[+heavy]**. They cannot appear as the weak node of a foot, as this would involve conflicting feature specifications, a situation not allowed in a declarative system.

A better approach, however, is to encode feature co-occurrence restrictions separately by means of *templates*, which are general descriptions of phrase structure rules. Advantages include the following: linguistic universals and language-specific rules are separated better, generalizations across rules can be stated just once, and readability of grammars is improved.

In the present example, we need the following templates:

```
[heavy=A] --> ([] / [heavy=A]).
[heavy=A] --> [branching=A], [branching=A].
[+branching] --> [], [].
```

The rule compiler applies every template to every rule with which it unifies. The first template is applied to every phrase structure rule that introduces the *weak-strong* pattern. The second and third templates are applied to every phrase structure rule that introduces a binary-branching structure, whether *weak-strong* or *strong-weak*, because there is no slash in the template. However, the feature unifications specified in templates are instantiated only to the extent that categories are defined for the relevant features. So, if only the categories *syl* and *rime* are defined for the feature *heavy*, and the feature *branching* is limited to nucleus and coda, the above templates derive the same effect as the phrase structure rule annotations they replace.

The syllable grammar also defines “spreading” of vocalic place features, which is how we model allophonic variation due to coarticulation. In our grammar, both vowels and consonants are defined for vocalic place features, which are complex feature structures of the following form:

```
voc:[+/-grv, +/-rnd, height=close/mid/open]
```

The idea is that *voc* features encode the primary articulation of vowels and glides, and the secondary articulation of consonants. The term “V-place” has recently been proposed by some other phonologists for roughly the same purposes. Vowels and glides are inherently specified for *voc* features, whereas nasals and obstruents are not. Liquids have inherent specifications as well; however, /l/ is “clear” (i.e., **[–grv]**) in onset position; “dark” (i.e., **[+grv]**) in coda position; and /r/ is unspecified for rounding.

To the extent that *voc* features are unspecified, a value is obtained for each feature through spreading. In YorkTalk, spreading of *voc* features is defined in the phrase structure rules on a case-by-case basis. In IPOX, this is done by means of templates, such that *voc* features appear to spread from *strong* nodes (where they

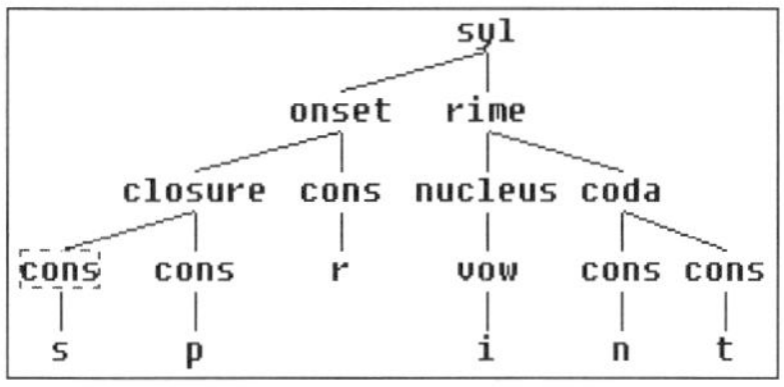

FIGURE 8.1. Syllable structure of /sprint/.

originate) to *weak* nodes.[4] As an example, consider onset /s/ in *sprint*, analyzed as shown in figure 8.1.

Having no inherent specification for *voc* features, /s/ (or, rather, the onset constituent that directly dominates /s/) must obtain values for these features through spreading. Since /r/ is unspecified for the feature *rnd*, /s/ obtains the value **[–rnd]** in accordance with /i/, just as it would in *spit*. However, because prevocalic /r/ is dark [OGC93, p. 216], /r/ is specified **[+grv]**, which is shared with /s/. With respect to this feature, /s/ in *sprint* is more like /s/ in *spot*. As a result, the *voc* features associated with /s/ in *sprint*, *spit*, and *spot* are all different. Phonetic interpretation is sensitive to these differences, which are reflected in the output waveforms as subtly but appropriately different frication spectra associated with /s/. (See Demo 1 on the CD-ROM.)

8.2.2 *Phonetic Interpretation 1: Temporal Interpretation*

At the next stage, the metrical-prosodic representation is assigned a temporal interpretation. As in YorkTalk, constituents are allowed to overlap with one another in various ways. Unlike YorkTalk, temporal interpretation in IPOX is determined by the compiler, and the user is prevented from incorporating ad hoc temporal interpretation constraints. All the developer can and must do is provide statements for durations assigned to constituents, which can be treated as terminals as far as phonetic interpretation is concerned, and the extent of overlap between adjacent nodes in the metrical-prosodic representation.

These statements are interpreted differently depending on whether a constituent occurs within a syllable. Across syllables, a *left-to-right strategy* is used, in which the duration of a constituent is the sum of the durations of each of the individual syllables within that constituent minus the amounts of overlap specified at the boundaries between those syllables. Within syllables, a *co-production strategy* is used, in which the duration of a constituent equals the duration assigned to the prosodic head of that constituent, and weak nodes are overlaid on their strong sister nodes at the left or right edge, depending on whether the configuration is *weak-strong* or *strong-weak*, respectively.

---

[4]In actuality, we are dealing with nondirectional feature sharing. However, if a feature is inherently specified on a vowel, and is shared with a consonant, then it "appears" to have been spread from the vowel onto the consonant.

When the *coproduction strategy* is used, the following conventions are observed:

1. If no duration is specified for a weak node, the duration is equal to the amount of overlap between the weak node and its strong sister node.
2. Alternatively, if no overlap is specified between a weak and a strong node, the amount of overlap equals the duration of the weak node.
3. A negative overlap quantity specifies the amount of nonoverlap (i.e., separation) between a weak node and its strong sister node.

As an example of a duration rule, the following statement asigns a duration of 680 ms to a nucleus marked as **[+long]** (a long vowel or a diphthong):

```
nucleus:[+long] => 680.
```

In the phonology, long vowels and diphthongs are modeled as branching nuclei, dominating both a vowel and an off-glide constituent (e.g., /iy/ for a front high long vowel; /ay/ for a front low-to-high diphthong). The above statement, however, effectively makes the nucleus a terminal element of the structure as far as phonetic interpretation is concerned. In other words, the phonetic interpretation of long vowels and diphthongs is holistic rather than compositional, even though they are assigned internal structure in the phonological analysis.

As an example of a rule for overlap, the following statement fixes the overlap between onset and rime at 200 ms:

```
onset + rime => 200.
```

Because the metrical relation between onset and rime is *weak-strong*, the onset is overlaid on the rime for the first 200 ms.

The various possibilities offered by the coproduction model are exemplified by the temporal interpretation of the syllable *sprint*, shown in the screen clip in figure 8.2.

As can be seen, consonants in the onset are temporally interpreted rather differently from coda consonants. Specifically, onset constituents do not have inherent durations: their durations follow from statements about nonoverlap between sister constituents within an onset. Coda constituents do have inherent durations, as well as statements about nonoverlap between sister constituents within a coda.

The phonetic exponents of a constituent are calculated relative to the start time and duration of the constituent. This does not mean, however, that these exponents are confined to the space allocated by temporal interpretation. Thus, durations of constituents cannot be equated with concatenative segment durations, nor can

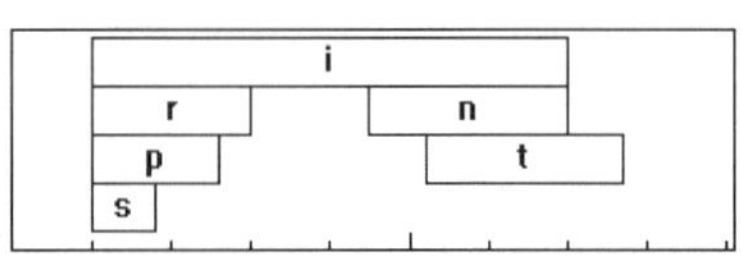

FIGURE 8.2. Temporal interpretation of /sprint/.

the boundaries in figure 8.2 be understood as boundaries between segments in the orthodox sense. On the other hand, the relation between perceptual segment durations and temporal interpretation is not totally arbitrary in actual practice, even if it is in principle. For example, the "distance" between onset and coda in a syllable is usually a fairly good measure of the perceptual duration of the vowel with which they are coproduced (see chapter 9 of this volume).

8.2.3 Phonetic Interpretation 2: Parametric Interpretation

Temporal interpretation is only one aspect of phonetic interpretation. In addition, values for the synthesizer parameters must also be determined in order to provide a complete phonetic interpretation of the abstract phonological structures we initially computed. Parametric phonetic interpretation is done by evaluation of a set of phonetic exponency rules, which have the following general form:

```
Category:Features => Exponents.
```

These rules are evaluated for each node in the metrical-prosodic representation. The nodes of the tree are visited in a *top-down* fashion. In this way, holistic properties of a constituent are computed first, to be worked out in finer detail by lower-level constituents. As an example, consider the generation of F_0 contours for metrical feet within a single phrase. The phonetic exponency of a foot for this parameter is a simple linear high-to-low fall. At lower levels of prosodic structure, individual consonants and vowels contribute small details to the complete F_0 specification (e.g., so-called consonantal perturbations), similar to the F_0 algorithm briefly discussed in [PB88, pp. 176–177].

Across syllables, the order of interpretation is *left-to-right*. The motivation for this is that in connected speech each syllable needs to be "glued" to the previous one (see also subsection 8.3.1). Within syllables, however, the order of interpretation is *head-first*. In this way, weak nodes may adapt their parameters to those of their strong sister nodes because the parameters of the strong sister node (the head) are already known. The phonetic motivation for this is that consonants coarticulate with vowels, either directly or indirectly through other consonants, but not vice versa.

The results of evaluating phonetic exponency rules are added to the *phonetic exponency database*. Each entry in the database has the following general form:

```
<Parameter, Time1, Time2, Value1, Value2>
```

As an example, consider the following exponency rule for the second formant F2 for fricatives in onset position:

```
cons:[-coda, -son, +cnt] =>
     A = END,
     B = F2_END,
     C = F2_VAL,
     D = F2_LOCUS,
     E = F2_COART,
     F = f2(0.2*A),
```

```
G = f2(A+(B*A)),
f2(0.2*A, 0.5*A, 0.9*A, A, A+(B*A)) = (F, C, C, D+E*(G-D), G).
```

In this rule, the feature specification `cons:[-coda, -son, +cnt]` specifies a fricative consonant is dominated by an onset constituent. The `[-coda]` feature is inherited from the onset node dominating *cons*. The letters `A` to `G` are variables for times and parameter values, which are local to the rule. The built-in macro `END` returns the duration of the current constituent. `F2_END`, `F2_VAL`, `F2_LOCUS`, and `F2_COART` are calls to user-defined lookup tables; the current constituent is passed as an argument to the call. `f2(0.2*A)` and `f2(A+(B*A))` query the phonetic exponency database for a value for F2 at a particular point in time (at 20 percent of the duration and at slightly more than 100 percent of the duration, respectively); if such a value cannot be inferred from the database, the system returns a default value. `f2(0.2*A, 0.5*A, 0.9*A, A, A+(B*A))` specifies the points in time between which values for F2 are to be interpolated, and `= (F, C, C, D+E*(G-D), G)` specifies the corresponding values for F2. The term `D+E*(G-D)` implements a locus equation of the form $Locus + Coart * (Vowel - Locus)$ (see [AHK87, pp. 113–115].

The above rule implements the general structure of F2-exponents for fricatives in onset position. Lookup table entries are used to fill in the specifics of different types of fricatives in various contexts. For example, in the case of /s/ the lookup table for `F2_VAL` returns a value of 1700 or 1400, depending on whether the *voc* features define a spread or a rounded context, respectively.

By way of illustration, figure 8.3 shows the application of the above rule in different contexts. Figure 8.4 shows the formant structure and voicing for the syllable *sprint*.

8.2.4 Parameter Generation and Synthesis

If figure 8.4 seems messy, this is because all phonetic exponents of a structure are shown simultaneously. However, the phonetic exponency database is more than just a mixed bag of parameter tracks, as it also takes into account the way in which some constituents are overlaid on other constituents. Phonetic exponency in IPOX, as in YorkTalk, is very much like paint applied to a canvas in layers, obscuring

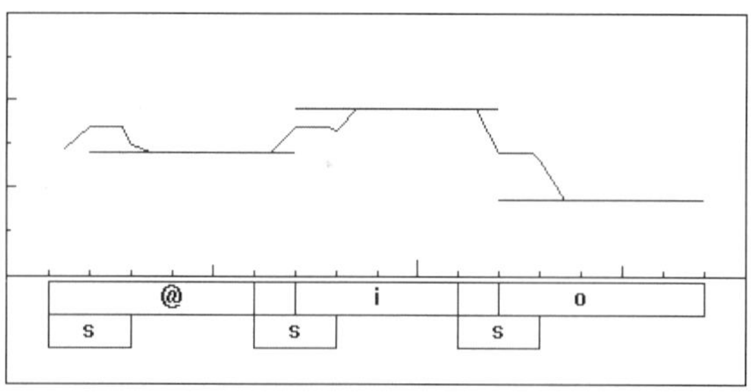

FIGURE 8.3. Phonetic exponency of /s/ for F2.

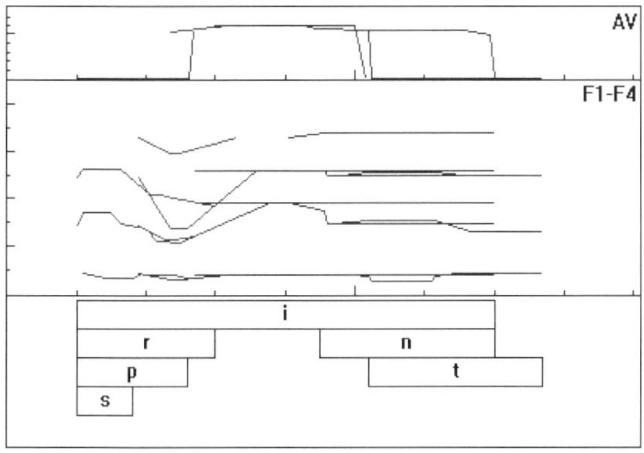

FIGURE 8.4. Phonetic interpretation of *sprint*.

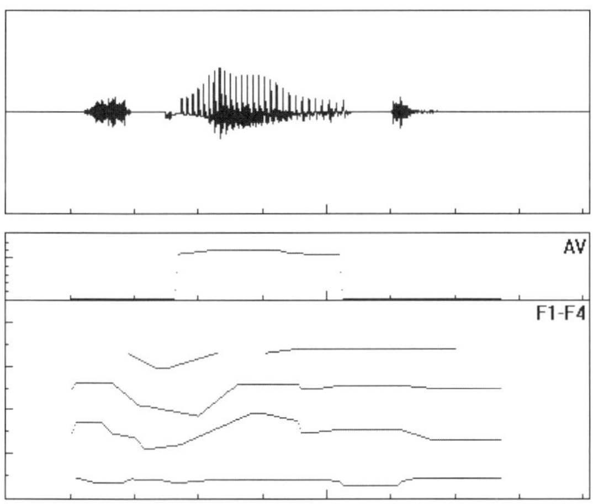

FIGURE 8.5. Parameter generation and synthesis of *sprint*.

earlier layers where it is applied and blending in with the background at the edges. Thus, it is possible to take the top-most value for a parameter at arbitrary points in time (such as every 5 ms), and use this value as input to the synthesizer. If the exponency rules have done their job well, the result is a continuous set of parameter tracks, as shown in figure 8.5, which also shows the waveform generated by the Klatt synthesizer.

Note that in our system it is not necessary to control each synthesis parameter at every point in time. For example, in figures 8.4 and 8.5 it can be seen that the fourth formant F4 is actively controlled only for onset /r/ and the coda nasal, and simply left at its default value (not shown) elsewhere. In fact, the default in IPOX is to do nothing. For this reason, and because of the declarative nature of the system, each exponency rule substantiates an independent empirical claim about the phonetic realization of a piece of phonological structure.

In a similar fashion, the lookup tables used by the exponency rules are sparse tables: a distinction made in one case (e.g., front versus back, spread versus round) need not be made in all cases. Thus, it is feasible to supply phonetic detail as seems necessary, or make sweeping generalizations as seems appropriate.

Also, the use of a specific synthesizer restricts the exponency rules to a specific set of parameters and the range of admissible values for each parameter. The system as a whole, however, is not dependent on a particular synthesizer configuration, and could easily be adapted to any synthesizer that accepts a parameter file as input, including an articulatory synthesizer or hybrid approach between acoustic and articulation-based synthesis (see, e.g., chapter 16 of this volume).

8.3 Polysyllabic Words

Phonetic interpretation in IPOX and YorkTalk is *compositional*, that is, the interpretation of a constituent is a function of the interpretation of its parts and the way in which they are combined. For example, the phonetic interpretation of a syllable consists of the phonetic interpretation of the rime, combined with the phonetic interpretation of the onset that is overlaid on the rime. Thus, within a syllable the mode of combination is head-first, as dictated by metrical structure, using the coproduction model of temporal interpretation. Across syllables, the mode of combination is slightly different (left-to-right, such that the onset is overlaid on the coda of the previous syllable). However, this does not change the fact that phonetic interpretation is compositional across syllables as well. That is, the phonetic interpretation of a polysyllabic utterance is simply the combination of any number of syllables, each of which is no different from its isolation form. On the face of it, this point of view seems problematic in that it raises the questions of how to deal with the special properties of intervocalic consonant clusters as well as how to handle vowel reduction and speech rhythm.

If we are to maintain that phonetic interpretation is compositional, the answer to these two questions must lie in the way in which syllables are combined to form utterances. The next subsections discuss our solutions to these problems.

8.3.1 *Ambisyllabicity*

One of the problems raised by compositionality is how to make sure that a single intervocalic consonant is properly coarticulated with its neighboring vowels. For example, in a word such as *bottle* /bot@l/ the /t/ should be interpreted as a coda with respect to the preceeding vowel, and as an onset with respect to the following vowel. Thus, the intervocalic /t/ must be parsed twice, once as the coda of the first syllable, once as the onset of the second syllable. In other words, the /t/ must be made *ambisyllabic*. This way we derive the coarticulation of /t/ with the preceding vowel /o/, just as in /bot/, as well as with the following vowel /@/, just as in /t@l/. However, without doing anything else, the resulting speech is simply the concatenation of two syllables, separated by a short pause. This type of juncture is appropriate if the two syllables belong to different prosodic phrases, but not word-internally. To avoid this, the onset /t/ is overlaid on the coda /t/ in temporal interpretation just enough to create the effect of a single intervocalic /t/.

More generally, different amounts of overlap are used in different contexts, depending on the nature of the intervocalic cluster as well as the prosodic configuration in which it appears [Col95]. In the case of two identical stops, different possibilities for overlap between syllables can be visualized as follows (see Demo 2 on the CD-ROM):

- very short closure

```
--Vowel-|-Closing-|---Closure---|-Release-|---
                              -Closure-|-Release-|-Vowel--
```

- normal intervocalic closure

```
--Vowel-|-Closing-|---Closure---|-Release-|---
                                 -Closure-|-Release-|-Vowel--
```

- long closure

```
--Vowel-|-Closing-|---Closure---|-Release-|---
                                       -Closure-|-Release-|-Vowel--
```

In IPOX and YorkTalk, ambisyllabicity is not restricted to single intervocalic consonants (as in most phonological theories). Within (Latinate parts of) words, intervocalic clusters are parsed with *maximal ambisyllabicity*. Compare: /win[t]@r/, /a[sp]rin/ and /si[st]@m/ (see Demo 3 on the CD-ROM).

By parsing the bracketed clusters as ambisyllabic, we derive the fact that in each of these cases the first syllable is heavy, and that the /t/ in /wint@r/ is aspirated, whereas in /sist@m/ it is not aspirated, as it occurs in the onset cluster /st/.

Various kinds of assimilation occurring in intervocalic clusters must be dealt with as well. We will not discuss assimilation in any detail here, but briefly sketch a general approach (for a detailed analysis see [Loc92]).

A declarative system does not allow feature-changing rules, as would be employed in traditional generative analyses of assimilation phenomena. However, it is possible to use underspecification in the analysis component, combined with feature-filling rules to further instantiate underspecified structures. As an example, consider the following rule for determining the amount of overlap between syllables with an ambisyllabic voiceless stop (in this rule, the value of *cns* is a complex feature structure defining consonantal place of articulation):

```
coda:[-voi, -son, -cnt, cns=A] +
   onset:[-voi, -son, -cnt, cns=A] => ...
```

If both coda and onset are fully specified for *cns* features, this rule merely checks for equality, and assigns the appropriate amount of overlap (indicated by . . .). However, if either the coda or the onset is underspecified for *cns* features, these features become shared, effectively assimilating the coda to the following onset or vice versa. Also, the cluster receives the temporal interpretation of a single ambisyllabic consonant.

In actual practice, though, cases of full assimilation are fairly rare, and a slightly more subtle approach is needed, in which the feature structures of assimilated consonants are similar but not identical, and in which the temporal interpretation of an assimilated cluster is slightly different from ambisyllabic consonants. Also, it will be necessary to use different assimilation rules for different morphophonological and prosodic contexts.

8.3.2 Vowel Reduction and Speech Rhythm

Connecting syllables as discussed in the previous subsection helps to smooth the boundaries between them. It does not, however, produce a very satisfying result if all syllables in a word or utterance have the same durations as their isolated monosyllabic counterparts. Therefore, to model the rhythm of connected speech, syllables that appear in a polysyllabic utterance are assigned a "compression factor," depending on the position of a syllable in a metrical foot, and the internal makeup of a syllable.

This compression factor is taken into account during temporal interpretation, such that various constituents within a syllable are compressed accordingly. In parametric interpretation, a distinction is made between absolute and relative timing. For example, fricatives and sonorants are relatively timed (i.e., all references to points in time are expressed as proportions of the duration of a constituent), but stop consonants use absolute timing. Because of this, a stop consonant in a compressed syllable occupies a larger proportion of the syllable than in a syllable that is not compressed. In other words, compression is not linear.

Also, rules for overlap between constituents take into account compression with one major exception: Overlap between onset and rime is always 200 ms (this is an arbitrary constant), even if constituents within the onset are compressed. Thus, the notional start of the vowel is always located 200 ms after the start of the syllable. As a consequence, vowels are more sensitive to compression than their neighboring consonants.

The effect of syllable compression is illustrated in figure 8.6 for the second syllable of *bottle*: the left panel shows the temporal and phonetic interpretation of this syllable without compression; the right panel shows the same syllable compressed to 62 percent of its duration.

Because compression is not linear, the two pictures are strikingly different. In the compressed version, the onset overlaps a much greater portion of the rime, and the vowel is almost totally eclipsed. Also, as a result of compression, the formant transitions are qualitatively different. This can be seen rather clearly in the transitions of the second formant F2. In both versions the coda /l/ is coarticulated

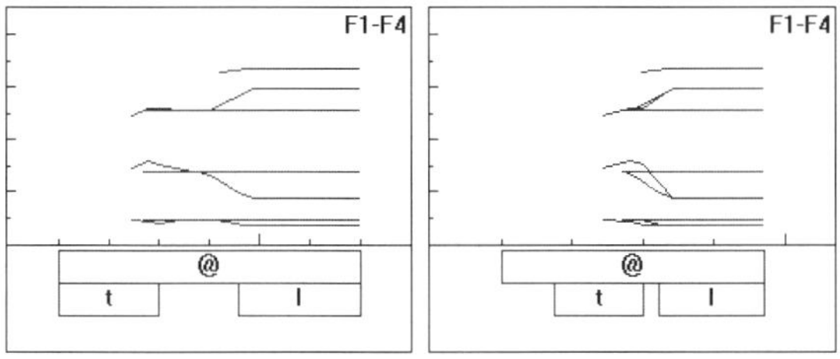

FIGURE 8.6. /t@l/ without (left) and with (right) compression.

with the vowel. However, in the compressed version the onset /t/ is coarticulated with the coda /l/ and only indirectly with the vowel. Perceptually, this creates the effect of a *syllabic sonorant* (see Demo 4 on the CD-ROM).

It is important to note that although the vowel appears to have been deleted, its presence in the background is still notable in the formant transitions, which would have been different if we had selected another vowel. Thus, the question is raised as to whether an analysis of vowel reduction/deletion in terms of syllable compression can be generalized to full vowels as well. We have preliminary evidence that this is the case.

As an example, consider the unstressed prefix *sup* in words such as *suppose* and *support*. In received pronunciations of these words, it appears that the vowel has been deleted. However, we have synthesized successful imitations of these words, using the full vowel /^/ (as in *but*). If the first syllable in *suppose* is compressed to 60 percent, then the vowel obtains a slightly reduced quality. With compression to 52 percent, the vowel is eclipsed, resulting in the pronunciation *s'ppose* (see Demo 5 on the CD-ROM).

A segmental analysis of vowel elision would seem to predict that a version of the word *support* in which the first vowel is deleted is phonetically similar, if not identical, to the word *sport*. By contrast, an analysis in terms of compression predicts subtle, but notable differences in temporal as well as spectral structure. Specifically, it is correctly predicted that /p/ is aspirated in *s'pport*, but not in *sport* (see Demo 6 on the CD-ROM).

Even more challenging are the apparent changes in vowel quality in a stem such as *photograph* when it appears in a "stress-shifting" Latinate derivation. In isolation, the main stress is on the first syllable, and the vowel in the second syllable is reduced. In *photography*, however, the main stress is on the second syllable, and the vowels in the first and third syllables undergo reduction. Finally,

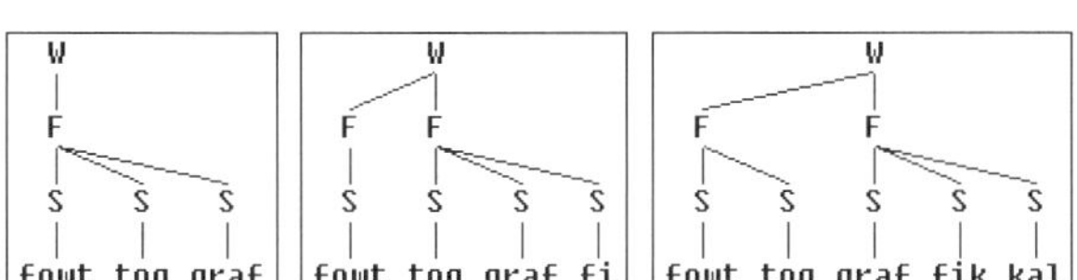

FIGURE 8.7. *Photograph*, *photography* and *photographical*.

in *photographical* the main stress is on the third syllable, and reduced vowels are found in the second and final syllable.[5] In a segmental system, we would need to posit alternations with schwa for each vowel in the stem *photograph*. In a declarative system like IPOX, we could use a feature such as **[+reduced]** to trigger a special set of phonetic exponency rules. However, often this does not appear to be necessary. We have succesfully synthesized these three words using full vowels (/ow/ as in *blow*, /o/ as in *pot*, and /a/ as in *sad*) in analysis as well as phonetic interpretation. The prosodic analysis trees assigned to these words by IPOX are shown in figure 8.7. By varying syllable compression in accordance with the metrical-prosodic structure, we obtain the expected alternations between full and reduced vowels (see Demo 7 on the CD-ROM). Also, the reduced vowels are appropriately different in quality depending on the "underlying" vowel, which we think is an additional advantage with respect to an analysis using alternation with schwa.

The evidence presented above suggests that in our system vowel reduction is the natural consequence of variations in speech rhythm that are independently motivated. In order to further substantiate this claim, however, we need to examine alternations between full and reduced vowels more systematically.

8.4 Connected Speech

As mentioned in section 8.1, the current version of IPOX lacks interfaces between morphosyntactic structure on the one hand, and metrical-prosodic structure on the other. In this section we briefly discuss the kind of interface we have in mind, illustrated with an analysis of the sentence "*This is an adaptable system.*" Also, we discuss how, with a few adaptations to the prosodic grammar, we have been able to generate this sentence, despite the lack of a syntax-prosody interface.

The main problem posed by this sentence is a severe mismatch between morphosyntactic and prosodic structures. This is illustrated in the metrical represen-

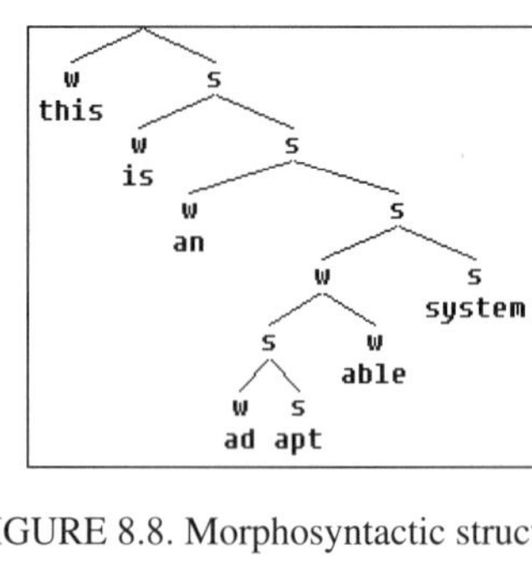

FIGURE 8.8. Morphosyntactic structure.

[5]Note that in the novel, but well-formed derivation *photographicality* we would find a full vowel for the affix *-al*.

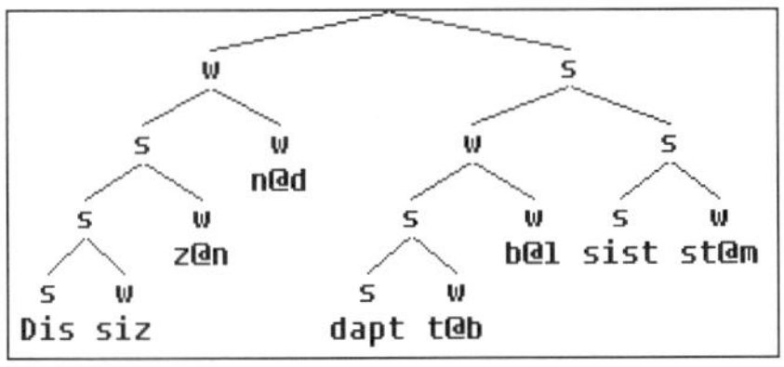

FIGURE 8.9. Prosodic structure (metrical tree).

tations in figures 8.8 and 8.9 (in these screen clips, category labels are not shown, and metrical structure is shown by labeling nodes with w(eak) or s(trong)). The structure in figure 8.8 is obtained by parsing orthographic input using simple IPOX grammars for English syntactic and morphological structure, in which phrase structure rules are annotated with metrical structure. The structure in figure 8.9 is obtained by parsing phoneme-based input using a prosodic grammar for English. (The internal structure of syllables is not shown.)

Although the two structures are radically different, they are systematically related, and a general solution would need to take both into account. In generative phonology, this relation is usually described in procedural terms: unstressed function words and stray initial syllables initially remain "unparsed," and are "adjoined" at a later stage to the preceding foot. In declarative theory, such a solution is not available, so the two structures must be produced in parallel. In our current example, the structure in figure 8.9 is arrived at by requiring that heavy syllables such as /dapt/ and /sist/ appear as the head of a foot, whereas light syllables are weak nodes of a foot (except phrase-initially). Such a restriction, however, is warranted only in the Latinate part of the lexicon [Col94], and should not be generalized to phrase-level prosodic structure. However, setting the grammar up this way allowed us to experiment with the generation of short sentences.

Figure 8.10 shows the prosodic structure of our sentence again, this time in the "headed" format of figures 8.1 and 8.8. Figure 8.11 shows the generated formant structure and F0, as well as the waveform generated from this structure (see Demo 8 on the CD-ROM).

This experiment illustrates that generating connected speech is not qualitatively different from generating a two-syllable word, although for a longer utterance we would need to worry about prosodic phrasing as well as be able to supply a more varied intonation. In the present setup, intervocalic clusters are ambisyllabic as much as possible, and each metrical foot receives the same linear falling F_0. In future work, we envisage incorporation of a more sophisticated treatment of F_0

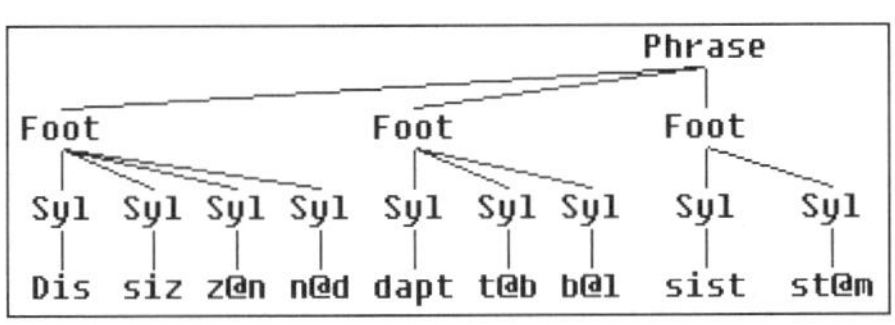

FIGURE 8.10. Prosodic structure (headed tree).

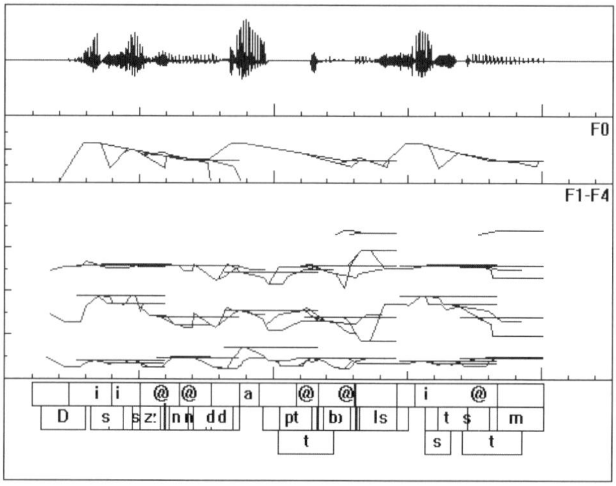

FIGURE 8.11. Phonetic interpretation.

generation along the lines of [Pie81], whose autosegmental approach to intonation is close in spirit to the methods employed in IPOX.

8.5 Summary

We have described an architecture for analyzing a sentence syntactically, morphologically and prosodically, and computing phonetic parameters for the Klatt synthesizer from such an analysis. A number of phenomena that are unrelated in a more conventional system based on rewrite rules, such as coarticulation, unstressed vowel shortening, centralization of unstressed vowels, syllabic sonorant formation, and elision of unstressed vowels before syllabic sonorants, are modeled in IPOX as natural concomitants of the all-prosodic view of phonological structure and phonetic interpretation. In the future we shall improve the phonetic quality of our English and Dutch phonetic parameters, as well as address the prosody-syntax interface and the phonetics of intonation more thoroughly.

An inspection copy of the IPOX system is available on the World Wide Web from one of the following locations:

```
ftp://chico.phon.ox.ac.uk/pub/ipox/ipox.html
http://www.tue.nl/ipo/people/adirksen/ipox/ipox.html
```

The software and documentation are freely available for evaluation and non-profit research purposes only. The authors reserve the right to withdraw or alter the terms for access to this resource in the future. Copyright of the software and documentation is reserved ©1994, 1995 by Arthur Dirksen/IPO and John Coleman/OUPL.

Acknowledgments: This chapter has benefited from comments by two anonymous reviewers. The research of Arthur Dirksen has been made possible by a fellowship of the Royal Netherlands Academy of Arts and Sciences.

References

[AHK87] J. Allen, M. S. Hunnicut, and D. KLatt. *From Text to Speech: The MITALK System*. Cambridge University Press, Cambridge, 1987.

[Col92] J. S. Coleman. "Synthesis-by-rule" without segments or rewrite rules. In *Talking Machines: Theories, Models, and Designs*, G. Bailly, C. Benoit, and T. R. Sawallis, eds. Elsevier, Amsterdam, 211–224, 1992.

[Col94] J. S. Coleman. Polysyllabic words in the YorkTalk synthesis system. In *Phonological Structure and Phonetic Form: Papers in Laboratory Phonology III*, P. A. Keating, ed. Cambridge University Press, Cambridge, 293–324, 1994.

[Col95] J. S. Coleman. Synthesis of connected speech. To appear in *Work in Progress* No. 7. Speech Research Laboratory, University of Reading, 1–12, 1995.

[Dir93] A. Dirksen. Phonological parsing. In *Computational Linguistics in the Netherlands: Papers from the Third CLIN meeting*, W. Sijtsma and O. Zweekhorst, eds. Tilburg University, Netherlands, 27–38, 1993.

[DQ93] A. Dirksen and H. Quené. Prosodic analysis: the next generation. In *Analysis and Synthesis of Speech: Strategic Research Towards High-Quality Text-to-Speech Generation*, V. J. van Heuven and L. C. W. Pols, eds. Mouton de Gruyter, Berlin, 131–144, 1993.

[Loc92] J. K. Local. Modelling assimilation in nonsegmental, rule-free synthesis. *Papers in Laboratory Phonology II: Gesture, Segment, Prosody*, G. J. Docherty and D. R. Ladd, eds. Cambridge University Press, Cambridge, 190–223, 1992.

[OGC93] J. P. Olive, A. Greenwood and J. Coleman. *Acoustics of American English Speech: A Dynamic Approach*. Springer-Verlag, New York, 1993.

[Ple81] J. B. Pierrehumbert. Synthesizing intonation. *J. Acoust. Soc. Amer.* 70(4):985–995, 1981.

[PB88] J. B. Pierrehumbert and M. E. Beckman. *Japanese Tone Structure*. MIT Press, Cambridge, MA, 1988.

Appendix: Audio Demos

Demo 1 - Coarticulation:
Spreading of vocalic place features in the phonology is reflected in phonetic interpretation by subtle differences in frication spectra associated with /s/: spit, spot, sprint.
Demo 2 - Syllable overlap:
Three versions of /bot@l/ bottle, with ambisyllabic /t/, generated with different amounts of syllable overlap:
Demo 3 - Ambisyllabicity:
Intervocalic clusters are parsed with maximal ambisyllabicity. In the following words, the bracketed clusters are ambisyllabic: /win[t]@r/, winter; /si[st]@m/, system; /a[sp]rin/, aspirin. Note that /t/ is aspirated in winter, but not in system.

Demo 4 - Syllable compression I:
Again, three versions of /bot@l/ bottle, this time with different amounts of compression for the first and second syllable: Note that when the second syllable /t@l/ is compressed to 62 percent, the vowel is almost fully eclipsed, creating the impression of a syllabic sonorant.
Demo 5 - Syllable compression II:
Two versions of /sp̂owz/ suppose, with different amounts of compression for the unstressed prefix /sp̂/.
Demo 6 - Syllable compression III:
A segmental analysis of vowel elision would seem to predict that "s'pport" is phonetically identical to "sport." Our analysis in terms of syllable compression correctly predicts subtle (and less subtle) differences: Note that /p/ is aspirated in "s'pport" but not in "sport."
Demo 7 - Syllable compression IV:
The three words below have been synthesized using full vowels (/ow/ as in blow, /o/ as in pot, and /a/ as in sad) in analysis as well as phonetic interpretation. By varying syllable compression in accordance with metrical-prosodic structure, we obtain the expected alternations between full and reduced vowels: /fowtograf/, "photograph"; /fowtografi/, "photography"; /fowtografikal/, "photographical."
Demo 8 - Connected speech:
Our first attempt at generation of a full sentence with IPOX.
/Disiz@n@dapt@b@lsist@m/, "This is an adaptable system."

9

A Model of Timing for Nonsegmental Phonological Structure

John Local
Richard Ogden

ABSTRACT Usually the problem of timing in speech synthesis is construed as the search for appropriate algorithms for altering durations of speech units under various conditions (e.g., stressed versus unstressed syllables, final versus non-final position, nature of surrounding segments). This chapter proposes a model of phonological representation and phonetic interpretation based on Firthian prosodic analysis [Fir57], which is instantiated in the YorkTalk speech generation system. In this model timing is treated as part of phonetic interpretation and not as an integral part of phonological representation. This leads us to explore the possibility that speech rhythm is the product of relationships between abstract constituents of linguistic structure of which there is no *single* optimal distinguished unit.

9.1 Introduction

One of the enduring problems in achieving natural-sounding synthetic speech is that of getting the rhythm right. Usually this problem is construed as the search for appropriate algorithms for altering durations of segments under various contextual conditions (e.g., initially versus finally in the word or phrase, in stressed versus unstressed syllables). Van Santen [van92] identifies a number of different approaches employed to control timing in synthesis applications and provides an overview of their relative strengths and weaknesses. He refers to these as (i) sequential rule systems [Kla87]; (ii) lookup tables; (iii) binary (phone) classification trees [Ril92]; and (iv) equation techniques, based directly on duration models [CUB73]. All the approaches he discusses rest on the assumption that there is a basic unit to be timed, that it is some kind of (phoneme-like) segment, and that rhythmic affects are assumed to "fall out" as a results of segmental-level timing modifications. Van Santen's own, sophisticated approach to modeling duration (employing sums-of-products models) has produced improved results in segmental timing for the AT&T synthesis system [van94]. However, as he acknowledges, units other than the segment are required to make synthetic speech sound natural.

In large part, the pervasive problem of rhythm in synthesis can be seen to arise from the adherence by researchers to representations which are based on concatenated strings of consonant and vowel segments that allocate those segments uniquely to a given syllable. Campbell and Isard [CI91] have suggested that a more effective model is one in which the syllable is taken as the distinguished timing unit and segmental durations are accommodated secondarily to syllable durations. Our view is that rhythm requires statements of relations between units in particular pieces of linguistic structure. Any successful account of the rhythmic organization of language requires representations that will permit the expression of hierarchical structure of varying domains. In the approach sketched here, we reject string-based data structures in favor of hierarchically structured, nonsegmental representations that admit structure sharing. These provide for the felicitous expression and representation of relationships necessary to generate synthetic versions of polysyllabic utterances that faithfully mimic the rhythmic organization of natural speech without any single optimal distinguished unit of timing. Timing, then, is seen as "prosodic."

In the first part of the chapter, we briefly discuss rhythm and the phonetic interpretation necessary to model it. We then present a timing model based on temporal constraints that operate within and between syllables. In the second part, we compare the temporal characteristics of the output of the YorkTalk system with Klattalk [Kla] on the one hand and the naturalistic observations of [Fow81] on the other. We show that it is possible to produce similar, natural sounding temporal relations by employing linguistic structures that are given a compositional parametric and temporal interpretation [Loc92, Ogd92].

9.2 Syllable Linkage and Its Phonetic Interpretation in YorkTalk.

9.2.1 Principle of Parametric Interpretation

YorkTalk instantiates a version of Firthian prosodic analysis (FPA) [Fir57, Loc92]. Firthian analysis specifically rejects phonological and phonetic segmentation. It provides an analysis of phonological contrasts in terms of abstract categories that have their definition in terms of the relations into which they enter. These phonological categories have no implicit phonetic content, unlike those of conventional generative phonology (e.g., [CH68, Ken94]). In FPA special analytic consideration is given to structure at various levels of linguistic statement. One consequence of this is that the "same" phonological category may be given a different phonetic interpretation, depending on its place in phonological, morphological, lexical, interactional or, for instance, syntactic structure [Sim92, Spr66, Hen49, Hen52, Kel89, Kel92]. Recent acoustic evidence for such structural differentiation can be found in [Man92], where it is shown that there are timing and spectral differences between long nasal portions that represent (i) n - n (as in "win no's") and (ii) n - ð (as in "win those") where at a broad phonetic level the two pieces could be represented as [n:]. ([Loc92] also provides similar evidence in articulatory and acoustic domains.)

In speech synthesis there is a logical division between a level of symbolic representation and a level at which acoustic/articulatory description is handled. In FPA there is a strict separation of phonetic and phonological information. As a consequence of this, FPA requires that temporal information be treated as a part of phonetic interpretation and not as an integral part of phonological representation (see [Car57]). The YorkTalk speech generation system makes use of this central principle of FPA. This provides for phonological representations that do not change in order to produce phonetically different tokens of a given lexical item. Therefore, there is no "derivation" of the phonetics from the phonology as there is in conventional phonological accounts (including the more recent nonlinear models). The approach to phonology, then, is declarative (nonderivational). The model we have outlined, with its two distinct levels of representation and with categories that can be given more than one phonetic interpretation and in which the languages of phonetics and phonology are different, could be unconstrained. The phonetic interpretation of phonological structures in YorkTalk is constrained by the Principle of Compositionality (see [Par84]), which states that the "meaning" of a complex expression is a function of the meanings of its parts and the rules whereby the parts are combined (see [Fir57, Whe81]). Thus the phonetic interpretation of a complete structure (e.g., an utterance or a syllable) is a function of the phonetic interpretation of its constituent parts (see [Col92]). This means that any abstract feature, or bundle of features, at a particular place in a phonological representation is always interpreted in the same way.

YorkTalk's nonsegmental phonological representations are constructed by metrical and phonotactic parsers, which parse input into structures consisting of feet, syllables, and syllable constituents.

9.2.2 *Structured Phonological Representations*

In YorkTalk's phonological structures, the rime is the head of the syllable, the nucleus is the head of the rime and the strong syllable is the head of the foot (see [Col92]). Every node in the graph is given a head-first temporal and parametric phonetic interpretation. A coproduction model of coarticulation [Fow80] is implemented in YorkTalk by overlaying parameters. Because the nucleus is the head of the syllable, the nucleus and syllable are coextensive. By fitting the onset and coda within the temporal space of the nucleus, they inherit the properties of the whole syllable. Where structures permit, constituents are shared between syllables as shown below (ambisyllabicity). The temporal interpretation of ambisyllabicity is the temporal and parametric overlaying of one syllable on another ([Loc92, Ogd92]). Figure 9.1 gives a partial phonological representation of the two syllables of the nonsense word si:sə. In this figure ϕ labels metrical feet and σ labels phonological syllables. We assume a constituent structure for syllables, which branches into onset, rime, and within rime into nucleus and coda. Joining terminal branches provides an ambisyllabic representation of the intersyllabic consonant.

Some flavor of our approach can be gained from the statement below, which shows part of the formal phonetic interpretation of a coda node whose features

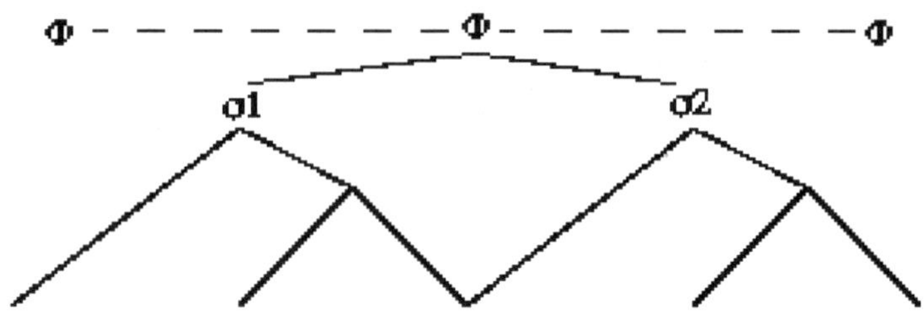

FIGURE 9.1. Partial phonological graph of utterance si:sə.

will match those for a "liquid." The main body of the statement consists of (time, value) pairs.

```
f2= exponency(coda(Liquid)
(Start-(0.1*Dur), Value1)
(Start+(0.1*Dur),0.8*Value1)
(Start+(0.2*Dur),0.76*Value1)
(Start+(0.3*Dur),0.74*Value1)
(End, 0.74*Value1), (Start=750, End=900))
```

Time statements refer to Start, End and Dur. Start refers to a reference point for the start of the coda constituent. End refers to the end of the coda constituent and therefore the end of the nucleus and syllable constituents. Dur is End-Start. Start and End are calculated by applying temporal constraints to the syllable structure as a whole. Start and End do not demarcate "segment" boundaries. Apparent segment boundaries are the result of sudden changes arising from particular combinations of such parameter strips. Thus, YorkTalk does not model segments. Segments are a product of phonetic interpretation of structures, and, because phonetic interpretation is carried out on phonological structures, segments are not an available unit for association with timing. Moreover, there is evidence from various sources that timing is most appropriately driven by nonterminal (i.e., nonsegment) constituents (e.g., [CI91, Col92, VCR92, Smi93]).

9.3 The Description and Modeling of Rhythm

Abercrombie [Aber64] provides a seminal description of rhythmic-quantity configurations in disyllabic feet in English. In particular he draws attention to two different kinds of rhythmic patterns, which can be observed in initially accented disyllables. The first, which he labels "short long," is found in disyllabic structures where the first syllable can be analyzed as "light" (i.e., a rime with a phonologically short nucleus and nonbranching coda). The second kind of rhythmic patterning, "equal-equal," is found in disyllabic structures having a "heavy" first syllable (i.e., a rime with a phonologically short nucleus and branching coda or a phonologically long nucleus irrespective of coda structure). This analysis provides for an account of the observed differences in word pairs such as *whinny* versus *windy* ([wɪNDɪ]) and *filling* versus *filing*. Notice that the first pair of words allow us to see that the phonetic exponents of a syllable's strength and weight (which in large part determine its rhythmic relations) include more features than traditional "suprasegmentals" and are not simply local to that syllable. The final vocalic portions in these words have different qualities depending on whether the first syllable is light or heavy: For some speakers the final [i] vowel of *windy* is closer than that of *whinny*, whereas for other speakers the situation is reversed. For some speakers the final vowel of *windy* is diphthongal whereas the final vowel of *whinny* in monophthongal [Loc90].

In YorkTalk, such rhythmical effects are primarily modeled by the temporal interpretation function "squish"—a unit of temporal compression.

Polysyllabic utterances are compositionally constructed from single syllable structures. Achieving the correct duration of a syllable and, more importantly, achieving the correct duration of syllables in relation to each other is the key to producing the correct rhythm. To achieve these relations we make use of two key concepts: distance and squish. Squish expresses a ratio between a polysyllabic versus a monosyllabic foot. Distance is a measure of separation of onset (end) and coda (start) in syllables. Thus for any syllable:

$$\text{syllable squish} = \frac{\text{distance of squished syllable}}{\text{distance of syllable in a monosyllabic utterance}}$$

The value of the distance is directly related to: (1) the position of the syllable within the foot (a strong syllable, foot-medial syllable, or foot-final syllable); (2) whether the preceding syllable slot is empty; (3) the value of the rime head features *[long, voi]* and the nucleus feature *[height]*; and (4) the position of the foot in the utterance. In the same way that the headedness of the structures directs the parametric interpretation, so it also directs the temporal interpretation. So, for example, the first syllable ($\sigma 1$), si:s, in Figure 9.1 is interpreted as below:

> as a monosyllable: SyllDur: 680, OnsDur: 200, CodaDur: 250; Squish is 1; Distance is 230;
>
> as first syllable of si:s/schwa: SyllDur: 335, OnsDur: 120, CodaDur: 155; Squish is 0.61; Distance is 140.

Figure 9.2 pictures informally the structural inheritance of temporal constraints within and between syllables.

The metrical parser determines word-accentual patterns and groups syllables into feet. Within the foot we distinguish three types of syllable: strong, middle, and final. This has consequences for the temporal constraints that apply during temporal interpretation. The assignment of a syllable as strong, middle, or final depends in part on whether it is in a Graeco-Germanic or Latinate foot [Col93, Ogd93].

The YorkTalk model makes a number of predictions about durational characteristics of strong syllables in different structural contexts. We treat strong syllables

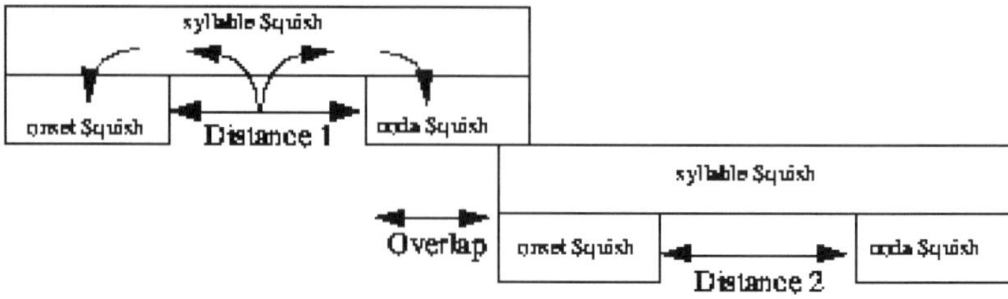

FIGURE 9.2. Inheritance of temporal constraints for syllables.

that are final in the utterance in the same way as we treat monosyllabic utterances (effectively no temporal compression). Strong syllables with succeeding weak syllables, however, are "squished." Weak syllables in middle and final positions in the foot are interpreted with different temporal characteristics; if they are medial they are more temporally compressed than if they are final.

9.3.1 *Effects Modeled by Squish*

The account of squish we have given so far has been purely mechanical, that is, we have focused exclusively on the aspects of implementation rather than the uses to which squish can be put. In this section we shall look in a little more detail at the sorts of phonetic effects we achieve with squish.

Syllabic Laterals and Nasals

One of the principles of the phonology of YorkTalk is that no information is deleted or destroyed. This presents us with "deletion," which refers to the problem by which, according to traditional descriptions, linguistically important segments are missed under certain conditions. One of these "deletion phenomena" is that of syllabic laterals at the end of words such as "animal," "bottle," "fiddle," "gristle," for which a vowel is sometimes said to have been deleted and the consonant syllabified. Another view of the same material is that syllabicity is associated with different phonetic events, and that in one case it is associated with vocalicity, and in another with laterality and darkness or nasality. In YorkTalk's phonology, words such as these are represented as a schwa + lateral/nasal in the rime. Here the structure is crucial: those syllables which are last in the foot and whose rimes contain əl have a very small value for distance (where "small" means about 50 ms, and "large" for a long vowel such as i: means something like 200 ms). Thus the syllable as a whole has a very high squish value. (Small distance implies a high squish. High squish means a great deal of temporal compression, which means a low number for the value of squish. We shall try not to use too many of these high/low metaphors.) The effect of this is that the laterality or nasality starts early and so no vocalic portion can be discerned either auditorily or acoustically; yet this impression of a syllabic consonant is achieved entirely without deleting anything and without changing the parameters for either the schwa-vowel or the lateral at the end. In other words, squish allows us to produce a new sound by using old ones. The use of temporal relations to achieve effects like this has been explored by Browman and Goldstein [BG89] and Coleman [Col93].

Deletion of Vowels

According to the traditional segmental account, as a word like "support" is said at a greater tempo, so the vowel of the first syllable reduces and is finally deleted. "Support" should then be homophonous with "sport"; however, this is demonstrably not the case, because in the plosive of "sport" there is no aspiration, whereas in "support" the plosive *is* aspirated because it is initial in a strong syllable and is not adjacent to a phonological unit whose exponent is friction (as it is in "sport"). By squishing the first syllable of "support" it is possible to produce something like [spɔt], but the plosive will be aspirated because structurally it is not in a fricative/plosive cluster. Furthermore, the initial friction has the correct resonance, that is, the resonance suitable for a syllable whose nucleus is central in quality. In "sport" both the friction and the plosion are appropriate for a syllable whose nucleus is back and round (see [Col93, Man92]).

An Example of Syllable Squish : Man, Manage, Manager

This example is chosen to illustrate the difference in the temporal interpretation of the syllable written "nage." (YorkTalk uses maximal ambisyllabicity [Loc92], so that the syllables are (orthographically): man, man-nage, man-nage-ger.)

"Man," being a monosyllable in isolation (final in the utterance), has a squish of 1. In "manage" the distance for the second syllable is 85 ms, and the squish is 0.71, the distance that applies to a foot-final syllable that is preceded by a strong light syllable and contains a voiced rime. In "manager" the distance for the second syllable is 70 ms and the squish is 0.6, the distance that applies to a *foot-medial* syllable that is preceded by a strong light syllable and contains a voiced rime. Note that the position in the foot (final, medial) is crucial. The spectrograms for these three words as generated by the synthesizer YorkTalk are presented for comparison in figures 9.3, 9.4, and 9.5. Note in particular the durations of nasal portions and voiced portions. The durations of the utterances are 500 ms, 515

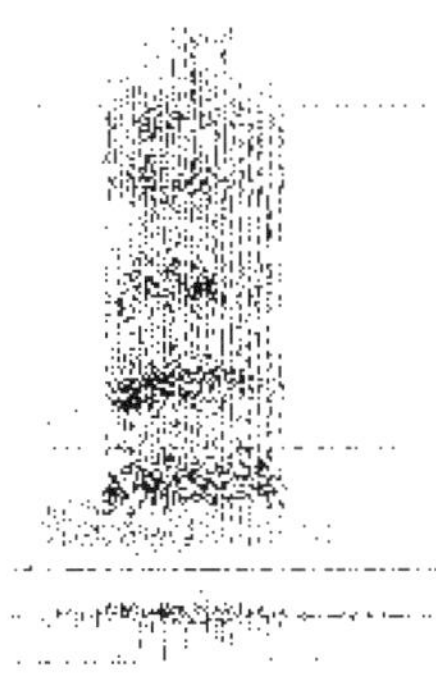

FIGURE 9.3. Spectrogram of synthetic "man."

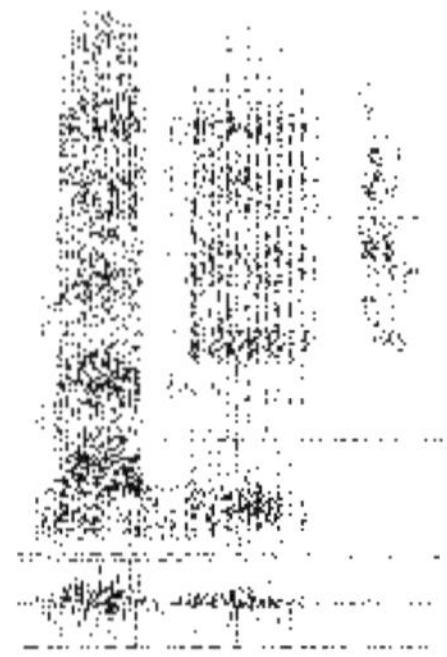

FIGURE 9.4. Spectrogram of synthetic "manage."

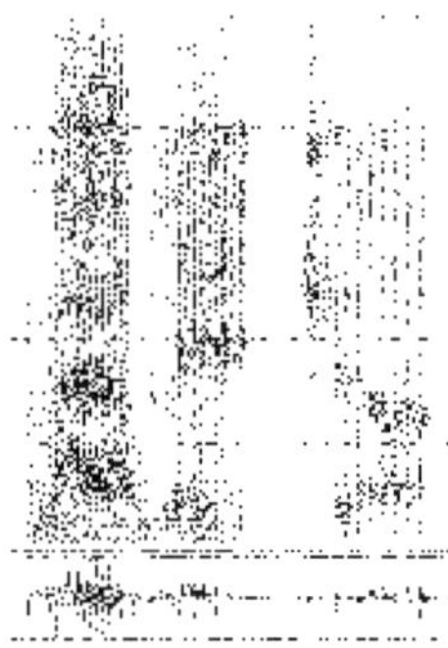

FIGURE 9.5. Spectrogram of synthetic "manager."

ms and 565 ms for "man," "manage," and 'manager," respectively. The synthesis could be interpreted as consistent with the "liberal view" of isochrony [Fow83] and perhaps the view that timing in "isochrony" languages is foot timing rather than stress timing [Wii91], although isochrony is not explicitly programmed into YorkTalk. YorkTalk (like real speech) gives the impression of isochrony without actually being isochronous.

Temporal Relations Between Feet

Temporal relations between feet are handled in the same way as temporal relations between syllables, i.e., by using squish and distance. At present, YorkTalk interprets utterances composed of up to three feet: $\phi 1$, $\phi 2$, $\phi 3$. The first two are called the first feet ($\phi 1$ and $\phi 2$). The last foot ($\phi 3$) is compressed in the same way as an utterance consisting of one foot, whereas the first feet are squished to produce a faster tempo than the last foot. A single first foot is squished more than two first feet.

An Example of Foot Squish: "Humpty Dumpty"

"Humpty Dumpty" is parsed as a two-foot utterance, with both feet of the form σs σw (figure 9.6).The distances for the syllables in the first foot are less than

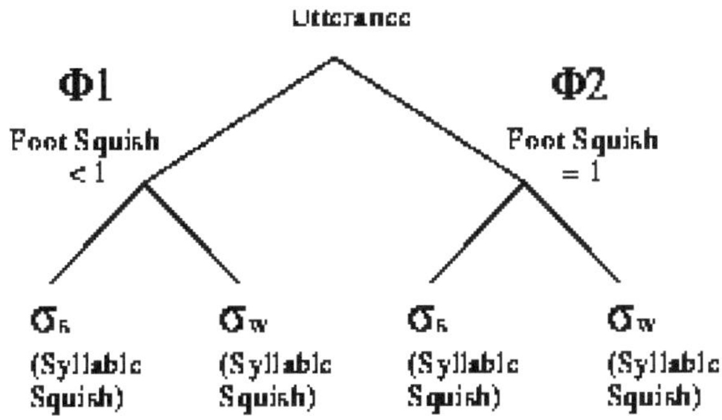

FIGURE 9.6. Temporal relations in the multiple-foot utterance Humpty Dumpty.

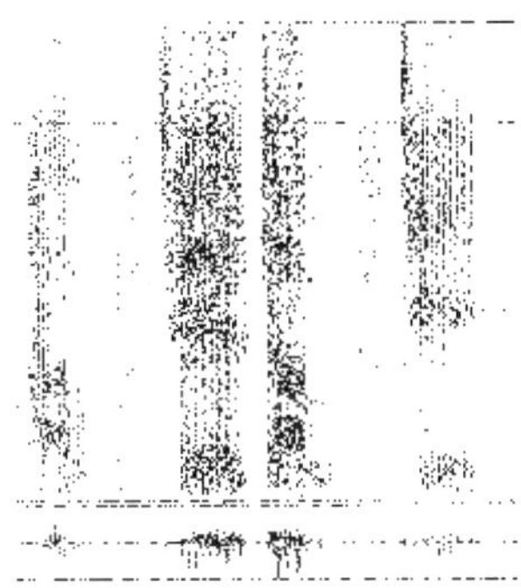

FIGURE 9.7. Spectrogram of synthetic "Humpty Dumpty."

the corresponding distances for similar syllables in the last foot of an utterance. So the distance for the syllable "hump" is 80 ms, and for the syllable "dump" it is 100 ms. The effect of this is that the squish for "hump" is lower than that for "dump," and thus the first foot is produced at a faster tempo than the second foot. Essentially, the syllable squish is squished again at the foot level. Note that without phonological structure, the implementation of foot-level timing would be very much more complex. In YorkTalk, because the structure is available, it is possible to compress a syllable and all its contents, or a foot and all its contents, without treating the lowest level units separately. The YorkTalk spectrograph for "Humpty Dumpty" is shown in figure 9.7.

9.4 Comparison of the Output of YorkTalk with Natural Speech and Synthesis

Figure 9.8 presents a comparison of natural American English [Fow81], natural British English [LO94], American synthesis (Klattalk), and YorkTalk British English synthesis. Lack of suitable published results prevents us from comparing

	Fowler	Klatt	British English	YorkTalk
si:sə	0.51	0.49	0.4	0.3
si:səsə	0.47	0.46	0.35	0.3
səsi:	0.9	0.84	0.98	1
səsəsi:	1.04	0.78	0.95	1

FIGURE 9.8. Mean duration of [i:] in polysyllables expressed as a proportion of monosyllabic [i:] in [si:].

	Klatt	YorkTalk	British English
si:sə	1:1.06	1:1.3	1:1.1
si:səsə	1:0.47:1.05	1:0.3:1.13	1:0.4:1.2
səsi:	0.29:1	0.3:1	0.3:1
səsəsi:	0.27:0.27:1	0.2:0.29:1	0.2:0.3:1

FIGURE 9.9. Duration of vocalic portions as a proportion of the [i:].

our durational measures with those generated by other nonsegmental approaches to synthesis.

Figure 9.8 expresses the ratios of vowel durations of the words elicited by Fowler as proportions of the vowel duration of the stressed vowel in the word. Fowler gives no durations for the unstressed vowels, so that a complete comparison is not possible. It can be seen that YorkTalk models the relations between vowel durations quite accurately, without any explicit notion of "vowel segment." The actual differences in duration give an impression of a difference in tempo, rather than a difference in rhythm.

There are notable differences between the natural American English observed by Fowler and the natural British English we have observed. Most particularly, the degree of final lengthening is greater in British than in American English. The actual (i.e., nonrelative) duration of the i: vowels is shorter in the polysyllabic words in American English than in British. The ranges of variation in duration expressed proportionally for natural British English encompass those of Fowler's data. They also include the proportion generated by YorkTalk the duration of i: when preceded by weak syllables, but final in the word, is virtually identical to that of i: in a monosyllable. Despite durational differences in absolute terms between natural speech and the YorkTalk synthetic speech, the rhythmical relations between syllables are maintained in YorkTalk.

As can be seen, the relative durations of the vocalic portions generated by YorkTalk closely model those of the natural British polysyllabic words (Figure 9.9). This is achieved without the use of segment or distinguished timing unit at any stage of the generation process. Fowler's account of the durational differences in stressed vowels is to use the formula of [LR73], which employs a best-fit model to estimate anticipatory and backward-shortening parameters. The formula also requires intrinsic duration, predicted duration, and the number of previous and subsequent unstressed syllables. No linguistic structure whatsoever is modeled by this formula because the syllables are treated as being linearly arranged with no hierarchical structure. Fowler suggests changing the formula to add a parameter for word-final lengthening, based on her observations. Klatt's durations are achieved by applying eleven additive and multiplicative rules, which may variously lengthen or shorten a segment or leave a segment unchanged. Klatt's rules make use of some linguistic structure (such as segment type, position in the word, position in the utterance, etc.), but the structure is not such an integral part of linguistic representation for Klattalk as it is for YorkTalk. For instance, Klattalk has no feet, and all the linguistic features are expressed in the segments, whereas YorkTalk distributes the phonological features across all parts of its structures [Col92, Ogd92]. The YorkTalk model provides a unified temporal description that allows us to capture relations between the phonetic exponents of parts of structure simply.

9.5 Conclusion

There is more than one way to generate synthetic speech whose durations mimic those of natural speech. Techniques may or may not use some preferred unit of timing. In generating synthetic speech whose rhythm is right for the language being modeled, it is important to model relations between syllables rather than to concentrate almost exclusively on individual syllables or segments, or on segmental durations.

References

[Abe64] D. Abercrombie. Syllable quantity and enclitics in English. In *Honour of Daniel Jones*, D. Abercrombie, D. B. Fry, P. A. D. MacCarthy, N. C. Scott and J. L. Trim, eds. Longman Green, London, 216–222, 1964.

[BG89] C. P. Browman and L. M. Goldstein. Towards an articulatory phonology. *Phonololgy Yearbook* 3: 219–252, 1989.

[CI91] W. N. Campbell and S. D. Isard. Segment durations in a syllable frame. *J. Phonetics* 19:37–47, 1991.

[Car57] J. C. Carnochan. Gemination in Hausa. In *Studies in Linguistic Analysis*, Special Volume of the Philological Society, 2nd edition, 49–81, 1957.

[CH68] N. Chomsky and M. Halle. *The Sound Pattern of English*. Harper & Row: New York, 1968.

[CUB73] C. H. Coker, N. Umeda, and C. P. Browman. Automatic synthesis from ordinary English text. *IEEE Transactions on Audio and Electroacoustics*, AU-21, 3: 293–298, 1973.

[Col92] J. C. Coleman. The phonetic interpretation of headed phonological structures containing overlapping constituents. *Phonology Yearbook* 9(1):1–44, 1992.

[Col93] J. C. Coleman. Polysyllabic words in the YorkTalk synthesis system. In *Papers in Laboratory Phonology III*, P. Keating, ed. Cambridge University Press, 293–324, 1993.

[Fir57] J. R. Firth. A synopsis of Linguistic Theory. In *Studies in Linguistic Analysis*, Special Volume of the Philological Society, 2nd edition, 1–32, 1957.

[Fow80] C. A. Fowler. Coarticulation and theories of extrinsic timing. *Journal of Phonetics* 8:113–133, 1980.

[Fow81] C. A. Fowler. A relationship between coarticulation and compensatory shortening. *Phonetica* 38:35–50, 1981.

[Fow83] C. A. Fowler. Converging sources of evidence for spoken and pereived rhythms of speech: Cyclic production of vowels in sequences of monosyllabic stress feet. *Journal of Experimental Psychology: General* 112:386–412, 1983.

[Hen49] E. J. A. Henderson. Prosodies in Siamese. *Asia Major* 1:198–215, 1949.

[Hen52] E. J. A. Henderson. The phonology of loanwords in some South-East Asian languages. *Transactions of the Philological Society* 131–158, 1952.

[Kel89] J. Kelly. Swahili phonologcal structure: A prosodic view. In *Le Swahili et ses Limites*, M. F. Rombi, ed. Editions Recherche sur les Civilisations, Paris, 25–31, 1989.

[Kel92] J. Kelly. Systems for open syllabics in North Welsh. In *Studies in Systemic Phonology*, P. Tench, ed. Pinter Publishers, London and New York, 87–97, 1992.

[Ken94] M. Kenstowicz. *Phonology in Generative Grammar.* Basil Blackwell, Oxford, 1994.

[Kla] D. H. Klatt. *Klattalk: The conversion of English text to speech.* Unpublished manuscript, Massachusetts Institute of Technology, Cambridge, MA.

[Kla87] D. H. Klatt. Review of text-to-speech conversion for English. *Journal of the Acoustical Society of America* 82(3):737–793, 1987.

[LR73] B. Lindblom and K. Rapp. Some temporal regularities of spoken Swedish. *Papers in Linguistics from the University of Stockholm* 21:1–59, 1973.

[Loc90] J. K. Local. Some rhythm, resonance and quality variations in urban Tyneside speech. In *Studies in the Pronunciation of English: A Commemorative Volume in Honour of A C Gimson*, S. Ramsaren, ed. Routledge, London, 286-292, 1990.

[Loc92] J. K. Local. Modelling assimilation in a non-segmental rule-free phonology. In *Papers in Laboratory Phonology II*, G. J. Docherty and D. R. Ladd, eds. CUP, Cambridge, 190–223, 1992.

[LO94] J. K. Local and R. A. Ogden. Temporal exponents of word-structure in English. *York Research Papers in Linguistics.* YLLS/RP 1994.

[Man92] S. Y. Manuel, S. Shattuck-Hufnagel, M. Huffman, K. N. Stevens, R. Carlson, and S. Hunnicutt. Studies of vowel and consonant reduction. In *Proceedings of ICSLP* 2:943–946, 1992.

[Ogd92] R. A. Ogden. Parametric interpretation in YorkTalk. *York Papers in Linguistics* 16:81–99, 1992.

[Ogd93] R. A. Ogden. *European Patent Application 93307872.7 - YorkTalk.* 1993.

[Par84] B. H. Partee. Compositionality. In *Varieties of Formal Semantics*, F. Landman and F. Veltman, eds. Foris, Dordrecht, 281–312, 1984.

[Ril92] M. D. Riley. Tree-based modeling for speech synthesis. In *Talking Machines: Theories, Models, and Designs*, G. Bailly and C. Benoit, eds. Elsevier, North-Holland, Amsterdam, 265–273, 1992.

[Sim92] A. Simpson. The phonologies of the English auxiliary system. In *Who Climbs the Grammar Tree?* R. Tracy, ed. Niemeyer, Tuebingen, 209–219, 1992.

[Smi93] C. L. Smith. Prosodic patterns in the coordination of vowel and consonant gestures. Paper given at the Fourth Laboratory Phonology Meeting, Oxford, August, 1993.

[Spr66] R. K. Sprigg. Vowel harmony in Lhasa Tibetan: Prosodic analysis applied to interrelated vocalic features of successive syllables. *Bulletin of the School of Oriental and African Studies* 24:116–138, 1966.

[van92] J. P. H. van Santen. Deriving text-to-speech durations from natural speech. In *Talking Machines: Theories, Models, and Designs*, G. Bailly and C. Benoit, eds. Elsevier, North-Holland, Amsterdam, 275–285, 1992.

[van94] J. P. H. van Santen. Assignment of segmental duration in text-to-speech synthesis. *Computer Speech & Language* 8:95–128, 1994.

[VCR92] J. P. H. van Santen, J. Coleman, and M. Randolph. Effects of postvocalic voicing on the time course of vowels and diphthongs. *Journal of the Acoustical Society of America* 92:2444, 1992.

[Whe81] D. Wheeler. *Aspects of a Categorial Theory of Phonology.* Graduate Linguistics Student Association, University of Massachusetts at Amherst, 1981.

[Wii91] K. Wiik. On a third type of speech rhythm: Foot timing. In *Proceedings of the Twelfth International Congress of Phonetic Sciences*, Aix-en-Provence, 3:298–301, 1991.

10

A Complete Linguistic Analysis for an Italian Text-to-Speech System

Giuliano Ferri
Piero Pierucci
Donatella Sanzone

ABSTRACT A morpho-syntactic analyzer was constructed as part of a project aimed at developing an automatic text-to-speech system for the Italian language. This chapter describes the most important steps and the related problems involved in the automatic linguistic analysis of the Italian language. The process of analysis can be divided into three main phases: morphological analysis, phonetic transcription, and morpho-syntactic parsing. The morphological analyzer processes each word by means of context-free grammars to obtain all morphologic and syntactic features. Then, the phonetic transcription module locates the stressed syllables of each word and performs grapheme to phoneme conversion. Finally, syntactically related words are marked to form a suitable basis for pitch contour determination.

10.1 Introduction

During the past few years remarkable improvements have been made in the development of high-quality text-to-speech systems. Advanced digital signal processing techniques make it possible to generate highly intelligible and natural-sounding speech. Nevertheless, in spite of its potential quality, synthetic speech still sounds unnatural and is difficult to listen to, especially for long sentences. The reason for this could be poor modeling of the prosody-related phenomena that occur in natural speech. Several limitations of current systems contribute to this behavior, as for instance semantic/pragmatic and discourse factors that influence F_0 patterns. In this chapter a different class of limitations is addressed: Often automatic synthesizers are not able to correctly determine sufficient linguistic information from the input text to specify prosodic patterns properly.

As far as the sentence level is concerned, this lack of appropriate prosodic modeling produces a large degree of uncertainty about breath group determination in the input text, which very often results in the incorrect placement of prosodic phrase breaks and in a machine-like prosodic pattern generation. At the word level, the prosodic pattern determination requires the production of a proper fundamental frequency movement around the stressed syllable. The reliability of automatic lexical stress placement plays a fundamental role in this. At the phonetic level, the intelligibility of synthetic speech is also strongly dependent on a reliable grapheme to phoneme conversion process. In some languages—for instance, Italian—this process presents some ambiguities that can be partly solved using grammatical and/or morphological analysis. The above comments show the importance of reliable processing of the input text within a text-to-speech system.

A text-to-speech system could be logically divided into three main modules: linguistic analysis, prosodic pattern determination, and speech signal generation. The latter can be based on a parametric model of the vocal apparatus, such as the formant technique or LPC coding, or on time-domain processing techniques such as overlap and add methods. This stage could be taught as a language-independent component of the overall system, given that proper acoustic characterization for each language is made available. The prosodic pattern determination process seems to exhibit some sort of dependence on the language to be synthesized, although sometimes groups of languages can be treated in a similar way. To this extent, for instance, the Italian language shows several similarities with other European languages [Col90]. Linguistic analysis is generally acknowledged as the more language-dependent component of a text-to-speech system. This analysis carries out two main tasks: (1) letter to phoneme conversion is accomplished, forming the basis for the following acoustic unit selection; and (2) morphologic and syntactic analyses are executed and prosodic phrase boundaries and sentence level accentuation are determined.

Depending on the language concerned, linguistic processing involves different degrees of difficulty and complexity, resulting from the interaction of the several sources of knowledge. The main linguistic problems for the Italian language include the following:

1. Determination of morphological characteristics. An adjective or a noun can have up to four inflections based on its gender and number, whereas verbs can have up to a hundred different inflections based on mood, tense, and person.
2. Stressed syllable determination. Lexical stresses can fall on one of the last six syllables. There is neither graphic representation (except in final position) nor general deterministic rules to help in this task.
3. Ambiguities related to phonetic transcription [Sal90]. There are no reliable rules to assign the opening/closing degree of constriction for the stressed mid-vowels and voicing/unvoicing determination of `s` and `z`.

These problems can be solved using a *morphologic approach,* based on the determination of all basic morphologic components for each word. Each component brings specific linguistic characteristics to the word to which it refers. The analysis recognizes each component and can specify its syntactic and morphologic features as well as its phonetic transcription. The overall system is organized as follows:

1. morphological analysis, including the determination of all morphologic features (gender and number or mood, tense, and person) and syntactic features (noun, adjective, verb, and so on) of each word
2. phonetic transcription determination, with each grapheme being translated into a phoneme

3. morpho-syntactic analysis, in which some types of word sequences are recognized. This information can be used to assign phrasal intonation contours

This chapter describes in detail the three steps of analysis. In the first section there is the description of the morphologic phase. The next section deals with the phonetic transcription process. The third section describes the morpho-syntactic analyzer. Evaluations and remarks about future work conclude the chapter.

10.2 The Morphologic Analysis

The aim of the morphologic analysis is to determine, for each word, the lemma from which it is derived, its syntactic category (e.g., verb, noun, adjective, article), and its morphologic category (e.g., masculine, singular, indicative). Phonetic transcription related to each syntactic interpretation is also retrieved. This process will be described in detail in the next section.

10.2.1 The Definition of the Problem

In most works referring to the English language, morphology is considered only as a part of the syntactic parser. However, the morphology of Italian must be previously analyzed because it is more complex: There are more rules in Italian than in English and these rules present many exceptions.

From a morphologic point of view, a word can be seen as a complex entity made up of many subunits showing *formation* and *transformation* processes [Cho69]: New words can be generated by turning *base* words into *derived* words. For example:

	⇒	alteration	⇒	`cart-acc-ia` *(waste paper)*
`carta` *(paper)*	⇒	composition	⇒	`cart-a-monet-a` *(paper money)*
	⇒	prefixation	⇒	`in-cart-are` *(to wrap in paper)*
	⇒	cliticization	⇒	`in-cart-ar-lo` *(to wrap it in paper)*

These rules make the set of Italian words potentially unlimited. Therefore it is convenient to make use of an approach in which the lexicon of the system contains only the elementary lemmata of the language. All correct Italian words can be obtained by applying derivation rules.

These rules can be easily described by means of a context-free grammar in which each word results from the concatenation of the *stem* of a lemma with *affixes*, *endings*, and *enclitics*. This grammar can *analyze* a given word, by giving all the possible lemmata from which it is derived, and it can also *generate*, from a given lemma, all the current Italian words deriving from it. In this chapter, the first application will be assessed because it is the only one needed in a TTS system.

10.2.2 About the Lexicon

Data are organized according to the morphological structure of Italian words and divided into *prefixes*, *stems*, *suffixes* (both *derivative* and *alterative*), *endings*, and *enclitics*, which are stored with their linguistic characteristics. Data are stored in several tables (one for each morphological type of component); each table contains elements and related linguistic information allowing word classification in terms of morphologic and syntactic features. Phonetic transcription of each component is stored in the table.

The stems are grouped in a table with five attributes:

1. the invariable part of the lemma, called the stem (the access key in the table)
2. the lemma
3. the phonetic transcription of the stem
4. the class of endings associated with the lemma. A class of endings is the set of all the endings related to a given class of words. For example, each of the regular verbs of the first conjugation has the same ending: hence there exists a class called `dv_1coniug` containing all these endings.
5. the syntactic category of the lemma, for example, the information that *to have* is an *auxiliary transitive verb*

The following is an example of a stem table entry:

stem	lemma	transcription	ending class	syntactic category
`cas`	`casa`	ˈkaz	`dn_cosa`	`noun_common`

The set of the classes of endings forms a table with four attributes:

1. the name of the class (the access key in the table)
2. one ending belonging to the class
3. the phonetic transcription of the ending
4. the morphologic category associated with this ending

The following is an example of some endings table entries:

ending class	ending	transcription	morphologic category
`da_bello`	a	`a`	`fem_sing`
`da_bello`	e	`e`	`fem_plur`
`da_bello`	i	`i`	`mas_plur`
`da_bello`	o	`o`	`mas_sing`

The class *da_bello* contains all the endings needed to inflect all the nouns behaving like the word `bello` (*beautiful* or *handsome*).

The affixes can be divided into *prefixes* preceding the stem of the lemma and *suffixes* following the stem of the lemma. The set of prefixes is a table with only two attributes:

1. the prefix
2. the phonetic transcription of the prefix

It is important to note that prefixes do not change the morphologic and syntactic features of a lemma. For example, the prefix `s` transforms the transitive verb `vestire` (*to dress*) into the intransitive verb `svestire` (*to undress*). The following is an example of some prefix table entries:

prefix	transcription
`ri`	ri
`s`	s

The set of derivative suffixes is a table with five attributes:

1. the stem of the suffix (the access key to the table)
2. the suffix
3. the phonetic transcription
4. the ending class of the suffix
5. the syntactic category of the suffix

Unlike prefixes, suffixes can change both morphologic and syntactic features of the original word: They change verbs into nouns or adjectives (*deverbal suffixes*), nouns into verbs or adjectives (*denominal suffixes*), and adjectives into verbs or nouns (*deadjectival suffixes*). The second attribute is chained to the stem of the original lemma to obtain the derived lemma. For example, from the stem of the lemma `monte` (*mountain*), which is a noun, the suffix `uoso` forms `mont-uoso` (*mountain-ous*). The following is an example of a suffix table entry:

suffix stem	suffix	transcription	ending class	syntactic category
`uos`	`uoso`	ˈwoz	`da_bello`	`qualif_adj`

In the Italian language there is another peculiar type of suffixation process: the *evaluative suffixation* [Sca86], or *alteration*. The alterations change the morphologic and semantic features of the altered word, but not its syntactic category. Alteration is a typical feature of the Italian language and it is completely absent in other languages such as English or French. For example, it is possible to change the single word `gatto` (*cat*) into the single word `gattino`, for which the English language requires two words (*little cat*). The set of alterations is a table with four attributes:

1. the stem of the alteration (the access key in the table)
2. the phonetic transcription
3. the ending class of the alteration
4. the semantic type of the alteration

The following is an example of an alternations table entry:

alteration	transcription	ending class	semantic category
`in`	ˈin	`da_bello`	`diminutive`

Another typical derivation in Italian is *cliticization*. The enclitics are pronouns linked to the ending of a verb. For example, the English words *bring it* are translated into only one Italian word, `portalo`. Enclitics are always not stressed. The set of enclitics is a table with three attributes:

1. the enclitic (the access key in the table)
2. the phonetic transcription
3. the morphologic category of the enclitic

The following is an example of an enclitics table entry:

enclitic	transcription	morphologic category
lo	lo	mas_sing

Two more sets of data have been purposely defined to handle fixed sequences of words, such as idioms and proper names.

The set of the most common Italian idioms has been structured as a table with four attributes:

1. the most significant word of the idiom (the access key in the table)
2. the idiom
3. the phonetic transcription
4. the syntactic category

Using an idiom table it is possible to identify the idiom without performing the analysis of each of the component words. For example, `in modo che` (*in such a way as*) is an idiom used in the role of a conjunction:

keyword	idiom	transcription	syntactic category
modo	in modo che	inmodo'ke	conj_subord

The set of proper names is stored in a table with five attributes:

1. the most significant word of the proper name (the access key in the table)
2. the proper name
3. the phonetic transcription
4. the syntactic category of the name
5. the morphologic category of the name

keyword	proper name	transcription	synt.	morph.
banca	banca d'Italia	baŋkadi'talja	name_bank	fem_sing

Finally, two more tables are defined: the first is used by the preanalyzer to detect abbreviations, and the second could be modified by the end user to add dynamically new words to the system vocabulary.

Each table is accessed using an indexed search to improve performances. The lexicon used is composed of 91 prefixes, 7088 stems, 178 derivative suffixes, 46 alterative suffixes, 1911 endings grouped in 112 classes, 32 enclitics, 1112 proper nouns, and 91 idioms. The system can recognize about 11,000 lemmas corresponding to more than 100,000 inflected words.

10.2.3 The Morphologic Analyzer

The building of a lemma depends on several morphologic components, which follow well-defined composition rules. These rules can be described using *formal grammars* [Cho69]. The morphologic analysis is based on the following context-free grammar [Rus87]:

WORD ⇒ {prefix}^{n} <RADIX> <REM>
RADIX ⇒ <stem> {suffix}^{n}
REM ⇒ {ending}^{m} {enclitic}^{n}
m=0,1,2; n=0,1,2,3, ...

Each word is analyzed from left to right and is split into one stem, preceded by one or more prefixes (if any) and followed by one or more suffixes (if any); by an ending; and, if the word is a verb, by one or more enclitics (if any). Compound names can be recognized by applying grammar rules twice. All linguistic features depend on the above breakdown: grammatical characteristics are given by stems and suffixes, whereas morphologic characteristics (gender and number or mood, tense, and person) are given by endings. The parser analyzes each word and produces, for each form analyzed, the list of all its characteristics:

1. the lemma it derives from
2. its syntactic characteristics
3. its morphologic characteristics (none for invariable words)
4. the list of alterations (possibly empty)
5. the list of enclitics (possibly empty)

The analyzer provides all possible interpretations for each word. For example:

`porta` *(door)*	⇒ `port` (stem, noun) + `a` (ending, feminine sing.)
`porta` *(he brings)*	⇒ `port` (stem, verb) + `a` (ending, 3rd sing., simple pres.)
`porta` *(bring!)*	⇒ `port` (stem, verb) + `a` (ending, 2nd sing., imperative)

This approach presents some problems because the rules of the formal grammar are too general. Many morphologically valid compositions could be accepted and among these it is necessary to select only correct interpretations. Therefore it is important to insert specific controls to *filter* the morphologic analysis results.

Two kinds of *filters* are implemented:

1. controls carried out during morphologic component selection, using special information included in tables of lexicon in order to direct the searching only toward *valid* compositions [San93]
2. controls applied at the end of the morphologic analysis phase in order to select only the correct ones from among all *morphologically* possible candidates

The second type of filter is the most important. The most selective filter of this category checks the exact interpretation of every word, verifying its presence in a table that contains every lemma recognizable by the system. The inflected word is accepted if and only if the lemma is in the basic lexicon. The following are some entries of the basic lexicon:

lemma	syntactic category
`abbreviare`	transitive verb
`bello`	qualified adjective

All inflections of the verb `abbreviare` (*to shorten*) are recognized, as are the masculine, feminine, singular and plural forms of adjective `bello` (*beautiful*).

10.3 The Phonetic Transcription

In the previous section, we presented several morphologic tables, each one containing information about phonetic transcription. In fact, the grapheme to phoneme conversion is accomplished during morphologic analysis, but in this section we will keep them separate.

10.3.1 The Definition of the Problem

The idea of using a morphologic parser to determine the correct phonetic transcription of words comes from some observations about problems on the stress assignment of Italian words.

Phonetic transcription is generally not a very difficult task in the Italian language, but there is a good deal of ambiguity that can affect the intellegibility and quality of synthetic speech, especially for the automatic lexical stress assignment problem. In Italian there is no explicit graphic indication in a written text of the location of the stressed vowel (except for a few cases of final stress, such as `città` – *town*). The lexical stress can fall on any of the last six syllables for verbs followed by enclitics and on any of the last four syllables in the remaining cases. Stress position depends more on morphological and grammatical characteristics than on syllabic structure. Moreover, rule-based approaches often fail because of the great number of exceptions. These considerations suggest treating every word as a potential exception, and assigning to each entry of the lexicon the related phonetic transcription.

Each morphological component is labeled with its own phonetic transcription (sometimes more than one) to solve each ambiguity. If a word is not found in the system's vocabulary, a simple set of rules is applied to assign phonetic transcription.

10.3.2 Automatic Stress Assignment

Exhaustive rules to determine the exact stress position are difficult to find [Bal91], and in some cases cannot be defined (for instance, for the Italian language).

By exploiting the morphologic approach and assigning the phonetic transcription to lexical particles explicitly, the stress position is consequentially localized for each one. The stress can fall only upon fixed components. The following scheme describes the different possibilities:

1. Prefixes are never stressed.

2. Stems are always stressed (on one of the last two syllables).
3. Endings can be either stressed or unstressed.
4. Derivative and alterative suffixes can be either stressed or unstressed.
5. Enclitics are never stressed.

When every basic morphologic component is retrieved, the stress position of every one is known, and so the stress position on the whole word is found (eventually more than one). The primary lexical stress can be localized on the right-most stressed syllables [Mar90]. For example:

`casa` *(house)*	⇒ ˈkaz+a	= ˈkaza
`casetta` *(little house)*	⇒ ˈkaz+ˈet:+a	= kaˈzet:a
`casettina` *(very little house)*	⇒ ˈkaz+ˈet:+ˈin+a	= kazet:ˈina

Other significant advantages can be found in stress localization on clitics verbs: adding enclitics (atonic pronouns) at the end of the word moves stress leftward. For example:

`porta` *(bring)*	⇒ ˈpɔrt+a	= ˈpɔrta
`portalo` *(bring it)*	⇒ ˈpɔrt+a+lo	= ˈpɔrtalo
`portaglielo` *(bring it to him)*	⇒ ˈpɔrt+a+ʎ:e+lo	= ˈpɔrtaʎ:elo

10.3.3 Opening and Closing Vowels

The morphologic approach (storing the phonetic transcription for each morphologic component) completely solves the problems related to automatic stress assignment. Furthermore, in the Italian language there are five graphemic symbols for vowels `a`, `e`, `i`, `o`, and `u`; some of them, namely the vowels `e` and `o`, can be pronounced with an opening or closing degree of constriction. Even if this contrast is dependent on regional dialect, there is general agreement on the evidence for this phenomenon in what is sometimes called "Standard Italian." For consistency, we have choosen to follow the pronunciation recommended by Canepari [Can92]. The use of a morphologic-based phonetic transcription allows us to resolve the open/closed ambiguities for `e` and `o`.

This distinction should be more carefully considered in the case of homographs, in which different pronunciations may correspond to different meanings. For example, the word `pesca` could be pronounced in two different ways: [ˈpeska] (*fishing* or *he fishes*) and [ˈpɛska] (*peach*). The system assigns a phonetic transcription to every possible syntactic interpretation. In the above example, `pesca` is broken down into a stem (`pesc`) followed by an ending, but the stem has two phonetic transcriptions associated with it, corresponding to both pronunciations.

10.3.4 Voiced and Unvoiced Consonants

The same problem described for the mid-vowels `e` and `o` arises for the consonants `s` and `z`. Using the same approach, problems related to the voicing determination of consonants `s` and `z` are addressed.

Determination of voicing degree for `s` could be performed using rules [Lep78]: for example, at the beginning of the word, `s` is generally unvoiced (such as in `salire` [sa'lire] – *ascend*). The same applies if `s` appears just after a consonant (such as in `penso` ['pɛnso] – *I think*) or before an unvoiced consonant (such as in `pasto` ['pasto] – *meal*). Otherwise, `s` is voiced before a voiced consonant (for example, `snello` ['znɛl:o] – *slim*).

Problems arise when `s` is in intervocalic position. Usually `s` is voiced, such as in `isola` ['izola] (*island*). But there are some exceptions, as in the case of affixed and compound words and in numerical sequences. For example, the word `ri-salire` [risa'lire] (*re-ascend*) is obtained by a prefixation process: the phonetic transcription of the stem includes unvoiced `s`. In this case the complete prefixed word is pronounced using unvoiced `s` even if `s` is intervocalic. The same exception applies for the clitics beginning with `s`. Some rules [San93] are also applied when different word components are joined, which sometimes results in transforming the original phonetic transcription. For example, the word `smettere` (*to stop*) is composed by a prefix (`s`), a stem (`mett`), and an ending (`ere`): The phonetic transcription associated with the prefix is an unvoiced `s`, but because it precedes a voiced consonant, it is modified to the related voiced phoneme [z].

More evident problems are present in the pronunciation of the consonant `z` because there are no fixed rules to determine the correct voicing decision, and a voicing error would produce unpleasant acoustic effects. For example, the words `pezzo` ['pɛts:o] (*piece*) and `mezzo` ['mɛdz:o] (*means*) contain the same sequence of graphemic symbols (`ezzo`), each one with a different phonetic transcription.

10.4 The Morpho-Syntactic Analysis

The morpho-syntactic analyzer performs the analysis of contiguous words, allowing for:

- elimination of syntactic ambiguities
- determination of syntactically related sequences of words, to be grouped together from a prosodic point of view

Isolated words with many syntactic interpretations can take on, in some specific word sequences, only one syntactic meaning. For example, the word `stato` (*state, condition, status, stayed*) takes on only one meaning in the sequence `sono stato promosso` (*I've been promoted*). Furthermore, the above same sequence of words is recognized as a verbal phrase: so it is possible, for example, to assign a specific intonation contour, and no pauses will be inserted between the words of the same group. The determination of related contiguous words can be performed in three different steps: The first step occurs before morphologic analysis, the second one is executed just after morphologic analysis by recognizing regular structures, and the third is a syntactic parser.

10.4.1 The Preanalyzer

The preanalyzer simplifies morphologic analysis by recognizing all the fixed sequences of words in the sentence. There are four types of sequences retrieved in this phase:

1. idioms
2. proper names
3. dates
4. numbers

The determination of idioms does not require morphologic analysis. For example, to analyze a sequence of words such as `in modo che` (*in such a way as*), it is not necessary to know that `in` is a preposition, `modo` is a noun and `che` could be a conjunction or a pronoun; the only useful information is that this sequence takes on the role of a conjunction.

For proper names is also necessary to know that, for example, `Giovanni Agnelli` or `Banca Nazionale del Lavoro` are single entities, that the first one is the name of a male person, and the second is the name of a bank.

Idioms and proper names are recognized by means of a string-pattern-matching algorithm. For the idioms only, this algorithm is triggered when the most significant word of the idiom (generally the noun) appears in the input sentence. This mechanism improves the system performances because idioms often begin with articles or prepositions that appear frequently in the sentence. It would not be convenient to perform the comparison every time the analyzer met such words.

Date expressions such as `sabato 21 gennaio 1995` (*Saturday, January 21, 1995*) are considered as single entities to simplify the subsequent phases of analysis. They are recognized using the following context-free grammar:

DATE	$\Rightarrow$	<name proper day> <DATE 1>
DATE	$\Rightarrow$	<DATE 1>
DATE	$\Rightarrow$	<DATE 2>
DATE 1	$\Rightarrow$	<number day> <name proper month>
DATE 1	$\Rightarrow$	<number day> <DATE 2>
DATE 2	$\Rightarrow$	<name proper month> <number year>

Numbers are also recognized by means of another context-free grammar, translating strings into numbers. In this way it is possible to evaluate in the same manner numbers such as `1995` and strings such as `millenovecentonovantacinque` (*one thousand nine hundred ninety-five*).

NUMBER	$\Rightarrow$	<NUM1>
NUMBER	$\Rightarrow$	<`'mille'`>
NUMBER	$\Rightarrow$	<`'mille'`> <NUM1>
NUMBER	$\Rightarrow$	<NUM1> <`'mila'`>
NUMBER	$\Rightarrow$	<NUM1> <`'mila'`> <NUM1>
NUM1	$\Rightarrow$	<NUM2>
NUM1	$\Rightarrow$	<NUM3>
NUM1	$\Rightarrow$	<NUM4>

NUM2	⇒	<units> <NUM3>
NUM3	⇒	<`'cento'`>
NUM3	⇒	<`'cento'`> <NUM4>
NUM4	⇒	<units>
NUM4	⇒	<tens>
NUM4	⇒	<tens> <units>

10.4.2 The Morpho-Syntactic Analyzer

The morpho-syntactic analyzer uses the results from the previous stages to detect regular syntactic structures. The following structures are recognized:

1. compound tenses of verbs
2. comparative and superlative degree of adjectives
3. mixed numeric expressions

Compound tenses of verbs are described by means of a context-free grammar: when the analyzer finds a past participle or a present progressive form of a verb, the rules of the grammar are applied. The grammar recognizes structures containing one or more adverbs between the auxiliary verb and the past participle or gerund form of the main verb. The following phrases are recognized by the grammar: `io sono stato ancora chiamato` or `io sono ancora stato chiamato` (*I have been called again*). Note also that in the Italian language adverbs may be located between the auxiliary and the past participle or between the two past participles.

COMP TENSE	⇒ <v tran aux> {adverb}^{n} <PAST> {REM1}
COMP TENSE	⇒ <v intran aux> {adverb}^{n} <PAST> {REM1}
COMP TENSE	⇒ <v intran aux> {adverb}^{n} <REM>
COMP TENSE	⇒ <v intran fraseol> {adverb}^{n} <REM2>
COMP TENSE	⇒ <v intran servil> {adverb}^{n} <REM1>
PAST	⇒ <v tran(past part)>
PAST	⇒ <v intran(past part)>
REM	⇒ <v intran aux p part> {adverb}^{n} <PAST>
REM1	⇒ {adverb}^{n} <v tran infinity>
REM1	⇒ {adverb}^{n} <v intran infinity>
REM2	⇒ <v tran gerund>

When a rule is successfully applied, the morphologic categories of the verbs are changed and the attribute *active/passive* is specified.

In the previous example (`io sono stato ancora chiamato`) the analyzer accomplishes two tasks: First, it recognizes a sequence of five words that will be pronounced with the same overall intonation contour and without pause between words; and second, it solves an ambiguity. In fact, the word `ancora` can assume three different meanings (with different pronunciations):

- adverb (aŋˈkora - *again, still*)
- noun (ˈaŋkora - *anchor*)

- verb (ˈaŋkora - *to anchor*)

but in this context it can assume only the first meaning.

Qualificative adjectives can be used in the comparative or superlative degree. In this case, the adjective is preceeded by the adverb `più` (*more*) or `meno` (*less*) and, for the superlative degree only, by the definite article.

This kind of syntactic structure can be recognized by a context-free grammar. It is applied when the analyzer finds the words `più` o `meno` followed by a qualificative adjective. By using this grammar it is possible to recognize expressions such as `più bello` (*more beautiful*), `il più bello` (*the most beautiful*), and `il libro più bello` (*the most beautiful book*). In the Italian language, in fact, the noun can appear between the article and the adverb of the superlative.

SUPERL REL	⇒	<art determ> <COMPARATIVE>
SUPERL REL	⇒	<art determ> <noun> <COMPARATIVE>
COMPARATIVE	⇒	<'più'> <adj qualif>
COMPARATIVE	⇒	<'meno'> <adj qualif>

Mixed numeric expressions such as `2 miliardi 400 milioni` (*2 billion and 400 million*) can appear frequently in Italian text (in newspaper articles, for example). Another context-free grammar allows the recognition of such structures and translating them in numeric form. The rules of this grammar will be applied when the parser finds words such as `miliardo` (*billion*), `milione` (*million*), and `mila` (*thousand*).

NUM COMP	⇒	<num> <'miliardo'> {NUM1} {NUM2} {num}
NUM COMP	⇒	<NUM1> {NUM2} {num}
NUM COMP	⇒	<NUM2> {num}
NUM1	⇒	<num> <'milione'>
NUM2	⇒	<num> <'mila'>

10.4.3 The Syntactic Parser

The last stage of linguistic analysis performs syntactic parsing in order to reach two main goals: (1) to solve the remaining syntactic ambiguities, and (2) to assign intonation group boundaries.

The richness of information available from the previous phases of analysis can allow the use of very sophisticated parsers. Many techniques could be used here. For example, a chart-parsing algorithm could be successfully applied [Rus91] or a rather different probabilistic method could be followed [MM91]. But the approach choosen in the first implementation is very similar to the one described by O'Shaugnessy [OSh89], even if the dictionary used is quite large.

The first task for the parser is to segment a sentence into very small phrasal units (*phonological words*), each containing a few words. A phonological word is uttered with a stress only on the main lexical word and without inner word boundaries. In some special cases a phonological word can be composed by one or more punctuation marks. In the next step, the parser classifies each phonological word depending on the syntactic meanings of the selected lexical words. Then, the parser can connect sequences of phonological words into *intonation groups*. An intonation group is a sequence of words uttered without pauses and using a specific prosodic pattern. Last, the parser classifies the intonation groups depending on the chosen phonological words. The specific prosodic pattern can be selected depending on this intonation group classification.

The following is a more detailed description of the algorithm used. First of all, the lexical words are divided into five different classes:

[**F**] – Function words (articles, prepositions, conjunctions);

[**C**] – Content words (nouns, adjectives, verbs, adverbs);

[**A**] – Ambiguous words (words with more than one possible interpretation: function and content);

[**V**] – Verbal words (words with the syntactic meaning only of a verb);

[**T**] – Punctuation marks;

The parser operates from left to right through each sentence. The ambiguities are solved using rules depending on the different syntactic characteristics of the current word and the class of the previous word. A *finite state automaton* segments the sentence, locating the end of each phonological word, in which a sequence of content words is broken up by a function word. Each phonological word is classified by its elements according to the following rules:

[**NG**] – noun group (a noun or an adjective and its preceding function words except prepositions);

[**VG**] – verb group (a verb optionally preceded by modal and auxiliary verbs);

[**AG**] – adverbial group (one or more adverbs eventually followed by nouns and/or adjectives);

[**PN**] – prepositional noun group (a preposition followed by a NG);

[**PV**] – prepositional verb group (a preposition followed by a VG);

[**PA**] – prepositional adverbial group (a preposition followed by an AG);

[**F**] – function group (a function word or a punctuation mark).

After this classification, the parser has to locate the intonation group boundaries. The end of an intonation group could be found before a VG phonological word or before an F phonological word. By using this approach, pauses will be inserted before verbs (emphasizing them) and before the sentence-ending punctuation marks. Conversely, pauses will rarely occur before PN or PV phonological words. However, if sequences of many phonological words without pauses are detected, a boundary is inserted before PN or PV group.

Finally, the intonation groups are classified depending on their structure and the punctuation marks. Each intonation group receives a different prosodic pattern depending on its morphology and its role in the sentence. There are several prosodic pattern categories, referred, for example, to beginning, ending and continuing intentions of the sentence. Several intonation contours are available for each category: for example, ending contours can be affirmative, exclamatory, or interrogative.

10.5 Performance Assessment

The system was tested using a set of 17,809 words, morphologically, syntactically, and phonetically labeled by hand. This text comes from Italian newspapers, mostly treating the economy and politics. It had been chosen as a subset of a larger corpus used in a previous project to assess the statistical distribution of words in the Italian language, for speech recognition applications [MM91]. After the initial test, 1,784 words of this test corpus were not successfully recognized. These words were analyzed by hand and added to the appropriate vocabulary tables.

Coding was carried out in the C programming language, in order to make the system portable across computing platforms. For the test mentioned above, processing time was estimated as the average time required to complete the analysis of a word in the input text. Performance is reported for a Risc machine (IBM RS/6000 mod. 250 equipped with a PowerPC processor) and for a PC-class machine, namely a 66 MHz clocked 486-DX2 processor. The average analysis time required for a word was 2.8 msecs in the first environment and 5.8 msecs in the second one. Furthermore, this measure represents a small contribution to the overall CPU requirement of the text-to-speech system, especially when compared to the modules that are responsible to the speech signal generation. Using a *simplified overlap and add* technique, one of the fastest available for speech signal generation, the average ratio between the CPU requirements of the linguistic analysis module versus the signal generation module is in the range between 1:10 and 1:20 (depending on the required output sampling frequency).

No generally accepted assessment method has been found in the technical literature to test the effectiveness of syntactic marking provided by the system for the F_0 contour assignment task. Thus, only informal tests have been carried out on a sixty-sentence corporus of newspaper text, read by a professional speaker. The analysis showed that more than 95% of sentences were classified in a way that is consistent with the behavior of the natural F_0 contour of the speaker.

10.6 Conclusion

A complete system for the automatic linguistic analysis for an Italian text-to-speech system has been presented. The system is based on a morphology-driven lexicon for the Italian language, containing information about phonetic transcription. Furthermore, a set of context-free grammars, representing the rules of word, idiom, and sentence formation, is used. This system has been proven to be effective in managing the tasks of morphologic analysis, phonetic transcription, and morpho-syntactic parsing. The output of the system is suitable to feed all the subsequent modules that are necessary to complete the process of automatic speech synthesis from written text. The implementation assessed the system as a viable and very efficient way to perform linguistic analysis in a text-to-speech system for the Italian language.

Future extensions of the current system include the development of *intelligent* tools for vocabulary management, which would be able to locate words that are currently not recognized by the system and to give to the system administrator the results of a preliminary analysis to be completed by hand. Another area of interest is in the development of syntactic disambiguation algorithms based on statistical techniques, similar to what is currently used for automatic speech recognition systems, to raise the recognition score in large vocabulary applications. Furthermore, the development of a special prosodic processor able to efficiently manage all the input coming from the morpho-syntactic analysis stage seems to be a natural consequence of the adopted approach. Finally, the suitability of this system for other languages, and thus for multilingual applications, needs to be assessed.

References

[Bal91] M. Balestri. A coded dictionary for stress assignment rules in Italian. In *Proceedings of 2nd European Conference on Speech Communication and Technology*, Genoa, Italy, 1169–1171, 1991.

[Can92] L. Canepari. *Manuale di Pronuncia Italiana*. Zanichelli, Bologna, Italy, 1992.

[Cho69] N. A. Chomsky. *L'analisi del Linguaggio Formale*. Boringhieri, Turin, Italy, 1969.

[Col90] R. Collier. Multi-lingual intonation synthesis: Principles and applications. In *Proceedings of ESCA Workshop on Speech Synthesis*, Autrans, France, 273–276, 1990.

[Lep78] G. C. Lepschy. *Saggi di Linguistica Italiana*. Il Mulino, Bologna, Italy, 1978.

[Mar90] P. Martin. Automatic assignment of lexical stress in Italian. In *Proceedings of ESCA Workshop on Speech Synthesis*, Autrans, France, 149–152, 1990.

[MM91] G. Maltese, and F. Mancini. A technique to automatically assign parts-of-speech to words taking into account word-ending information through a probabilistic model. In *Proceedings of 2nd European Conference on Speech Communication and Technology*, Genoa, Italy, 753–756, 1991.

[OSh89] D. D. O'Shaughnessy. Parsing with a small dictionary for applications such as text to speech. *Computational Linguistics* 15(2):97–108, June 1989.

[Rus87] M. Russo. A generative grammar approach for the morphologic and morphosyntactic analysis of Italian. In *Proceedings of 3rd European Conference of ACL*, Copenhagen, Denmark, 32–37, 1987.

[Rus91] T. Russi. Robust and efficient parsing for applications such as text-to-speech conversion. In *Proceedings of 2nd European Conference on Speech Communication and Technology*, Genoa, Italy, 775–778, 1991.

[Sal90] P. L. Salza. Phonetic transcription rules for text-to-speech synthesis of Italian. *Phonetica* 47:66–83, 1990.

[San93] D. Sanzone. *Un Sistema di Analisi Morfologica per la Lingua Italiana*. Degree thesis in Mathematic Science, University of Rome, 1993.

[Sca86] S. Scalise. *Generative Morphology*. Foris Publications, Dordrecht, Holland – Riverton, CA, 1986.

11

Discourse Structural Constraints on Accent in Narrative

Christine H. Nakatani

ABSTRACT This chapter examines the relationship between discourse structure and intonational prominence or pitch accent. It is argued, based on the distribution of pitch accent in an unrestricted spontaneous narrative, that accent function must be interpreted against a dynamic background of linguistic factors, including grammatical function, lexical form, and discourse structure as formulated within the computational discourse modeling framework presented by Grosz and Sidner [GS86]. A generalization emerges from analysis of the spontaneous narrative that accent functions as a marker of the attentional status of entities in a discourse model. The results of this study may be applied in message-to-speech synthesis systems to enable richer and more meaningful prosodic variation.

11.1 Introduction

The hierarchical structure of discourse influences accent decisions in systematic ways. Previous studies have shown, for example, that the accentuation of referring expressions can be correlated with discourse structural properties [Ter84], and that taking discourse structure into account can improve the performance of pitch accent assignment algorithms [Hir93]. On the other hand, these same studies as well as others show that accent is determined partly by lexical form and partly by grammatical function, among other linguistic factors (cf. [Bro83, Fuc84, Alt87, TH92]). Our account of the distribution of pitch accent in a spontaneous narrative unifies and reconciles previous findings by defining how accent combines with lexical form and grammatical function to mark the attentional status of entities in a discourse model. Our interpretation of accent function in discourse involves two basic claims: first, the meanings conveyed by choices of lexical form and syntactic structure are separate but interacting; and second, the role of accentuation must be interpreted against the background of these choices in linguistic expression. The new analyses of accent function are formally cast within the computational discourse modeling framework of Grosz and Sidner [GS86].

In section 11.2, we describe our analyses of an unrestricted spontaneous narrative. In section 11.3, we formulate a discourse-based interpretation of accent function to account for findings of the study, and we present the framework of attentional state modeling that we assume in our analyses. In section 11.4, we dis-

...and Masson was a con man
he's one of those slick greasy kind of people who
weasel their way into your life
and Eissler—being an old man—wanted to unload this
cuz he wanted to retire and die okay
analysts are very very strange people
they're very closely knit in other words
they will not talk to you if you have not been analyzed
they will not talk to anybody who is not an analyst alright...

FIGURE 11.1. Excerpt from the spontaneous narrative monologue (on CD-ROM).

cuss the discourse functions of accent in six linguistic configurations. We discuss related work in section 11.5 and conclude in section 11.6.

11.2 The Narrative Study

An excerpt of the spontaneous narrative monologue analyzed in this study is provided on the accompanying CD-ROM (see Appendix). The text transcription of the excerpt appears in figure 11.1. The monologue consists of 20 minutes of American English speech, obtained using sociolinguistic interview techniques.[1]

Several factors contributing to accent decisions can be interrelated using the notion of *information status*, or *given/new* [Pri81, Pri88]. Simply put, information that is being communicated for the first time in a discourse is considered new to the discourse or *discourse-new*, whereas information previously mentioned is given or *discourse-old*. Previous research has shown there is a general tendency for given information to be unaccented and new information to be accented [Bro83]. It is also generally thought that the information status of pronoun referents is given, and the information status of proper names and full noun phrases is new (but see section 11.4). Finally, there is a general tendency for grammatical subjects to represent given information and grammatical direct objects to represent new information [Pri88]. Thus, the general claim about the accentuation of given/new information predicts that (1) pronouns are unaccented and full noun phrases and proper names are accented, and (2) subjects are unaccented and direct objects are accented. The narrative study provides data challenging both of these basic hypotheses.

[1]The narrative was collected by Virginia Merlini for the purpose of studying American gay male speech and was made available by Mark Liberman at the University of Pennsylvania Phonetics Laboratory.

11.2.1 Analysis

In the narrative study, 481 animate noun phrase referring expressions were analyzed for accentuation, grammatical function (e.g., subject, direct object, object of preposition), and form of referring expression (e.g., proper name, pronoun, definite/indefinite noun phrase). We define two accentuation classes, *prominent* and *nonprominent*. The accentuation of an expression is said to be prominent if its head and/or an internal modifier is marked by either H* or a complex pitch accent in Pierrehumbert system of English intonation [Pie80]. An expression not meeting this criterion is said to be nonprominent.[2] For the present analyses, we therefore conflate L* accented expressions and deaccented expressions in the nonprominent category.[3] Accentuation judgments were made by the author by listening and by examining the pitch tracks, amplitude waveform, and spectrographic information in some cases. All speech analysis was performed using Entropic Research Laboratories Waves+ software [Tal89] on Sun workstations.

11.2.2 Results

Our analyses consider the role of accentuation in relation to grammatical function and lexical form of referring expression. We focus on the two most frequent grammatical functions (subject versus direct object) and the two largest classes of lexical forms (pronouns versus explicit forms, i.e., proper names, definite NPs, and indefinite NPs), based on distributions in our narrative.

Overall results in table 11.1 show that explicit forms are generally prominent and pronouns generally nonprominent, providing indirect support for the hypothesis that new information (conveyed by explicit forms) is generally prominent and given information (conveyed by pronouns) is generally nonprominent. Although this trend is significant ($p < .00001$, $\chi^2 = 69.5$, $df = 1$), the accentuation on 20% (40/200) of the referring expressions in table 11.1 is contrary to the information status thought to be conveyed by the form of the referring expression itself.[4]

Overall results also show that subjects are somewhat more likely to be prominent than direct objects, although this trend is not significant ($p < .3$, $\chi^2 = 1.3$, $df = 1$). These distributions nonetheless run counter to the prediction that subject position, which is the preferred grammatical function for old information, is less likely to hold a prominent expression than is direct object position.

---

[2] In analyzing proper names, an additional complication arises due to the independent lexical status of first and last names in English. For the study, a proper name was considered prominent if at least one of the first and last names bore an H* or complex pitch accent. Similarly, we do not address issues concerning NP-internal accent placement. The narrative contains relatively few internally modified complex NPs.

[3] A more systematic study of the possible discourse uses of L* versus deaccenting remains as future work. A total of six tokens, all explicit forms, were labeled with the L* accent. Based on this small sample, we could not clearly distinguish between the discourse functions of L* and deaccenting.

[4] Terken [Ter84] has reported similar distributions for task-oriented monologues.

TABLE 11.1. Percentage of prominent referring expressions.

Percent Prominent	Subject		Direct Object		Total	
	%	N	%	N	%	N
Pronouns	23%	25/111	7%	1/15	21%	26/126
Explicit forms	91%	49/54	55%	11/20	81%	60/74
Total	45%	74/165	34%	12/35	43%	86/200

Closer examination of the data reveals an asymmetric interaction of grammatical function and lexical form. Subjects are prominent only 23% of the time as pronouns, but are regularly made prominent as explicit forms ($p < .00001$, $\chi^2 = 68.3$, $df = 1$). Direct objects, on the other hand, are prominent 55% of the time as explicit forms, but are rarely made prominent as pronouns ($p < .005$, $\chi^2 = 8.9$, $df = 1$). Clearly, grammatical function and lexical form alone cannot be used to reliably predict accentuation for the classes of subject pronouns and direct object explicit forms. The notion of discourse-old/discourse-new status is also of little value in predicting accentuation for these problematic classes because all of the pronouns in the narrative, subject or otherwise, refer to discourse-old entities, whereas fewer than one-half of the prominent direct object explicit forms refer to discourse-new entities.

11.3 A Discourse-Based Interpretation of Accent Function

We propose that the distributions of pitch accent in the narrative study can be explained by the communicative functions that accent serves in various linguistic configurations, as summarized in table 11.2.[5] The proposed functions of accent share in common the fact that each serves to manipulate the dynamic record of the activated or salient entities in a discourse model. This interpretation of accent function provides an overarching framework that unifies and reconciles previous findings on the meaning of accents, while contributing original analyses of the problematic accentuation classes uncovered in the narrative study.

Our analyses rely on notions of discourse segmental (or global) as well as utterance-based (or local) levels of discourse structure. Grosz and Sidner [GS86] propose three interrelated components of discourse structure: *intentional structure*, *linguistic structure*, and *attentional state*. Briefly, a discourse is comprised of *discourse segments* whose hierarchical relationships are determined by intentional structure and realized by linguistic structure. Discourse processing proceeds at two

[5]Terminology in table 11.2 is defined in subsection 11.3.1. *Cb* stands for backward-looking center; *Cp* stands for preferred center.

TABLE 11.2. Discourse functions of accent.

Prominent Expressions	
Linguistic Factors	*Discourse Function*
SBJ pronoun	Shift local attention to new *Cb*
SBJ explicit form	Introduce new global referent as *Cp*
DOBJ explicit form	Introduce new global referent

Nonprominent Expressions	
Linguistic Factors	*Discourse Function*
DOBJ explicit form	Maintain referent in global focus
DOBJ pronoun	Maintain non-*Cb* referent in secondary local focus
SBJ pronoun	Maintain *Cb* referent in primary local focus

levels, global and local. The global level concerns relationships among discourse segments, whereas the local level concerns relationships among utterances within a single segment. This discourse structure theory takes the modeling of speaker intention to be central to discourse interpretation and generation, and allows for the direct integration of planning mechanisms [Loc94]. Thus it provides a theoretical basis for the design of collaborative spoken-language systems in which message-to-speech synthesis systems may be embedded.

11.3.1 Modeling Attentional State

In Grosz and Sidner's [GS86] framework, the notion of salience is formalized by attentional focusing mechanisms that are claimed to underlie discourse processing in general. The attentional state component dynamically records the entities and relations that are salient at any point in the discourse. The present analyses require rudimentary definitions of four attentional statuses provided by the Grosz and Sidner model: primary local focus or *Cb*, secondary local focus, immediate global focus, and nonimmediate global focus.

The global level of attentional state is modeled as a last-in first-out *focus stack* of *focus spaces*, each containing representations of the entities and relations salient within the discourse segment corresponding to the focus space [Gro77]. The focus space at the top of the focus stack is termed the *immediate focus space*. Pushes and pops of focus spaces obey the hierarchical segmental structure of the discourse. An empty focus space is pushed onto the stack when a segment begins; entities are recorded in the focus space as the discourse advances until the discourse segment closes and its focus space is popped from the focus stack. Those entities represented in the top focus space on the stack are in immediate global focus. Entities represented elsewhere on the stack are in nonimmediate global focus.

It is important to note that when an embedded segment opens, a new focus space is pushed onto the stack on top of that of its embedding segment. Upon the close of an embedded segment, therefore, the focus space of the *embedding*

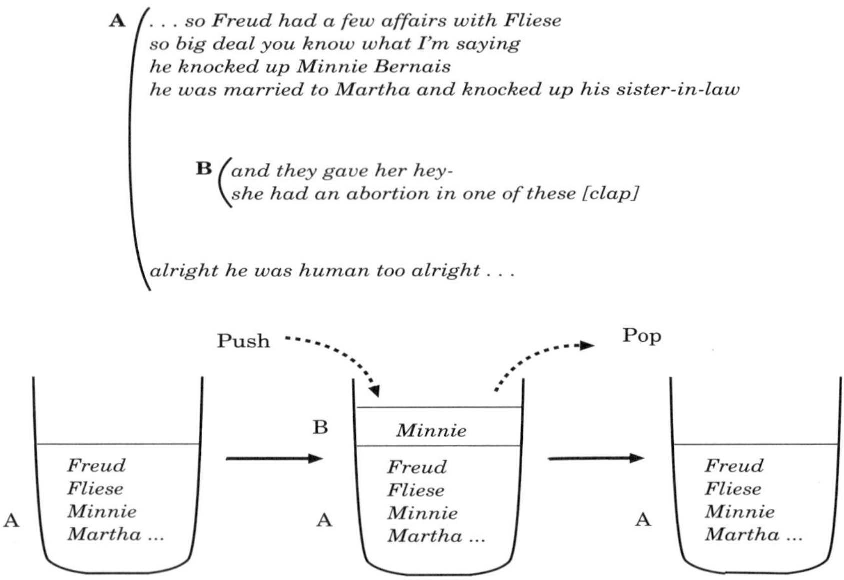

FIGURE 11.2. Illustration of global focusing mechanisms.

segment becomes the immediate focus space, and entities within it are said to be accessible.

Figure 11.2 illustrates the motivating phenomenon for the global focusing mechanisms; when the pronoun *he* is encountered after a pop, the entities in the preceding embedded segment are no longer on the focus stack and are therefore not available for pronoun reference. The pronoun instead refers to an entity in the focus space of the outer segment that is resumed.

The local level of attentional state is modeled by *centering* mechanisms [Sid79, JW81, GJW83, GJW95]. All of the salient discourse entities realized in an utterance are recorded in the partially ordered set of *forward-looking centers* (*Cf* list) for that utterance. The *Cf* list specifies a set of possible links forward to the next utterance. The ranking of the *Cf* list members reflects their relative salience in the local discourse context at the time of the uttering of that particular utterance. Each utterance also has a single *backward-looking center* (*Cb*), which is the most central and salient discourse entity that links the current utterance to the previous utterance. The *Cb* of an utterance U_n, $Cb(U_n)$, is defined as the highest-ranking member of the *Cf* list of the prior utterance, $Cf(U_{n-1})$ that is realized in U_n. In the current state of centering theory, the *Cf* list members are ordered based on grammatical function and surface position [GGG93, GJW95]. The highest ranking member of the *Cf* list is called the *preferred center* (*Cp*), and is the most likely candidate to become the *Cb* of the following utterance.

Figure 11.3 illustrates how centering operates on a sequence of utterances that is centrally about Freud, who remains the *Cb* throughout. As noted, centering

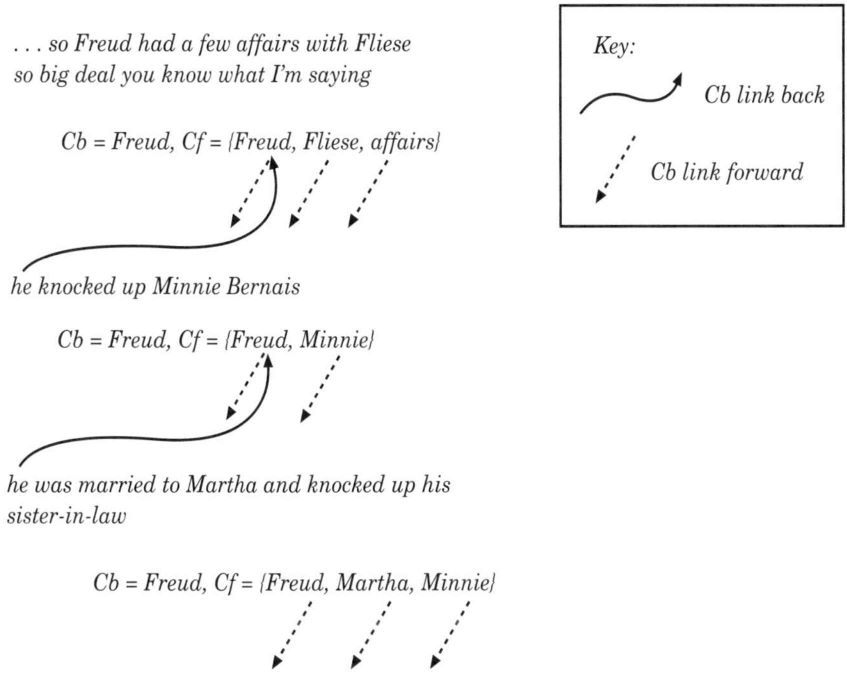

FIGURE 11.3. Illustration of local focusing (centering) mechanisms.

constructs are computed within each discourse segment and are reinitialized at the start of a new segment.

Roughly speaking, Grosz and Sidner's [GS86] global and local attentional state mechanisms distinguish two levels of global salience and two levels of local salience. A discourse entity may be globally salient by virtue of its being represented on the focus stack in either the immediate focus space or a non-immediate focus space. Entities in the immediate focus space are claimed to be relatively more accessible than those in focus spaces deeper in the stack. At the local level, the *Cb* is claimed to be more salient than non-*Cb* members of the *Cf* list. Psychological evidence for this claim is provided by anaphor resolution experiments [H-D98, GGG93].

We emphasize that, unlike most existing theories of information status or of focus of attention, this framework assumes a fundamental qualitative distinction between global and local salience. Further, it characterizes notions of discourse salience relative to a dynamically unfolding record of mutual beliefs established during the discourse. This contrasts with theories of given/new information status that rest upon the notion of shared knowledge of discourse participants (e.g., Prince 1981's taxonomy of shared knowledge [Pri81]), or that stipulate successive grades on a single scale of givenness (e.g., Gundel et al. scale of familiarity [GHZ89]). Prince's 1988 reexamination of her taxonomy of shared knowledge in terms of the notions of discourse-new and discourse-old seems to address issues of global focusing within a single discourse segment; there is no analog in either Prince's 1981 or 1988 studies [Pri81, Pri88] of segment-to-segment–based manipulations of attentional state.

11.3.2 Discourse Analysis of the Narrative

Our narrative is essentially a purposeful monologue in which the speaker tries to persuade the interviewer to read a biography of Sigmund Freud. The speaker describes the life stories of various personae from this biography, explaining how they became involved in the Freudian circle of analysts and their activities within it. The genre of narrative may exhibit complex turns of topics and events, many of which the listener may not be able to predict before the time of hearing. Whereas structural choices of syntactic configuration and lexical form of referring expression can be readily detected and widely agreed upon, analysis of the topic or discourse structure poses greater difficulties and uncertainties. Nevertheless, analysis of a completed narrative often reveals clear structure underlying the speaker's presentation of information at various levels of linguistic analysis.

We undertook a discourse structure analysis of the narrative sample based on speaker intention [GS86] following procedures defined and utilized by Grosz and Hirschberg [GH92]. The narrative was divided into hierarchically structured *discourse segments*, each of which was assigned a *discourse segment purpose* that describes the communicative goal for that segment, which was intended by the speaker to be recognized by the hearer [GS86]. Cue phrases or discourse markers also served as cues to segmental structure. Finally, the global and local focusing statuses of the referring expressions in the narrative were determined based on a discourse segmentation by the author (following procedures described in Grosz and Hirschberg [GH92]) and centering constructs (determined by hand, by using the segmentation analysis and the centering rules as defined in Grosz et al. [GJW95].

Although numerous theoretical and practical issues in the area of discourse analysis require further research, we take the approach that it is fruitful to bootstrap from rudimentary analyses guided by available theory and practice, making use of independent evidence from intonational correlates of discourse structure and meaning [GH92].

11.4 Discourse Functions of Accent

We present detailed analyses of the discourse structural properties of referring expressions in six linguistic configurations, namely those specified in table 11.2.[6]

[6]Two additional configurations are logically possible, namely nonprominent subject explicit forms and prominent direct object pronouns. The narrative contained six cases of

We propose that accentuation serves to signal changes in the attentional state of a discourse in systematic ways.

11.4.1 Manipulations of Local Focus: Pronouns

We analyzed the twenty-five cases of prominent subject pronouns and found six cases of emphasis or contrast and three cases requiring limited inference to determine the pronoun referent. The remaining cases can be viewed within the framework of attentional state modeling as falling into two major classes of center shifts.

For one class (seven cases), accentuation signaled reference to a previous discourse center that was not salient in the immediate discourse context. For this class, intonation marks a global center shift, which accompanies a focus space pop upon the completion of an embedded discourse segment. Accent in these cases signals reference to a previous center that was not realized in the immediately preceding utterance, and therefore was not a member of the *Cf* list of the immediately preceding utterance. Prominent pronouns signaling global shifts occurred as the first grammatical subject following the completion of the embedded discourse segment.

For the other class (nine cases), accentuation marked a local shift in attention away from the current discourse center to a new discourse entity that was indeed salient, but not centrally so, in the immediate discourse context. In contrast to the first class, the pronoun referent always occurred in the *Cf* list of the immediately preceding utterance, but never as the *Cb*.

An excerpt of the narrative displayed in figure 11.4 illustrates these two kinds of center shifts signaled by accent on subject pronouns. In figure 11.4, the second accented pronoun, *HE*, illustrates the first class and marks a global shift in attention back to the outer segment. This accented pronoun realizes the previously established *center* of the outer segment, namely *Masson*. The first accented pronoun, *SHE*, illustrates the second class of local shifts, establishing *Anna Freud* as the *center* of the embedded subsegment.

In contrast to the class of prominent subject pronouns, the majority of nonprominent subject pronouns referred to the *Cb* of the last utterance (73%). A smaller

---

the former (out of 54 subject explicit forms total) and one of the latter (out of 15 pronoun direct object forms total). These distributions led us to formulate the following constraint:

> If grammatical function and form of referring expression convey conflicting *given/new* statuses, then accentuation must "reinforce," or agree with the (preferred) *given/new* status conveyed by the form of referring expression. A corollary of this hypothesis is that for the cases where a referring expression of a certain form is realized as its preferred grammatical function (e.g., pronouns as subjects, proper names as nonsubjects), the speaker is free to use accenting to convey linguistic information apart from *given/new* status, such as topic shift, emphasis, or contrast [Nak93].

Further empirical investigation is needed to test these specific hypotheses for a variety of speakers and genres.

So Masson became the new curator.
He flies to London and, you know,
he's already met Anna Freud and therefore
he has access to the secret cupboard of Freudian letters
and naturally Anna assumed that uh

> **SHE [H*]** *was a brilliant woman too –*
> *she did more a lot of work in child psy- psychiatry and psychoanalysis*

assumed that **HE [H*]** *would keep this information*
you know within the confines of the psychoanalytical group

Well, as Masson was studying these letters he realized...

FIGURE 11.4. Examples of H*-accented prominent pronouns (in boldface capitals) signaling local and global shifts in centers.

subclass of nonprominent subject pronouns (14%) referred to the *Cb* of a neighboring segment on the focus stack or of the segment recently in immediate global focus. Interestingly, eight out of the twelve cases in this subclass were instances of subject pronouns that continued the *Cb* of an outer segment upon the closing of an embedded segment. We conjecture, therefore, that although there may be a tendency to signal with accent a global shift in centers as defined above, it is clearly not necessary to do so. Psychological testing is needed to determine whether accent plays a facilitative role in cuing center shifts of this sort. For the remaining 13% (eleven cases) of nonprominent subject pronouns, which do not continue or resume a *Cb*, three tokens occur in repairs, two in dialogue tags, and one in response to an interruption from the interviewer.

The occurrences of direct object pronouns fell into three subclasses: intersentential anaphora (three cases), for which pronominalization of an NP in a subordinate clause is licensed by established syntactic rules; multiple pronouns (five cases), for which the direct object pronoun occurs in addition to a subject pronoun realizing the *Cb*; and objects of verbs of perception (five cases), for which it is hypothesized that the lexical semantics of the verb (e.g., *see, approach*) actually render the direct object as the *Cp* or most likely candidate to become the most salient entity in the continuation of the discourse (see [GJW95]). The final token of the nonprominent direct object pronouns occurs in a repair, whereas the single prominent direct object pronoun token occurs in a contrastive context. We conclude that centering theory, enhanced with intersentential rules and provisions for lexically marked classes of verbs, is sufficient to account for the attentional role of lack of accent on direct object pronouns.

11.4.2 Manipulations of Global Focus: Explicit Forms

Analysis of the direct object explicit forms shows that accentuation is generally determined by the global focus state. Eighty percent of these expressions are cases of first mention since the last discourse segment boundary, although three-quarters of these are also references to discourse-old entities. In our narrative, intonational prominence marks the introduction of entities into a segment's focus space; crucially, these entities do not occur in the sister segment or immediately embedding segment. In contrast, lack of intonational prominence marks the (re)introduction of entities recently in global focus immediately following a focus space pop or push. In the case of a pop, the reintroduced entity must have been previously introduced in the embedding segment that becomes the immediate focus space after the pop. In the case of a push, the entity must have been represented in the focus space of a sister segment. In short, we interpret the choice of accentuation on direct object proper names as a reflection of the global salience of the corresponding entity. References to entities that are either in a neighboring focus space on the focus stack or in the most recently popped focus space do not require accentual prominence. References to entities that lack salience at the global level of attentional state, on the other hand, are reintroduced into a focus space by prominent expressions. Finally, we remark that the two cases that were not first mentions since a discourse segment boundary both occurred during the speaker's reading from a book; we cannot account for these last two cases at this time.

Examples of prominent and nonprominent direct object explicit forms appear in figures 11.5 and 11.6, respectively. In figure 11.5, a new character, Eissler, is introduced with a prominent proper name in direct object position. In figure 11.6, the central character of the preceding sister segment, Swales, is reintroduced with a nonprominent proper name at the start of a new segment that centrally concerns Anna Freud.

Applying the principles of our analysis, the accentuation of 90% of the direct object explicit forms in our narrative is correctly predicted. These results surpass those of alternate strategies, such as accenting only and all cases of discourse-

You wanna hear the story about PAUL MASSON?...
In his thirties, late thirties, he decided to be psychoanalyzed,
and he was, and he became very interested in Freud.
And so he read, and he jumped into the Freud world, the Freudian world, like I jump into a pot of pasta. Okay.

And he approached **PAUL [H*]** *[sic]* **EISSLER [H*]**,
head of the Freudian archives based in New York...

FIGURE 11.5. Introduction of new global referent: Example of prominent direct object explicit form (in boldface capitals).

Peter Swales.
He's another interesting character in this little drama.
Was a hippie in the sixties... And all of a sudden he realized that his interest was in psychoanalysis.
And again, like Masson, or Maason, some people call him Maason, jumped into it like I would in a... I don't know what, I said pasta before, I'll say something else. And,

But you see, ANNA FREUD didn't like **Peter [-] Swales [L*]**
(Is she dead?) No she was sh- yes she's dead...

FIGURE 11.6. Maintenance of immediate global referent: Example of nonprominent direct object explicit form (in boldface lower case).

initial mentions (65% correct) or accenting only and all cases of segment-initial mentions (55% correct).

The class of explicit forms in subject position shares the property that a majority of expressions are first mentions since the last segment boundary (81%). Of the 10 out of 54 cases that are not segment-initial, three occur in repairs and three in quoted contexts. This lends support to the generalization that explicit lexical forms are used to signal the global newness of a discourse referent. However, fully 91% of subject-explicit forms are prominent, compared to 55% for direct object–explicit forms. We claim that intonational prominence in these cases signals not only the global newness of a referent but also a shift in centers from that of the previous segment to the new subject referent. This is consistent with psychological results showing a delay in reading times when a proper name is used in subject position to realize an established *Cb* [GGG93]. Under this analysis, prominence on subject-explicit forms, like prominence on subject pronouns, allows the hearer to infer a local attentional shift.

11.5 Discussion

The hypothesized functions of accentuation include the marking of emphasis, contrast, topic structure, and information status ([Hal67, Lad80, Bro83, Fuc84, Ter84], inter alia). We relate our findings from the narrative study to a number of results from previous empirical investigations of accent function and related phenomena. Three major hypotheses from prior work are reviewed, namely that accentuation communicates (1) given/new information status, (2) topichood, and (3) relative salience.

11.5.1 Given/New

In a study of task-oriented speech, Brown [Bro83] confirmed a general tendency for given information to be unaccented and new information to be accented, where lexical items referring to referents previously mentioned in the discourse were considered "given." In her corpus, very few instances of discourse-old information were accented (less than 5%). Brown concluded that when syntax and lexical form convey conflicting information statuses, lexical form provides the reliable clue. Her analysis fails to assign grammatical function any principled role in interpreting accent choices.

Other researchers have observed that given information can be made prominent by accent in more regular ways than suggested by Brown's results. Fuchs [Fuc84] hypothesized that given items may be left accented to signify their givenness, or may be accented "to establish them as a point of relevance/'newness' with respect to the question of immediate concern at that point" (p. 144). Later work addressed from different linguistic perspectives the circumstances under which given information is accented (e.g., Horne [Hor91] on metrical-phonological constraints, Selkirk [Sel93] on syntactic factors, and Hirschberg [Hir93] on corpus-based exploration of the relative contributions of various factors). In a study of elicited speech in which the lexical and syntactic forms of utterances were controlled, Terken and Hirschberg [TH92] demonstrated that structural properties, such as the persistence of grammatical function and surface position, contribute to the determination of accent for given information. All of these studies have relied on notions of given/new that are nonhierarchical and unilevel. That is, segmentation of the discourses is linear or nonexistent, and there is no distinction between local and global salience in the above accounts.

11.5.2 Topichood

The role of accentuation in topic-marking was investigated by Terken [Ter84]. In his instruction-giving monologues, Terken found that the first introduction of a discourse topic was generally accented and subsequent references to the topic were often expressed as unaccented pronouns. In contrast, noninitial references to nontopical entities were often expressed as accented full forms. Terken's first observation accords with the large distributions of prominent subject-explicit forms (serving as topical or Cp introductions) and nonprominent subject pronouns (generally continuing or resuming the Cb). His second observation accords with our analyses of direct object–explicit forms. The vast majority of these refer to entities that were previously introduced into the discourse yet were not topics or Cbs.

Our interpretation of accent function takes Terken's account two steps further. First, we identify a significant correlation between grammatical position and accentuation. Second, we outline a more fine-grained proposal about factors determining accentuation on direct object–explicit forms, specifying a set of conditions on the global attentional state that appears to license the accenting as well as deaccenting of direct object–explicit forms in discourse. In short, the lack of intonational prominence on nontopical referents is determined by the recent global focusing history and is not due simply to their nontopical nature.

Terken [Ter84], Fuchs [Fuc84], Cahn [Cah90], and others have made observations similar to ours about the topic-shifting function of accent. For example, Cahn proposed that items may be H*-accented to reinstantiate the *Cb* after a discourse push or pop [Cah90, pp. 15–16], and Terken noted that "the use of accented pronouns immediately after topic introduction seems to be [a] way to confirm for the listener that there has been a topic shift" [Ter84, p. 70]. We cast these notions more precisely by identifying two distinct sets of conditions on the local and global attentional state that appear to license the accenting of subject pronouns in narrative discourse.

11.5.3 Relative Salience

The association between accent placement and Grosz and Sidner's [GS86] attentional structure was first theorized by Hirschberg and Pierrehumbert [HP86]. In particular, they claimed that "just as salience is always determined relative to some particular context, accent placement must be determined with respect to the segment in which the accentable item appears.... [It] is the signaling of salience relative to the discourse segment that produces the secondary effects of given-new distinction, topic-hood or contrastiveness, and the favoring of one reference resolution over another" [HP86, p. 14]. Later theoretical work more closely examined the relationship between accenting and local focusing [Cah90, Kam94]. These studies focused on the problem of accented pronouns, and so the analyses of the role of accentuation in attentional modeling are not as general as the one presented here.

The accentuation of proper names has also been investigated by prosody researchers. It has been suggested that proper names should often be unaccented because their very use presumes familiarity with the named entity on the part of the hearer (see discussion in [Lad80, p. 91]. On the other hand, Hirschberg [Hir93] noted that the accenting of given proper names in a large speech corpus could be explained by the proposal by Sanford and colleagues [SMG88] that proper names may be used to refocus the speaker's attention on previously established discourse entities that are not salient in the immediate discourse context. The preponderance of segment-initial mentions among explicit forms of all types seems to accord with this idea. Such reintroductions were found to often bear pitch accent in news speech [Hir93]. Our narrative data argue for an analysis in which the choice of lexical form is independent of, but related to, the choice of accentuation. Use of an explicit form indicates newness relative to the immediate focus space, whereas accentuation reflects newness relative to the neighboring focus spaces.

In this regard, our analysis of direct object proper names poses a problem for Grosz and Sidner's stack model of global focusing [GS86] because the focus space of a sister segment is popped from the focus stack before its sibling focus space is pushed. As mentioned in section 11.3, entities in a focus space that is popped from the focus stack are claimed to be unavailable for reduced reference. However, our proposed analysis can be reconciled with the Grosz and Sidner model if we relax this constraint and allow for entities in the most recent immediate focus space to remain globally salient. Davis and Hirshberg implemented precisely this modification in the global focusing mechanisms of a message-to-speech system [DH88, p. 142].

Finally, we point out that our analysis of direct object–explicit forms extends a previous hypothesis of Terken and Nooteboom that reference resolution proceeds differently for accented and unaccented expressions; namely, that listeners assume that an unaccented expression refers to a member of a "restricted set of activated entities" in the discourse context, whereas the interpretation of an accented expression is not constrained in this manner [TN87, p. 148]. The notion of activated entities is formalized in the narrative study in terms of the structured contents of the global focus stack. We note that our analyses show that Terken and Nooteboom's hypothesis applies to pronominal expressions as well, for which the relevant set of restricted entities is formally cast in terms of the centering constructs computed at the local level of discourse processing.

11.5.4 Summary

The discourse-based interpretation of accent function subsumes previous hypotheses concerning given/new information status, topichood, and relative salience. To summarize, we emphasize the following contributions of our study. First, we found that accentuation cannot simply be associated with form of referring expression, but rather makes an independent contribution to the structuring of information in discourse (cf. [Bro83]). Second, we utilized the distinction between local and global levels of attention to make precise two different notions of discourse salience. The generalization emerges that intonational prominence marks the newness of discourse entities at either the local or global level, or both (as for prominent subject explicit forms). Against the appropriate background of grammatical and lexical properties, accent may cue precise inferences for reference resolution and attentional state modeling.[7]

11.6 Conclusion

As speech synthesis technology becomes more widespread in language-understanding systems, it is critical to make maximal use of prosody to convey meaningful spoken messages. Psychological research [TN87, NK87] has shown that

[7] A similar inference-cuing role for accent is proposed in Hirschberg and Ward [HW91] for a very different reference phenomenon. Hirschberg and Ward found that accentuation patterns on anaphors in the source clause in VP-ellipsis constructions affected the interpretation of target clause anaphors in systematic ways: accenting served to flip preferences from either strict to sloppy or sloppy to strict readings of the target clause, *against* the background of *underlying* lexico-semantic preferences for either a strict or sloppy reading.

contextually appropriate intonation, including pitch accent placement, facilitates language comprehension. Current pitch accent assignment modules in speech synthesis systems [Hir93, HFLL93] have made measurable gains by heuristic discourse modeling. This study furthers our understanding of discourse structural constraints on accentuation for the genre of spontaneous narrative. We build upon previous work by developing a discourse-based interpretation of accent function in a computational framework amenable to implementation in spoken language systems. Results from corpus-based studies such as this may be best tested in message-to-speech systems, in which discourse structure and meaning can be directly encoded. The further development of pitch accent assignment algorithms that model the interactions among discourse structural constraints and other linguistic factors affecting accentuation should enable richer and more meaningful prosodic variation for speech synthesis.

Acknowledgments: This work was partially supported by a National Science Foundation (NSF) Graduate Research Fellowship, and NSF grants no. IRI-90-09018, no. IRI-93-08173, and no. CDA-94-01024 at Harvard University. The author thanks Barbara Grosz and Julia Hirschberg for many valuable discussions.

References

[Alt87] B. Altenberg. *Prosodic Patterns in Spoken English: Studies in the Correlation Between Prosody and Grammar for Text-to-Speech Conversion.* Lund University Press, Lund, Sweden, 1987.

[Bro83] G. Brown. Prosodic structure and the given/new distinction. In *Prosody: Models and Measurements*, A. Cutler and D. R. Ladd, eds. Springer-Verlag, Berlin, Germany, pp. 67–78, 1983.

[Cah90] J. Cahn. The effect of intonation on pronoun referent resolution. Unpublished manuscript, Massachusetts Institute of Technology Media Laboratory, Cambridge, MA, 1990.

[DH88] J. Davis and J. Hirschberg. Assigning intonational features in synthesized spoken directions. In *Proceedings of 26th Annual Meeting of the Association for Computational Linguistics*, Buffalo, NY, 1988.

[Fuc84] A. Fuchs. 'Deaccenting' and 'default accent.' In *Intonation, Accent and Rhythm*, D. Gibbon and H. Richter, eds. Walter deGruyter, Berlin, Germany, pp. 134–164, 1984.

[GGG93] P. C. Gordon, B. J. Grosz, and L. A. Gilliom. Pronouns, names, and the centering of attention in discourse *Cognitive Science* 17:311–347, 1993.

[GH92] B. J. Grosz and J. Hirschberg. Some intonational characteristics of discourse structure. In *Proceedings of the International Conference on Spoken Language Processing*, ICSLP, Banff, Canada, 1992.

[GHZ89] J. Gundel, N. Hedberg, and R. Zacharski. Givenness, implicature and demonstrative expressions in English discourse. *CLS 25. Parasession on Language in Context*, 89–103, 1989.

[GJW83] B. J. Grosz, A. K. Joshi, and S. Weinstein. Providing a unified account of definite noun phrases in discourse. In *Proceedings of 21st Annual Meeting of the Association for Computational Linguistics*, Cambridge, MA, 1983.

[GJW95] B. J. Grosz, A. K. Joshi, and S. Weinstein. Centering: A framework for modelling the local coherence of discourse. *Computational Linguistics* 21(2), 1995.

[Gro77] B. J. Grosz. The representation and use of focus in dialogue understanding. Technical Report 151, SRI International, Menlo Park, CA, 1977.

[GS86] B. J. Grosz and C. Sidner. Attention, intentions, and the structure of discourse. *Computational Linguistics* 12(3), 1986.

[Hal67] M. A. K. Halliday. Notes on transitivity and theme in English, Part 2. *Journal of Linguistics* 3:199–244, 1967.

[H-D98] S. B. Hudson-D'Zmura. *The Structure of Discourse and Anaphor Resolution: The Discourse Center and the Roles of Nouns and Pronouns*. Ph.D. Thesis, University of Rochester, Rochester, NY, 1988.

[HFLL93] M. Horne, M. Filipsson, M. Ljungquist, and A. Lindstrom. Referent tracking in unrestricted texts using a lemmatized lexicon: Implications for the generation of prosody. In *Proceedings of the Third European Conference on Speech Communication and Technology*. ESCA, Berlin, Germany, 1993.

[Hir93] J. Hirschberg. Pitch accent in context: Predicting intonational prominence from text. *Artificial Intelligence* 63:304–340, 1993.

[Hor91] M. Horne. Why do speakers accent 'given' information? In *Proceedings of the Second European Conference on Speech Communication and Technology*. ESCA, Genova, Italy, 1991.

[HP86] J. Hirschberg and J. Pierrehumbert. The intonational structuring of discourse. In *Proceedings of 24th Annual Meeting of the Association for Computational Linguistics*, New York, 1986.

[HW91] J. Hirschberg and G. Ward. Accent and bound anaphora. *Cognitive Linguistics* 2(2):101–121, 1991.

[JW81] A. K. Joshi and S. Weinstein. Control of inference: Role of some aspects of discourse structure centering. In *Proceedings of International Joint Conference on Artificial Intelligence*, 385–387, 1981.

[Kam94] M. Kameyama. Stressed and unstressed pronouns: Complementary preferences. In *Proceedings of the FOCUS and NLP Conference*, Germany, 1994.

[Lad80] D. R. Ladd. *The Structure of Intonational Meaning*. Indiana University Press, Bloomington, 1980.

[Loc94] K. Lochbaum *Using Collaborative Plans to Model the Intentional Structure of Discourse*. Ph.D. thesis, Harvard University, Cambridge, MA, 1994.

[Nak93] C. Nakatani. Accenting on pronouns and proper names in spontaneous narrative. In *Proceedings of the ESCA Workshop on Prosody*, Lund, Sweden, 1993.

[NK87] S. G. Nooteboom and J. G. Kruyt. Accent, focus distribution, and the perceived distribution of given and new information: An experiment. *Journal of the Acoustical Society of America* 82(5), 1987.

[Pie80] J. Pierrehumbert. *The Phonology and Phonetics of English Intonation*. Ph.D. thesis, Massachussetts Institute of Technology, Cambridge, MA, 1980.

[Pri81] E. Prince. Toward a taxonomy of given-new information. In *Radical Pragmatics*, P. Cole, ed. Academic Press, New York, 1981.

[Pri88] E. Prince. The ZPG letter: Subjects, definiteness, and information-status. In *Discourse Description: Diverse Analyses of a Fund Raising Text*, S. Thompson and W. Mann, eds. Elsevier Science Publishers, Amsterdam, Holland, 1988.

[Sel93] E. O. Selkirk. Sentence prosody: Intonation stress and phrasing. In *Handbook of Phonological Theory*, J. Goldsmith, ed. Basil Blackwell, Oxford, 1993.

[Sid79] C. Sidner. *Toward a Computational Theory of Definite Anaphora Comprehension in English*. MIT Technical Report AI-TR-537, 1979.

[SMG88] A. Sanford, K. Moar, and S. Garrod. Proper names as controllers of discourse focus. *Language and Speech* 31:43–56, 1988.

[Tal89] D. Talkin. Looking at speech. *Speech Technology* 4:74–77, 1989.

[Ter84] J. Terken. The distribution of pitch accents in instructions as a function of discourse structure. *Language and Speech* 27:269–289, 1984.

[TH92] J. Terken and J. Hirschberg. Deaccentuation and persistence of grammatical function and surface position. *Language and Speech* 37(2):125–145, 1994.

[TN87] J. Terken and S. G. Nooteboom. Opposite effects of accentuation and deaccentuation on verification latencies for given and new information. *Language and Cognitive Processes* 2(3/4):145–163, 1987.

Appendix: Audio Demo

The sound file contains the spontaneous narrative monologue analyzed in the study.

12

Homograph Disambiguation in Text-to-Speech Synthesis

David Yarowsky

ABSTRACT This chapter presents a statistical decision procedure for lexical ambiguity resolution in text-to-speech synthesis. Based on decision lists, the algorithm incorporates both local syntactic patterns and more distant collocational evidence, combining the strengths of decision trees, N-gram taggers and Bayesian classifiers. The algorithm is applied to seven major types of ambiguity in which context can be used to choose the pronunciation of a word.

12.1 Problem Description

In speech synthesis, one frequently encounters words and numbers for which pronunciation cannot be determined without context. Seven major types of these homographs are addressed here:

1. *Different Part of Speech*: The largest class of pronunciation ambiguity consists of homographs with different parts of speech: *Three lives were lost* versus *One lives to eat*. These cases can typically be resolved through local syntactic patterns.

2. *Same Part of Speech*: A word such as *bass* or *bow* exhibits different pronunciations with the same part of speech and thus requires additional "semantic" evidence for disambiguation.

3. *Proper Names* such as *Nice* and *Begin* are ambiguous in capitalized contexts, including sentence initial position, titles and single-case text.

4. *Roman Numerals* are pronounced differently in contexts such as *Chapter III* and *Henry III*.

5. *Fractions/Dates* such as *5/16* may be pronounced as *five-sixteenths* or *May 16th*.

6. *Years/Quantifiers*: Numbers such as *1750* tend to be pronounced as *seventeen-fifty* when used as dates, and *one-thousand seven-hundred fifty* when used before measure words, as in *1750 miles*. Related cases include the distinction between *the 727 pilot* and *727 people*.

7. *Abbreviations* may exhibit multiple pronunciations, such as *St.* (Saint or Street) and *Dr.* (Doctor or Drive).

Some homographs exhibit multiple categories of ambiguity. For example, *lead* has different pronunciations when used as a verb and noun, and between two noun senses (*He followed her lead* versus *He covered the hull with lead*). In general, we resolve the part-of-speech ambiguity first and then resolve the additional semantic ambiguity if present.

12.2 Previous Approaches

N-gram taggers [Jel85, Chu88, Mer90] may be used to tag each word in a sentence with its part of speech, thereby resolving those pronunciation ambiguities that correlate with part-of-speech ambiguities. The Bell Laboratories TTS synthesizer [SHY92] uses Church's PARTS tagger for this purpose. A weakness of these taggers is that they are typically not sensitive to specific word associations. The standard algorithm relies on models of part-of-speech sequence, captured by probabilities of part-of-speech bigrams or trigrams, to the exclusion of lexical collocations. This causes difficulty with cases such as *a ribbon wound around the pole* and *a bullet wound in the leg*, which have identical surrounding part-of-speech sequences and require lexical information for resolution. A more fundamental limitation, however, is the inherent myopia of these models. They cannot generally capture longer distance word associations, such as between *wound* and *hospital*, and hence are not appropriate for resolving many semantic ambiguities.

Bayesian classifiers [MW64] have been used for a number of sense disambiguation tasks [GCY94], typically involving semantic ambiguity. In an effort to generalize from longer distance word associations regardless of position, an implementation proposed in [GCY92] characterizes each token of a homograph by the 100 words nearest to it, treated as an unordered bag.[1] Although such models can successfully capture topic-level differences, they lose the ability to make distinctions based on local sequence or sentence structure. In addition, these models have been greatly simplified by assuming that occurrence probabilities of content words are independent of each other, a false and potentially problematic assumption that tends to yield inflated probability values. One can attempt to model these dependencies (as in [BW94]), but data sparsity problems and computational constraints can make this difficult and costly.

Decision trees [BFO84, BDDM91] can be effective at handling complex conditional dependencies and nonindependence, but often encounter severe difficulties with very large parameter spaces, such as the highly lexicalized feature sets frequently used in homograph resolution.

[1] Leacock et al. have pursued a similar bag-of-words strategy, using an information-retrieval-style vector space model and neural network [LTV93].

The current algorithm described in this chapter is a hybrid approach, combining the strengths of each of the three preceding general paradigms. The algorithm was proposed in [SHY92] and refined in [Yar94a]. It is based on the formal model of **decision lists** in [Riv87], although feature conjuncts have been restricted to a much narrower complexity, namely word and class trigrams. The algorithm models both local sequence and wide context well, and successfully exploits even highly nonindependent feature sets.

12.3 Algorithm

Homograph disambiguation is ultimately a classification problem in which the output is a pronunciation label for an ambiguous target word and the feature space consists of the other words in the target word's vicinity. The goal of the algorithm is to identify patterns in this feature space that can be used to correctly classify instances of the ambiguous word in new texts.

For example, given instances of the homograph *lead* below, the algorithm should assign the appropriate label /lid/ or /lɛd/.

Pronunciation	Context		
(1) lɛd	.. it monitors the	*lead*	levels in drinking
(1) lɛd	... median blood	*lead*	concentration was
(1) lɛd	.. found layers of	*lead*	telluride inside ..
(1) lɛd	... conference on	*lead*	poisoning in ...
(1) lɛd	.. strontium and	*lead*	isotope zonation ..
(2) lid	maintained their	*lead*	Thursday over ...
(2) lid	.. to Boston and	*lead*	singer for Purple
(2) lid	Bush a 17-point	*lead*	in Texas, only 3
(2) lid	his double-digit	*lead*	nationwide. The
(2) lid	the fairly short	*lead*	time allowed on ..

The following sections will outline the steps in this process, using the individual homographs *lead* (lid/lɛd) and *bass* (beɪs/bæs) as examples. The application of this algorithm to large classes of homographs such as fractions versus dates is described in section 12.4.

Step 1: Collect and Label Training Contexts

For each homograph, begin by collecting all instances observed in a large text corpus. Then label each example of the target homograph with its correct pronunciation in that context. Here this process was partially automated, using tools that included a sense disambiguation system based on semantic classes [Yar92].

For this study, the training and test data were extracted from a 400-million-word corpus collection containing news articles (AP newswire and *Wall Street Journal*), scientific abstracts (NSF and DOE), 85 novels, two encyclopedias and three years of Canadian parliamentary debates, augmented with e-mail correspondence and miscellaneous Internet dialogues.

Step 2: Measure Collocational Distributions

The driving force behind this disambiguation algorithm is the uneven distribution of *collocations* (word associations) with respect to the ambiguous token being classified. For example, the following table indicates that certain word associations in various positions relative to the ambiguous token *bass* (including co-occurrence within a $\pm k$ word window[2] exhibit considerable discriminating power.[3]

Position	Collocation	beɪs	bæs
Word to the right (+1 w)	bass *player*	105	0
	bass *fishing*	0	94
	bass *are*	0	15
Word to the left (-1 w)	*striped* bass	0	193
	on bass	53	0
	sea bass	0	47
	white bass	0	26
	plays bass	16	0
Within ±20 words ($\pm k$ w)	*fish* (in ±20 words)	0	142
	guitar (in ±20 words)	136	0
	violin (in ±20 words)	49	0
	river (in ±20 words)	0	48
	percussion (in ±20 words)	41	0
	salmon (in ±20 words)	0	38

The goal of the initial stage of the algorithm is to measure a large and varied set of collocational distributions and select those that are most useful in identifying the pronunciation of the ambiguous word.

In addition to raw word associations, the present study also collected collocations of lemmas (morphological roots), which usually provide more succinct and generalizable evidence than their inflected forms, and part-of-speech sequences, which capture syntactic rather than semantic distinctions in usage.[4] A richer set of

[2]Several different context widths are used. The ±20-word width employed here is practical for many applications. The issues involved in choosing appropriate context widths are discussed in [GCY92].

[3]Such skewed distributions are in fact quite typical. A study in [Yar93] showed that $P(pronunciation|collocation)$ is a very low entropy distribution. Certain types of content-word collocations seen only *once* in training data predicted the correct pronunciation in held-out test data with 92% accuracy.

[4]The richness of this feature set is one of the key reasons for the success of this algorithm. Others who have very productively exploited a diverse feature set include [Hea91], [Bri93], and [DI94].

positional relationships beyond adjacency and co-occurrence in a window is also considered, including trigrams and (optionally) verb-object pairs. The following table indicates the pronunciation distributions observed for the noun *lead* for these various types of evidence:

Position[5]	Collocation	lɛd	lid
+1 L	lead *level/N*	219	0
-1 W	*narrow* lead	0	70
+1 W	lead *in*	207	898
-1W,+1W	*of* lead *in*	162	0
-1W,+1W	*the* lead *in*	0	301
+1P,+2P	lead , *<NOUN>*	234	7
±k W	*zinc* (in ±*k* words)	235	0
±k W	*copper* (in ±*k* words)	130	0
-V L	*follow/V* + lead	0	527
-V L	*take/V* + lead	1	665

Step 3: Compute Likelihood Ratios

The discriminating strength of each piece of evidence is measured by the absolute value of the the log-likelihood ratio:

$$\text{Abs}(\text{Log}(\frac{P(Pronunciation_1|Collocation_i)}{P(Pronunciation_2|Collocation_i)}))$$

The collocation patterns most strongly indicative of a particular pronunciation will have the most extreme log-likelihood. Sorting evidence by this value will list the strongest and most reliable evidence first.

Note that the estimation of $P(Pronunciation_j|Collocation_i)$ merits considerable care. Problems arise when an observed count in the collocation distribution is 0, a common occurrence. Clearly the probability of seeing *zinc* in the context of the /lid/ pronunciation of *lead* is not 0, even though no such collocation was observed in the training data. Finding a more accurate probability estimate depends on several factors, including the size of the training sample, the nature of the collocation (adjacent bigrams, verb-object pairs, or wider context), our prior expectation about the similarity of contexts, and the amount of noise in the training data.

Several smoothing methods have been explored in this work, including those discussed in [GCY92]. The preferred technique is to take all instances of the

[5] Position markers include +1 (token to the right), -1 (token to the left), $\pm k$ (co-occurrence in $\pm k$-token window) and -V (head verb). Possible types of objects at these positions include W (raw words), P (parts of speech), and L (lemmas—a class of words consisting of different inflections of the same root, such as *take/V* = *takes, took, taken, take, taking*).

same raw frequency distribution (such as 2/0 or 10/1) and collectively compute a smoothed ratio that better reflects the true probability distribution. This is done by holding out a portion of the training data and computing the mean observed distribution there (e.g., 1.8/0.2) for all of the collocations that have the same raw frequency distribution in the first portion (e.g., 2/0). This mean distribution in the held-out data is a more realistic estimate of the distribution expected in independent test data, and hence gives better predictive power and better probability estimates than using the unsmoothed values.

Smoothed ratios are very sensitive to type of collocation being observed. A 1/0 observed ratio for adjacent content words has a smoothed ratio of .92/.08, whereas a 1/0 observed ratio for function-word collocations 5 to 50 words away has a smoothed ratio close to .5/.5, indicating that a 1/0 training distribution is essentially noise here, with little predictive value.

The process of computing distributions of the form $x/0$ (and also $x/1$, etc.) for all values of x can be simplified by the observation that the mapping from observed ratios to smoothed ratios tends to exhibit a log-linear relationship when other factors such as distance are held constant. This is shown in the figure below for observed $x/0$ distributions for adjacent content word collocations. The points and jagged line (A) are the empirically observed values in held-out data. Line B constitutes the least-squares fit of these data, which is a reasonable fit, especially given that the empirical values are poor estimates for large x due to limited sample points in this range.

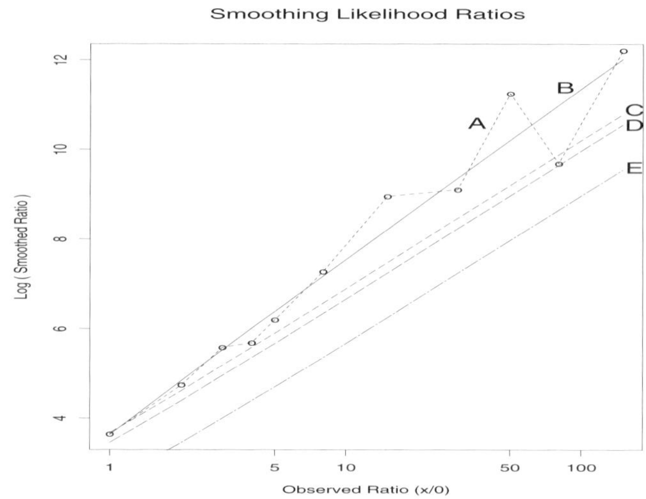

Satisfactory results may be obtained, however, by a much simpler smoothing procedure. Adding a small constant α to the numerator and denominator ($x/z \rightarrow (x+\alpha)/(z+\alpha)$) roughly captures the desired smoothing behavior, as shown in lines C ($\alpha = .085$), D ($\alpha = .1$), and E ($\alpha = .2$) of the figure. The constant α is determined empirically for the different types of collocation and distance from the target word. However, the value does not vary greatly for different homographs, so adequate performance can be achieved by reusing previous values rather than estimating them afresh from held-out training data.

Step 4: Sort by Likelihood Ratio into Decision Lists

Preliminary decision lists are created by sorting all collocation patterns by the absolute value of the smoothed log likelihood ratio, computed as described above. The following are highly abbreviated examples:

Decision List for *lead (noun)* *(highly abbreviated)*		
LogL	Evidence	Pron.
11.40	*follow/V* + lead	⇒ lid
11.20	*zinc* in $\pm k$ wrds	⇒ lɛd
11.10	lead *level/N*	⇒ lɛd
10.66	*of* lead *in*	⇒ lɛd
10.59	*the* lead *in*	⇒ lid
10.51	lead *role*	⇒ lid
10.35	*copper* in $\pm k$	⇒ lɛd
10.16	lead *poisoning*	⇒ lɛd
8.55	*big* lead	⇒ lid
8.49	*narrow* lead	⇒ lid
7.76	*take/V* + lead	⇒ lid
5.99	lead , *NOUN*	⇒ lɛd
1.15	lead *in*	⇒ lid

Decision List for *bass* *(highly abbreviated)*		
LogL	Evidence	Pron.
10.98	*fish* in $\pm k$ wrds	⇒ bæs
10.92	*striped* bass	⇒ bæs
9.70	*guitar* in $\pm k$	⇒ beɪs
9.20	bass *player*	⇒ beɪs
9.10	*piano* in $\pm k$	⇒ beɪs
9.01	*tenor* in $\pm k$	⇒ beɪs
8.87	*sea* bass	⇒ bæs
8.49	*play/V* + bass	⇒ beɪs
8.31	*river* in $\pm k$	⇒ bæs
8.28	*violin* in $\pm k$	⇒ beɪs
8.21	*salmon* in $\pm k$	⇒ bæs
7.71	*on* bass	⇒ beɪs
5.32	bass *are*	⇒ bæs

The resulting decision lists are used to classify new examples by identifying the highest line in the list that matches the given context and returning the indicated classification. This process is described in step 6.

Step 5: Optional Pruning and Interpolation

The decision lists created above may be used *as is* if we assume that the likelihood ratio for the jth entry in the list is roughly the same when computed on the entire training set and when computed on the residual portion of the training set for which the first $j-1$ entries have failed to match. In other words, does the probability that *piano* indicates the /beɪs/ pronunciation of *bass* change significantly conditional on not having seen *fish*, *striped*, *guitar*, and *player* in the target context?

In most cases the *global* probabilities (computed from the full training set) are acceptable approximations of these *residual* probabilities. However, in many cases we can achieve improved results by interpolating between the two values. The residual probabilities are more relevant, but because the size of the residual training data shrinks at each level in the list, they are often much more poorly estimated (and in many cases there may be no relevant data left in the residual on which to compute the distribution of pronunciations for a given collocation). In contrast, the global probabilities are better estimated but less relevant. A reasonable compromise is to interpolate between the two, where the interpolated estimate is $\beta_i \times global + (1 - \beta_i) \times residual$. When the residual probabilities are based on a large training set and are well estimated, β_i is small (the residual will dominate). In cases for which the relevant residual is small or nonexistent, β_i is large and the smoothed probabilities rely primarily on the better estimated global values. If all $\beta_i = 0$ (exclusive use of the residual), the result is a degenerate (strictly right-branching) decision tree with severe sparse data problems. Alternately, if one assumes that likelihood ratios for a given collocation are functionally equivalent at each line of a decision list, then one could exclusively use the global (all $\beta_i = 1$). This is clearly the easiest and fastest approach, as probability distributions do not need to be recomputed as the list is constructed.

Which approach is best? Using only the global proabilities does surprisingly well, and the results cited here are based on this readily replicable procedure. The reason is grounded in the strong tendency of a word to exhibit only one sense or pronunciation per collocation (discussed in step 3 and [Yar93]). Most classifications are based on an x versus 0 distribution, and although the magnitude of the log-likelihood ratios may decrease in the residual, they rarely change sign. There are cases for which this does happen, and it appears that some interpolation helps, but for *this* problem the relatively small difference in performance does not necessarily justify the greatly increased computational cost.

Two kinds of optional pruning can increase the efficiency of the decision lists. The first handles the problem of "redundancy by subsumption," which occurs when more general patterns higher in the list subsume more specific patterns lower down. The more specific patterns will never be used in step 6 and may be omitted. Examples of this include lemmas (e.g., *follow/V*) subsuming inflected forms (*follow*, *followed*, *follows*, etc.), and bigrams subsuming trigrams. If a bigram unambiguously signals the pronunciation, probability distributions for dependent trigrams need not even be generated because they will provide no additional useful information.

The second, pruning in a cross-validation phase, compensates for over-modeling of the training data (which appears to be minimal). Once a decision list is built it is applied to its own training set plus some held-out cross-validation data (*not* the test data). Lines in the list that contribute to more incorrect classifications than correct ones are removed. This also indirectly handles problems that may result from the omission of the interpolation step. If space is at a premium, lines that are never used in the cross-validation step may also be pruned. However, useful information is lost here, particularly for a small cross-validation corpus; these lines may have proved useful during later classification of the test data. Overall, a 3% drop in performance is observed, but more than a 90% reduction in space is realized. The optimum pruning strategy is subject to cost-benefit analysis. In the results reported below, all pruning except this final space-saving step was utilized.

Step 6: Using the Decision Lists

Once the decision lists have been created, they may be used in real time to determine the pronunciations of ambiguous words in new contexts.

From a statistical perspective, the evidence at the top of this list will most reliably disambiguate the target word. Given a word in a new context to be assigned a pronunciation, if we may base the classification only on a single line in the decision list, it should be the highest-ranking pattern that is present in the target context. This is uncontroversial and is solidly based in Bayesian decision theory.

The question, however, is what to do with the less-reliable evidence that may also be present in the target context. The common tradition is to combine the available evidence in a weighted sum or product. This is done by Bayesian classifiers, neural nets, IR-based classifiers, and N-gram part-of-speech taggers. The system reported here is unusual in that it does no such combination. *Only* the single most reliable piece of evidence matched in the target context is used.

There are several motivations for this approach. The first is that combining all available evidence rarely produces a different classification than using just the single most reliable piece of evidence, and when these differ it is as likely to hurt as to help. A study in [Yar94a] based on 20 homographs showed that the two methods agreed in 98% of the test cases. Indeed, in the 2% cases of disagreement, using only the single best piece of evidence worked slightly *better* than combining evidence. Of course, this behavior does not hold for all classification tasks, but it *does* seem to be characteristic of lexically based semantic classifications. This may be explained by the previously noted observation that in most cases, and with high probability, words exhibit only one sense in a given collocation [Yar93].

Thus for this type of ambiguity resolution, there is no apparent detriment, and some apparent performance gain, from using only the single most reliable evidence in a classification. There are other advantages as well, including run-time efficiency and ease of parallelization. However, the greatest gain comes from the ability to incorporate nonindependent information types in the decision procedure. A given word in context may match several times in the decision list, once each for its part of speech, lemma, inflected form, bigram, trigram, and possible word-classes as well. By using only one of these matches, the gross exaggeration of probability from combining all of these nonindependent log-likelihood ratios is avoided. These dependencies may be modeled and corrected for in Bayesian formalisms, but it is difficult and costly to do so. Using only one log-likelihood ratio without combination frees the algorithm to include a wide spectrum of highly nonindependent information without additional algorithmic complexity or performance loss.

12.4 Decision Lists for Ambiguity Classes

This algorithm may also be directly applied to large classes of ambiguity, such as distinguishing between fractions and dates. Rather than train individual pronunciation discriminators for *5/16* and *5/17*, etc., training contexts are pooled for all individual instances of the class. Because the disambiguating characteristics are quite similar for each class member, enhanced performance due to larger training sets tends to compensate for the loss of specialization.

12.4.1 Class Models: Creation

Decision lists for ambiguity classes may be created by replacing all members of the class found in the training data (e.g., *5/16* and *5/17*) with a common class label (e.g., *X/Y*). The algorithm described in section 11.3 may then be applied to these data.[6]

An abbreviated decision list for the fraction/date class is shown below:

Decision List for Fraction/Date Class		
LogL	Evidence	Pronunciation
8.84	*<NUMBER>* (X/Y)	$\Rightarrow$ *fraction*
7.58	(X/Y) *of*	$\Rightarrow$ *fraction*
6.79	*Monday* in $\pm k$ words	$\Rightarrow$ *date*
6.05	*Mon* in $\pm k$ words	$\Rightarrow$ *date*
5.96	(X/Y) *mile*	$\Rightarrow$ *fraction*
5.68	(X/Y) *inch*	$\Rightarrow$ *fraction*
4.22	*on* (X/Y)	$\Rightarrow$ *date*
3.96	*from* (X/Y) *to*	$\Rightarrow$ *date*

12.4.2 Class Models: Use

The use of these class decision lists requires one additional step: translating from the raw text (e.g., *5/16*) to its full pronunciation using the paradigm (e.g., *fraction*) selected by the decision list. In conjunction with the Bell Laboratories TTS speech synthesizer, the decision lists specify the chosen paradigm by using escape sequences surrounding the ambiguous form as output from the list (e.g., "`\!nfr 5/16 \!nfc`" for *fraction*).

Although the rules for this translation are typically straightforward and standard, a complication arises in the case of dates. American and British conventions differ regarding the order of day and month, pronouncing *3/7* in "*Monday, 3/7 at 5 PM*" as *March 7th* and *July 3rd*, respectively. It would seem reasonable to make this choice conditional on a global *British* or *American* parameter, set for the region of use. However, even if one decided to treat ambiguous dates conservatively (e.g., *three slash seven*), there is still considerable merit in pronouncing known fractions properly (e.g., *"3/7 of the"* as *three-sevenths* rather than *three slash seven*).

---

[6]There are advantages to filtering or weighting the training data such that each member of the class has roughly balanced representation. This causes the trained decision list to model the dominant common features of the class rather than the idiosyncrasies of its most frequent members.

12.4.3 Class Models: Incorporating Prior Probabilities

Clearly not every member of the class has the same inherent probability, independent of context. We can gain leverage by modeling these differences in prior probability. For example, the class ambiguity *year/quantifier* exhibits the following distribution of the prior probability of being a year, for numbers from 500 to 2000.[7]

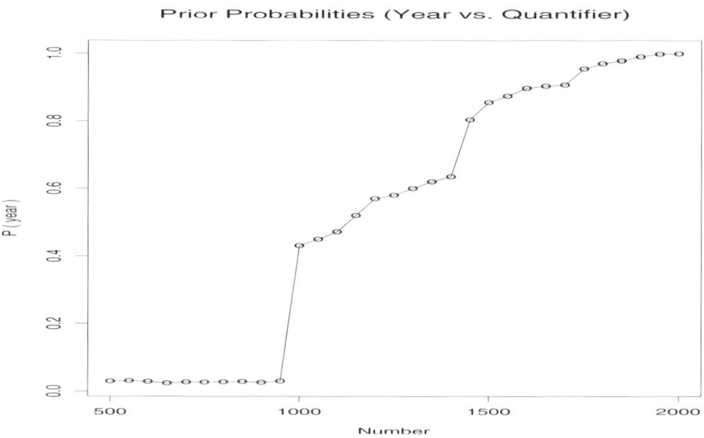

These prior probabilities may be used dynamically as follows. Train a decision list for the class assuming an uninformative prior. When applying the list, if the highest matching pattern based on context indicates the same pronunciation as the majority pronunciation based on the appropriate prior, return this result. If it indicates the minority pronunciation, find the highest matching pattern that indicates the majority pronunciation. If the difference in log likelihoods exceeds the log of the prior ratio, use the minority pronunciation.

12.4.4 Roman Numerals

Roman numerals are a case for which two-tiered class models may be productively used. The majority of Roman numerals (including II, III, VI, VII, VIII, IX, XII, XIII) exhibit the basic distinction between the uses *Chapter VII* and *Henry VII*. These are modeled in the abbreviated decision list below.

[7]The spurt at 1000 is due to the possibility of numbers greater than 1000 being written with a comma. This tendency is greatest for literary and news text, inconsistent in informal correspondence, and relatively rare in scientific text. The second spurt at roughly 1492 is due to a strong American bias in the training data's historical references.

Decision List for Roman Numerals (e.g., *VII*)		
LogL	Evidence	Pronunciation
9.63	<NEW-SENT> VII	⇒ *seven*
9.59	*king* (within $\pm k$ words)	⇒ *the seventh*
9.35	*Chapter* VII	⇒ *seven*
9.21	*Henry* VII	⇒ *the seventh*
9.16	*Edward* VII	⇒ *the seventh*
8.63	*Title* VII	⇒ *seven*
7.82	*Volume* VII	⇒ *seven*
7.65	*pope* (within $\pm k$ words)	⇒ *the seventh*
7.03	*Pius* VII	⇒ *the seventh*
6.57	*Mark* VII	⇒ *seven*
6.04	*Gemini* VII	⇒ *seven*
5.96	*Part* VII	⇒ *seven*
	○ ○ ○	
1.83	<PROP-NOUN> VII	⇒ *the seventh*

However, four Roman numerals exhibit an additional possible pronunciation. They include *IV* as /aɪ vi/ (for intravenous) and *I*, *V* and *X* (letters). For these cases, an initial decision list makes the primary distinction between these additional interpretations and the *numeric* options, based on such collocations as *IV drug, fluid, dose, injection, oral*, and *intramuscular*. If the *numeric* option is identified, the general Roman numeral list is consulted to determine if the final pronunciation should be as in *Article IV* or *George IV*. This two-tiered list maximizes the use of existing class models.

12.5 Evaluation

The following table provides a summary of the algorithm's performance on the classes of ambiguity studied.

System Performance		
	Prior	%
Type of Ambiguity (Examp.)	Prob.	Correct
Diff. Part of Speech (lives)	62	98
Same Part of Speech (bass)	72	97
Proper Names (Nice, Begin)	63	97
Roman Numerals (III)	75	97
Fractions/Dates (5/16)	59	94
Years/Quantifiers (1750)	67	93
Abbreviations (St., Dr.)	87	98
Average	69	96

A breakdown of performance on a sample of individual homographs follows:

Word	Pron1	Pron2	Sample Size	Prior Prob.	% Correct
lives	laɪvz	lɪvz	33186	69	98 [8]
wound	waʊnd	wund	4483	55	98
lead (N)	lid	lɛd	12165	66	98
tear (N)	tɛɚ	tɪɚ	2271	88	97
axes (N)	'æksiz	'æksɪz	1344	72	96
Jan	dʒæn	jɑn	1327	90	98
routed	ɹutɪd	ɹaʊtɪd	589	60	94
bass	beɪs	bæs	1865	57	99
Nice	naɪs	nis	573	56	94
Begin	bɪ'gɪn	beɪgɪn	1143	75	97
Chi	tʃi	kaɪ	1288	53	98
Colon	koʊ'loʊn	'koʊlən	1984	69	98
St.	seint	strit	624	74	99
in.	ɪntʃ	ɪntʃɪz	222	76	96
III	3	the 3rd	28146	70	98
IV	aɪ vi	*numeric*	2090	83	99
IV (*numeric*)	4	the 4th	1744	63	98
VII	7	the 7th	1514	76	98
Average			96558	69	**97**

Evaluation in each case is based on five-fold cross-validation using held-out test data for a more accurate estimate of system performance. Unless otherwise specified in the text, these results are based on the simplest and most readily replicable options in the algorithm above, and are hence representative of the performance that can be expected from the most straightforward implementation. Using more sophisticated interpolation techniques yields performance above this baseline. The sources of the test (and training) data are described in section 12.3, step 1.

12.6 Discussion and Conclusions

The algorithm presented here has several advantages, which make it suitable for general lexical disambiguation tasks that require attention to both semantic and syntactic context. The incorporation of word and optionally part-of-speech trigrams allows the modeling of many local syntactic and semantic constraints,

[8]As a standard for comparison, the PARTS tagger achieves 88% and 82% accuracy on these test data for *lives* and *wound*, respectively. A primary reason for the difference in performance is the lexicalization issue discussed in section 12.2.

whereas collocational evidence in a wider context allows for topic-based semantic distinctions. A key advantage of this approach is that it allows the use of multiple, highly nonindependent evidence types (such as root form, inflected form, part of speech, thesaurus category, or application-specific clusters) and does so in a way that avoids the complex modeling of statistical dependencies. This allows the decision lists to find the level of representation that best matches the observed probability distributions. It is a kitchen-sink approach of the best kind—throw in many types of potentially relevant features and watch what floats to the top. There are certainly other ways to combine such evidence, but this approach has many advantages. In particular, precision seems to be at least as good as that achieved with Bayesian methods applied to the same evidence. This is not surprising, given the observation in [LTV93] that widely divergent sense-disambiguation algorithms tend to perform roughly the same given the same evidence. The distinguishing criteria therefore become:

- How readily can new and multiple types of evidence be incorporated into the algorithm?
- Are probability estimates provided with a classification?
- How easy is it to understand the resulting decision procedure and the reasons for any given classification?
- Can the resulting decision procedure be easily edited by hand?
- Is the algorithm simple to implement, and can it be applied quickly to new domains?

The current algorithm rates very highly on all these standards of evaluation, especially relative to some of the impenetrable black boxes produced by many machine learning algorithms. Its output is highly perspicuous: the resulting decision list is organized like a recipe, with the most useful evidence first and in highly readable form. The generated decision procedure is also easy to augment by hand, changing or adding patterns to the list. The algorithm is also extremely flexible—it is quite straightforward to use any new feature for which a probability distribution can be calculated. This is a considerable strength relative to other algorithms, which are more constrained in their ability to handle diverse types of evidence. In a comparative study [Yar94b], the decision list algorithm outperformed both an N-Gram tagger and Bayesian classifier primarily because it could effectively integrate a wider range of available evidence types.

Overall, the decision list algorithm demonstrates considerable hybrid vigor, combining the strengths of N-gram taggers, Bayesian classifiers, and decision trees in a highly effective, general-purpose decision procedure for lexical ambiguity resolution.

Acknowledgments: This research was conducted in affiliation with the Linguistics Research Department of Bell Laboratories. It was also supported by an NDSEG Graduate Fellowship, ARPA grant N00014-90-J-1863 and ARO grant DAAL 03-89-C0031 PRI. The author would like to thank Jason Eisner and Mitch Marcus for their very helpful comments.

References

[BDDM91] P. Brown, S. Della Pietra, V. Della Pietra, and R. Mercer. Word sense disambiguation using statistical methods. In *Proceedings of the 29th Annual Meeting of the Association for Computational Linguistics*, Berkeley, 264–270, 1991.

[BFOS84] L. Brieman, J. Friedman, R. Olshen, and C. Stone. *Classification and Regression Trees*. Wadsworth & Brooks, Monterey, CA, 1984.

[Bri93] E. Brill. *A Corpus-Based Approach to Language Learning*. Ph.D. Thesis, University of Pennsylvania, Philadelphia, 1993.

[BW94] R. Bruce and J. Wiebe. Word-sense disambiguation using decomposable models. In *Proceedings of the 32nd Annual Meeting of the Association for Computational Linguistics*, Las Cruces, NM, 139–146, 1994.

[Chu88] K. W. Church. A stochastic parts program and noun phrase parser for unrestricted text. In *Proceedings of the Second Conference on Applied Natural Language Processing*, 136–143, 1988.

[DI94] I. Dagan and A. Itai. Word sense disambiguation using a second language monolingual corpus. *Computational Linguistics*, 20:563–596, 1994.

[GCY92] W. Gale, K. Church, and D. Yarowsky. A method for disambiguating word senses in a large corpus. *Computers and the Humanities*, 26:415–439, 1992.

[GCY94] W. Gale, K. Church, and D. Yarowsky. Discrimination decisions for 100,000-dimensional spaces. In *Current Issues in Computational Linguistics: In Honour of Don Walker*, A. Zampoli, N. Calzolari, and M. Palmer, eds. Kluwer Academic Publishers, Dordrecht, Holland, 429–450, 1994.

[Hea91] M. Hearst. Noun homograph disambiguation using local context in large text corpora. In *Using Corpora*, University of Waterloo, Waterloo, Ontario, 1991.

[Jel85] F. Jelinek. Markov source modeling of text generation. In *Impact of Processing Techniques on Communication*, J. Skwirzinski, ed. M. Nijhoff, Dordrecht, 1985.

[LTV93] C. Leacock, G. Towell, and E. Voorhees. Corpus-based statistical sense resolution. In *Proceedings, ARPA Human Language Technology Workshop*, Princeton, NJ, 260–265, 1993.

[Mer90] B. Merialdo. Tagging text with a probabilistic model In *Proceedings of the IBM Natural Language ITL*, Paris, France, 161–172, 1990.

[MW64] F. Mosteller and D. Wallace *Inference and Disputed Authorship: The Federalist*. Addison-Wesley, Reading, MA, 1964.

[Riv87] R. L. Rivest. Learning decision lists. *Machine Learning* 2:229–246, 1987.

[SHY92] R. Sproat, J. Hirschberg, and D. Yarowsky. A corpus-based synthesizer. In *Proceedings, International Conference on Spoken Language Processing*, Banff, 1992.

[Yar92] D. Yarowsky. Word-sense disambiguation using statistical models of Roget's categories trained on large corpora. In *Proceedings, COLING-92*, Nantes, France, 454–460, 1992.

[Yar93] D. Yarowsky. One sense per collocation. In *Proceedings, ARPA Human Language Technology Workshop*, Princeton, NJ, 266–271, 1993.

[Yar94a] D. Yarowsky. Decision lists for lexical ambiguity resolution: Application to accent restoration in Spanish and French. In *Proceedings of the 32nd Annual Meeting of the Association for Computational Linguistics*, Las Cruces, NM, 88–95, 1994.

[Yar94b] D. Yarowsky. A comparison of corpus-based techniques for restoring accents in Spanish and French text. In *Proceedings, 2nd Annual Workshop on Very Large Corpora*, Kyoto, Japan, 19–32, 1994.

Section III

Articulatory Synthesis and Visual Speech

13

Section Introduction. Talking Heads in Speech Synthesis

Dominic W. Massaro
Michael M. Cohen

This book documents that indeed "Progress in speech synthesis" is indeed being made. Just a little experience with the requirements of speech synthesis converts even the most optimistic to the realization of the tremendous endeavor that is required. Both a highly interdisciplinary approach and almost an unlimited supply of technological and human resources are necessary for starters. We can also expect that progress, although cumulative, will necessarily be gradual and too slow for many of us. We applaud the authors for their significant contributions and look forward to the continued progress of their promising research programs.

Beckman's tour de force in chapter 15 sensitizes us to the intricate relationship between basic research in speech science and speech synthesis technology. She illustrates this productive interplay of pure research and applied implementation in the case of intonation synthesis. It will be a challenge for our text-to-speech systems to arrive at the appropriate interpretation of "John doesn't drink because he's unhappy" given an analysis of the text semantics. And, of course, once we arrive at the proper interpretation, work remains to be done on both the proper acoustic [Cah90] and visual [Pel91] synthesis. She also demonstrates that the dynamics of speech articulation are better captured by higher-order properties such as timing rather than lower-order properties such as duration.

Beckman's call to more global descriptions is consistent with the framework described by Bickley, Stevens, and Williams in chapter 16. They describe procedures for the synthesis of segmental information on the basis of high-level parameters. This framework offers a productive compromise between terminal analog and articulatory speech synthesis because many of the high-level parameters are articulatory in nature: area of lip opening, average glottal area, and so forth. These high-level parameters are mapped via a set of equations into the lower-level parameters that actually control the synthesizer. Their examples of this techniques instills a degree of appreciation for this approach. We look forward to learning more about the positive aspects of this technique as well as an objective measure of the quality of the synthesis that results.

In chapter 17, Wilhelms-Tricarico and Perkell ambitiously attack the problem of biomechanical and physiologically based speech modeling. In particular, they appear to have successfully modeled the tongue and related control processes. It is heartening to see attention in this area (see also [Pel91]), given its importance as a cue for visual speech. In terms of the control processes, we hope to see further work to determine to what extent these processes are functionally hierarchical.

Perhaps one of the most significant developments in speech synthesis has been reuniting the free-floating voice with a talking head. The history of speech research and application has reasonably viewed speech as an auditory phenomenon. If a voice (or a speech synthesizer) speaks in the forest with no audience, is there speech? Our claim is that there isn't without a talking head to accompany it. More seriously, including a talking head to text-to-speech synthesizers offers the potential of a dramatic improvement in realistic synthesis, synthesis intelligibility, and end-user acceptability.

Perceptual scientists have documented that our sensory interactions in the world are seldom via a single modality. Rather, our experience is grounded in a multisensory interplay of all of our senses with a rich set of environmental dimensions. This multidimensional scenario also exists in language communication. Psychological experiments have revealed conclusively that our perception and understanding are influenced by the visible speech in the speaker's face and the accompanying gestural actions. These experiments have shown that the speaker's face is particularly helpful when the auditory speech is degraded due to noise, bandwidth filtering, or hearing impairment [Mas87, Sum91]. Although the influence of visible speech is substantial when auditory speech is degraded, visible speech also contributes to performance, even when paired with intelligible speech sounds. The importance of visible speech is most directly observed when conflicting visible speech is presented with intelligible auditory speech. One famous example resulted from the dubbing of the auditory syllable /ba/ onto a videotape of a talker saying /ga/. A strong effect of the visible speech is observed because a person will often report perceiving (or even hearing) the syllable /da/, /va/, or /ð/, but seldom /ba/ corresponding to the actual auditory stimulus.

A peculiar characteristic of bimodal speech is the complementarity of audible and visible speech. Visible speech is usually most informative for just those distinctions that are most ambiguous auditorily. For example, place of articulation (such as the difference between /b/ and /d/) are difficult via sound but easy via sight. Voicing, on the other hand, is difficult to see visually but is easy to resolve via sound. Thus, audible and visible speech not only provide two independent sources of information, these two sources are often productively complementary. Each is strong when the other is weak.

In addition to its applied value, synthetic speech has been central to the study of speech perception by human observers. Much of what we know about speech perception has come from experimental studies using synthetic speech. Synthetic speech gives the experimenter control over the stimulus in a way that is not always possible using natural speech. Synthetic speech also permits the implementation and test of theoretical hypotheses, such as which cues are critical for various speech distinctions.

It is believed that visible synthetic speech would prove to have the same value as audible synthetic speech. Synthetic visible speech could provide a more fine-grained assessment of psychophysical and psychological questions not possible with natural speech. For example, testing people with synthesized syllables intermediate between several alternatives gives a more powerful measure of integration relative to the case of unambiguous natural stimuli. In chapter 18, Le Goff, Guiard-Marigny, and Benoit address the question of which aspects of the speaking face are informative. Using video analysis of a face with painted lips, the authors are able to track the lips and to control the lips of a wire-frame model, first developed by Parke [Par74] and made more realistic in our laboratory. Perceivers were tested with just auditory speech under varying levels of white noise, or with the addition of a natural face, just synthetic lips, or the complete synthetic face (without the tongue).

They found a dramatic improvement in intelligibility with the addition of visual information. Moving lips help, the addition of the synthetic face helps more, and the natural face helps even more. The synthetic face had no tongue, and a future experimental question should be how close the synthetic face will be to a real face once a tongue is added. Guiard-Marigny, Adjoudani, and Benoit added a 3D model of the jaw and found some improvement in intelligibility relative to the synthetic lips alone, as shown in chapter 19. The jaw is controlled by a single data point corresponding to a dot on the speaker's chin. However, this improvement was not quite as large as that provided by the complete synthetic head in chapter 18. Future research comparing the jaw and the synthetic head will be of important value to the development of visible speech synthesis.

Our research indicates that, like audible speech, visible speech still does not duplicate the informative aspects of a real talking face. At this stage, it is difficult to predict the trajectory of visible speech synthesis. One issue might be whether articulatory or terminal analog synthesis offers the greatest potential. Adding the dimension of visible speech might provide a boost for articulatory synthesis because the articulators only have to be made visible to include this modality as input for the perceiver, rather than needing the additional step of a transformation to the acoustic signal, a process that has not yet been totally solved. On the other hand, with terminal analog synthesis, the investigators can concentrate on achieving a realistic animation without worrying about the physical hardware of living talkers.

It is also obvious that synthetic visible speech will have a valuable role to play in alleviating some of the communication disadvantages of the deaf and hearing-impaired. Analogous to the valuable contribution of using auditory speech synthesis in speech perception research, visible speech synthesis permits the type of experimentation necessary to determine (1) what properties of visible speech are used, (2) how they are processed, and (3) how this information is integrated with auditory information and other contextual sources of information in speech perception.

One applied value of visible speech is its potential to supplement other (degraded) sources of information. Visible speech is particularly beneficial in poor listening environments with substantial amounts of background noise. Its use is also important for hearing-impaired individuals because it allows effective spoken communication—the universal language of the community. Just as auditory speech synthesis has proved a boon to our visually impaired citizens in human machine interaction, visual speech synthesis should prove to be valuable for the hearing-impaired. Finally, synthetic visible speech had an important part of building synthetic "actors" [TT92] and played a valuable role in the exciting new sphere of virtual reality. We predict that the most progress in speech perception will be seen (no pun intended) in the continued refinement of artificial talking heads.

References

[Cah90] J. E. Cahn. *Generating Expression in Synthesized Speech.* MIT Media Lab Technical Report (revision of 1989 MS thesis), 1990.

[Mas87] D. W. Massaro. Speech perception by ear and eye: A paradigm for psychological inquiry. Lawrence Erlbaum Associates, Hillsdale, NJ, 1987.

[Par74] F. I. Parke. *A parametric model for human faces.* University of Utah Technical Report UTEC-CSc-75-047, 1974.

[Pel91] C. Pelachaud. *Communication and Coarticulation in Facial Animation.* University of Pennsylvania, Dept. of Computer and Information Science, Report MS-CIS-91-77, 1991.

[Sum91] Q. Summerfield. Visual perception of phonetic gestures. In *Modularity and the Motor Theory of Speech Perception*, I. G. Mattingly and M. Studdert-Kennedy, eds. Lawrence Erlbaum Associates, Hillsdale, NJ, 117–137, 1991.

[TT92] N. Thalmann and D. Thalmann. *Creating and Animating the Virtual World.* Springer-Verlag, Tokyo, 1992.

14

Section Introduction. Articulatory Synthesis and Visual Speech

Juergen Schroeter

14.1 Bridging the Gap Between Speech Science and Speech Applications

When asked to write an overview of a section of a book, one is faced with the problem of what service to provide besides the obvious attempt at gluing the papers in the section together and tying them to the other sections of the book. One option is to summarize current approaches to synthesis and "hot" issues being attacked in research. Eric Moulines [Mou92] satisfied this option in an excellent way in the predecessor of this book. The fact that there is little to add to Moulines's summary allows me to focus on a more specific, yet interdisciplinary issue: I have chosen to foster closer ties between basic research in speech production and applied research in speech synthesis. I will do this by highlighting problems in speech synthesis in need of solution, as well as pointing out recent findings in speech production research that are likely to impact speech synthesis. My approach will be somewhat analogous to the one of our contribution to the special session about the role of speech production in speech recognition at the June 1994 meeting of the Acoustical Society of America in Boston (written up [RSS95]).

Consider the history of research in speech synthesis from the early days of "Pedro the Voder" demonstrated at the 1939 World's Fair in New York City, over the many contributions made at Haskins Laboratories starting with the Pattern Playback machine built by Frank Cooper (e.g., [Mat74, Lib93]), and the famous book by Fant [Fan60], to the vast contributions by Klatt (see, e.g., [Ste92, Kla87]). From these and other important contributions, it seems obvious that speech synthesis and research in speech production are closely related: if we try to make a machine talk like a human, we had better know something about human speech production. Also, to understand how humans produce speech, it might be useful to test our models by letting them synthesize speech (or certain attributes of speech), and to compare the output of our models to what we measure in natural speech. These are the lines of thought in Mary Beckman's chapter, entitled "Speech Models and Speech Synthesis."

There is another side to the story, however, and this side is not mentioned in any chapter in this section and barely touched upon in section VIII, Systems and Applications. It is customer-driven by the ongoing revolutionary introduction of multimedia technology into our daily lives. Here, speech synthesis in the form of text-to-speech systems plays an increasing role in the communication between machines (computers) and humans, in particular when embedded in an information-retrieval system accessed over the telephone (see, e.g., section IV of [Rab94]). It is my strong belief that this role will expand and will reach each of our own desktop "boxes" soon (including, of course, PCs), if it didn't do so already. For example, games and gimmicks aside, I use TTS synthesis regularly to check out the flow of written sentences such as the ones I am typing right now.

The increasing exposure of the general public to computers creates expectations that, at present, even the best available technology isn't able to fulfill: completely natural-sounding synthetic speech generated from text. Going back to the potentially important role TTS systems could play in information technology, it is worth noting that most of the automatic voice responses you might hear today via telephone are playbacks of natural utterances, recorded in sentence-length (or longer) units. Very simple voice-response systems (e.g., those that play back digits responding to a request for a telephone number) concatenate several words spoken in isolation and do not make any modifications to pitch or duration. Recorded complete utterances are used even in systems in which content is updated very frequently. The significant costs associated with high-quality recordings of natural speech and of updating voice responses based on these recordings make a certain group of voice-response systems prime targets for next-generation TTS systems. In this effort, as the quality of TTS systems improves, more and more opportunities will open up. Hence, quality should be the number one issue in current research in speech synthesis.

What constitutes high-quality synthesis? Experience shows that the quality of TTS systems has several dimensions (see section VII on evaluation and perception). Although the ranking of the relative importance of these dimensions clearly depends on the specific application (e.g., a natural intonation contour over a paragraph of text is not important if we never synthesize more than one sentence at a time), it seems obvious that intelligibility generally ranks at the top, followed by "naturalness." Along these lines, it is worth noting that the increasing potency of today's computers allows us to improve the intelligibility of TTS systems by adding a visual (face/lip-reading) channel: Visual cues provided by synthesizing talking heads, as in the system described in chapter 19 by Guiard-Marigny et al. and in chapter 18 by Le Goff et al., are helpful in discriminating between confusable consonants such as nasals or between labial and alveolar stops. In general, however, I think that although we have made major inroads in intelligibility over the past decades, the issues related to naturalness need to be attacked more aggressively. Therefore, let's take a closer look.

What constitutes naturalness? Besides the obvious (and still somewhat premature) binary test of asking listeners to mark example utterances as "natural" or as "synthetic," it is much less obvious how to evaluate, let alone improve, naturalness. Again, there certainly are several dimensions to naturalness, possibly associated with different time scales. Whereas some dimensions are evaluated over a time frame that may encompass minutes (e.g., variables important for the natural-sounding synthesis of paragraph-length sections of speech), other perceptual dimensions may be evaluated for the naturalness of sentences, and others may be crucial for the naturalness of 100–200 ms segments of speech (e.g., dyads in concatenative synthesis). Finally, on the shortest scale (i.e., 5–20 ms), spectral measures borrowed from research in speech coding can help in assessing objectively (i.e., without employing human subjects) the quality of the "microscopic" building blocks of synthetic speech. Unfortunately, little is known about how humans assess the quality/naturalness of 100–500 ms segments of speech, let alone that of sentence-length utterances and beyond. It is interesting to note the apparent uniqueness of each of the individual components of the naturalness of a TTS system. The overall rating of a TTS system is likely to be low when any one of the task-relevant dimensions of naturalness (each dimension being associated with a different time scale) is low. For example, even with a complete copy-synthesis of a natural intonation of an example utterance, synthesized speech might still appear as not natural-sounding if it contains pops, tonal artifacts, and other short-term glitches. In fact, possible improvements made by a new intonation model might be masked by the generally poor microscopic quality of a given synthesizer. The same is true when a synthesizer is used that produces a perfectly natural waveform on a pitch-epoch by pitch-epoch basis, but is part of a TTS system that lacks a good pronounciation dictionary or intonational rules.

The question of what is more important, global or microscopic naturalness, is debatable (and to a large extent simply moot; see below). In any case, if we had objective measures for naturalness of time frames longer than 50 ms, we certainly could use these measures for improving unit-cutting procedures for concatenative TTS systems, as well as for evaluating segmental rules in rule-based TTS systems. In lieu of such long-term objective measures, we probably will have to continue relying on subjective evaluation methods for quite some time. However, given that good spectral measures exist for evaluating "microscopic" naturalness, it seems prudent to work one's way from the short time scales upward to the long time scales when trying to improve overall naturalness. Given that available spectral measures can be used to assure naturalness of short (10–50ms) speech frames, the next important milestone seems to be to solve the naturalness problem on segments of 100–500 ms. This also implies that one should start out by using the best available waveform synthesis algorithm before addressing, for example, problems in prosody. In any case, it is important to evaluate TTS systems embedded in an intended application. Then, the question of which system sounds more natural is pre-empted by the question of what system better fits the application, has a higher customer acceptance, and so forth.

How can research in speech production (e.g., one that is aimed at "articulatory synthesis") contribute toward solving the naturalness problem? This question is not easy to answer, given that high-quality synthesizers seem to adopt more and more pure signal-processing techniques such as PSOLA (see, e.g., [MC90]). Is this trend parallel to the trend in speech recognition about which a famous researcher once said (highly paraphrased!): "the more speech-knowledge we put into our recognizers, the worse they get!"?

Research in articulatory synthesis became significant in the early 1970s with the work on realistic (but still relatively simple) glottal models [IF72] and the work on articulatory models ([Mer73, Cok76] and others), and (maybe) peaked with the introduction of "task dynamics" by [BG86] and its practical implementation [SM89]. Although this kind of work has contributed significantly to our understanding of the speech production process, the progress made toward achieving synthesized speech of reasonable quality using these models has been painfully slow due to two closely related problems: (a) the models used for the glottis and the vocal tract have to be highly accurate, and (b) acquiring data necessary to improve these models is difficult and expensive. In addition, any method used to estimate control parameters for an articulatory synthesizer from just the speech signal (as an affordable alternative to (b)) is prone to errors (see [SS94]). In recent years, these problems have been alleviated somewhat by improved articulatory tracking techniques (e.g., [PCSMG92]), and better numerical simulation techniques (e.g., [AN92, INN92]). In this section, chapter 17 by Wilhelms-Tricarico and Perkell on a three-dimensional model of the tongue falls into this category. Although it is difficult to imagine that such computationally expensive models will ever find their way into commercially available speech synthesizers, it is highly likely that some of the results obtained by these models (and the experimental data upon which they are based) will do so. Let me explain why I believe this conjecture is true.

As pointed out above, accurate methods for the objective evaluation of 100–500 ms and longer "segments" of speech do not exist at present. Dynamic speech production models, however, should be able to produce natural-sounding transitions into and out of stops, between sonorants and fricatives, and so forth, by exploiting the inertia of the articulators (and therefore producing natural-sounding coarticulation). Hence, by combining models of speech production with (microscopic) spectral distance measures, we should be able to improve the naturalness of these transitions. This approach is outlined in chapter 16 by Bailly et al. Bickley and coworkers encapsulate knowledge of speech production in what they call "mapping relations" that derive (low-level) control parameters for the well-known Klatt formant synthesizer. This approach was also taken by early contributions of Coker [CF66, Cok76], who has made significant improvements to his system over recent years (personal communication, 1995). His articulatory, rule-based TTS addresses the problem of allophonic variations due to coarticulation and language demands.

The to-be-expected contributions of models of speech production to TTS systems go further than just solving the segment transition problem. Given that inventory acquisition for any TTS system is expensive, speaker mapping is a reasonable goal. Here, speaker mapping means transforming the synthesis inventory of one speaker to a hypothetical other speaker, or to another speaking style or regional accent by literally scaling the articulators (and the glottis) and by mapping the dynamic changes of the underlying control parameters.

In summary, by writing this introduction, I intended to highlight the links between research in speech production and the applied research in speech synthesis. People in both areas need to collaborate more to solve the naturalness problem in speech synthesis that I consider the next important milestone to reach. On the practical side, accurate models in speech production will help reduce the number of speakers (and speaking styles) needed to record TTS inventories.

References

[AN92] F. Alipour and J. Ni. Numerical simulation of unsteady flow in a glottal constriction with moving boundaries. *J. Acoust. Soc. Amer.* 92(4, Pt. 2):2391, 1993.

[BG86] C. P. Browman and L. Goldstein. Towards an articulatory phonology. *Phonology Yearbook*, 3:129–252, 1986.

[Cok76] C. H. Coker. A model of articulatory dynamics and control. *Proc. IEEE* 64(4):452–460, 1976.

[CF66] C. H. Coker and O. Fujimura. A model for specification of vocal tract area function. *J. Acoust. Soc. Amer.* 40:1271(A), 1966.

[Fan60] G. Fant. Acoustic theory of speech production. Mouton and Co., The Hague, 1960.

[IF72] K. Ishizaka and J. L. Flanagan. Synthesis of voiced sounds form a two-mass model of the vocal cords. *Bell System Techn. J.* 51(6):1233–1268, 1972.

[INN92] H. Iijima, M. Nobuhiro, and N. Nagai. Glottal impedance based on finite element analysis of two-dimensional unsteady viscous flow in a static glottis. *IEEE Trans. on Signal Processing* 40:2125–2135, 1992.

[Kla87] D. H. Klatt. Review of text-to-speech conversion for English. *J. Acoust. Soc. Amer.* 82(3):737–793, 1987.

[Lib93] A. M. Liberman. *Some assumptions about speech and how they changed.* Haskins Laboratories Status Report on Speech Research SR-113:1–32,1993.

[Mat74] I. G. Mattingly. Speech synthesis for phonetic and phonological models. In *Current Trends in Linguistics*, Vol. 12, T. A. Sebeok, ed. Mouton and Co., The Hague, 2451–2487, 1974.

[MC90] E. Moulines and F. Charpentier. Pitch-synchronous waveform processing techniques for text-to-speech synthesis using diphones. *Speech Comm.* 9:453–467, 1990.

[Mer73] P. Mermelstein. Articulatory model for the study of speech production. *J. Acoust. Soc. Amer.* 53(40):1070–1082, 1973.

[Mou92] E. Moulines. Synthesis models: A discussion. In *Talking Machines, Theories, Models, and Designs*, G. Bailly and C. Benoit, eds. Elsevier, North-Holland, Amsterdam, 7–12, 1992.

[PCSMG92] J. S. Perkell, M. H. Cohen, M. A. Svirsky, M. L. Matthies, I. Garabieta, and M. T. T. Jackson. Electromagnetic midsagittal articulometer systems for transducing speech articulatory movements. *J. Acoust. Soc. Amer.* 92(6):3078–3096, 1992.

[Rab94] L. R. Rabiner. Applications of voice processing in telecommunications. *Proc. IEEE* 82(2):199–228, 1994.

[RSS95] R. Rose, J. Schroeter, and M. M. Sondhi. The potential role of speech production models in automatic speech recognition. *J. Acoust. Soc. Amer.*, in print.

[SM89] E. L. Saltzman and K. Munhall. A dynamical approach to gestural patterning in speech production. *Ecological Psychology* 1(4):333–382, 1989.

[Son87] M. M. Sondhi and J. Schroeter. A hybrid time-frequency domain articulatory speech synthesizer. *IEEE Trans. Acoust., Speech, Signal Processing* ASSP-35(7):955–967, 1987.

[SS94] J. Schroeter and M. M. Sondhi. Techniques for estimating vocal-tract shapes from the speech signal. *IEEE Trans. Speech and Audio Proc.* 2(1):133-150, 1994.

[Ste92] K. N. Stevens. Speech synthesis methods: Homage to Dennis Klatt. In *Talking Machines, Theories, Models, and Designs*, G. Bailly and C. Benoit, eds. Elsevier, North-Holland, Amsterdam, 7–12, 1992.

15

Speech Models and Speech Synthesis

Mary E. Beckman

ABSTRACT Basic research in speech science over the last half century has benefited greatly from our endeavors to synthesize speech by machine. For example, developing programs for simulating the time course of fundamental frequency variation over sentences and longer utterances has been an indispensable research tool in our basic understanding of intonation. Synthesis systems, in turn, have directly benefited from being able to incorporate the models of linguistic control of F_0 originally built to test one or another theory of intonation. Models of temporal control are another area that can see important cross-fertilization of results and ideas between basic and applied research in synthesis. Current synthesis systems treat timing control by computing context-sensitive durations for phonetic segments, a method that integrates the use of statistical tools and large speech databases with the insights of several decades of smaller controlled laboratory experiments. Studies of articulatory kinematics suggest that closer attention to the spectral effects of articulator movement will be an important element in improving how well our synthesis systems capture the salient temporal correlates of stress and phrasing.

15.1 Theme and Some Examples

In 1991, Rolf Carlson edited a special issue of the *Journal of Phonetics* on the topic "Speech Synthesis and Phonetics." The issue was timed to commemorate the 200th anniversary of van Kempelen's mechanical synthesizer. However, the diversity of papers that he commissioned for it, and the characterization of three generations of synthesis work that he and Björn Granström wrote as an introduction for the papers, made the issue also a poignant memorial to the life of Dennis Klatt, a speech scientist and engineer untimely snatched from us, whose work embodied the theme of this chapter—namely, the opportunity that speech synthesis gives us to integrate a wealth of hard-won knowledge from many different areas of endeavor. Carlson and Granström stated the theme most eloquently in their introductory paper in that journal. They characterized how new ideas come into speech synthesis in the following way:

> [A] new idea is more like a ripe fruit to be harvested than a step into the unknown. Small steps push us in a certain direction and each step is a consequence of the past....We can see many examples of work that was done because it was simply the correct time to do it: a critical mass had been reached in the research community. ([CG91], p. 4)

Table 15.1 lists a few examples of the sort of harvest that Klatt himself described in his tutorial review of the history of text-to-speech systems for English [Kla87]. The first obvious example, going back half a century now, was the basic technique of formant synthesis, which built on decades of work on the acoustic theory of speech production, and on Fant's culminative implementation of this source-filter model in an electronic transmission-line analogue [Fan60]. By the early 1950s this work had led to the development of Cooper's Pattern Playback machine at Haskins Laboratories [CLB51], Lawrence's PAT synthesizer [Law53], and Fant's own OVE synthesizer [Fan53].

TABLE 15.1. Klatt's examples of the incorporation of basic models in the development of speech synthesis techniques and systems.

- Formant synthesis (e.g., Pattern Playback [CLB51], PAT [Law53], OVE I and II [Fan53, FM62]) built on work on the acoustic theory of speech production going back to [CK41] and even earlier.
- Formant-based synthesis-by-rule systems (e.g., [KG61, HMS64], Klatt's own Klattalk and MITalk systems [Kla82, AHK87]) built on work on acoustic correlates of segmental features (e.g., [PKG47, PB52, Fis54, DLC55]).
- Generalized rule-writing systems (e.g., [CG75, Her82]) built on work in generative phonology starting with [CH68], with more recent generations of these systems (e.g., [HKK85]) developed to accommodate more recent developments of the generative phonology model (e.g., [Gol76, Kah76]).
- Viable concatenative synthesis (e.g., [Oli77, SS86, SIMV90, PSP92, SO95]) made possible by basic research on statistical modeling of the digitized speech signal that gave rise to techniques such as LPC analysis-resynthesis [AH71, IS68].

Then, in the 1960s, there was the enormous amount of knowledge of acoustic phonetics incorporated in the first synthesis-by-rule systems. In his 1987 tutorial review of TTS systems for English, Klatt acknowledged the enormous debt that these systems owed to such scientists as Fischer-Jørgensen and Delattre by writing three large JASA pages summarizing their work as it was incorporated into these formant synthesis systems, and (in a more subtle tribute) by including on the accompanying LP record a sentence synthesized from a stylized spectrogram that Pierre Delattre had drawn for the Pattern Playback. Klatt included this as his first example of segmental synthesis by rule, describing it as the "Creation of a sentence from rules *in the head of Pierre Delattre*" ([Kla87], p. 784, my emphasis).

This synthesized sentence also illustrates the converse point I would like to emphasize today as well. Much of this early research in acoustic phonetics—for example, the locus model of consonant-vowel formant transitions—relied heavily on the availability of such tools as the Pattern Playback and OVE. The segmental model that provided the rules in Delattre's head got there in large part because of all the basic work on perceptual cues that the phoneticians and psychologists at Haskins and at other laboratories did in the late 1940s and 1950s using synthesized stimuli. In other words, it was not just that basic research in speech science provided knowledge essential for the development of speech synthesis technology. The basic research itself could not have occurred if there had not been this development in speech synthesis.

Later, in the 1970s, this push to develop formant-based synthesis-by-rule systems harvested the fruit of two decades of work in Generative Phonology. The representational framework and rewrite rules proposed by Chomsky and Halle [CH68] provided the basic computational model for the more generalized rule compilers that Carlson and Granström built into the Infovox system and that Hertz developed in her SRS system. Today, with Hertz's DELTA system, we are seeing the next generation of this sort of rule compiler, built to accommodate the new representational structures that the next generation of generative phonologists have proposed in order to accommodate tone and the effects of syllable structure.

One last example I would like to take from Klatt's article is something more akin to the first. At about the same time that the first generation of these formant rule compilers was being developed, important basic research on statistical models of the digitized speech signal by such people as Atal gave us LPC analysis-resynthesis. As Klatt pointed out in his article, this was the breakthrough necessary to finally have viable concatenative synthesis outside of a few limited domains such as telephone numbers, leading to Olive's diphone concatenation system for English. Today, we are seeing the full harvest of this new idea as many laboratories work to develop synthesis systems based on LPC diphone concatenation or related techniques. Just one example is the large-scale project at Bell Laboratories to apply to a host of other languages the TTS design that was built on Olive's original LPC-based diphone system for English (see [SO95]).

15.2 A Decade and a Half of Intonation Synthesis

This last example from Klatt's article again nicely illustrates how the benefit goes both ways. Because it allowed us to strip away the fundamental frequency contour and store a simple all-pole model of the accompanying spectra, LPC analysis-resynthesis provided a tool that has proved indispensable for harvesting the fruit of many decades of research on intonation in English and many other languages. In the last 15 years, we have made enormous progress in modeling fundamental frequency contours, in large part because the time was ripe in the relevant areas of linguistics—particularly in phonology and pragmatics—but also because of the development of this basic synthesis technique. We cannot do justice here to all the work that illustrates this point, and in any case, there are readily-available publications describing much of this work, including several of the chapters in this volume (e.g., chapter 32 by Möbius and chapter 33 by Higuchi et al.) and even an overview textbook in preparation ([Lad96]). So I will limit the current discussion to describing just three specific aspects of this progress, taken primarily from the work I know best—that of Pierrehumbert and her colleagues (see inter alia [Pie80, LP84, APL84, PB88, PS89, PH90]).

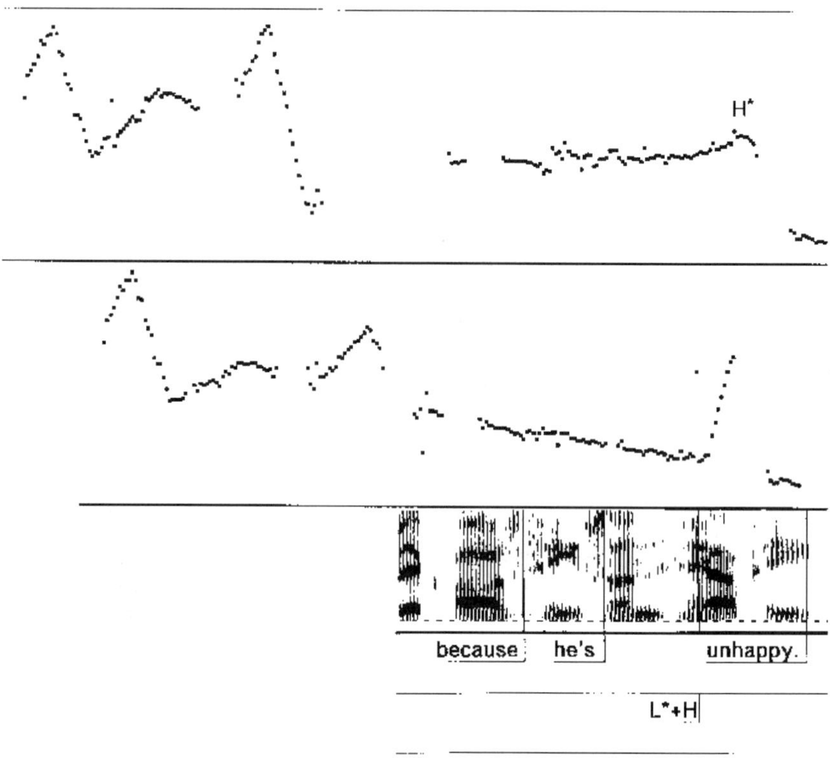

FIGURE 15.1. F_0 contours for example utterances illustrating two different tunes on the sentence *John doesn't drink because he's unhappy*. In this and the other figures in this section, salient aspects of tune are transcribed using Pierrehumbert's notation, that contrasts high "H" versus low "L" tones, and differentiates their functions with diacritics. Here, for example, the "*" indicates the tone of a pitch accent that is aligned to the stressed syllable, and the "+" joins the other tone in the bitonal accent to the starred tone.

The first illustration of how LPC analysis and resynthesis has aided our progress in understanding intonation is that we are now able to describe and even generate the full range of intonational contours that are possible in a notoriously difficult inventory of intonational contrasts. In English, a sentence can be uttered with any number of different tunes, and the choice of tune can have a profound effect on the meaning. It can change the presuppositions and even the truth value of the sentence, as in Jackendoff's famous example [Jac72], depicted in figure 15.1. The two panels show F_0 contours for two different synthesized renditions of the sentence *John doesn't drink because he's unhappy*. The synthesized utterances themselves are available on the accompanying CD-ROM (see Appendix). The first rendition presupposes that John does not drink, whereas the second implies the opposite: John does drink; he just is not a morose drunk. There are several differences between the two contours, including different intonational phrasings. The first rendition is broken into two intonational phrases at the comma, whereas the second rendition is just one phrase. One of the most salient aspects of the contrasting tunes, an aspect that had been impossible to capture in older TTS systems, is the difference between having a simple peak accent and having a more complex scooped accent shape on the stressed syllable in *unhappy*. The difference in pitch accent types is marked symbolically in the figure using Pierrehumbert's notation of H* versus L*+H, and it can be detected in the pitch tracks by comparing the timing of the F_0 peaks relative to the interruption of the F_0 trace at the voiceless segment [h]. We can synthesize this contrast now with the Bell Laboratories TTS system, because it incorporates Pierrehumbert's model of these qualitatively different pitch accent types.

Pierrehumbert's model is one of the best exemplars that I know of the theme of picking fruit when the time is right. It was the culmination of two major different strands of research. It built on decades of research by scores of linguists working on the phonology and pragmatics of English intonation, from O'Connor and Arnold [OA59], Bolinger [Bol58], and Halliday [Hal67] to Vanderslice and Ladefoged [VL72] (as interpreted by [Bru77]), Sag and Liberman [SL75], and many others. It also incorporated several important insights into the structure of pitch range variation that built on experimental work by many phoneticians and other speech scientists (e.g., [Col75, Mae76, Tho80]) combined with more qualitative observations by Africanists (e.g., [Cle81]).

Before her model was incorporated with Olive's diphones in the current Bell Laboratories TTS system, Pierrehumbert first tested it by building a program of rules for generating, from a symbolic transcription of the desired tune, a synthetic fundamental frequency contour to recombine with the LPC spectral coefficients for a natural utterance [APL84]. The second set of example utterances accompanying this chapter on the companion CD-ROM is from a demonstration of that program presented at a meeting of the Acoustical Society of America [PA84]. The utterances were generated by taking the LPC spectral coefficients of a natural utterance and recombining them with five different rule-generated intonational patterns. Figure 15.2 shows the pitch tracks and a symbolic representation of the contours as they would be transcribed in the ToBI framework [PBH94]. In all five intonation patterns, the accentuation is the same; there are two pitch accents—a prenuclear accent on the stressed syllable of *really* and the nuclear accent on the stressed syllable of *illuminating*.

The first and last patterns are ones that most American English speakers would readily recognize as contrasting with each other. The last contour, with low pitch targets on the two accented syllables and then a two-phase rise to the end of the utterance, is a common tune for a syntactically unmarked yes-no question in American English, whereas the first contour, with simple high targets on both accented syllables and a following fall, is very typical of the way that an American English speaker would read aloud a declarative sentence if put in front of a microphone in a sound booth.

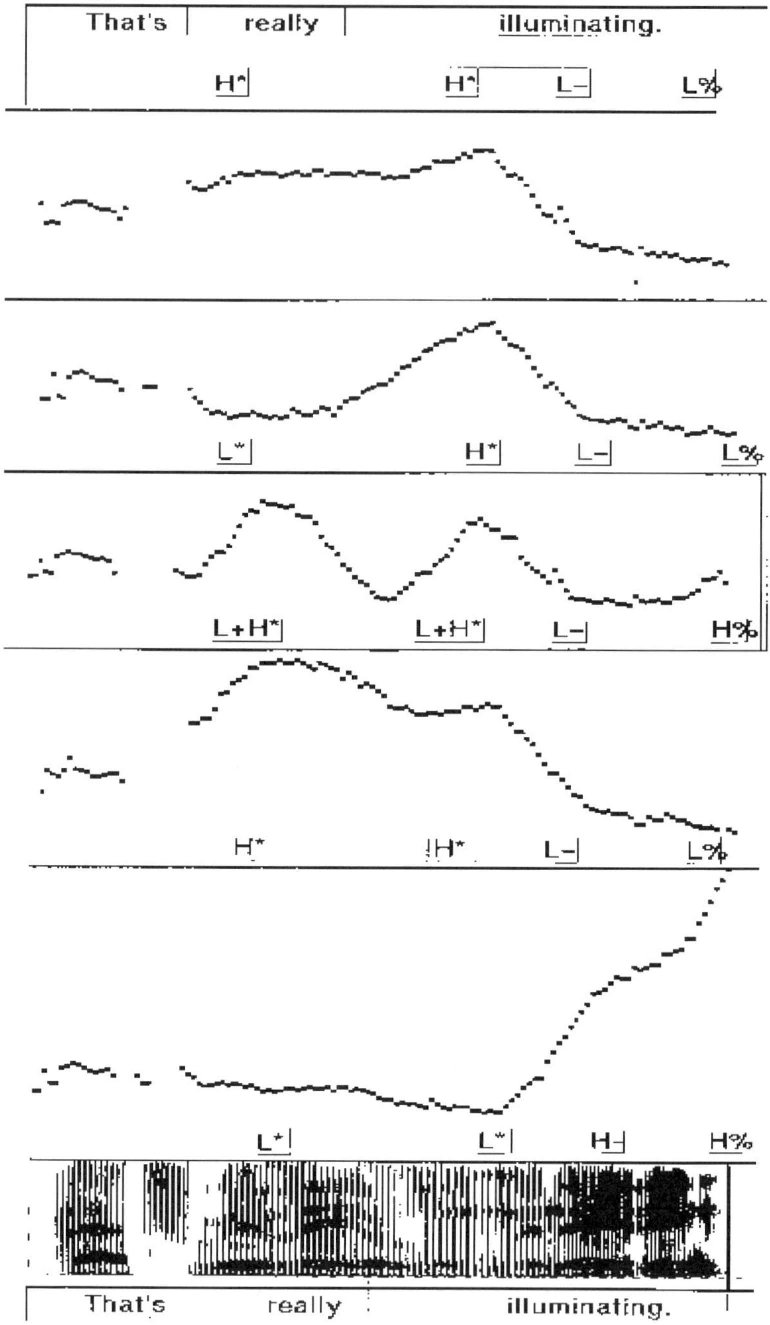

FIGURE 15.2. Synthetically generated F_0 contours for five tunes for the sentence *That's really illuminating* (reproduced, with permission, from [PA84]). As in Figure 15.1, the starred "H" and "L" tones (and the tones joined to them with "+" in the case of the third example) constitute pitch accents. The other tones are phrase accents ("H-" and "L-"), which specify the pitch in the region following the nuclear stress, and boundary tones ("H%" and "L%") specifying whether the pitch rises again at the edge of the intonational phrase. (See [PH90] and [PBH94] for details.)

The second contour, which is identical to the first except for a low rather than a high target for the first pitch accent, is the surprise-redundancy contour that Sag and Liberman [SL75] discuss.

The rising accents in the third contour are characteristic of lab-style declarative sentences for general Australian English speakers [FH93]. Database work by Ostendorf and her colleagues[OPS95] suggests that this accent type is not uncommon also in American newscasters style. Notice that the second peak in this contour is *downstepped*. That is, the backdrop pitch range has been deliberately reduced relative to the pitch range at the start of the intonational phrase. The synthesis rules generate this reduction of the pitch range automatically for sequences of accents involving this accent type.

The fourth contour also shows this downstep, but on a sequence of accent types that is otherwise indistinguishable from the two high accents in the first contour. This downstepping pattern is much more common in lab speech for Southern British English speakers, whereas in American English it can sound slightly pedantic. Pierrehumbert and Hirschberg [PH90] have characterized the difference in meaning between the two contours as a contrast between the following pragmatic functions. The H* H* sequence in the first contour simply highlights something as information to be added to the shared background that the speaker wants the hearer(s) to understand as relevant to the conversation, whereas the downstepping H* !H* sequence in the fourth contour does that and also asks the hearer to make an explicit connection between the currently highlighted information and some already assumed knowledge. This pragmatic function is important for narrative flow in story-telling or instruction-giving; hence perhaps the feeling of the pedant for American English speakers when the H* !H* L- L% pattern is produced in other contexts. In models that do not recognize the first and fourth contours as having different pitch accent types, however, what often happens is that the difference is instead ascribed to another part of the intonation system. So, for example, when Mattingly [Mat68] adapted to American English the formant synthesis rules that he had previously developed with Holmes and Shearme for British English [HMS64, Mat66], one of the things he did was to reduce the amount of baseline declination, to capture the fact that this downstepping contour is rather less likely in lab speech produced by American English speakers. Pierrehumbert's model, by contrast, explicitly separates this specification of a reduction in pitch range that is triggered by the choice of tune from the specification of more general manipulations of the backdrop pitch range such as declination. Thus the Bell Laboratories text-to-speech system can be used to generate both contours, so that the choice of contour type for any given pragmatic context is under the direct control of the user, and the difference between general American English and Southern British English in average backdrop pitch range that results from the different frequencies of use in the different dialects of the downstepped pitch accent can be modeled directly.

The discovery of such backdrop pitch range manipulations that are related to or triggered by the choice of local specifications in the tune is the second aspect of our progress in modeling intonation that illustrates my point about speech models and speech synthesis. It is an important discovery because rules such as downstep turn up in many languages, including languages that are otherwise very different from English. For example, Japanese is prosodically very unlike English, and the Japanese intonation system in particular shows several striking differences from the English system.

First, Japanese does not have the wide choice of pitch accent types illustrated in figure 15.2. Instead there is only one type—a fall in pitch on the accented syllable. Also, there is a crucial difference in what pitch accents do. In English, pitch accent placement is an important part of the sentence stress. Every well-formed intonational phrase must have at least one pitch accent, placed on the lexically stressed syllable of some word in the phrase, and putting a pitch accent on the lexically stressed syllable of that word marks the word as something new or important in the discourse, as something of importance. This means that to predict where the accents will go in a sentence, a text-to-speech system needs much more than a dictionary; it also needs a model of the larger discourse context (e.g., [Hir93] and chapter 11 of this volume). In Japanese, on the other hand, the pitch accents are predictable without any very elaborate model of the discourse structure, because instead of occurring just on prominent words, pitch accents occur (or do not) if they are specified (or not) in the dictionary. In other words, there are words that are inherently accented—such as the surname Fujimura /huzi'mura/—and there are words that are inherently not accented—such as the surname Fujinuma /huzinuma/. (Here the apostrophe after the second syllable in the phonemic transcription for Fujimura marks that syllable as the accented one, where the pitch falls.) So it is easy to find pairs of utterances that are structurally the same—they have the same number of intonational phrases, and the words play the same role in the discourse context—but one sentence may have many pitch accents and the other none at all. It depends predictably on what words are in the sentence.

On the other hand, Japanese is very much like English in that it also differentiates downstep (which in Japanese is a successive reduction in pitch range triggered at *each* accent) from the more general declination that occurs whether or not there are any accents in the phrase. Moreover, in Japanese, for which the accents are specified in the dictionary, there can be huge differences in the amount of downtrend between pairs of otherwise very similar sentences, if one sentence has many lexically accented words and the other has few or none. Earlier systems for synthesizing Japanese intonation contours (e.g., [FN79, BHF83]) did not recognize this important potential component of the downtrend, triggered by the local lexically predictable occurrence of pitch accent. Instead, all of the downtrend was modeled using baseline declination, as in Mattingly's [Mat68] or Klatt's [Kla82] rules for English. And those of us working on these synthesis systems for Japanese were sometimes baffled by the fact that our rules for declination sometimes produced far too much declination and sometimes produced far too little in a way that did not seem to be related to anything obvious like the length of the utterance or the number of phrases in it.

By the time that we were working on the intonation model described in [PB88], however, we could take advantage of important basic research such as [SS83] and

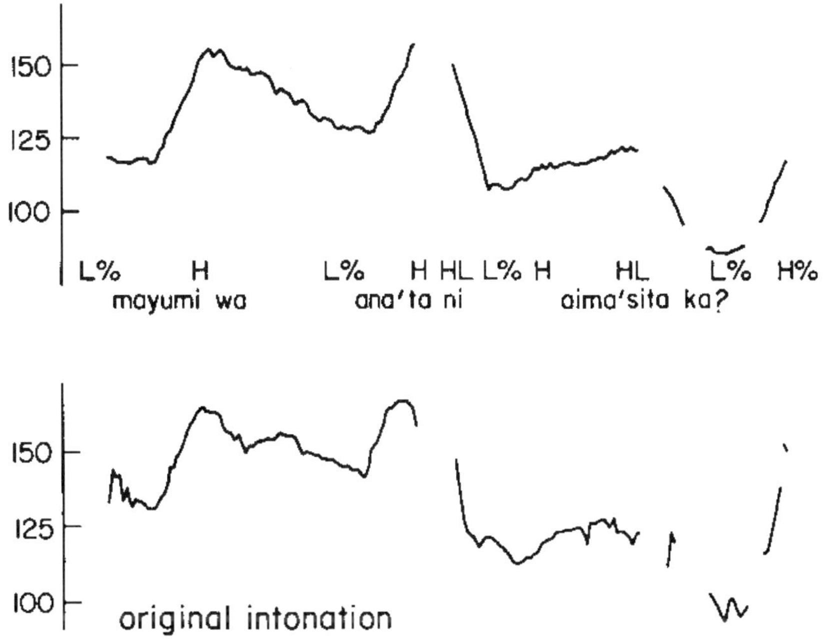

FIGURE 15.3. Rule-generated copy (top) and original F_0 (bottom) for utterance of /majumi-wa anata-ni aimasita ka/ 'Did Mayumi see you?' (Adapted, with permission, from [PB88], p. 177.)

[Pos84], which pointed the way toward separating downstep from declination and other components of downtrend. And because of the LPC-analysis-resynthesis technique, we could test this model in a program for generating synthetic F_0 contours for natural utterances without having to first build rules for formant values and durations. The third section of the accompanying example utterances on the CD-ROM demonstrates this program. There are two versions of an utterance stripped of its natural fundamental frequency contour and then recombined once with its original F_0 and once with the rule-generated copy. Figure 15.3 shows the two F_0 contours. The parameters for the model are described in chapter 7 of [PB88]. They include a speaker-specific downstep constant for the proportional reduction of pitch range automatically triggered by the accent on /ana'ta/ *you*. (See [PB88], pp. 207–208, for a table listing all of the parameter values used in generating the synthetic contour and a schematic figure depicting the tone targets generated by rule from those values.) The effect of the downstep can be seen by comparing the peak height relationships between /ana'ta-wa/ and the preceding and following phrases. The declination between /majumi-wa/ and /ana'ta-ni/ is negligible by comparison to the reduction in pitch range between /ana'ta-wa/ and /aima'sita ka/.

The third illustrative aspect of our progress in modeling fundamental frequency contours also involves pitch range (see figure 15.4). In both the Japanese and the English intonation synthesis systems, the user can specify a different backdrop

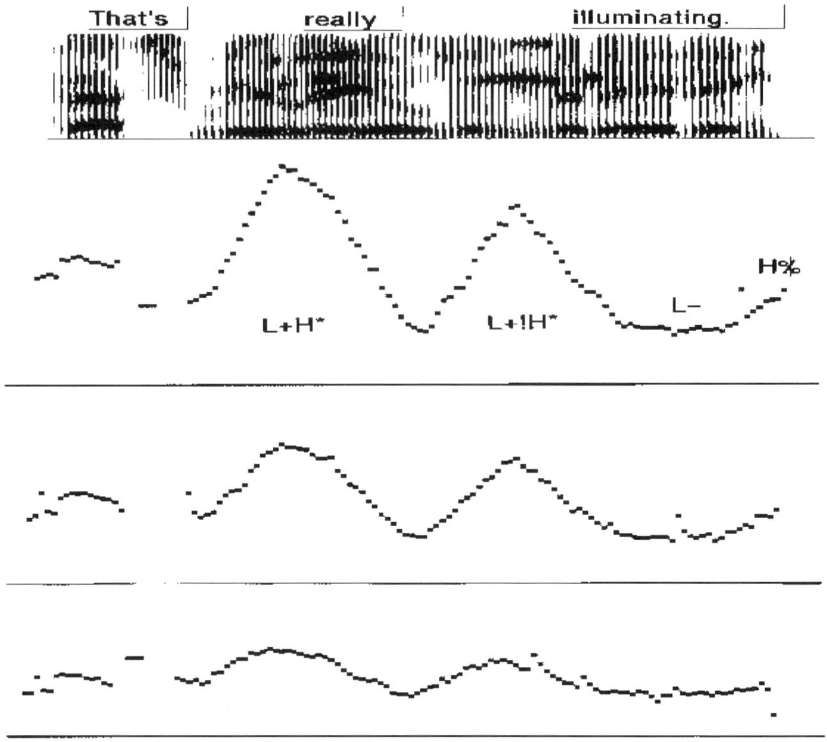

FIGURE 15.4. Synthetically generated F_0 contours for the third tune in figure 15.2, in three different overall pitch ranges. (Reproduced, with permission, from [PA84].)

pitch range for a phrase, and have the downstep ratio scaled automatically to the new pitch range. The next set of demonstration utterances on the accompanying CD-ROM demonstrates this aspect of the English system. It shows the third tune in figure 15.2, generated with three different pitch-range specifications. The downstepped L+!H* peak is reduced relative to the first L+H* peak by the same proportion of the overall pitch range.

Being able to specify this kind of overall variation in pitch range is very important because it can affect the interpretation of the intonational contour. For example, it can cue almost opposite meanings of the tune illustrated in figure 15.5. The two panels of the figure show fundamental frequency contours for two renditions of the sentence *Bob's going out with Anna*, which are also included on the accompanying CD-ROM. Both renditions have the same tune, which is related to the tune in the second utterance in figure 15.1. That is, this tune has the same "scooped" L*+H accent on the word with main stress followed by the same fall to a L- phrase tone, and then a rise to a H% boundary tone at the end of the utterance. Hirschberg and Ward have shown that the sentence produced with this contour has two different interpretations (the utterances in figure 15.5 are from their 1992 study [HW92]).

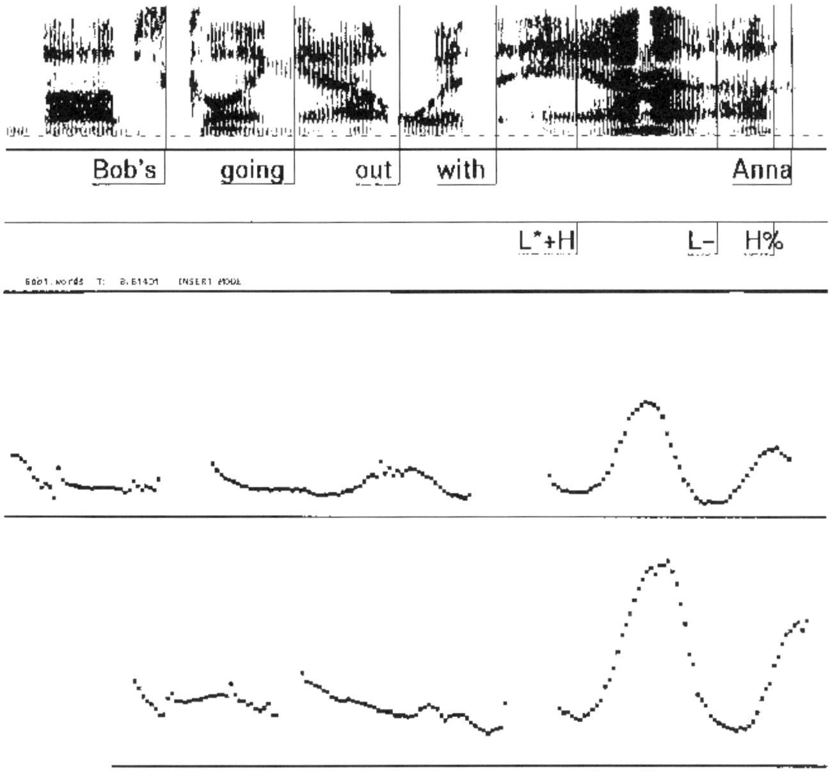

FIGURE 15.5. Two utterances of the sentence *Bob's going out with Anna* produced with the L*+H L- H% rise-fall-rise tune that can be interpreted (top panel) as uncertainty about the pragmatic appropriateness of the information conveyed or (bottom panel) as incredulous amazement about the information. (Reproduced, with permission, from the corpus recorded by [HW92].)

Imagine that Joe and John are having a conversation about a mutual acquaintance whose girlfriend has left him after putting $10,000 worth of charges on his credit card. If Joe asks, "Has Bob learned to pick more appropriate friends?" and John responds with the utterance depicted in the first panel of figure 15.5, this would imply that John is not sure whether Anna is any improvement over the previous girlfriend. This is the "uncertainty" interpretation of the contour. Joe might then respond with the second rendition in figure 15.5, to express his amazement at Bob's lack of judgment. This is the "incredulity" interpretation of the contour.

However, although the intonation pattern is qualitatively the same in the two cases—both are this rise-fall-rise tune—there are differences in such more continuous phonetic dimensions such as voice quality, amplitude, and pitch range [HW92]. The second interpretation, the incredulity one, tends to be louder and faster and have a more expanded pitch range. Hirschberg and Ward studied these other differences using a technique pioneered by Nakatani and Schaffer [NS78]:

begin with an LPC analysis of two natural utterances exemplifying the contrast; synthesize stimuli with all possible combinations of the original F_0 contour, the original durations and amplitudes, and the original LPC spectral coefficients; and then play these stimuli to listeners, to gauge which parts of the stimulus the listeners use in identifying the meaning. Doing this for the two renditions of the rise-fall-rise contour, Hirschberg and Ward found that the expansion of overall pitch range is the main cue differentiating the incredulity interpretation from the uncertainty interpretation.

There have been some studies showing that pitch range is used this way in Japanese as well. For example, Miura and Hara [MH95] have studied a contour with a H% boundary tone, which can be a simple yes-no question or a rhetorical question conveying the same kind of incredulity as the second interpretation of the rise-fall-rise contour in English. They also used LPC analysis-resynthesis to show that the most salient difference between the two interpretations is expanded pitch range. I suspect that the separate control of overall pitch range that is necessary to capture these phenomena in English and Japanese will turn out to be something that we need to incorporate into synthesis systems for many other languages as well. Work on French by Touati [Tou93] and on Korean by Jun and Oh [JO94] suggests that this will be true of those two languages at least.

Other recent work on Japanese by Maekawa and his colleagues [Maek91, MSKH93], by Kubozono [Kub93], Venditti [Ven94], Tsumaki [Tsu94], and others explores also how this kind of manipulation of overall pitch range interacts with downstep in signaling other relationships between sentences and their larger discourse contexts. For example, pragmatic narrow focus and some syntactic structures seem to block downstep. This blocking of downstep is typically accompanied by "reset," an expansion of the overall pitch range that also applies in comparable position in contrasting sentences with no accents—i.e., in sentences in which downstep would not have applied because the tonal trigger is not there in the lexical string. Many of these phenomena are reminiscent of things influencing pitch accent placement in English. These studies suggest that the next generation of synthesis models for Japanese must begin to incorporate some explicit discourse model to capture how downstep and pitch range manipulation work in longer narratives and in a larger range of communicative situations.

To summarize these three examples of recent progress in F_0 modeling, then, we can say that the harvest has been on both sides of the fence. The development of LPC analysis-resynthesis made it possible to develop programs for simulating the time course of fundamental frequency variation over sentences and longer utterances. This has been an indispensable research tool in our basic understanding of intonation. Text-to-speech systems, in turn, have directly benefited from being able to incorporate the models of linguistic control of F_0 originally built to test one or another theory of intonation. This close interaction between basic scientific models and speech technology has thus led us that much closer to our ultimate goal of going beyond synthesizing sentences from text to generating more complex discourses from an abstract specification of the underlying informational structure. Since this has brought me to talking about the future, what I would like to do now is to focus on another aspect of speech synthesis, in order to speculate a bit about where I think the next big new idea will be, about an aspect of speech for which the fruit of a lot of basic research is there on the tree, nearly ripe, and ready to be picked—namely, timing.

15.3 Models of Time

Our current models treat timing by specifiying durations for a succession of internally unanalyzed unit intervals, typically for a quasi-alphabetic unit corresponding roughly to the phoneme. That is, our model of time is still the same model that phoneticians such as Lehiste [Leh59], Lindblom [Lin63], and Koshevnikov and Chistovich [KC65] adopted in order to be able to relate measurements from spectrograms to the more familiar units of linguistic analysis. The model implicitly assumes something like a string of invariant target states output by the grammar and a clock ticking away in a separate module of the rule system to govern the intervals between motor impulses for triggering the successive states. The sequence of specified lengths of the unit intervals then becomes an independent primary phonetic parameter, analogous to the sequence of fundamental frequency values in a pitch track. This model led to many important findings about durational patterns associated with segmental and prosodic contrasts—for example, the finding that, other things being equal, low vowels are longer than high vowels; that closure durations for voiceless stops are considerably longer than those for voiced stops whereas vowels before voiceless coda consonants are shorter than vowels before voiced codas; that vowels in phrase-final or accented syllables are longer than vowels in phrase-medial or unaccented syllables; and so on—all of the effects that Klatt incorporated into his simple, elegant rules for specifying segment durations in Klattalk [Kla79]. When we go back and read the chapter of Lehiste's book [Leh70] summarizing these effects, or Klatt's paper updating that summary with more work on segment durations in English [Kla76], we are strikingly reminded how direct was the connection between the basic research and the synthesis rules.

The progress in synthesis that we have made since that time has been on two main fronts. We have harnassed speech recognition technology to make automatic segmentation tools in order to gather larger and larger corpora of durational measurements (e.g., [Leu85] as used in [PZ89], and chapter 25 of this volume). We also have devised ever more sophisticated statistical methods in order to get more and more accurate predictions from these corpora for the sizes of the effects and the interactions among them (see, e.g., [KTS90, CI91, van94]). The result is that the duration rules in any good TTS system today generate durations that usually fall well within the limits of the random variability that we see in measurements from natural speech. But there are still places in which the timing of utterances generated by these rules just does not sound natural, even cases in which the rules are generating values hardly different from values observed in natural speech. Why?

I think the answer is that we have taken segment duration as far as it will go as a satisfactory phonetic measure of timing, and that now we should look over the fence again into the neighboring garden of speech science, where an important thing has happened. We now have tools that let us sample articulatory states at decent recording intervals without bombarding the talker with dangerously high levels of X-rays—tools such as the magnetometer [NM93, PCSMG92]. With these tools, speech scientists have been studying the articulatory kinematics underlying the best-known durational effects for English and a few other languages, and the results of these studies suggest some ideas about what we must do to model aspects of timing that do not seem to be adequately captured when we simply manipulate segment durations.

The upper part of figure 15.6 presents an example that shows a set of jaw height traces from a study by Summers [Sum87], using a corpus of monosyllabic English nonsense words such as *bop* (/bap/) and *bob* (/bab/), which minimally contrasted in the voicing of the coda consonant and were embedded in sentences in which the target was either accented or deaccented. The sequences of labial consonants around low vowels was chosen so as to maximize the differential involvement of the jaw in the articulation of the neighboring segments. That is, since the lower lip is coupled to the jaw and the tongue body also rests on the jaw, talkers tend to raise the jaw to help the lips in making the labial consonant constrictions and to lower the jaw to help lower the tongue body in the low vowel. Thus we can get some idea of the articulatory kinematics underlying the duration pattern by tracking the time course of jaw lowering and then raising over the consonant–vowel–consonant (CVC) sequence.

The waveforms in the lower part of figure 15.6 show that when timing is gauged in the usual way by measuring the acoustic duration of the vowel interval, the effect of being followed by a voiceless consonant is indistinguishable from the effect of being in a deaccented syllable. The vowel in an accented *bop* is shortened relative to that in the accented *bob* by almost exactly as much as is the vowel in a deaccented *bob*. However, the jaw trace is very different in the two cases. The opening and closing movements in the deaccented *bob* are much slower than in the accented *bop*, so that the jaw does not open as wide. Since jaw height in the vowel is our indirect measure of tongue height, we might expect slower first formant transitions into the coda consonant in the deaccented *bob*, and a bit of undershoot for the first formant target. And that is what Summers did find: the deaccented *bob* and the accented *bop* showed different timing patterns for the formants, even though the two vowels were the same duration. A comparable study by de Jong [deJ95] of the patterns for tongue body and lip movements in words with alveolar codas following rounded back vowels suggests that the vowel length effects associated with accenting and deaccenting and with coda consonant voicing can be differentiated by the movement patterns of all major articulators. Work by Edwards, Beckman, and Fletcher [EBF91] shows dramatic differences in articulator dynamics between the effects of accent and of position in phrase as well. (See [NMK88, BE94] and the references cited in these papers for differential dynamics associated with other durational effects.)

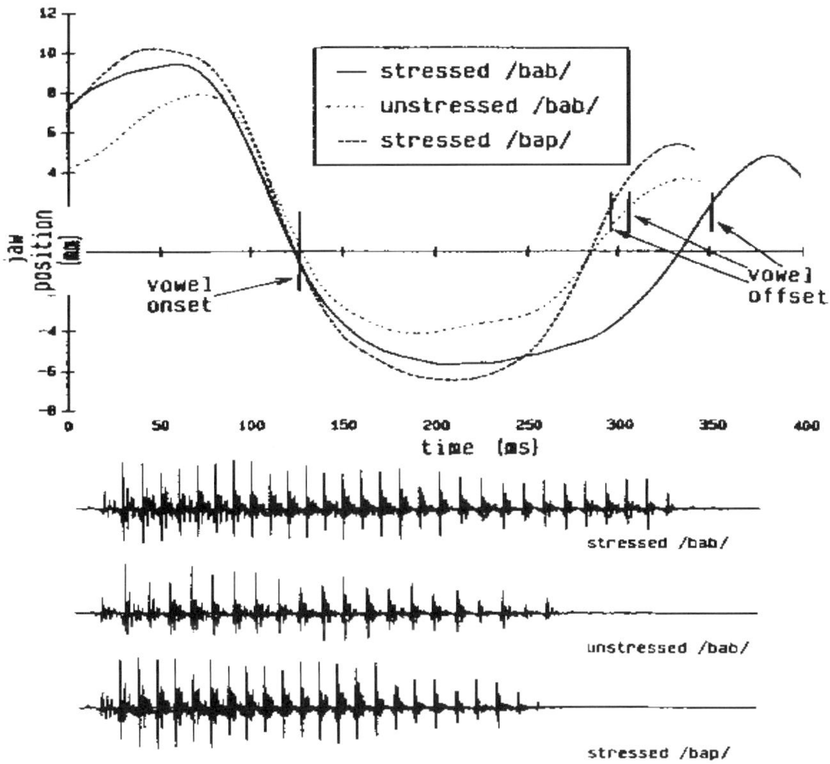

FIGURE 15.6. In upper panel, jaw height traces for representative tokens of an accented *bob* (solid trace), an unaccented *bob* (dotted trace), and an accented *bop* (dashed trace); lower panel shows acoustic waveforms for these same utterances. (Adapted, with permission, from [Sum87].)

What these studies suggest is that we need to stop thinking about these effects just in terms of duration—as if duration were some simple acoustic parameter such as fundamental frequency. Instead we need to start thinking of these as differences in timing per se. We have to start modeling the distinctive "acoustic signature" [Sum87] of each effect, the different time courses of the changes in the acoustic pattern, and the resulting undershoot of the spectral target, if there is any.

In order to model these timing differences in TTS systems as accurately as we now model segment durations, we are probably going to ultimately need either true articulatory synthesis systems capable of capturing directly the underlying articulatory dynamics, as in [Mae90, Mae91], or we will need to develop rules for parametric acoustic synthesis systems that are intelligently constrained by the articulatory dynamics, as in [SB91]. It is very encouraging to see the vitality of research in both of these areas, in continually ongoing work in many laboratories, and of a volume sufficient to constitute a sizable percentage of the papers at the last two of the biennial Seminars on Speech Production [Scu91, Lof95] as well as several papers at the Second ESCA/IEEE Workshop on Speech Synthesis (e.g. chapters 17 and 16 of this volume).

A difficult problem that we will have to address in building either sort of system is the fact that the mapping between articulation and acoustics is highly nonlinear. This central point of Stevens's Quantal Theory [Ste72, Ste89] has been reconfirmed in a number of other laboratories using a number of different models of the mapping. Figure 15.7 gives just one example of this work, a nomogram from a simulation study by Carré and Mrayati [CM91]. The articulatory model is a simple tube closed at one end and divided into four regions in which cross-sectional area can be varied, with consequences predicted by perturbation theory [MCG88]. The symbols connected by lines diagram the acoustic consequences of equal-sized changes in area for the relevant regions, to simulate linear articulatory trajectories from /i/ to each of the other oral vowels of French. As the graph shows, the spectral consequences of these linear articulatory changes are not linear. The lines in the F1-F2 space do not always change in the same direction along the entire articulatory continuum, and the symbols for equal-sized articulatory steps are not evenly spaced along each line. The most obvious case is in going from the constricted front and expanded back cavity of the /i/ to the constricted back and more neutral front cavity of the /o/; here the first formant turns around from increasing to decreasing values halfway through the trajectory, and the consequence of the last step is a change in the F1-F2 space that is nearly four times the size of the consequence of the articulatorily equal steps on either side of the F1 turnaround. Although less dramatic, however, other trajectories also show nonlinearities. Even the /i/ to /u/ trajectory, which is a fairly straight line, has very unevenly spaced symbols, indicating that a given amount of backing and rounding has a much smaller effect on the formant values in the beginning part of the trajectory than it does at the end of the trajectory. This work suggests that we will have to pay much closer attention to the spectral dynamics when we shorten durations to model the effects of deaccenting and the like on the high back vowel than on the high front vowel.

Of course, there is much basic work yet to be done before we can use this sort of knowledge to constrain formant synthesis rules. While our cache of basic articulatory studies is expanding rapidly, there is still a lot we do not know about what the articulatory patterns are, or about which aspects of the patterns we can ignore and which need to be modeled fairly closely in order for the spectral dynamics to sound natural and fluent.

There is also another important problem in reaping the knowledge we are rapidly gaining from this work on articulatory dynamics. The kind of fine spectral detail suggested in figure 15.7 is exactly what we are trying to avoid having to model by rule when we do diphone concatenation. We will need to be rather ingenious to devise efficient means for incorporating these effects into concatenative synthesis systems without jettisoning the advantage of the LPC model. One solution might be to combine concatenation with formant synthesis, as proposed by Pearson et al. [PMHH94]. This would let us concentrate our effort on just those parts of the signal where prosodic effects on spectral dynamics are the most salient and most

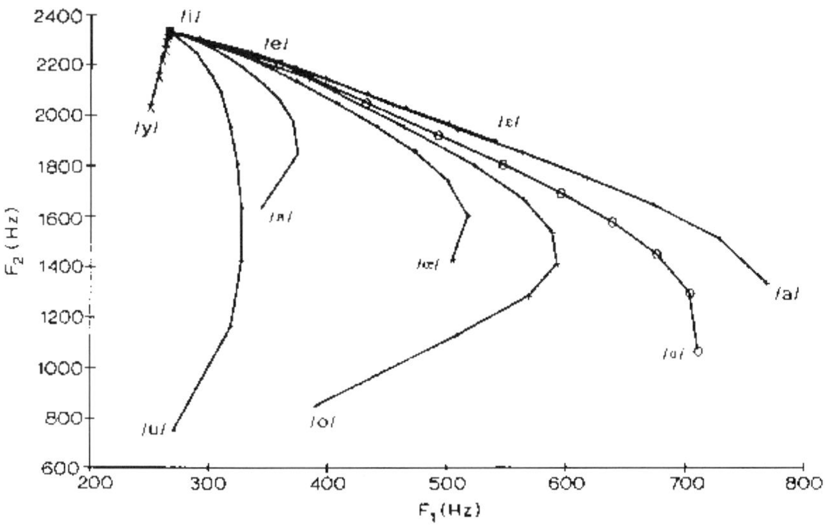

FIGURE 15.7. Nomogram of acoustic consequences in $F_1 - F_2$ space of varying relevant regions in a four-region model to simulate equal-sized steps along a trajectory between /i/ and each of the other simple vowels of French. (Adapted, with permission, from [CM91].)

important to model accurately, parts such as the transitions into and out of accented vowels.

Another solution may be to selectively stretch or shrink different parts of the concatenative unit to capture the salient dynamic differences. There is relevant work by Macchi et al. [MSW90] and by van Santen et al. [RC92], suggesting that we might be able to apply dynamic time warping to this problem. This work also makes clear, though, that we are going to have to think of these effects in terms of control units that can span any number of phoneme segments. Here we might benefit from work to incorporate more sophisticated phonological theories to describe and delineate larger prosodically governed control units (e.g., [Col92 and chapters 8 and 9 of this volume).

On the other hand, there is something that we can do already, and, in fact, can do rather easily with our current statistical methods. We can begin by modeling the dynamics of those aspects of the signal that are controlled directly by rule even in concatenative synthesis—namely, the amplitude and fundamental frequency contours. Figure 15.8 gives an example of what I have in mind. This is a figure from [vH94]. (See also [SP90] and other work cited there.) The figure shows the shape of the F_0 peak and following fall for an L+H* nuclear accent before an L- L% phrase tone sequence in sentence-final monosyllables with three different kinds of onset consonant—sonorants versus voiced obstruents versus voiceless obstruents. The F_0 starts much higher after the voiceless obstruents (as we would expect from work on intrinsic F_0 effects such as [HF53, Hom79]). What is interesting, however, is that this affects the timing of the F_0 trace throughout the syllable nucleus; the F_0

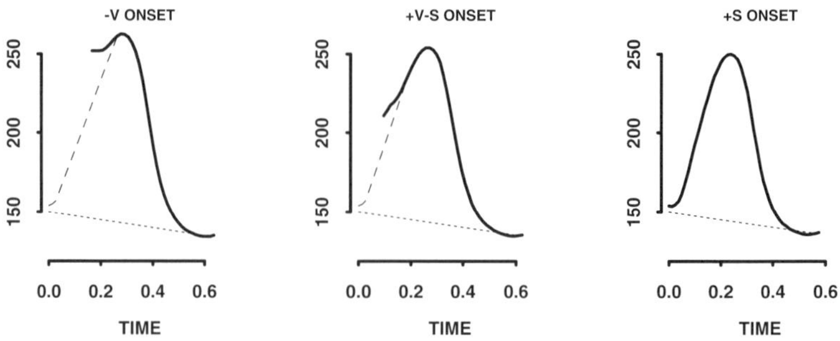

FIGURE 15.8. Average F0 shapes in monosyllabic nonsense words in nuclear accented position of the L+H* L- L% rise-fall contour. Reprinted with permission from [vSH94].

peak is relatively earlier and much more of the subsequent fall is realized within the sonorant portion of the syllable. I suspect that the failure to model this kind of interaction between the tune and the accompanying segments is an important source of perceived error in duration. That is, when we hear these nuclear-accented syllables as being too long or too short, we may be attending not to the duration per se, but to the failure to model these kinds of differences in F_0 dynamics.

This brings me to my last comment on this topic. Our time and energy are not infinite, and given the nature of current funding of synthesis research, some of that time and energy must be spent in developing and refining TTS systems that can be marketed by the companies for which we work. At some point, we do have to ask ourselves where to concentrate our resources in incorporating these models of speech timing into our synthesis systems. Which of the effects, and which aspects of the effects, are most perceptible and would have the most salient consequences if we did not capture the dynamics correctly? In looking at the literature on speech timing, one sometimes sees statements that this or that effect was so small as to be below the just noticable difference (JND) for a durational differences. I think that the psychoacoustic literature on JNDs for durations is completely irrelevant here. For the most part, we have got the segment durations right. Instead, we need to look to studies of the perception of timing per se, of the sensitivity of listeners to the dynamics of the changes for amplitude, F_0, or spectral shape. Psychoacousticians have recently begun to address these questions (e.g., [PRHMZ92, AY94]), and with some care in how we apply these results, we soon should be able to harvest from that orchard, too.

15.4 Valediction

The theme of this chapter has been the interaction of basic speech models and speech synthesis. I have mentioned a few of the classic examples of the way in which new ideas come into synthesis from many other areas of speech science, and have discussed at some length a current example and a possible future example of this abundant harvest. In the course of developing these examples, I have mentioned the work of many kinds of scientist—work by mathematical psychologists and psychoacousticians, by physicists and electrical engineers and computer scientists, by phonologists and speech physiologists. This fact alone—the diversity of our training—illustrates my theme as well as any specific result I have discussed. Speech synthesis is what it is today because people such as Dennis Klatt were constantly looking to apply knowledge from many different disciplines. It is a hybrid fruit of a flower that has been cross-pollinated from trees blooming in many neighboring gardens. It is important to insure that future generations of speech scientists and synthesis experts understand this: that we must all continually look across the fence to the flowers on trees next door. We must foster this outlook by our own example: by reading broadly, and by trying to cast our work in ways that will let us publish in journals addressed to audiences somewhat outside of our own specializations. We never know whence the next new idea will come, and so we need to open our labs to visiting researchers and postdocs from other disciplines, so that our students can learn to talk to people who have been trained very differently from them. And, finally, we every once in a while need to have interdisciplinary workshops such as the Second ESCA/IEEE Workshop on Speech Synthesis, where we can get together ourselves to talk with other people doing research different from our own.

References

[AH71] B. S. Atal and S. L. Hanauer. Speech analysis and synthesis by linear prediction of the speech wave. *J. Acoust. Soc. Amer.* 5:637–655, 1971.

[AHK87] J. Allen, S. Hunnicut, and D. H. Klatt. *From Text to Speech: The MITalk System.* Cambridge University Press, Cambridge, 1987.

[APL84] M. A. Anderson, J. B. Pierrehumbert, and M. Y. Liberman. Synthesis by rule of English intonation patterns. In *Proceedings of the 1984 IEEE International Conference on Acoustics, Speech and Signal Processing*, 2.8.2–2.8.4, 1984.

[AY94] K. Aikawa and R. A. Yamada. Comparative study of spectral representations in measuring the English /r/-/l/ acoustic-perceptual dissimilarity. In *Proceedings of the 1994 International Conference on Spoken Language Processing*, 2039–2042, 1994.

[BE94] M. E. Beckman and J. Edwards. Articulatory evidence for differentiating stress categories. In *Phonological Structure and Phonetic Form: Papers in Laboratory Phonology III*, P. A. Keating, ed. Cambridge University Press, Cambridge, 7–33, 1994.

[BHF83] M. E. Beckman, S. Hertz, and O. Fujimura. SRS pitch rules for Japanese. *Working Papers of the Cornell Phonetics Laboratory* 1:1–16, 1983.

[Bol58] D. Bolinger. A theory of pitch accent in English. *Word* 7:199–210, 1958.

[BP86] M. E. Beckman and J. B. Pierrehumbert. Intonational structure in English and Japanese. *Phonology Yearbook* 3:255–310, 1986.

[Bru77] G. Bruce. *Swedish Word Accents in Sentence Perspective*, Travaux de l'institut de linguistique de Lund, 1977.

[CG75] R. Carlson and B. Granström. A phonetically oriented programming language for rule description of speech. In *Speech Communication*, vol. 2, G. Fant, ed. Almqvist and Wiksell, Uppsala, 245–253, 1975.

[CG91] R. Carlson and B. Granström. Speech synthesis development and phonetic research: A personal introduction. *J. Phonetics* 1:3–8, 1991.

[CH68] N. Chomsky and M. Halle. *The Sound Pattern of English*. Harper and Row, New York, 1968.

[CI91] W. N. Campbell and S. D. Isard. Segment durations in a syllable frame. *J. Phonetics* 19:37–47, 1991.

[CIMV90] W. N. Campbell, S. D. Isard, A I. C. Monoghan, and J. Verhoeven. Duration, pitch and diphones in the CSTR TTS system. In *Proceedings of the 1990 International Conference on Spoken Language Processing*, Kobe, Japan, 825–828, 1990.

[CK41] T. Chiba and M. Kajiyama. *The Vowel: Its Nature and Structure*. Kaiseikan, Tokyo, 1941.

[CLB51] F. S. Cooper, A. M. Liberman, and J. M. Borst. The interconversion of audible and visible patterns as a basis for research in the perception of speech. In *Proceedings of the National Academy of Sciences*, Washington, DC, 37:318–325, 1951.

[Cle81] G. N. Clements. The hierarchical representation of tone features. In *Harvard Studies in Phonology 2*, G. N. Clements, ed. Harvard Department of Linguistics, Cambridge, MA, 1981.

[CM91] R. Carré and M. Mrayati. Vowel-vowel trajectories and region modeling. *J. Phonetics* 19:433–443, 1991.

[Col75] R. Collier. Physiological correlates of intonation patterns. *J. Acoust. Soc. Amer.* 58:249–255, 1975.

[Col92] J. Coleman. "Synthesis-by-rule" without segments or rewrite rules. In *Talking Machines: Theories, Models, and Designs*, G. Bailly and C. Benoit, Elsevier, North-Holland, Amsterdam, 1992.

[deJ95] K. de Jong. The supraglottal articulation of prominence in English: Linguistic stress as localized hyperarticulation. *J. Acoust. Soc. Amer.*, 97:491–504, 1995.

[DLC55] P. Delattre, A. M. Liberman, and F. S. Cooper. Acoustic loci and transitional cues for consonants. *J. Acoust. Soc. Amer.* 27:769–774, 1955.

[EBF91] J. Edwards, M. E. Beckman, and J. Fletcher. The articulatory kinematics of final lengthening. *J. Acoust. Soc. Amer.* 89:369–392, 1991.

[Fan53] G. Fant. Speech communication research. *Ing. Vetenskaps Akad. Stockholm* 24:331–337, 1953.

[Fan60] G. Fant. *Acoustic Theory of Speech Production*. Mouton, The Hague, 1960.

[Fis54] E. Fischer-Jørgensen. Acoustic analysis of stop consonants. *Phonetica* 2:42–59, 1954.

[FH93] J. Fletcher and J. Harrington. Characteristics of Australian intonation. Presentation at the Third ToBI Workshop, Ohio State University, 27–30 June 1993.

[FM62] G. Fant and J. Martony. *Speech Synthesis*. Quarterly Progress and Status Reports, Speech Transmission Laboratories, Royal Institute of Technology, Stockholm, 2:18-24, 1962.

[FN79] H. Fujisaki and S. Nagashima. A model for the synthesis of pitch contours of connected speech. Annual Report, Engineering Research Institute, Faculty of Engineering, University of Tokyo, 28:53–60, 1979.

[Gol76] J. Goldsmith. *Autosegmental Phonology*. Doctoral dissertation, Massachusetts Institute of Technology, Cambridge, MA, 1976.

[Hal67] M. A. K. Halliday. *Intonation and Grammar in British English*. Mouton, The Hague, 1967.

[Her82] S. Hertz. From text to speech with SRS. *J. Acoust. Soc. Amer.* 72:1155-1170, 1982.

[HF53] A. S. House and G. Fairbanks. The influence of consonant environment on the secondary acoustical characteristics of vowels, J. Acoust. Soc. Amer. 25, 105-113, 1953.

[Hir93] J. Hirschberg. Pitch accent in context: Predicting intonational prominence from text. *Artificial Intelligence* 63(1–2):309–340, 1993.

[HKK85] S. Hertz, J. Kadin, and K. Karplus. The Delta rule development system for speech synthesis from text. *Proc. of the IEEE*, 73:1589–1601, 1985.

[HMS64] J. N. Holmes, I. G. Mattingly, and J. N. Shearme. Speech synthesis by rule. *Language and Speech* 7:127–143, 1964.

[Hom79] J.-M. Hombert. Consonant types, vowel quality, and tone. In *Tone: A Linguistic Survey*, V. A. Fromkin, ed. Academic Press, New York, 1979.

[HW92] J. Hirschberg and G. Ward. The influence of pitch range, duration, amplitude and spectral features on the interpretation of the rise-fall-rise intonation contour in English. *J. Phonetics* 20:241–251, 1992.

[IS68] F. Itakura and S. Saito. Analysis-synthesis telephony based on the maximum likelihood method. In *Proceedings of the Sixth International Congress on Acoustics*, Tokyo, Japan, paper C-5-5, 1968.

[Jac72] R. Jackendoff. *Semantic Interpretation in Generative Grammar*, MIT Press, Cambridge, MA, 1972.

[JO94] S.-A. Jun and M. Oh. A prosodic analysis of three sentence types with "WH" words in Korean. In *Proceedings of the 1994 International Conference on Spoken Language Processing*, Tokyo, Japan, 323–326, 1994.

[Kah76] D. Kahn. *Syllable Based Generalizations in English Phonology*. Doctoral dissertation, Massachusetts Institute of Technology, Cambridge, MA, 1976.

[KC65] N. A. Koshevnikov and L. A. Chistovich. *Speech: Articulation and Perception*, U.S. Dept. of Commerce translation, JPRS 30-543, 1965.

[KG61] J. Kelly and L. Gerstman. An artificial talker driven from phonetic input. *J. Acoust. Soc. Amer., Suppl. 1*, 33:S35, 1961.

[Kla76] D. H. Klatt. Linguistic uses of segmental duration in English: Acoustic and perceptual evidence. *J. Acoust. Soc. Amer.* 59:1208–1221, 1976.

[Kla79] D. H. Klatt. Synthesis by rule of segmental durations in English sentences. In *Frontiers of Speech Communication Research*, B. Lindblom and S. Öhman, eds. Academic Press, London, 287–299, 1979.

[Kla82] D. H. Klatt. The Klattalk text-to-speech system. In *Proceedings of the 1982 IEEE International Conference on Acoustics, Speech and Signal Processing*, Paris, France, 1589–1592, 1982.

[Kla87] D. H. Klatt. Review of text-to-speech conversion for English. *J. Acoust. Soc. Amer.*, 82:737–793, 1987.

[KTS90] N. Kaiki, K. Takeda, and Y. Sagisaka. Statistical analysis for segmental duration rules in Japanese speech synthesis. In *Proceedings of the 1990 International Conference on Spoken Language Processing*, Kobe, Japan, 17–20, 1990.

[Kub93] H. Kubozono. *The Organization of Japanese Prosody*. Kurosio, Tokyo, 1993.

[Lad80] D. R. Ladd. *The Structure of Intonational Meaning*. Indiana University Press, Bloomington, 1980.

[Lad96] D. R. Ladd. *Intonational Phonology*. Cambridge University Press, Cambridge, in press.

[Law53] W. Lawrence. The synthesis of speech from signals which have a low information rate, In *Communication Theory*, W. Jackson, ed. Butterworths, London, 460–469, 1953.

[Leh59] I. Lehiste. *Acoustic-Phonetic Study of Internal Open Juncture*. Supplement to *Phonetica*, 5, 1959.

[Leh70] I. Lehiste. *Suprasegmentals*. MIT Press, Cambridge, MA, 1970.

[Leu85] H. C. Leung. *A Procedure for Automatic Alignment of Phonetic Transcriptions with Continuous Speech*. S.M. thesis, Massachusetts Institute of Technology, Cambridge, MA, 1985.

[Lin63] B. Lindblom. Spectrographic study of vowel reduction. *J. Acoust. Soc. Amer.* 35, 1773–1781, 1963.

[Lof95] A. Löfqvist, guest editor. *Speech Production Models and Data III*, Special issue of *J. Phonetics* 23(1–2), 1995.

[LP84] M. Y. Liberman and J. P. Pierrehumbert. Intonational invariance under changes in pitch range and length. In *Language Sound Structure: Studies in Phonology Presented to Morris Halle*, M. Aranoff and R. T. Oehrle, eds. MIT Press, Cambridge, MA, 1984.

[Mae76] S. Maeda. *A Characterization of American English Intonation*. Doctoral dissertation, Massachusetts Institute of Technology, Cambridge, MA, 1976.

[Mae90] S. Maeda. Compensatory articulation during speech: Evidence from the analysis and synthesis of vocal-tract shapes using an articulatory model. In *Speech Production and Speech Modelling*, W. J. Hardcastle and A. Marchal, eds. Kluwer, Dordrecht, Holland, 1990.

[Mae91] S. Maeda. On articulatory and acoustic variabilities. *J. Phonetics* 19:321–331, 1991.

[Maek91] K. Maekawa. Perception of intonational characteristics of WH and non-WH questions in Tokyo Japanese. In *Proceedings of the 12th International Congress of Phonetic Sciences*, vol. 4, Aix-en-Provence, France, 202–205, 1991.

[Mat66] I. G. Mattingly. Synthesis by rule of prosodic features. *Language and Speech* 9:1–13, 1966.

[Mat68] I. G. Mattingly. *Synthesis-by-rule of General American English*, Supplement to *Status Report on Speech Research, Haskins Laboratories*, SR-15/16, pages 1–223, 1968.

[MCG88] M. Mrayati, R. Carré, and B. Guérin. Distinctive regions and modes: A new theory of speech production. *Speech Commun.* 7:257–286, 1988.

[MH95] I. Miura and N. Hara. Production and perception of rhetorical questions in Osaka Japanese. *J. Phonetics* 23(3):291–303, 1995.

[MSKH93] K. Maekawa, T. Sato, S. Kiritani, and H. Hirose. Tookyoo-hoogen no bunreberu de no pitti geraku kindenzugakuteki kenkyuu [Electromyographic study of phrase-level pitch falls in the Tokyo dialect]. In *Proceedings of the March 1993 Meeting of the Acoustical Society of Japan*, Tokyo, Japan, 235–236, 1993.

[MSW90] M. J. Macchi, M.F. Spiegel, and K. L. Wallace. Modeling duration adjustment with dynamic time warping. In *Proceedings of the 1990 International Conference on Acoustics, Speech, and Signal Processing*, 333–336, 1990.

[NM93] N. Nguyen and A. Marchal. Assessment of an electromagnetic system for the investigation of articulatory movements in speech production. *J. Acoust. Soc. Amer.* 94(2):1152–1155, 1993.

[NMK88] S. Nittrouer, K. Munhall, J. A. S. Kelso, B. Tuller, and K. S. Harris. Patterns of interarticulator phasing and their relationship to linguistic structure. *J. Acoust. Soc. Amer.* 84:1653–1661, 1988.

[NS78] L. H. Nakatani and J. A. Schaffer. Hearing "words" without words: Prosodic cues for word perception. *J. Acoust. Soc. Amer.* 63:234–245, 1978.

[OA59] J. D. O'Connor and G. F. Ward. *Intonation of Colloquial English*. Longman, London, 1959.

[Oli77] J. P. Olive. Rule synthesis of English from diadic units. In *Proceedings of the 1977 IEEE International Conference on Acoustics, Speech and Signal Processing*, Hartford, CT, 568–570, 1977.

[OPS95] M. Ostendorf, P. J. Price, and S. Shattuck-Hufnagel. *The Boston University Radio News Corpus*. Report No. ECS-95-001, Electrical, Computer and Systems Engineering Department, Boston University, Boston, 1995.

[PA84] J. Pierrehumbert and M. Anderson. A computer program for synthesizing English intonation. *J. Acoust. Soc. Amer.*, Suppl. 1, 76:S3, 1984.

[PB52] G. E. Peterson and H. L. Barney. Control methods used in a study of the vowels. *J. Acoust. Soc. Amer.* 24:175–184, 1952.

[PB88] J. B. Pierrehumbert and M. E. Beckman. *Japanese Tone Structure*. MIT Press, Cambridge, MA, 1988.

[PBH94] J. Pitrelli, M. E. Beckman, and J. Hirschberg. Evaluation of prosodic transcription labeling reliability in the ToBI framework. In *Proceedings of the 1994 International Conference on Spoken Language Processing*, Yokohama, Japan, 123–126, 1994.

[PCSMG92] J. Perkell, M. Cohen, M. Svirsky, M. Matthies, I. Garabieta, and M. Jackson. Electro-magnetic midsagittal articulometer (EMMA) systems for tranducing speech. *J. Acoust. Soc. Amer.* 92:3078–3096, 1992.

[PH90] J. Pierrehumbert and J. Hirschberg. The meaning of intonation contours in the interpretation of discourse. In *Intentions in Communication*, P.R. Cohen, J. Morgan, and M. E. Pollack, eds. MIT Press, Cambridge, MA, 271–311, 1990.

[Pie80] J. Pierrehumbert. *The Phonology and Phonetics of English Intonation*. Doctoral dissertation, Massachusetts Institute of Technology, Cambridge, MA, 1980.

[PKG47] R. K. Potter, G. A. Kopp, and H. C. Green. *Visible Speech*. van Nostrand, New York, 1947.

[PMHH94] S. Pearson, H. Moran, K. Hata, and F. Holm. Combining concatenation and formant synthesis for improved intelligibility and naturalness in text-to-speech systems. In *Proceedings of the Second ESCA/IEEE Workshop on Speech Synthesis*, New Paltz, NY, 69–72, 1994.

[Pos84] Poser, W. J. (1984). *The Phonetics and Phonology of Tone and Intonation in Japanese*. Doctoral dissertation, Massachusetts Institute of Technology, Cambridge, MA, 1984.

[PRHMZ92] R. D. Patterson, K. Robinson, J. Holdsworth, D. McKeown, C. Zhang, and M. Allerhand. Complex sounds and auditory images. In *Auditory Physiology and Perception*, Y. Cazals, L. Dermany, and K. Horner, eds. Pergamon, Oxford, 1992.

[PS89] J. B. Pierrehumbert and S. A. Steele. Categories of tonal alignment in English. *Phonetica* 46:181–196, 1989.

[PSP92] T. Portele, B. Steffan, R. Preuß, W. F. Sendlmeier, and W. Hess. HADIFIX – A speech synthesis system for German. In *Proceedings of the 1992 International Conference on Spoken Language Processing*, Banff, Canada, 1227–1230, 1992.

[PZ89] J. F. Pitrelli and V. W. Zue. A hierarchical model for phoneme duration in American English. In *Proceedings Eurospeech '89*, Paris, France, 324–327, 1989.

[SB91] K. N. Stevens and C. Bickley. Constraints among parameters simplify control of Klatt formant synthesizer. *J. Phonetics* 19:161–174, 1991.

[Scu91] C. Scully, ed. *Speech Production: Models, Methods and Data*. Special issue of *J. Phonetics* 19(3–4), 1991.

[SL75] I. Sag and M. Liberman. The intonational disambiguation of indirect speech acts. In *Papers from the Eleventh Regional Meeting, Chicago Linguistics Society*, Chicago, 487–497, 1975.

[SO95] R.W. Sproat and J.P. Olive. A modular architecture for multi-lingual text-to-speech. In *Talking Machines II: More Theories, Models, and Designs*, J. P. H. van Santen, R. Sproat, J. P. Olive, and J. Hirschberg, eds. Springer-Verlag, 1996 (this volume).

[SP90] K. Silverman and J. Pierrehumbert. The timing of prenuclear high accents in English. In *Papers in Laboratory Phonology I: Between the Grammar and Physics of Speech*, J. Kingston and M. E. Beckman, eds. pages 72–106. Cambridge University Press, Cambridge, 72–106, 1990.

[SS83] S. Sagisaka and H. Sato. *Secondary Accent Analysis in Japanese Stem-Affix Concatenations (Transactions of the Committee on Speech Research S83-05).* Acoustical Society of Japan, Tokyo, 1983.

[SS86] S. Sagisaka and H. Sato. Composite phoneme units for the speech synthesis of Japanese. *Speech Commun.* 5:217–223, 1986.

[Ste72] K. N. Stevens. Quantal nature of speech. In *Human Communication: A Unified View*, E. E. David, Jr. and P. B. Denes, eds. McGraw-Hill, New York, 1972.

[Ste89] K. N. Stevens. On the quantal nature of speech. *J. Phonetics* 17:3–45, 1989.

[Sum87] W. V. Summers. Effects of stress and final-consonant voicing on vowel production: Articulatory and acoustic analyses. *J. Acoust. Soc. Amer.* 82:847–863, 1987.

[Tho80] N. Thorsen. Neutral stress, emphatic stress, and sentence intonation in Advanced Standard Copenhagen Danish. *Annual Report of the Institute of Phonetics*, University of Copenhagen, 14:121–205, 1980.

[Tou93] P. Touati. Prosodic aspects of political rhetoric. In *Proceedings of an ESCA Workshop on Prosody* (Working Papers of the Lund University Department of Linguistics, 41), 168–171, 1993.

[Tsu94] J. Tsumaki. Intonational properties of adverbs in Tokyo Japanese. In *Proceedings of the 1994 International Conference on Spoken Language Processing*, Yokohama, Japan, 1727–1730, 1994.

[van94] J. P. H. van Santen. Using statistics in text-to-speech system construction. In *Proceedings of the Second ESCA/IEEE Workshop on Speech Synthesis*, ESCA, New Paltz, 240–243, 1994.

[Ven94] J. J. Venditti. The influence of syntax on prosodic structure in Japanese. In *Papers from the Linguistics Laboratory, Working Papers in Linguistics*, Ohio State University, 44:191-223, 1994.

[vH94] J. P. H. van Santen and J. Hirschberg. Segmental effects on timing and height of pitch contours. In *Proceedings ICSLP '94*, Yokohama, Japan, 719–722, 1994.

[VL72] R. Vanderslice and P. Ladefoged. Binary suprasegmental features and transformational word-accentuation rules. *Language* 48:819–838, 1972.

[vRC92] J. P. H. van Santen, M. A. Randolph, and J. C. Coleman. Effects of postvocalic voicing on the time course of vowels and diphthongs. *J. Acoust. Soc. Amer.* 92:2444, 1992.

Appendix: Audio Demos

16

A Framework for Synthesis of Segments Based on Pseudoarticulatory Parameters

Corine A. Bickley
Kenneth N. Stevens
David R. Williams

ABSTRACT Procedures are described for rule-based synthesis of segmental aspects of an utterance using a formant synthesizer controlled by high-level parameters. The rules for synthesis of consonants are described in terms of: (1) the formation and release of consonantal constrictions; and (2) the actions of the secondary articulators of the glottis, the velopharyngeal opening, and active pharyngeal expansion or contraction. Examples of synthesis based in part on these rules are given. Advantages of using high-level parameters for rule-based synthesis are discussed.

16.1 Background and Introduction

The production of speech can be described as a patterned sequence of relatively slow movements or adjustments of articulatory structures punctuated by movements that create discontinuities in the spectrum of the sound. Each of these latter movements is a consequence of the creation of a narrow constriction or closure in the airway by the lips, the tongue blade, or the tongue body. The slower movements are generated by the tongue body, by rounding of the lips, and by adjustments of vocal-fold stiffness to produce changes in fundamental frequency. Also in the class of slower movements with nonabrupt acoustic consequences are adjustment of the velopharyngeal opening, the glottal opening, and the volume of the pharyngeal region of the vocal tract (for obstruent consonants). In English these latter adjustments are usually made in conjunction with the discontinuity-forming movements.

In developing models for speech production, then, we can think of three classes of control parameters:

1. Parameters relating to movements of the tongue body, mandible, lip rounding, and vocal-fold stiffness

2. Parameters specifying the formation or release of narrow constrictions in the oral cavity

3. Parameters describing the cross-sectional areas of velopharyngeal and glottal orifices and active expansion or contraction of the vocal-tract volume behind a constriction

Consonant production can be efficiently represented in terms of closures and openings of vocal-tract orifices. This representation of consonants, based on a speech production model, is well founded on theoretical grounds, and, we argue, has advantages over ones that use parameters that are primarily acoustic in nature. The class 2 and class 3 parameters listed above provide a natural and concise means of specifying the relevant closings and openings for consonants as well as associated adjustments of the secondary articulators. The acoustic consequences of changes in the three types of parameters interact, so that there is not always a one-to-one relation between the settings of the parameters in one domain and attributes of the sound output. These interactions are handled automatically in the system described in this chapter.

The articulatory movements can be modeled by a few parameters, the time course of each of which represents information that is relevant for calculating the myriad of acoustic parameters needed in a synthesizer of the type described in [KK90]. We have reported previously on a scheme for speech synthesis based on this idea of a small set of parameters, some of which are articulatory in nature and others of which are acoustic. The synthesis scheme uses a set of mapping relations to transform a small set of high-level (HL) parameters into the many low-level (LL) parameters of a Klatt-type synthesizer [SB91]. This method of specifying HL parameters and then automatically calculating LL parameters has proved effective in generating high-quality synthetic speech (see [WBS92, BSW94] for examples).

In this chapter, we describe the inventory of high-level control parameters that are used in the current implementation of a speech synthesizer based on the above speech-production principles and we review some of the mapping relations from HL parameters to the LL parameters that control the formant synthesizer. We then present examples of speech synthesized using this set of parameters to control the synthesizer. Finally, we describe a framework for generating the control parameters by rule for consonants in various phonetic environments, given a linguistic specification of the utterance to be synthesized.

16.2 Control Parameters and Mapping Relations

There are 10 HL parameters; they specify the time variation of the cross-sectional areas of the glottis, the velopharyngeal port, the consonantal constrictions formed in the oral tract, and the active expansion or contraction of the vocal-tract volume behind a constriction, as well as parameters related to the natural frequencies of the vocal tract and the fundamental frequency [SB91]. The relation between the parameters and the vocal-tract shape is schematized in figure 16.1. The parameters *f1*, *f2*, *f3*, *f4*, and *f0* are acoustic counterparts of the parameters in class 1 described above. The areas *al* and *ab* are class 2 parameters, and *an*, *ag*, and *ue* are class

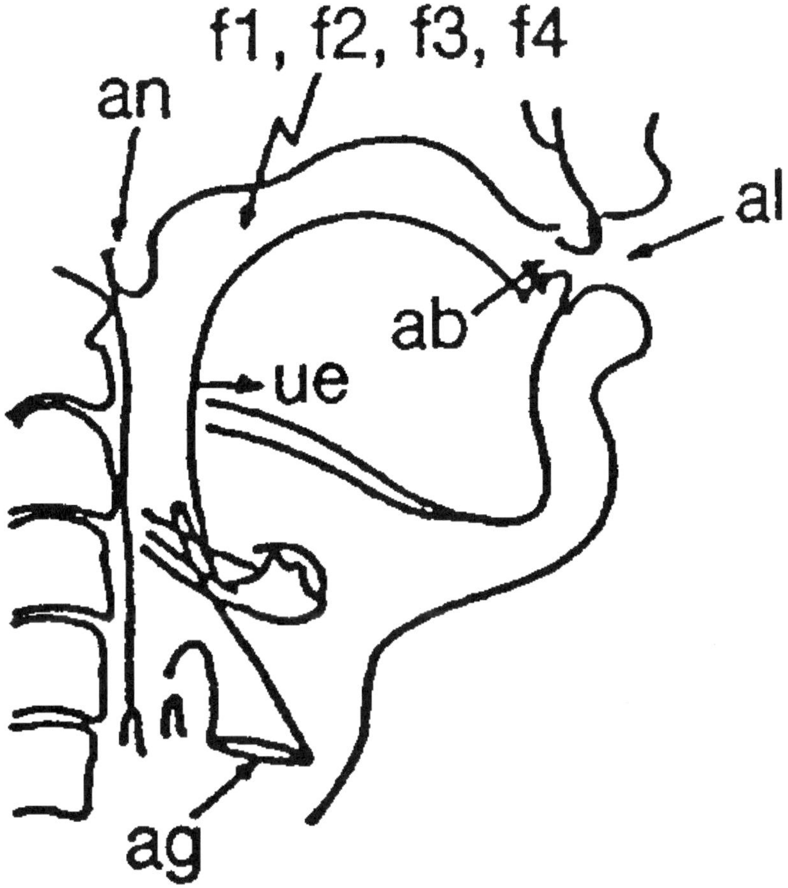

FIGURE 16.1. The HL parameters of *al*, *ab*, *an*, and *ag* shown above represent the orifice area of lip opening, constriction formed by the tongue blade, opening to the nasal cavities, and average glottal area, respectively. The formant parameters *f1*, *f2*, *f3*, and *f4* are used to indicate indirectly the vocal-tract configuration. The parameter *ue* corresponds to the rate of active expansion or contraction of the vocal-tract volume behind a constriction. The HL parameter *f0* (fundamental frequency) is not shown in the figure.

3 parameters. It should be noted that the formant parameters *f1*, *f2*, *f3*, and *f4* represent the natural frequencies of the vocal tract in the absence of an increased glottal opening *ag* or velopharyngeal opening *an*, and in the absence of local constrictions formed by *al* and *ab*. In this sense, then, these parameters are "virtual" formants. As will be observed later, this method of specifying vocal-tract shape has advantages when developing rules for the synthesis of consonants.

The HL parameters are transformed in the synthesizer into a larger set of LL parameters that control a Klatt-type synthesizer [KK90], as shown in figure 16.2. The transformation is achieved by a set of equations or mapping relations that

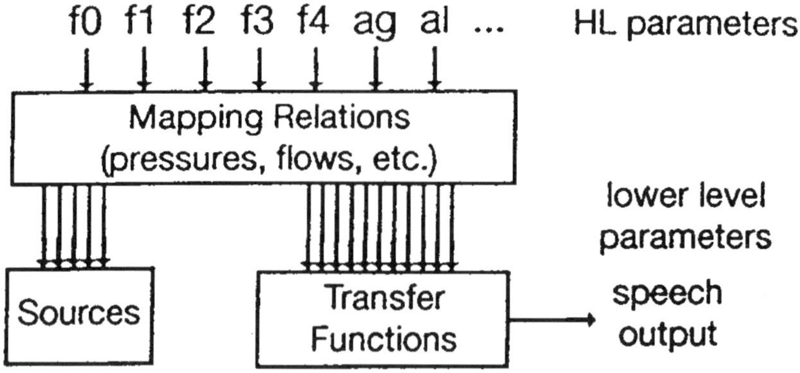

FIGURE 16.2. The HL parameters are mapped into intermediate parameters of airflows and pressure in the oral cavity, which in turn are mapped into the low-level parameters that control the sources and transfer functions of KLSYN88, a Klatt-type synthesizer [KK90].

relate the orifice sizes and formants to the characteristics of the periodic glottal source, aspiration noise, frication noise, and poles and zeros of the vocal-tract transfer function.

Thus each of the LL parameters is calculated from a set of equations in which the HL parameters are the arguments. To illustrate this process, we review the calculation of the source parameters *AF* (amplitude of frication noise), *AH* (amplitude of aspiration noise generated at the glottis), and *AV* (amplitude of glottal vibration source) for obstruent consonants. In the derivation of these LL source parameters, the first step is to calculate the pressures and flows at the supraglottal and glottal orifices. The calculation is based on a model that relates flow through an orifice to the pressure drop across the orifice, and includes yielding walls and a yielding glottis [Rot68, Ste93b]. The orifice sizes are specified by the HL parameters, which vary with time in a way that depends on the place of articulation and voicing characteristics of the consonant. The supraglottal constriction area is given by *al* or *ab* for a lip or tongue-blade constriction; it is calculated from the formant parameters when the major constriction is formed by the tongue body. The minimum of these constriction areas (*acx*) is used in the model for the calculation of pressures and flows. The glottal area (*agx*) is given by *ag*, modified by forces due to an increased supraglottal pressure. When the constriction areas and intraoral pressure P_m have been calculated from the model, the values of the LL source parameters are calculated from the following equations:

$$\begin{aligned} AF &= 20\log[\mathrm{K}_f P_m^{1.5} acx^{0.5}] \\ AH &= 20\log[\mathrm{K}_h (P_s - P_m)^{1.5} agx^{0.5}] \\ AV &= 20\log[\mathrm{K}_v (P_s - P_m)^{1.5}] \end{aligned}$$

where K_f, K_h, and K_v are constants that are adjusted to give the correct relative amplitudes of the sources, and P_s is the subglottal pressure. The LL parameters *AF*, *AH*, and *AV* are, respectively, the relative amplitudes (in dB) of the frication noise source, the aspiration noise source, and the glottal vibration source. The above equations are based on theoretical and empirical observations of source amplitudes and their relations to pressures, flows, and orifice sizes (cf. [Ste71, Sha85, Lad62]). The equation for *AV* is further modified to account for a decreased amplitude as *ag* is increased (corresponding to a spread glottis) and to reflect cessation of vocal-fold vibration when *agx* exceeds a certain threshold and when the transglottal pressure $P_s - P_m$ decreases below a threshold value. The mapping relations also contain equations that modify the open quotient *OQ* and the spectral tilt *TL* as functions of the glottal area. These equations are again based on theoretical models of vocal-fold behavior, as well as empirical data [Ste95].

In addition to these equations relating *KL* source parameters to vocal-tract orifices, there are equations that specify (1) changes in the formant bandwidths as a function of the glottal opening and velopharyngeal opening; (2) modification of the first-formant frequency by the glottal opening, tongue-blade or lip opening, and velopharyngeal opening; (3) the frequencies and bandwidths of a pole-zero pair as a function of the velopharyngeal opening; and (4) the amplitude of frication noise excitation of individual formants depending on the place in the vocal tract where noise is generated (as determined by the activity of *al* or *ab* or as inferred from the HL formant parameters).

16.3 Examples of Synthesis from HL Parameters

Here we illustrate the process of synthesizing consonants involving the HL area parameters *al* and *ab* (areas of lip and blade opening) in a vowel–consonant–vowel (VCV) sequence, which has concurrent adjustment of the velopharyngeal opening (*an*), the cross-sectional area of the glottis (*ag*) and the rate of change of the volume of the vocal-tract cavity (*ue*). We also describe the adjustment of supraglottal constrictions and glottal opening for a sequence of two obstruent consonants.

In the case of a VCV sequence such as that in [ʌpɔ] as in "upon," the parameter specifications are simple: once the class 1 parameters have been established, only one class 2 parameter (*al*) and one class 3 parameter (*ag*) need to be specified. The *al* and *ag* parameters for generating this sequence are shown in figure 16.3(a). The lip opening decreases rapidly to zero, remains at zero for about 80 ms, and then increases rapidly. The glottal opening *ag* increases gradually through the closure interval, reaches a maximum near the release time, and then decreases to a modal value. The formant parameters (not shown) follow trajectories appropriate for a labial consonant.

For the sequence [ʌmɔ] the HL parameter tracks are similar to those needed for "upon," with the parameter *ag* set to the modal value, and with the addition of the specification of the parameter *an* for the nasalization of the consonant, as shown in figure 16.3(b). A sequence such as [ʌbɔ] can be synthesized using HL parameter tracks, which are again similar to those for [ʌpɔ], with differences in the parameter

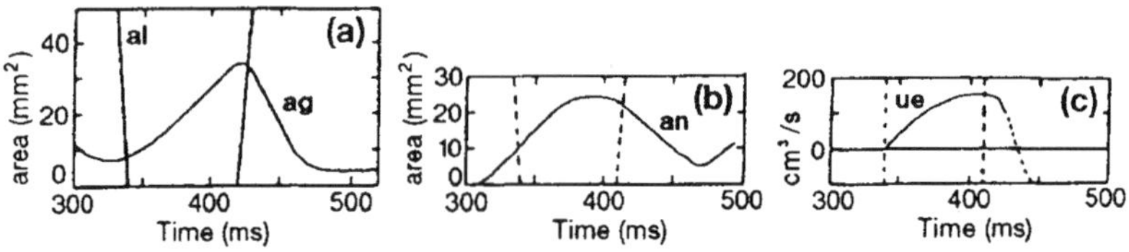

FIGURE 16.3. HL parameters as specified by rule for synthesis of various labial consonants: (a) areas of lip and glottal openings for [ʌpɔ]; (b) area of velopharyngeal opening for [ʌmɔ] with *al* shown by dashed lines; (c) parameter *ue* (vocal-tract expansion) for [ʌbɔ] with vertical dashed lines indicating times of consonantal closure and opening. See text for details.

ag (maintained at a value to indicate modal voicing throughout the consonantal closure), and the addition of the parameter *ue* to enable vocal-fold vibration to continue during part or all of the constriction interval. A typical time course for *ue* is given in figure 16.3(c). Variations on this pattern for *ue* would effect more or less voicing during the constriction interval.

Synthesis of a consonant cluster such as [əsvɑ] in "curious volume" requires only the specification of a narrowing of the tongue-blade constriction (HL parameter *ab*) followed by the lip constriction (*al*). For this cluster, the glottal opening (*ag*) is formed during the first member of the sequence, and then returns to a modal value when the second constriction is formed. (Several examples of the synthesis of consonant clusters have been presented earlier [BSW94].)

These examples illustrate some of the simplifications that are achieved in specifying the control parameters when HL parameters are used for the synthesis of consonants. Once the locations of a consonantal closure and release are established, the trajectories of the rapidly changing parameter indicating the area of the consonantal constriction at the closure and at the release (i.e., *al* for lip and *ab* for tongue-blade constrictions) can be specified. These are the class 2 parameters listed in section 16.1 above. Changes of the class 3 parameters describing the movements of secondary articulators (i.e., parameters *ag*, *an*, and *ue*) must be timed relative to the closing and opening landmarks, as shown in figure 16.3. Perceptual data obtained from listener judgments of stimuli in which these class 2 and class 3 consonantal parameters are manipulated over a range of values indicate that some latitude is permitted in the specification of these parameters [Wil94].

The use of the HL virtual formant parameters (i.e., the class 1 parameters noted above) also leads to a simplification of consonant synthesis. We note that the HL formant parameters are identical for the three utterances /ʌpɔ/, /ʌbɔ/, and /ʌmɔ/, because the movements of the tongue body and hence the parameters *f1*, *f2*, *f3*, and *f4* are the same for all three sequences. The actual low-level formant parameters that control the Klatt synthesizer must be, of course, different for the three utterances. The glottal spreading in the vicinity of the landmarks for /p/ causes an increase in the first-formant frequency and bandwidth, but these shifts are taken care of in the mapping equations that transform the HL parameters into LL parameters. Likewise, for the consonant /m/, manipulation of the velopharyngeal opening area automatically causes a shift in the first-formant frequency and in formant bandwidths, as well as the introduction of an additional pole-zero pair. But the HL formant parameters can be the same for both /p/ and /m/, since the HL formants are defined to be the natural frequencies of the vocal tract before the effects of a glottal opening or a velopharyngeal opening are taken into account.

The parameters used to synthesize the cluster /sv/ in /əsvɑ/ illustrate another type of simplification that is achieved through the use of HL parameters. The time course of the area parameters *ab* and *al* for the individual consonants in the cluster are essentially the same as they would be if the consonants were to occur singly in intervocalic position. The parameters *ab* for /s/ and *al* for /v/ are simply concatenated in such a way that the increase of *ab* at the tongue-blade release is coincident with the decrease of *al* for the lip closure. The HL parameter *ag* follows the normal open-closing pattern for /s/ but, because the calculated intraoral pressure maintains an increased level through the time of the labial constriction, the actual glottal opening remains wide during a portion of the labial constriction, and glottal vibration does not resume until later in the labial region. Again we have a rather complex series of acoustic events that are a consequence of a rather simple concatenation of the stylized patterns of HL parameters.

16.4 Toward Rules for Synthesis

The examples in the previous section have illustrated that relatively stylized patterns of HL parameters can be used for synthesis of many of the consonants, and that some of these patterns are largely independent of the environment in which the consonants occur. This synthesis experience has led us to develop a framework for the rules that are needed to generate the HL parameters when a linguistic description of the utterance is given. We assume that this description is given in terms of a phonetic transcription (or transcription in terms of hierarchical arrangements of features), together with a specification of syllables as having a nuclear pitch accent, a nonnuclear pitch accent, no accent but not reduced, reduced, and having a particular boundary tone. We concentrate in this chapter on rules for generating the segmental component of the synthesis for consonant segments in various phonetic environments.

To illustrate our progress in the development of the rules, we describe the synthesis of the phrase "once upon a midnight." The phonetic string /wʌnsəpɔnəmɪdna^{y}t/ is shown at the top of table 16.1, without the prosodic markings. A subset of the relevant features for this sequence is displayed under the phonetic units. The feature markings are those used by [Ste93a]. The phonetic transcription and prosodic markings are used to generate a series of landmarks. Each [+consonantal] phonetic segment corresponds to two landmarks: one at the time of the formation of the constriction and one at the time of the release. A vocalic segment is indicated by one landmark for monophthong vowels, and by two landmarks for diphthongs

TABLE 16.1. The phonetic string corresponding to "once upon a midnight" and relevant feature markings.

	w	ʌ	n	s	ə	p	ɔ	n	ə	m	ɪ	d	n	aʸ	t
consonantal	−	−	+	+	−	+	−	+	−	+	−	+	+	−	+
continuant			−	+		−		−		−		−	−		−
sonorant			+	−		−		+		+		−	+		−
lips						+				+					
tongue blade			+	+				+				+	+		+
round	+					−				−					
anterior			+	+				+				+	+		+
spread glottis				+		+						−			−

(one for the main vowel and one for the offglide). For a glide, one landmark is specified. Each of these landmarks is tagged with a set of descriptors indicating the values of the relevant features for the segment corresponding to the landmark. A set of timing rules governs the placement of the landmarks.

In this example, the timing of the landmarks is determined by hand. The locations of the landmarks for the utterance are shown by arrows at the top of figure 16.4. The time variation of the four area parameters for the synthesis of the sentence is displayed in the upper two panels of figure 16.4. (The parameter *ue* is not used in this sentence.) The bottom panel displays the HL formant parameters *f1* and *f2*.

For each landmark, the HL parameters are specified by rules that depend on the features associated with the landmark and with adjacent landmarks. For the synthesis of this sentence, the parameter tracks for the HL formant parameters and for the *f0* parameter are entered by hand, except for times immediately adjacent to consonantal landmarks. The parameters for the four areas and for *ue* (when required) are, however, determined by rule. These are the parameters corresponding to class 2 and class 3 in the listing in section 16.1.

In general, the rules for the HL parameters *al* and *ab* share many characteristics. For example, there is a series of alternations of rapid closures and releases for stop consonants, with the closure interval for singleton stops or nasals being 80–90 ms, shorter times when the segment is in a cluster (/n/ in "once") and when the consonant is flapped (/n/ in "upon a"), and longer times for a homorganic sequence (/dn/ in "midnight"). The rate of closure and release of *al* and *ab* for stops is the same (50 cm^2/s). For fricatives the area reaches a broad minimum of about 0.1 cm^2, and the rate of movement is slower. The rule-based trajectory for a fricative preceded by a nasal is illustrated by the parameter *ab* in the time interval 220–320 ms. It is noted that in the schwa vowel in the sequence "once upon," the parameter *ab* reaches a value of only 60 mm^2 before *al* decreases (325 ms), so that this short vowel is never free of the influence of a consonantal constriction.

For each pair of consonant landmarks, a class 3 parameter (*an*, *ag*, or *ue*) is usually needed. In this example, there are four nasal consonants, so there are four peaks in the *an* parameter. For example, in the first vowel, the rules prescribe that *an* begins to increase in the first vowel (at 50 ms), reaches a peak of 30 mm^2 at the time the alveolar closure is made (150 ms), and then decreases to zero (at 190 ms)

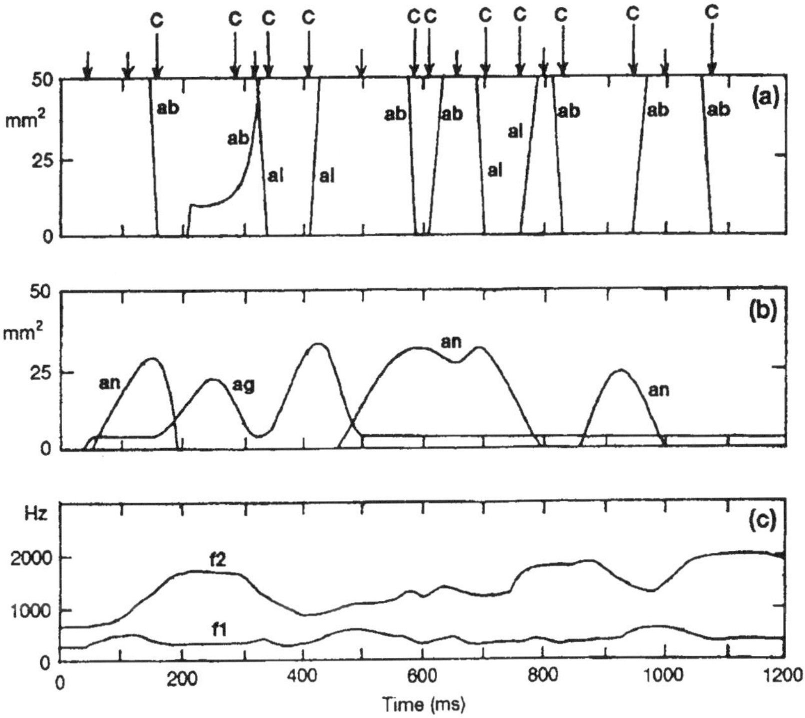

FIGURE 16.4. (a) HL tracks *al* and *ab* for "once upon a midnight" (top panel); (b) HL tracks *ag* and *an* (middle); (c) HL tracks *f1* and *f2* (bottom). (The HL parameter *f0* is not shown in this figure.) The locations of landmarks are indicated by arrows at the top of the figure. The paired abrupt consonantal landmarks are labeled C.

just before the release of the tongue blade to form the fricative. For the sequence "upon a mid," *an* begins to increase about 30 ms following the /p/ release, decreases after the /n/ release (610 ms), but never achieves velopharyngeal closure before increasing again, so that it reaches a second maximum near the closure for /m/ (at about 670 ms). Similarly, the parameter *ag* begins to increase (160 ms) before *ab* increases (210 ms) to form the fricative following the nasal in "once upon." A voiceless fricative requires an increased glottal opening relative to the modal value needed for the /n/. The parameter *ag* must also increase for the voiceless /p/, so that it reaches a maximum near the /p/ release (420 ms). Because the rate of change of *ag* is limited, *ag* is unable to return to its modal value in the unstressed vowel in "once upon," and consequently this vowel is produced with a somewhat spread glottis.

As we have observed, the combined set of parameters in figure 16.4, when processed by the mapping relations as in figure 16.2, lead to a specification of the LL parameters that control the Klatt synthesizer. The result of the synthesis is a highly intelligible phrase.

As noted above, the HL formant parameters were entered by hand for this utterance, except near the consonant landmarks, where *f2* and *f3* were required to pass through prescribed regions appropriate for each consonant. Rules for generating the formant parameters from a feature-based description of the type given in table 16.1 are currently under development. These rules will be based heavily on past synthesis-by-rule efforts with a formant synthesizer [Kla87].

This work was supported in part by grant #MH52358 from the National Institutes of Health.

References

[BSW94] C. A. Bickley, K. N. Stevens, and D.R. Williams. Synthesis of consonant sequences using a Klatt synthesizer with higher-level control. *J. Acoust. Soc. Amer.* 95(2):2815, 1994.

[KK90] D. H. Klatt and L.C. Klatt. Analysis, synthesis, and perception of voice quality variations among female and male talkers. *J. Acoust. Soc. Amer.* 87(2):820–857, 1990.

[Kla87] D. H. Klatt. Review of text-to-speech conversion for English. *J. Acoust. Soc. Amer.* 82:737–793, 1987.

[Lad62] P. Ladefoged. Subglottal activity during speech. In *Proceedings of the Fourth International Congress of Phonetic Sciences.* Mouton, The Hague, 73–91, 1962.

[Rot68] M. Rothenberg. The breath stream dynamics of simple-released-plosive production. *Bibliotheca Phonetica, No. 6.* S. Karger, Basel, 1968.

[SB91] K. N. Stevens and C. A. Bickley. Constraints among parameters simplify control of Klatt formant synthesizer. *J. Phonetics* 19:161–174, 1991.

[Sha85] C. Shadle. *The Acoustics of Fricative Consonants.* RLE Technical Report 506, Massachusetts Institute of Technology, Cambridge, MA, 1985.

[Ste71] K. N. Stevens. Airflow and turbulence noise for fricative and stop consonants: Static considerations. *J. Acoust. Soc. Amer.* 50:1180–1192, 1971.

[Ste93a] K. N. Stevens. Lexical access from features. In *Speech Technology for Man-Machine Interaction*, P. V. S. Rao and B. B. Kalia, eds. Tata McGraw-Hill, New Delhi, 21–46, 1993a.

[Ste93b] K. N. Stevens. Models for the production and acoustics of stop consonants. *Speech Comm.* 13:367–375, 1993b.

[Ste95] K. N. Stevens. *Phonetics*, forthcoming.

[WBS92] D. R. Williams, C. A. Bickley, and K. N. Stevens. Inventory of phonetic contrasts generated by high-level control of a formant synthesizer. In *Proceedings of the International Conference on Speech and Language Processing*, Banff, Canada, 571–574, 1992.

[Wil94] D. R. Williams. Modeling changes in magnitude and timing of glottal and oral movements for synthesis of voiceless obstruents. *J. Acoust. Soc. Amer.* 95(2):2815–2816, 1994.

17

Biomechanical and Physiologically Based Speech Modeling

Reiner F. Wilhelms-Tricarico
Joseph S. Perkell

ABSTRACT Improvements in speech synthesis may be achieved through an increased understanding of the actual physiology and control of speech production. Toward this end, a three-dimensional dynamic finite-element tongue model is described, which is the first component of a research project directed at a physiologically based computer simulation of speech production.

17.1 Introduction

It has been hoped for decades that speech synthesis based on articulatory geometry and dynamics would result in a breakthrough in quality and naturalness of speech synthesis, but this has not happened. It is now possible to generate very high quality synthetic speech, such as with the Klatt synthesizer, by modeling only the properties (spectral, etc.) of the *output* signal. Although physiological and physically based rules have been developed for reducing the large number of degrees of freedom of the Klatt synthesizer [SB91], we speculate that there is potential for even further improvement in speech synthesisers, which would be realized with the deeper understanding gained from modeling of the actual properties of the vocal-tract and the neural control system.

Our goal is to build a computational vocal-tract model that approximates the individual three-dimensional geometry of the vocal tract and attempts to simulate the individual's movements during speech, taking the biomechanics and muscle physiology into account as much as possible. The output of the model, parameters describing speech kinematics and acoustics, can then be made to emulate the original measurements of kinematic and acoustical parameters of the individual speaker. Therefore, the model will be a tool for quantitative investigations of the relation between the physical and physiological properties of the vocal tract plant and the neural controller. Although this research is not primarily grounded in speech synthesis, we believe that speech synthesis can benefit from the improved understanding of underlying mechanisms that will come from biomechanical articulatory modeling and understanding of speech motor control.

17.2 Articulatory Synthesizers

Most previous articulatory models have been two-dimensional geometric representations of the midsagittal vocal tract. It is often assumed that studies of limb motor control benefit from understanding the mechanics of the limbs; this idea has not yet taken hold in studies of speech motor control. Several attempts have been made to overcome the merely kinematic nature of articulatory models by introducing *some kind* of dynamics, and a number of "articulatory synthesizers" have been developed (cf. [Hen67, Mer73, Cok76, Mae82, SS87, SM89]). However, even the most advanced of these efforts has not incorporated the actual three-dimensional anatomy, biomechanics of the vocal tract structures, and the kind of hierarchical nonlinear control mechanism that will almost certainly be required.

Current research on articulatory modeling has begun to appreciate the usefulness of the three-dimensional information that magnetic resonance imaging (MRI) and other methods can provide. Several reconstructions of three-dimensional vocal tract configurations in vivo have been presented, and work on three-dimensional models of the vocal tract is in progress in several laboratories. From such research it becomes possible to take into account the structure of individual vocal tracts. Such geometric information, together with information on tissue properties, forms the basis for modeling realistic dynamic behavior. The dynamic three-dimensional tongue model described below is a step in this direction.

17.3 A Finite Element Tongue Model

To model computationally the movement (including deformation) of a soft tissue body such as the tongue, the body is subdivided into finite elements that have simple shapes. The methods are applicable to the lips as well as the soft palate. For example, in previous work, hexahedral elements with 8 nodes (irregular bricks) have been used to represent the tongue (see [Wil94]).

17.3.1 Modeling Soft Tissue

For the description of motion of the body, two representations are used. An arbitrary configurational state is used as the reference configuration. The deformed configuration is described relative to this reference configuration. The time-variant mapping of the position of each point with respect to the reference configuration onto its position in the deformed configuration constitutes the movement and deformation of the body. The state of motion (including deformation) can be represented by a displacement vector field and displacement velocity vector field relative to the reference configuration. Stress and force fields are also formulated as functions of time and location of the points with respect to the reference configuration. The equations of motion that govern the deformation process can be formulated in terms of the displacement and velocity fields. In the numerical approximation, the displacement and velocity fields within each element are piece-wise continuously interpolated between the field values at the element nodes. With this discretization process, the theoretically infinite number of equations of motion are approximated by a finite number of equations, which are obtained by lumping the force fields at the element nodes. In the 8-node elements, trilinear interpolation methods are utilized. For 27-node elements, which are planned for future extensions of the model, triquadratic interpolation methods are used. In both cases, the interpolation is continuous to at least first order and continuous at the interelement boundaries.

The result of the discretization is a nonlinear second-order system of ordinary differential equations with nonconstant coefficients, in which the variables are the displacement vectors and the displacement velocities for all degrees of freedom in the system. There are three degrees of freedom for most nodes. This number may be reduced, for nodes that move only along a surface or nodes whose motion is completely prescribed, for example, by virtue of being attached to the jaw.

The system of equations relates the forces and displacement accelerations at each node. The control of this system is achieved by specifying external muscle activation levels for each muscle in the modeled body of soft tissue. In particular, the human tongue consists of some passive tissues and at least 12 different muscle pairs, which are partially interwoven. It is modeled as a continuum in which the fine details of the tissue structures are not included. Briefly, the following simplifications are made:

- Individual interdigitating bundles are not modeled. Instead the continuum is assumed to possess several spatial directions in which tensile stress can be generated independently.
- Nonmuscular tissues (such as glands and fat) are assumed to be isotropic.
- Nerve fibers and blood vessels are ignored.
- The tissue is modeled as completely incompressible.

Muscle fibers are represented as continuous, directed fields in which active tensile stress can be produced. Figure 17.1 shows an example of such a representation. During the movement, the local orientation of the muscle fibers is taken into account by updating the direction of the muscle fiber fields during the computation. At each point in the tongue body, several muscle fiber fields with different directions can overlap to represent interdigitation of muscles. The tissue is modeled as incompressible; this condition is maintained for each finite element during the simulation by computing an element-wise constant hydrostatic pressure field (in the case of the 8-node element) assuring the volume constancy of the element. In the case of the triquadratic element, the pressure fields will be computed with linear terms in each spatial direction.

Muscle Models

Since an accurate model of tongue muscle tissue that includes its biomechanical properties is not available, a pragmatic phenomenological muscle model was

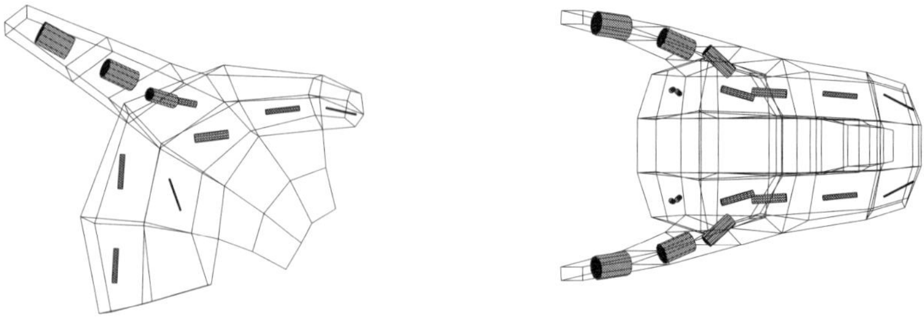

FIGURE 17.1. Display of the fiber directions of the Styloglossus muscle. Left: lateral projection; right: view from the top. The tongue tip is on the right. The direction of muscle fibers is symbolized by cylinders. The diameter of the cylinders is proportional to the fiber density parameters in the model.

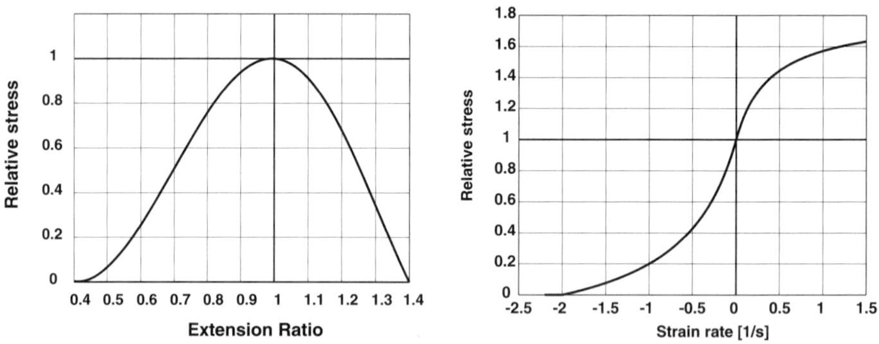

FIGURE 17.2. Left: a stress mobilization function as a function of the muscle fiber extension ratio (stretch); right: the dependence of the active stress on the rate of contraction for a constant stretch of 1. At negative strain rates the muscle fibers are contracting. The strain rate is positive if the muscle fibers are elongated by external forces. The total stress along the fibers, which is a function of activation, strain, and strain rate, is obtained by multiplying the two curves and by multiplying the result with an activation level and with a constant maximal stress.

adopted. The stress in a muscle tissue has two components: active and passive stress. The passive stress is modeled as a nonlinear-elastic, linear-viscoelastic response of the tissue to deformation.

The active component is computed in a stress production model that takes into account the elongation and the rate of elongation or shortening of the muscle fibers. The left half of figure 17.2 shows, for one arbitrary level of activation, the dependence of the stress on the strain[1] along the direction of the muscle fibers. The function has a maximum at the relative rest length of the fibers. The right half of the figure shows the dependency of the stress on the strain rate in the direction of the muscle fibers. On the abscissa, zero corresponds to rest; the negative range represents contraction. The positive range represents the behavior when the fibers are stretched due to external forces.

Because the changing muscular activation levels modify the constitutive equations of the muscle tissue, they indirectly influence the stress field in the continuum, which is computed based on the instantaneous strain and rate of strain. This results in a modification of the node forces by the muscle activation levels. The motion of the approximated continuum is obtained by specifying initial conditions (position and velocity) for each node and integrating the equations of motion numerically. Thus, the varying muscle activity levels constitute a multidimensional parametric control.

17.3.2 Outline of the Research on the Tongue Model

The presented tongue model has been used as a first test case for some of the methods for simulating the behavior of soft tissue articulators. Several compromises and simplifications were introduced to obtain a model that keeps the computational burden within manageable limits so that the model could be tested on state-of-the-art small computers. Its main limitations are that only eight of the muscles of the tongue are represented and that the spatial discretization into finite elements is still too approximate. Currently, the input to the model, the activation levels of the muscles, have to be specified manually. Figure 17.3 shows a simulation example of a tongue depression, for which three muscles were activated. Although the feasibility of the methods can be shown, the model does not yet simulate realistic tongue movements. More realistic simulations will be possible after several refinements. It is necessary to represent more accurately the anatomy of the tongue and floor of the mouth; and to include a moving mandible and hyoid bone; and to include modeling the interactions between the tongue body and the constraining walls of the oral cavity. A higher number of elements will be used to allow better definition of individual muscles and differentiation between different tissue types. From a strictly biomechanical perspective, a proposed number of 150-200 elements may be considered too coarse a representation; however, this number may provide enough detail to further the objectives of the overall modeling effort.

Following these improvements, a master model will be developed that allows subject-specific customization by use of MRI and other data, so that simulations with the model can be compared to data from the individual speaker. In the master model, the shapes of the tongue and other organs will be complemented with

---

[1] Stress is measured in the same units as pressure, namely as force per area (Newton/m^2 or dyne/cm^2). Strain is a measure of deformation, measured as the ratio of change of length divided by reference length. Stretch is defined as the ratio of length divided by reference length. The strain rate is the rate of change of strain or stretch, in units of 1/s.

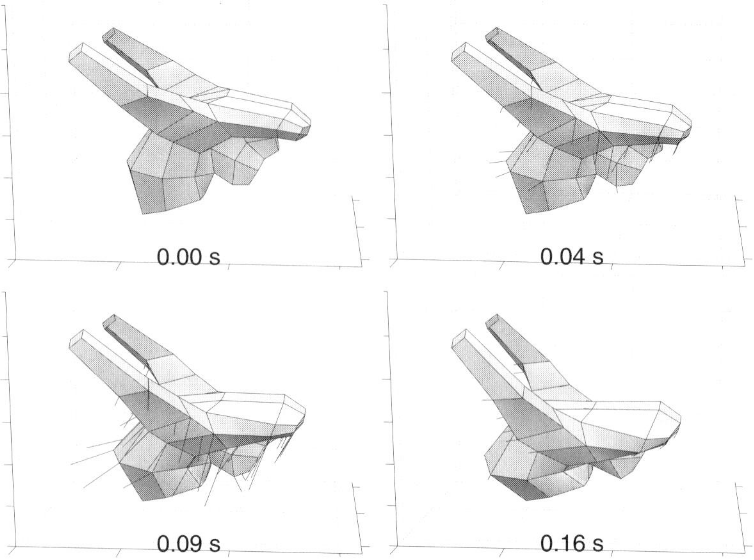

FIGURE 17.3. First 160 ms of a lowering gesture computed with the following constant activations, switched on at time 0 s: genioglossus anterior 0.2, genioglossus post 0.1, and hyoglossus 0.8. The tongue reaches a new equilibrium at about 250 ms after some damped rocking movement.

muscle fiber directions from anatomical drawings. With current MRI methods such detailed anatomical data cannot be obtained for an individual. The combination of gross shape information of soft tissue articulators from MRI data with small-scale structural information from anatomical drawings, such as Miyawaki's [Miy74] sectional drawings of the tongue, can be achieved by using a warping mapping (the thin-spline mapping method, see [Boo91, WW95]) that deforms the drawings such that their outer contour fits with tongue shapes obtained from MRI data.

Another important step is an assessment of biomechanical parameters. To approximate values of constitutive parameters such as maximal muscle stress (force per unit cross-sectional area) and maximal velocity of contraction, information from the literature is used. For example, it is generally agreed that tongue tissue is not significantly different from other skeletal muscle. Thus, if necessary, missing information can be provided from other, probably related, skeletal musculature (for examples of laryngeal musculature, see [KHO81]). In addition, mechanical experiments on subjects are planned in order to augment and refine information obtained from the literature about parameter values that determine the biomechanical behavior. For example, the displacements of a number of points on the tongue tip are tracked optically while the tip is displaced with an external force transducer. Other experiments may record a combination of EMG (electromyogram), movement, and force. These experiments will include a variety of subject tasks, such as pushing the tongue tip against a force transducer with various effort levels. The resulting information will be used for tuning the muscle model.

The complete vocal tract model will approximate the anatomy and biomechanical properties of an individual speaker's vocal tract. Its time-variant input is a vector of muscular activation levels, and its outputs are various kinematic variables describing the changing vocal tract configuration. In further extensions, an acoustic output will be added. The biomechanical model allows, at various stages, the computation of energy flows, muscular efforts, muscle lengths, and contraction or extension rates. Further, in the completed biomechanical model it will be possible to compute the moments in time when, as a result of muscle control signals, collisions between the tongue and the hard palate occur, and it will be possible to compute vocal tract parameters such as the forces exerted on the hard palate. Such physical observables may serve as a basis for studies of effort of articulation, for example, for different speaking rates and degrees of reduction. Certain physical observables of the system also can be converted into (hypothesized) sensory information. This information can be used in designing and tuning a planned control system, as will be outlined below.

17.4 The Controller

Having a sufficiently realistic simulation of the biomechanics of the vocal tract does not solve any interesting questions by itself. On the contrary, it will confront us with an ill-posed problem of motor control that has been solved by the co-evolution of the human central nervous system (CNS) and the speech apparatus. It would be desirable to apply neurophysiological principles in designing the control structure, but there is currently very little that can be used as a foundation for implementing quantitative neurologically based models of speech motor control. This is a long-range possibility and may be relevant for research on speech pathology in the future. However, the availability of a biomechanical vocal tract model makes it possible to investigate more general speech motor control models because assumptions about the controlled plant are replaced by explicit properties of the biomechanical model.

It is generally agreed that the CNS planning of movement is organized by orchestrating muscle groups as synergies that, in the example of limb movement, may contain several agonist and antagonist pairs, rather than by controlling individual muscles. The results by Munhall, Löfqvist and Kelso [MLK94], which showed that laryngeal articulation could be influenced by perturbing the lip movements, support further the idea that "coordinative structures" exist on levels that deal with multiple, biomechanically uncoupled articulators. The key to understanding such structures of motor control seems to be a unification of sensory and motor information processing. Online orosensory information modifies the execution of the motor plan, as is demonstrated in speech production in studies in which the synergisms involving upper and lower lip and jaw movements are investigated using perturbations (see [AGC84, KTV84]), or recently in the above-cited similar studies of lip larynx coordination [MLK94]. It has been concluded in these studies that the motor program includes the planning of how sensory information is used at the lowest levels to cause compensatory adjustments among multiple articulators.

The organization and linking of articulators may involve online use of both internal feedback and peripheral sensory information in the form of a "sensory template" that is specific for a motor control task. A sensory template is defined, according to Burgess [Bur92] as "a central representation of the sensory receptor discharge that would be expected to occur during a movement if the movement is executed according to the plan." Sequences of articulatory and acoustic goals may be accomplished by a hierarchical control system that dynamically sets up synergisms between agents that are less coupled (physically and neurologically) in higher control tiers than in lower control tiers. Such a system achieves a reduction of the number of degrees of freedom to be controlled at each succeedingly higher level.

It should be noted that such an understanding is not in contradiction with the idea that speakers figure out individually how to organize their multiarticulator system for producing the "correct" vowel *acoustics*: Johnson, Ladefoged and Lindau [JLL93] found that midsagittal vocal tract configurations are consistent within a speaker but vary between speakers; the interspeaker variability could not be explained by anatomy, gender, or any other factors. One may conclude that the domain of the actual underlying goal may differ, depending on the type of speech sound, with some goals for consonants being specified, at least partially, in articulatory terms, and goals for vowels in acoustic terms.

To aid in the preprogramming of speech movements, it has been postulated that speakers acquire an internal model and use it to predict the sensory results of the neural controller's actions. This *forward model* emulates the causal flow of information from the controller to the resulting action, including the sensory response that follows from the actions (see [JR92]). The controller is structured such that it can convert its input, consisting of acoustic and articulatory goals, into commands that act upon the vocal tract. The resulting actions produce orosensory, proprioceptive, tactile, and auditory sensations. On the various levels of this forward model, the difference between ascending sensory signals and predicted sensory signals is a source of information that may be used to adjust the performance of the controller. Such adjustments do not necessarily take place with closed-loop, moment-to-moment use of feedback. It is hypothesized that the feedback is used to maintain ("validate" or "calibrate") the internal model. For a more detailed discussion of these ideas, see [Per94].

17.4.1 *Considerations About the Structure of the Controller*

Even at the current stage of development of the vocal tract model, it is possible to think about the general structure of a control mechanism that will compute the muscular activation time functions to steer the biomechanical model. This structure is not understood in terms of actual neurological functionality but rather as "software" that should simulate plausible functional components of the neurological controller. The controller transforms input signals that are described in terms of desired acoustical and/or articulatory goals into muscular activations, which cause movement of the biomechanical plant.

As in other models of speech production, the tasks that presumably underlie the resulting speech movements have to be postulated for the current approach. Ideas and decisions about subdividing the controller into levels, and into subunits that act in parallel at each level, interact strongly with decisions about the representation of the task space.

The purpose of the controller, as described "in a nutshell" by Perkell [Per94] is to control multiple constrictions to determine the aerodynamics and acoustics of the vocal tract, multiple articulator movements for each constriction, and multiple muscles for each articulatory movement. This hierarchy can be expressed by assuming three levels of the controller. The lowest level incorporates control structures that generate synergistic muscle actions that result in simple gestural movements, such as raising the tongue blade or rounding the lips. The next level orchestrates the "elemental gestures" (see also [Fuj93])[2] of the lower level to perform articulatory tasks that can be best described as creating vocal tract constrictions with certain characteristics (manners), for example creating an appropriate constriction for a vowel or producing a bilabial closure or a dento-alveolar constriction for a fricative. The third level, which orchestrates both lower levels, receives input signals that are described in terms of both desired acoustical consequences of articulation and/or as articulatory goals directly. The selection of acoustic goals and articulatory goals at the highest level comprises the translation of a hypothetical symbolic representation of speech into control actions.

Jordan and Rumelhart's distal learning strategy will be used as a paradigm for the implementation of each level of the controller, starting at the lowest level. The psychological and neurophysiological idea of the internal model, or efferent copy, appears in this strategy as a forward model. The tentative general structure, shown in figure 17.4 has two components. One component (the controller *C*) maps from "intentions" (i) to motor commands (u), and the other component, called the forward model (*FM*), from motor commands to predicted sensations ($\hat{s}$). The forward model is trained using the difference between predicted sensation and actual sensation (s), which arise in the plant (the biomechanical system - *P*) as the result of the controlled actions. The composite system (*C* and *FM*) is trained using the difference between the desired sensations (d) and the actual sensation. See [JR92] for further details. In this context, figure 17.4 is a sketch of the first-level controller. Because the biomechanical plant (*P*) is a dynamic system, the internal model of the lowest control level will also be a dynamic system. The plant transforms the motor control input u and its current state x into two types of sensory results, s, and $h(x,u)$, which makes the plant's state x (consisting of all displacements and velocities) par-

[2] The elemental gestures can be understood similar to Fujimura's elemental gestures in the C/D model: Elemental gestures correspond one-to-one to articulatory dimensions which are specified by a combination of (i) articulator (e.g., tongue tip, tongue dorsum); (ii) action (e.g., raising, retraction); and (iii) manner (e.g., lateral, rounded).

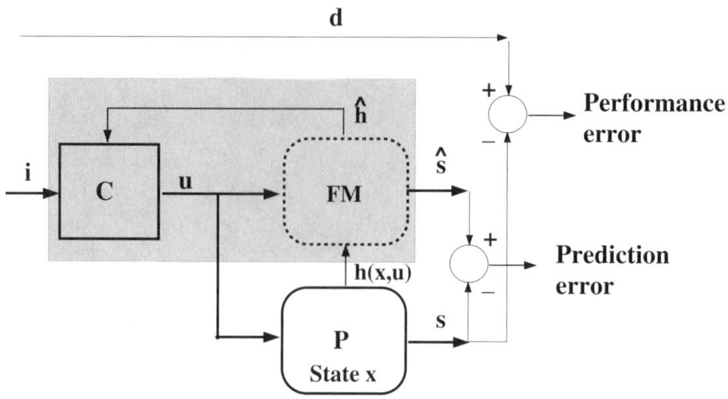

FIGURE 17.4. The composite learning strategy (adopted and modified from [JFA93]). The controller (C) transforms intentions (i) into motor commands (u). The forward model (FM) predicts $\hat{s}$ as the sensory result of the action u on the biomechanical plant. The difference between the predicted sensation $\hat{s}$ and the actual sensation s, the prediction error, is used to optimize the forward model during training. The forward model is sensitive to the state of the plant; it receives input $h(x, u)$ that contains information about the state (for example, muscle length information). Once the forward model is sufficiently accurate, the performance error (that is, the difference between desired sensations, d, and sensations, s), is then used to train the controller. The performance error is transformed into a control error by using a conjugate forward model in which the signal flow is inverted. The controller receives estimated information about the state of the plant $\hat{h}$, which is generated in the forward model, based on its partial representation of the plant. This enables the controller to take the state of the plant into account in movement planning.

tially observable by the forward model. The signal $h(x,u)$ may contain measures of the length and rate of change of the muscles. The forward model learns to map motor commands (u) and current observables of the state ($h(x,u)$) into estimated sensory output ($\hat{s}$). It further learns to predict the development of relevant plant state-related functions, shown as $\hat{h}$. Once the forward model is trained, the actual controller (C) relies on the estimated information ($\hat{h}$) about the state of the plant. This amounts to "internalizing" a feedback loop in the controller-forward model composite system. In applications on arm control models, similar control systems have been implemented as partially recurrent neural networks (see [JFA93]). In the cited example, the involved forward model receives as input the full state information of the controlled system. Particularly in the case of the tongue, there are very many degrees of freedom. The assumption about the role of the forward model may thus be replaced by the following view: The purpose of the internal model is not to simulate the plant, rather to "mimic" certain aspects of it. It needs to represent sufficient information to allow the prediction of the result of an action onto the plant. This information is represented by the observable ($h(x,u)$).

It is unreasonable to assume that a single forward model could be generated in one step for use in control of the complex biomechanical model of the entire vocal tract. In view of the complexity of the biomechanical model and the speech motor

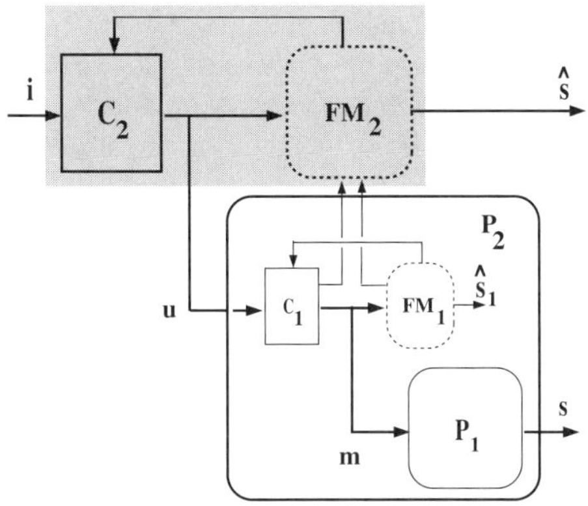

FIGURE 17.5. The second level of the controller, consisting of the subcontroller C_2, generates higher-level motor commands u and controls the plant P_2, which is an encapsulated system consisting of the biomechanical model and the lowest-level controller C_1. The biomechanical plant P_1 is driven by muscle activation functions m, which are generated by the lower-level controller C_1. To train the forward model 2, either the difference between the predicted sensation $\hat{s}$ and the predicted sensation $\hat{s}_1$, or the difference between predicted sensation $\hat{s}$ and actual sensation s can be used. The controller C_2 is trained based on comparing the predicted sensations $\hat{s}$ with desired sensations associated with the intentions i.

control task, it will certainly be necessary to subdivide the overall control problem on each level into smaller ones. Subdivision on the lowest level is particularly sensible because the biomechanical plant consists of parts (e.g. the tongue body, tongue blade, mandible, lips, velum) that can act quasi-independently. Subdivision at the next higher level is motivated by the possibility for control of constrictions at different locations along the vocal tract with different manners. Another motivation for subdivision into subcontrollers for special tasks, in particular on the second level, is seen in recognized structures for interarticulatory coordination such as between jaw and lips. Jordan and Jacobs [JJ93] have extended their previously proposed competing experts paradigm to constitute a hierarchical architecture [the "hierarchical mixture of experts" (HME) paradigm] that allows such a subdivision of control problems. The subdivision process is in principle automatic, but can be aided by initial biases which follow from "natural" subdivisions of the controlled plant into previously identified articulatory structures (on the lowest level), as outlined above, and coordinative structures (on the middle level).

By starting to build the controller from the bottom to the top, from simple movement models to complex movement models, each level of the controller is designed to achieve a reduction of complexity and degrees of freedom for the higher level controllers. For example, subcontrollers in the second level do not have to care about and operate without knowledge of individual muscle states, because that knowledge is incorporated into the lowest level of control. The internal model built into the lowest level of control includes a partial (but sufficient) representation of the biomechanical model. The next higher level operates on an "encapsulated" lower level, and its internal model includes a representation of the effects of combined generalized motor commands (such as "tongue tip raising" and "tongue back lowering") on sensory output, but only to the extent that it is relevant for the second level. Stated differently, the higher levels of the controller receive more general intentions, issue more global motor commands, and have actions that result in more abstract sensations. Figure 17.5 sketches this scheme for the two lower levels. The controller of the second level C_2 controls the augmented plant P_2, which consists of the lower-level controller C_1 and the biomechanical model. Before controller C_2 can be trained properly, the forward model FM_2 of the middle level needs to be trained to include a partial representation of the augmented system P_2, including the prediction of state-related information of the controller in P_2. The third level of the proposed hierarchy will operate on a plant formed by encapsulating the presented structure, and augmenting it further by adding another level of acoustical ouput resulting from computations in an extended biomechanical and acoustical model.

17.5 Conclusions

We have presented the outline of a biomechanical vocal tract model for speech motor control research, and we have described the first realized component, a preliminary three-dimensional finite element model of the tongue. The purpose of the model is to approximate realistically the anatomy and relevant physiological aspects of the speech production apparatus of individual speakers, in order to make explicit the biomechanical constraints of the vocal tract system on speech production. This approach amounts to posing a control problem of considerable complexity, but one that actually approximates the problem faced by the CNS. For solving this problem, several possible solutions may exist. Because there are not enough facts established to design a control system based on neuro-physiological insights, we propose an alternative strategy that is based on a model of learning (the distal teacher framework), general insights on limb and speech motor control (coordinative structures), and efficient models of information processing and representation (hierarchical mixture of experts).

References

[AGC84] J. H. Abbs, V. L. Gracco, and K. J. Cole. Control of multimovement coordination: Sensorimotor mechanisms in speech motor programming. *J. Motor Behavior* 16(2):195–231, 1984.

[Boo91] F. L. Bookstein. *Morphometric Tools for Landmark Data: Geometry and Biology*. Cambridge University Press, Cambridge, 1991.

[Bur92] P. R. Burgess. Equilibrium points and sensory templates. *Behavioral and Brain Sciences* 15(4), 1992. Open Peer Commentary to Bizzi et al. (1992), Does the nervous system use equilibrium-point control to guide single and multiple joint movements?

[Cok76] C. H. Coker. A model of articulatory dynamics and control. In *Proceedings IEEE* 64:452–460, 1976.

[Fuj93] O. Fujimura. C/D model: A computational model of phonetic implementation. In *DIMACS Proceedings*, E. S. Ristad, ed. American Mathematical Society, 1993.

[Hen67] W. L. Henke. Preliminaries to speech synthesis based on an articulatory model. In *Proceedings of the 1967 IEEE Boston Speech Conference*, 170–177, 1967.

[JFAss] M. I. Jordan, T. Flash, and Y. Arnon. A model of the learning of arm trajectories from spatial targets. *J. Cognitive Neuroscience*, in press.

[JJ93] M. I. Jordan and R. A. Jacobs. *Hierarchical Mixtures of Experts and the EM Algorithm*. Computational Cognitive Tech. Rep. 9301, Massachusetts Institute of Technology, 1993.

[JLL93] K. Johnson, P. Ladefoged, and M. Lindau. Individual differences in vowel production. *J. Acoust. Soc. Amer.* 94(2):701–714, 1993.

[JR92] M. I. Jordan and D. E. Rumelhart. Forward models: Supervised learning with a distal teacher. *Cognitive Science* 16:307–354, 1992.

[KHO81] Y. Kakita, M. Hirano, and K. Ohmaru. Physical properties of the vocal fold tissue: Measurements on excised larynges. In *Vocal Fold Physiology*, K. N. Stevens and M. Hirano, eds. chapter 25. University of Tokyo Press, Tokyo, 1981.

[KTVBF84] J. A. S. Kelso, B. Tuller, E. Vatikiotis-Bateson, and C. A. Fowler. Functionally specific articulatory cooperation following jaw perturbations during speech: Evidence for coordinative structures. *J. Experimental Psychology: Human Perception and Performance* 10(6):812–832, 1984.

[Mae82] S. Maeda. A digital simulation method of the vocal tract system. *Speech Comm.* 199–229, 1982.

[Mer73] P. Mermelstein. Articulatory model for the study of speech production. *J. Acoust. Soc. Amer.*, 53(4):1070–1082, 1973.

[Miy74] K. Miyawaki. A study of the musculature of the human tongue. *Ann. Bul. Research Institute of Logopedics and Phoniatrics, Univ. Tokyo* 8:23–50, 1974.

[MLK94] K. G. Munhall, A. Löfqvist, and J. A. S. Kelso. Lip-larynx coordination in speech: Effects of mechanical perturbations to the lower lip. *J. Acoust. Soc. Amer.* 95(6):3605–3616, 1994.

[Per94] J. S. Perkell. Articulatory processes. *A Handbook of Phonetic Science*, J. Hardcastle and J. Laver, eds. 1994, in press.

[SB91] K. N. Stevens and C. A. Bickley. Constraints among parameters simplify control of Klatt formant synthesizer. *J. Phonetics* 19:161–174, 1991.

[SM89] E. L. Saltzman and K. G. Munhall. A dynamical approach to gestural patterning in speech production. *Ecological Psychology* 1:333–382, 1989.

[SS87] M. M. Sondhi and J. Schroeter. A hybrid time-frequency domain articulatory speech synthesizer. *IEEE Transactions on Acoustics Speech and Signal Processing* ASSP-37(7):955–967, 1987.

[Wil94] R. Wilhelms-Tricarico. Physiological modeling of speech production: Methods for modeling of soft-tissue articulators. *J. Acoust. Soc. Amer.* 97(5):3085–3098, 1995.

[WWT94] C.-M. Wu and R. Wilhelms-Tricarico. Tongue structural model: Integrating MIR data and anatomical structure into a finite element model of the tongue. In *Proceedings of the XIII International Congress of Phonetic Sciences ICPHS 95*, Stockholm, Sweden, 490–493, 1995.

Appendix: Video Demo

Tongue protrusion movie.

This movie was generated by activating some of the muscles of the tongue model with the following activation levels (in percent of maximal strength): Genioglossus anterior, 30%; genioglossus posterior, 30%; styloglossus, 40%; and transversalis and verticalis, 90%. The strong contraction of the transversalis and verticalis muscles compresses the tongue in directions that are approximately perpendicular to the length axis of the tongue. Because the total volume of the tongue is held constant, the tongue body lengthens. The combined effect of the contraction of the verticalis muscles, genioglossus anterior, and styloglossus results in bucking up of the tongue body and formation of a groove on the tongue back near the tip of the tongue.

Tongue lowering movie.

This movie shows a tongue depression gesture that was generated by activating the genioglossus posterior and anterior slightly (20%) and the hyoglossus with 90% of its maximal strength. The hyoglossus muscles are on both sides of the tongue and pull it toward the hyoid bone, back and down. In the model, the hyoid bone is represented only as a fixed plane at the bottom of the model. Because the model overshoots its new equilibrium point after activating the muscles, a dampened oscillatory movement can be observed.

18

Analysis-Synthesis and Intelligibility of a Talking Face

Bertrand Le Goff
Thierry Guiard-Marigny
Christian Benoît

ABSTRACT Analytic measurement of visual parameters relevant to the labial production of speech as well as real-time, 3D-computer-animated models of the lips and of the face have been implemented on two coupled computers, so that synthetic lips alone or a whole facial model can mimic on line (or play back) the actual gestures of a natural speaker. The geometric measurements performed on the speaker's lips and jaw are made through image processing of the front and profile view of the speaker's face. Data are transmitted to a graphics computer through a control interface, which delivers the proper parameters to control the animation of the 3D models. The lip model uses five control parameters; the facial model uses one extra: jaw lowering. At present, the tongue is not controlled. We present the real-time techniques used for analysis, animation of the 3D models, and synchronization of the two processes. Finally, we evaluate the bimodal intelligibility of speech under five levels of acoustic degradation by added noise. We compare the intelligibility of the speech signal presented alone, with the lip model, with the facial model, and with the original speaker's face. Our results confirm the importance of visual information in the perception of speech: the whole natural face restores two-thirds of the missing auditory intelligibility when the acoustic transmission is degraded or missing; the facial model (tongue movements excluded) restores half of it; and the lip model alone restores one-third of it.

18.1 Introduction

Parametric models of the face aim at providing the viewer with the maximum quantity of information carried by the speaker's face through a small number of commands. In Japan, Aizawa et al. [AHS89] developed a system for analysis-synthesis of faces that allows a low-bit rate transmission of the differences between adjacent images at 10 images per second (ips) and of the parameters relevant to the facial expressions based upon the facial action coding system (FACS [EF78]). In North America, Terzopoulos and Waters [TW91] used a contour analysis technique to detect specific deformations of the face in order to animate a parametric model of the face. In France, Saulnier et al. [SVG94] elaborated a system for real-time detection of the global movement of human heads as well as of their most salient expressions and lip gestures for tele-virtuality purposes. Here we present a system for analysis-synthesis of speaking faces in which a small number of parameters allow lip and jaw gestures to be accurately reconstructed. First, we present the models of the lips and of the face that we used, as well as their control parameters. We then describe the acquisition technique used for parameter measurement based on video processing of a speaker's face. Finally, we discuss the performance of our analysis-synthesis system in terms of the intelligibility of the visual information, and we compare it as a function of the kind of visual display used.

18.2 The Parametric Models

The high-resolution model of the lips we used is presented in chapter 19 paper by Guiard-Marigny et al. in this volume. It is controlled through five parameters that are easy to measure on a speaker's face: width (A), height (B) of the internal lip contour, lip contact protrusion (C), upper lip ($P1$) and lower lip ($P2$) protrusion.

The model of the face that we used was first designed by Parke [Par74] (figure 18.1). It has been implemented on a SGI graphics computer and improved for speech production by Cohen and Massaro [CM93, CM94]. It is animated through controls related to physiological gestures, e.g., "raise chin," "raise lower lip," "jaw thrust," and so on. Our objective was to animate this model as well as possible from the above-mentioned parameters of the lip model. A control interface has thus been developed to predict the original commands of the face model from only six parameters that are easy to measure on a speaker's face. Five parameters were used in the control of the lip model. An extra one was necessary for the face model, namely the chin vertical displacement (M). The interface used mostly linear combinations of parameters. It allows different face and animation styles to be generated, e.g., large versus narrow face, hypo- versus hyperspeech, and the like. The original lips were replaced with the high-resolution lip model in the face model. This greatly improves the control of the face model with our six parameters.

18.3 Video Analysis

The speaker is filmed from front and side by two video cameras (figure 18.2). His lips are made up with a blue color (chroma-key blue makeup). Dots are also painted on his chin and on the goggles he wears. This eases image processing, which is based on chrominance rather than on luminance. A chroma-keyer transforms the blue color into saturated black so that the lip vermilion area and the chin dots are darker than any other pixel on the screen. This makes automatic analysis much simpler and more accurate. After digitizing the video frames on a gray scale, the software developed by Lallouache [Lal91] allows the contours of the lips and of the chin dot to be extracted on each field, i.e., 50 ips. Then, geometric

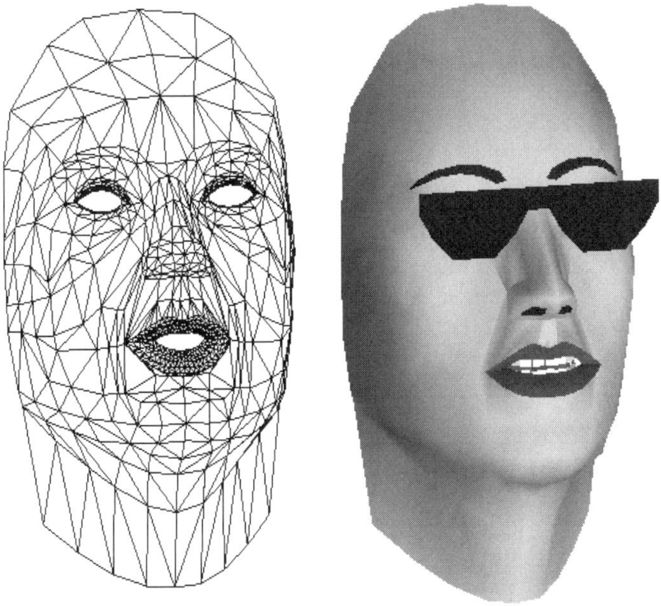

FIGURE 18.1. Modified version of Parke's model. Left: wireframe structure; right: Gouraud-shaded rendering

measurements are made on each contour, as seen on the bottom panel of figure 18.2. A version of this labiometric software, delivering highly accurate measurements, is installed on a PC-based workstation connected to a VCR. A real-time version of this software has also been installed on a SGI computer directly connected to the cameras [ALGAB94]. Due to hardware limitations of the VideoStarter Board used, there is a maximum input rate of 13 ips.

18.4 Real-Time Analysis-Synthesis

The real-time analysis-synthesis system presented in figure 18.3 is currently implemented on two SGI computers connected by Ethernet. One computer is used for video analysis and the other for image synthesis. The video signal comes either from the cameras through the preprocessing devices or from a VCR. The command parameters are transmitted at 12.5 ips. The graphics computer thus receives a new set of parameters every 80 milliseconds (ms). Optionally, the parameters can be linearly interpolated between two images so that the models are synthesized at a video rate (25 ips). In parallel, the acoustic signal is directly digitized on the graphics computer where it is delayed in order to resynchronize the audio and video signals. The audio-visual delay due to the whole analysis-synthesis process is on the order of 200 ms. Of course, it can be dramatically decreased depending on the hardware used.

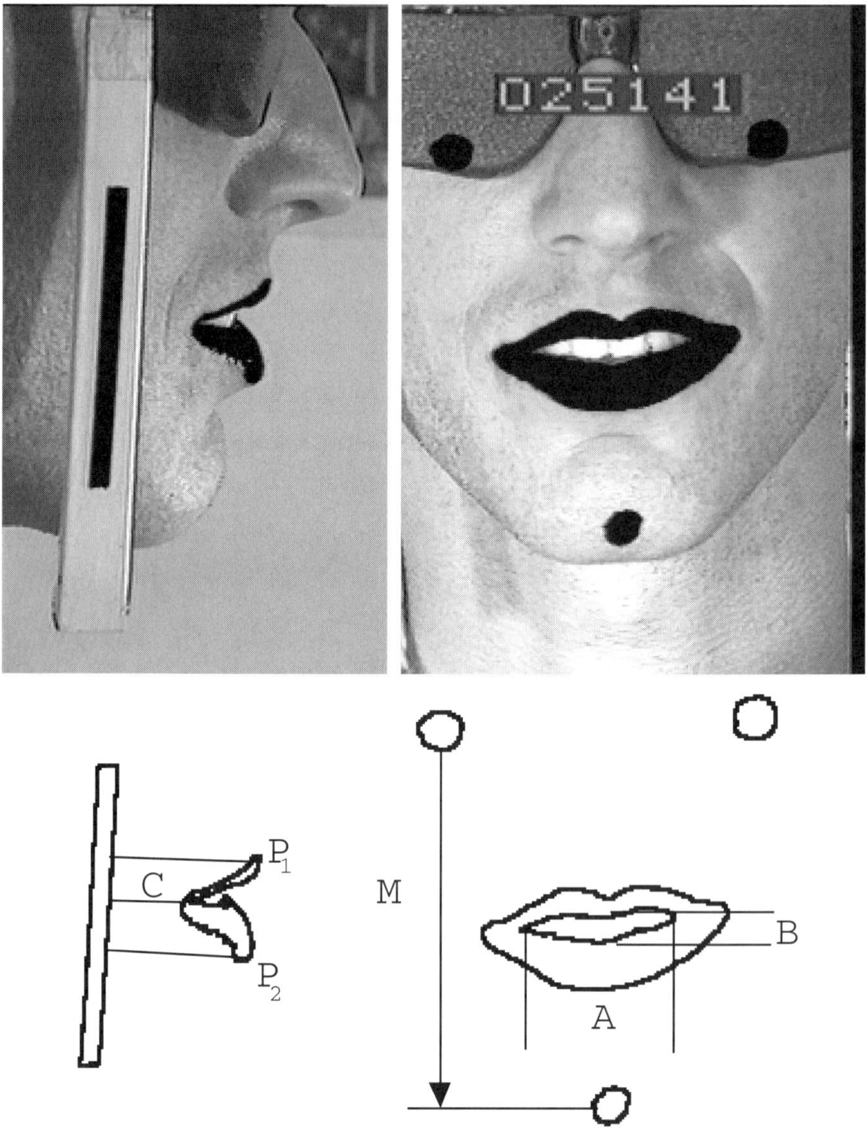

FIGURE 18.2. Speaker filmed from front and side (profile view): original image (top panel), and geometric analysis (bottom panel).

18.5 Intelligibility of the Models

Extending the experiment by Benoît et al. [BMK94], the audiovisual intelligibility of the face model and of the lip model have been quantified under five conditions

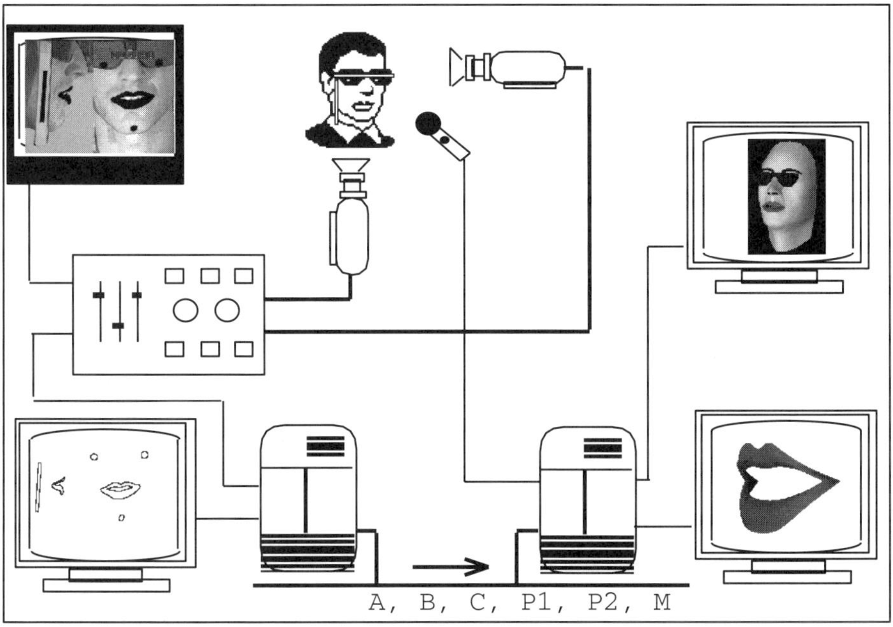

FIGURE 18.3. Schematic of the real-time analysis-synthesis process for the animation of a speaking face.

of acoustic degradation. These five conditions are ranging from −18 dB to +6 dB, by steps of 6 dB (−18, −12, −6, 0, +6).

18.5.1 *Preparation of the Stimuli*

The speech material consisted of the natural acoustic utterances of a French speaker and of four kinds of visual display: (1) no video, (2) natural face, (3) synthetic lips, and (4) synthetic face. The two synthetic models were animated from parameter files so that no delay affected the original synchrony between audio and video.

The corpus was made of VCVCV nonsense words. V was one of the three French vowels /a/, /i/ or /y/. C was one of the six French consonants /b/, /v/, /z/, /ʒ/, /R/ or /l/. The test words were embedded in a carrier sentence of the form "C'est pas VCVCVz ?".

1. **No video**: The 18 different sentences were first digitized. They were then acoustically degraded by the addition of white noise, at five S/N levels, by 6 dB steps. There were overall 90 audio stimuli. A pseudorandom order was used for presentation. Ten extra stimuli were appended before the actual test so that subjects could adapt to the test conditions.

   These acoustic stimuli served as a reference to the next three experimental conditions for which the natural face or the synthetic models were simply synchronized to the audio part.

2. **Natural face**: The original video recording of the speaker was digitized and compressed on a PC through a VIDEIS board. The front view of the lower part of the face, from the neck to the middle of the bridge of the nose, was displayed on a 15-inch monitor. The actual width of the whole face on the screen was roughly 10 cm. The video rate was 25 ips (PAL format) with a VHS-like quality. Audio stimuli were post-synchronized with the image display. A visual-alone condition was added to the five audiovisual conditions. This subtest had 108 stimuli.
3. **Synthetic lips**: The lip model was Gouraud shaded and animated at 50 ips on the 19-inch monitor of an SGI Elan. It was displayed at a 20-degree angle view from the sagittal plane. The actual width of the lips on the screen was roughly 10 cm. The audio file controlled the display of the model, one image being calculated every 320 audio samples.
4. **Synthetic face**: The face model was Gouraud shaded and animated at 25 ips on the 19-inch monitor of an SGI Elan. It was displayed at a 20-degree angle view from the sagittal plane. The actual width of the whole face on the screen was roughly 10 cm. The digital audio files controlled the display of the model, one image being calculated every 640 audio samples.

Procedure : Fourteen normal-hearing French subjects took part in the experiment. The order of presentation of the four subtests was balanced across the subjects. Each subtest lasted 20 minutes. Each subject ran no more than two subtests per half-day. Subjects answered through a keyboard in the "natural face" test. They answered with the mouse on the screen in the other tests. Subjects were requested to respond to both the vowel and the consonant as much as they could guess it. A "?" response was tolerated, however.

18.5.2 Global Intelligibility

A test word was first considered correct only if both the vowel and the consonant were correctly identified. As for auditory and visual intelligibility of natural stimuli, the results obtained in this experiment are in agreement with those by Benoît et al. [BMK94]. Adding the video image of the natural face shows a dramatic gain in intelligibility over presenting the audio alone, as seen in figure 18.4. The lip model and the face model also contribute strongly to improve the intelligibility of auditory speech. Scores in lipreading conditions are not presented in figure 18.4 simply because they are strictly identical to those obtained under the most degraded acoustic condition (S/N = -18 dB). In fact, no acoustic cues could even be detected at this noise level. Figure 18.4 shows that the synthetic lips account for one-third of the intelligibility carried by the whole natural face, whatever the acoustic degradation. The synthetic face accounts for the two-thirds of it.

The contribution of the synthetic lips and of the synthetic face to visual speech intelligibility is certainly impressive when considering the large contribution of the whole natural face. Five parameters are sufficient to animate the lip model alone. Even without the teeth, the tongue, the chin, and the skin, the intelligibility carried by the lip model is striking, and this is obtained with a very small quantity of information. As for the face model, a sixth parameter by itself almost doubles

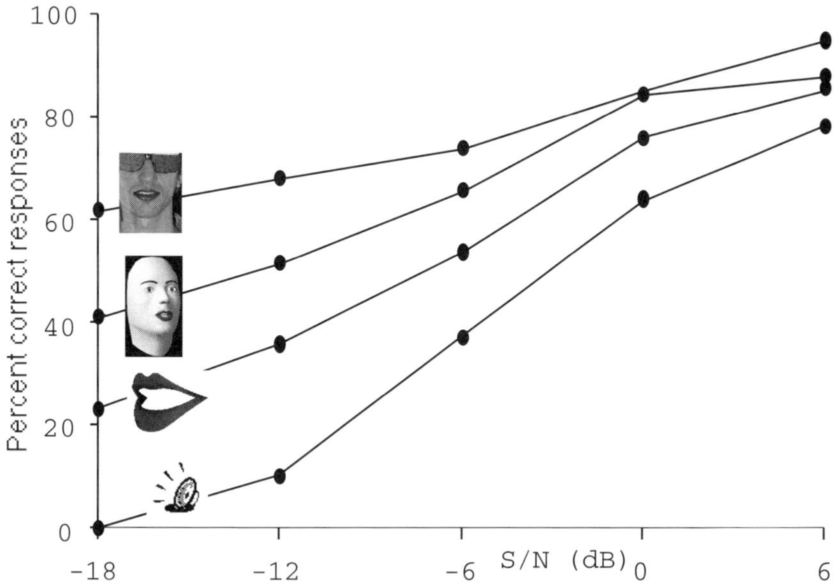

FIGURE 18.4. Intelligibility scores obtained by 18 subjects in the identification of 18 stimuli, as a function of acoustic degradation, depending on the mode of presentation: audio alone, audio plus the lip model, audio plus the face model, audio plus the whole natural face (from bottom to top).

the visual intelligibility provided by the synthetic lips to the perceiver. Here again, there is no control of the tongue, but the teeth and the chin are animated, and the structure of the face is coherently displayed. The visual information provided by the teeth and the chin allows subjects to disambiguate confusions among spread vowels (/i/ versus /a/), which are largely mixed up through the lips alone.

Sumby and Pollack [SP54] proposed an index of the visual contribution to the missing auditory information: (I[AV]-I[A])/(1-I[A]) where I[AV] and I[A] are the audio-visual and audio intelligibility scores in a given S/N condition. Figure 18.5 shows the evolution of this index along the acoustic degradation at the three S/N conditions where all differences in intelligibility are significant, i.e., between -18 dB and -6 dB, for the three audio-visual conditions. The index is remarkably constant over the acoustic conditions of degradation.

Overall, the whole natural face restores two-thirds of the missing information when the acoustics is degraded or missing; the facial model (tongue movements excluded) restores one-half of it; and the lip model restores one-third of it. This is strong evidence that a very low bit rate of information (five or six parameters 25 times per second) is sufficient to transmit a great deal of the visual information carried on by the speaker's natural face, even though tongue gestures are not yet controlled.

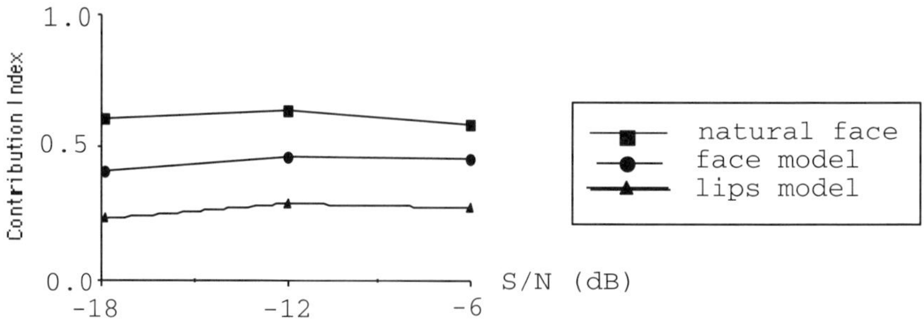

FIGURE 18.5. Contribution index of the visual information to missing acoustic information.

TABLE 18.1. Confusion matrices of consonants, irrespective of the response on the vowel (S/N = −12 dB). Stimuli are presented in rows. Percepts are presented in columns. Scores are out of a posssible 42.

Audio only

	b	ʒ	l	R	v	z	?
b	**11**	5	1	3	5	1	16
ʒ	3	**10**		7	1	5	16
l	6	3	**1**	7	4		21
R	4	1	2	**13**	3	3	16
v	6	1	2	7	**5**	1	20
z	4	7	2	1	3	**2**	23

Natural face

	b	ʒ	l	R	v	z	?
b	**38**				4		
ʒ		**35**	1	1	3	2	
l		3	**25**	10	1	2	1
R		6	3	**25**	1	7	
v	3	3		1	**34**		1
z		9	1	1	2	**28**	1

Lip model

	b	ʒ	l	R	v	z	?
b	**39**	1			1		1
ʒ		**10**	6	17	4		5
l			**16**	21	1		4
R	1	2	10	**21**	4		4
v	18		4	1	**13**		6
z	1	4	3	7	5	**18**	4

Face model

	b	ʒ	l	R	v	z	?
b	**36**		1		3		2
ʒ		**30**	3	3		3	3
l		2	**14**	21	1	1	3
R	2		14	**19**	1	3	3
v	7	2	4	4	**21**	1	3
z		6	6	5	7	**15**	3

18.5.3 Consonant Confusions

Consonant confusions are presented in table 18.1 at S/N = −12 dB where differences are at their maximum.

Whatever the consonant, there is a very strong disambiguation due to visual information. The disambiguation power follows the same hierarchy as that of global intelligibility, from the lips alone to the natural face, through the synthetic face.

- /b/ is the consonant best identified audio-visually, although it is given as a response to many /v/ stimuli, especially with the lip model (43%). The absence of teeth is obviously the reason for these confusions.

- /ʒ/ identification is not improved when vision of the lip model is added to audio. However, /ʒ/ is rather well identified with the synthetic face or with the natural face. With the lip model, there are many confusions between /ʒ/ and /R/ in spread-vocalic context. Moreover, not only is /ʒ/ never identified in a /i/ context, but it leads subjects to identify the vowel /i/ as an /a/ (/iʒiʒi/ is perceived as /aRaRa/). Adding the chin and the teeth disambiguate both the carrier vowel and the carrier consonant.

- The two liquids /l/ and /R/ are mixed up in all conditions. A main reason is obviously that there is no tongue associated with the lip model, and that the tongue is not controlled in the face model. Even with the human face, confusions occur in /i/ and /y/ contexts, when the lip opening is too small for subjects to see the vertical tongue movement characteristic of /l/.

- /z/ is auditorily identified below chance, and /ʒ/ is then the most frequent response. This is obviously due to the background noise used. Surprisingly enough, our lip model helps subjects to disambiguate /z/ and /ʒ/ better than the face model. Nevertheless, a significant amount of /ʒ/ responses to /z/ stimuli remains when a natural face is presented. In fact, these confusions occur only in the /y/ context. /y/ has such an important coarticulatory effect that all consonants (except /b/ and /v/) look very similar when surrounded by two /y/'s. Therefore, audiovisual confusion of consonants presented in a /y/ context is mostly based on auditory similarities.

18.5.4 Vowel Confusions

The vowel confusions are presented in table 18.2 at S/N = -12 dB where differences are at their maximum. Complementarity between audition and vision is clearly seen in table 18.2. In the auditory mode, /a/ is seldom confused with /i/ or /y/, whereas /i/ and /y/ are largely confused with each other. Conversely, /y/ is seldom confused with /a/ or /i/ in the (audio-)visual mode, whereas there are many confusions between /a/ and /i/ in the audio-visual mode.

As stated above, there is almost no auditory confusion between /a/ and /i/. However, when the lip model is simultaneously displayed, an /a/ response is given to an /i/ stimulus in 50% of the cases, whereas /a/ is almost always identified. This effect remains to a smaller extent with the natural face, where an /i/ stimulus still leads to /a/ responses in 12% of all cases. Of these latter confusions, 70% are observed with /izizi/ (perceived as /azaza/ by half of the subjects.) In fact, this is mostly due to the individual utterances /izizi/ and /azaza/ selected as stimuli for our experiment. Benoît et al. [BLMA92] showed that /a/ has a smaller lip and jaw opening and takes the shape of an /i/ when surrounded by /z/. They also noticed that /z/ has the same shape when surrounded by /i/ or /a/. Those statistical observations

TABLE 18.2. Confusion matrices of vowels, irrespective of the response on the consonant (S/N = −12 dB). Stimuli are presented in rows. Percepts are presented in columns. Scores are out of a possible 84.

Audio only

	a	i	y	?
a	**52**	2	4	26
i	4	**25**	34	21
y	2	12	**48**	22

Natural face

	a	i	y	?
a	**82**	1		1
i	10	**72**		2
y			**84**	0

Lip model

	a	i	y	?
a	**76**	3	2	3
i	35	**35**	8	6
y		1	**80**	3

Face model

	a	i	y	?
a	**75**	9		
i	2	**82**		
y	1	1	**82**	

were obtained from a multidimensional analysis of 10 utterances of /izizi/ and /azaza/. We looked back at the original data. It turns out that the /azaza/ used in our intelligibility test is among those with the largest lip and jaw opening. The /izizi/ here selected is also the one with the largest lip opening (and with an average jaw opening.) Differences in lip opening between our two stimuli /izizi/ and /azaza/ are in the range of 1 mm. Corresponding differences in the jaw opening are in the range of 0.5 mm. It is thus not surprising that this hyperarticulated /azaza/ has been correctly identified, whereas half of the subjects perceived the hyperarticulated /izizi/ as another /azaza/ when they had the opportunity to see the natural face. This effect is emphasized with the lip model, in which subjects cannot see the (even small) jaw movements. More surprisingly, the face model allows subjects to correctly identify /i/ on the one hand, and to respond with an /i/ percept to an /a/ stimulus. Those errors occur when /a/ is coarticulated with labial consonants in two-thirds of the cases. The fact that /izizi/ is not here perceived as /azaza/ is probably due to an insufficient control of the chin displacements by the face model.

18.6 Conclusion

A first prototype of an analysis-synthesis model of a speaking face runs in real-time at the ICP. The synthesis module uses only five anatomical parameters to animate the lip model developed at the ICP (Institut de la Communication Parlée). It uses six parameters to animate Parke's model of the face in which the ICP lip model has been integrated. Those parameters are easily measured on the front and profile views of a speaker's face through video analysis. The intelligibility carried on by these parameters well complements the intelligibility of the acoustic signal. The quantity of information to be transmitted is very low, on the order of a few hundred bits per second. Therefore, many applications can be foreseen in the area of multimodal telecommunications as well as in the automatic animation of synthetic actors. Furthermore, this analysis-synthesis process allows the influence of a given parameter to be tested in order to better understand the psychophysics of speech perception by the eye, because each control parameter can be individually modified, either automatically or by hand, before being applied to one model or the other. When used off-line, the analysis module allows temporal modifications to be processed, for instance, so that the influence of rate could be tested in the bimodal perception of speech.

Acknowledgments: This study was supported by the French CNRS and by the European ESPRIT-BRA project No. 8579 "MIAMI."

We thank Christian Abry, Ali Adjoudani, Omar Angola, Alain Arnal, Marie-Agnès Cathiard, Mike Cohen, Sonia Kandel, Tahar Lallouache, and Jean-Luc Schwartz for technical and scientific support.

References

[Adj93] A. Adjoudani. Elaboration d'un modèle de lèvres 3D pour animation en temps réel. In *Mémoire de D.E.A. Signal Image Parole*, Institut National Polytechnique, Grenoble, France, 1993.

[AHS89] K. Aizawa, H. Harashima, and T. Saito. Model-based analysis-synthesis image coding (MBASIC) system for a person's face. *Signal Processing: Image Communication* 1:139–152, 1989.

[ALGAB94] O. Angola, B. Le Goff, T. Guiard-Marigny, A. Adjoudani, C. Benoît, and M. Cohen. Analyse-synthèse de visages parlants. In *Proceedings of 20emes Journées d'Etude sur la Parole*, Societe Française d'Acoustique, Tregastel, France, 1994.

[BMK94] C. Benoît, T. Mohamadi, and S. Kandel. Effects of phonetic context on audio-visual intelligibility in French. *J. Speech & Hearing Res.* 37:1195–1203, 1994.

[BLMA92] C. Benoît, M. T. Lallouache, T. Mohamadi, and C. Abry. A set of French visemes for visual speech synthesis. In *Talking Machines: Theories, Models and Designs*, G. Bailly and C. Benoît, eds. Elsevier Science Publishers B.V., North-Holland, Amsterdam, 485–504, 1992.

[CM93] M. M. Cohen and D. W. Massaro. Modeling coarticulation in synthetic visual speech. In *Proceedings of Computer Animation93*, Magnenat-Thalmann and Thalmann eds. Geneva, Switzerland, 1993.

[CM94] M. M. Cohen and D. W. Massaro. Development and experimentation with synthetic visible speech. *Behavioral Research Methods, Instrumentation, & Computers* 26:260–265, 1994.

[EF78] P. Ekman and W. Friesen. *Facial Action Coding System.* Consulting Psychologists Press, Palo Alto, CA, 1978.

[GAB94] T. Guiard-Marigny, A. Adjoudani, and C. Benoît. A 3D model of the lips. In *Proceedings of the 2nd ETRW on Speech Synthesis*, New Platz, NY, 1994.

[Gui92] T. Guiard-Marigny. Animation en temps réel d'un modèle paramétrisé de lèvres. In *Mémoire de D.E.A. Signal Image Parole*, INP, Grenoble, France, 1992.

[Lal91] M. T. Lallouache. *Un Poste "Visage-parole" Couleur. Acquisition et Traitement Automatique des Contours des Lèvres*. Doctoral thesis, de l'Institut National Polytechnique de Grenoble, 1991

[LeG93] B. Le Goff. Commandes paramétriques d'un modèle de visage 3D pour animation en temps réel. In Mémoire de D.E.A. Signal Image Parole, Institut National Polytechnique, Grenoble, France, 1993.

[Par74] F. I. Parke. *A parametric model for human faces*. Ph.D. Dissertation, University of Utah, Department of Computer Sciences, 1974.

[SP54] W. H. Sumby and I. Pollack. Visual contribution to speech intelligibility in noise. *J. Acoust. Soc. Amer.* 26:212–215, 1954.

[Sum79] Q. Summerfield. Use of visual information for phonetic perception. *Phonetica*, 36:237–344, 1979.

[SVG94] A. Saulnier, M. L. Viaud, and D. Geldreich. Analyse et Synthèse en temps réel du Visage pour la Télévirtualite. In *Actes de la Conférence Imagina '94*, Institut National de l'Audiovisuel, 173–182, 1994.

[Ter91] D. Terzopoulos and K. Waters. Techniques for realistic facial modeling and animation. In *Computer Animation '91*, N. Magnenat-Thalmann and D. Thalmann, eds. Springer-Verlag, Tokyo, 59–74, 1991.

Appendix: Video Demos

The CDROM contains video demos of our system.

19

3D Models of the Lips and Jaw for Visual Speech Synthesis

Thierry Guiard-Marigny
Ali Adjoudani
Christian Benoît

ABSTRACT 3D models of the lips and jaw have been developed in the framework of an audio-visual articulatory speech synthesizer. Unlike most of the regions of the human face, the lips are essentially characterized by their border contours. The internal and external contours of the vermilion zone can be fitted by means of algebraic equations. The coefficients of these equations must be controlled so that the lip shape can be adapted to various speakers' conformations and to any speech gesture. To reach this goal, a 3D model of the lips has been worked out from geometrical analysis of the natural lips of a French speaker. Our lip model was developed to adjust a set of continuous functions best fitting the contours of 22 reference lip shapes. Only five parameters are necessary to predict all the equations of the lip model. For the jaw, we used a 3D wire-frame structure of a human skull. Two translations and one rotation can be applied for the animation of the jaw. We developed an analysis-synthesis package that allows both lip and jaw motion kinematics to be predicted from video analysis of a real speaker face.

19.1 Introduction

Even though the auditory modality is dominant in speech perception, it has been shown that the visual modality increases speech intelligibility. This is observed when there is a background noise [SP54, Nee56, Erb69, Erb75, BMJ74, Sum79, BMK94], when the message is linguistically complex, or when the language used is not familiar to the perceiver [RMG87]. These are the reasons why synthetic faces have been developed to enhance the intelligibility of speech synthesizers.

McGrath [McG85] in English, and then LeGoff et al. [LGB95] in French showed that human lips alone carry more than half the visual information provided by a whole natural face. McGrath [McG85] also showed that vision of the teeth somewhat increases the intelligibility of a message: it disambiguates sounds differing in jaw position such as "bib" versus "bab" These results are evidence that a synthetic face must be made of a "good" model of the lips first and then of a "good" model of the jaw. In this perspective, we first developed two 3D parametric models: one of the lips and one of the jaw. The lip model is based on parametric equations of

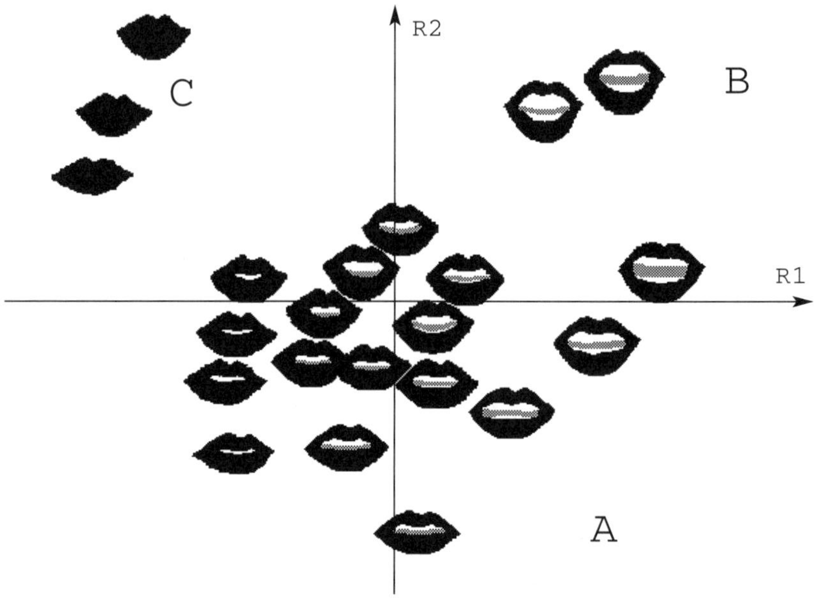

FIGURE 19.1. Projection of the front views of 22 basic lip shapes used as a reference database, and of their three most characteristic parameters in a factorial plane. *A*, *B* and *C* are depicted in figure 19.2.

the contours of the vermilion zone. The jaw model is a rigid bone structure that can be controlled by means of two translations and one rotation. Our efforts are preliminary steps toward the design of a 3D model of all the speech articulators of the human face, whether they are directly visible or not.

19.2 The 2D Model of the Lips

Contrary to the other regions of the human face, the lips are mainly characterized by their contours. The lip model presented here was thus based on the identification of algebraic equations best fitting the actual contours of a (French) speaker's lips. A 2D lip model was first designed by Guiard-Marigny [Gui92] from the front views of 22 basic lip contours, as shown in figure 19.1. Those shapes (so-called "visemes") were first identified by [BLMA92] from a multidimensional analysis of a French speaker's facial gestures. Guiard-Marigny [Gui92] predicted a good approximation of the internal and external lip contours in the coronal plane by means of a limited number of simple mathematical equations. To do so, he split the vermilion contours into three regions, as shown in the right part of figure 19.2. The same kind of polynomial and sinusoidal equations were used to describe both the internal and external lip contours. The speaker's lips were considered symmetrical so that only the right part of the lips was calculated. For each of the 22 visemes, 15

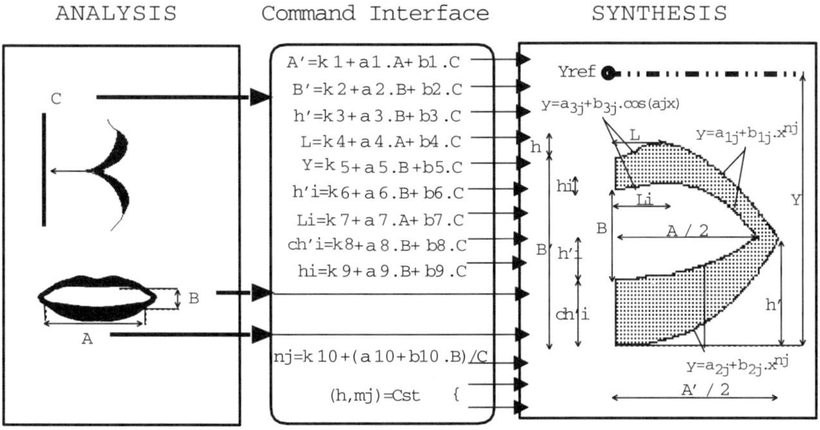

FIGURE 19.2. Schematic of the analysis (2 × 2D)–synthesis (2D) process. All equation coefficients of the lip contours were experimentally obtained by best fitting the modeled contours with the real ones.

coefficients were necessary for the equations to best fit the natural contours. The number of coefficients was then decreased by iteratively predicting one coefficient from another or a set of others, based on phonetic knowledge. Figure 19.3 gives an example of a correlation that was optimized between two coefficients measured on the front view, after having introduced a coefficient from the profile view. This decreased the dispersion of data due to some protruded and some spread shapes. Ultimately, the 2D model is controlled through only three parameters: the width (A) and the height (B) of the internal lip contour, and the lip contact protrusion (C). These anatomical distances can be automatically measured on a speaker's face, as shown in Figure 19.2.

19.3 The 3D Model of the Lips

To derive a 3D model from the above described 2D model, Adjoudani [Adj93] used the same technique as that used for the 2D model. Adjoudani wanted to identify the equations of the lip contours that best fit the projection of the natural contours in the axial plane. Adjoudani first obtained those contours by manually matching the front and the profile contours, as shown in figure 19.4. The axial plane was selected because of the strong influence of the jaw on the lip shape. An example of the reconstructed curves from the viseme /a/ is given in figure 19.5. In order to render the volume of the lips, Adjoudani identified three intermediate contours between the internal and the external contours. He obtained 10 polynomial equations. An iterative process allowed Adjoudani to predict all the necessary coefficients of these equations from five parameters. Those control parameters are the above-mentioned three parameters that command the 2D model and two extra

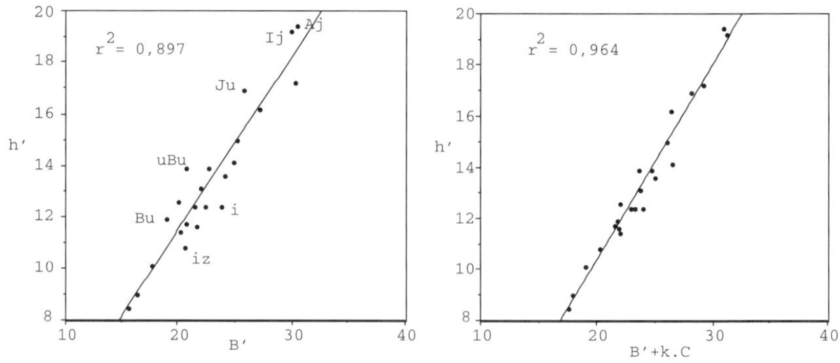

FIGURE 19.3. Improvement of a correlation between parameters of the model measured in the coronal plane (h' = vertical distance between the bottom and the corner of the lips; B' = height of the external contours) after having introduced an extra parameter measured in the axial plane (C = lip contact protrusion) in order for rounded visemes (consonants in a /y/ context and vowels in a /ʒ/ context) and spread visemes (/i/ with or without a /z/ context) to get closer to the average relationship between h' and B'.

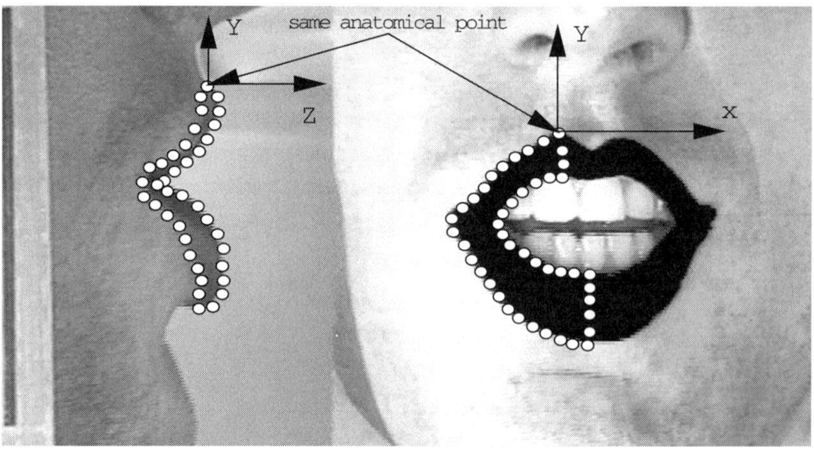

FIGURE 19.4. Matching of the front and the profile contours to obtain 3D contours.

parameters: the protrusion of the upper lip and that of the lower lip. We do not assume here that there are five degrees of freedom in the human lip gestures. Our goal is not to find the smallest set of independent parameters that may describe all lip gestures. Rather, our goal is to create an easy-to-use model of the lips that can be controlled from easily measured parameters on a real speaker's face and is easy to predict by rules for a text-to-speech system.

Finally, this set of five parameters allows any lip shape to be reconstructed with a fair approximation of a visible speech sequence uttered by our reference speaker (or another). Figure 19.6 displays the "wire-frame" structure and the final rendered image of our 3D lip model. Figure 19.7 shows the real lips of the speaker and the

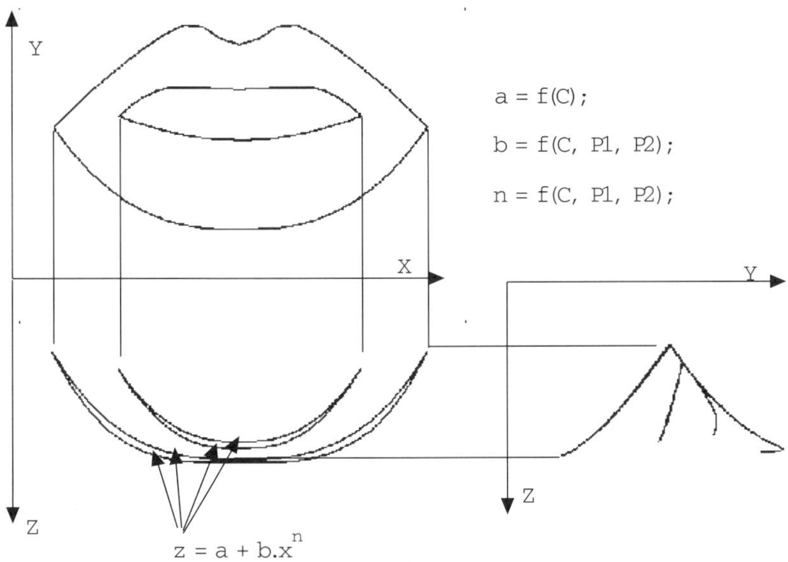

FIGURE 19.5. Identification of the lip contour equations in the axial plane $z = f(x)$; matching of these contours with those first studied in the coronal plane $y = f(x)$; and projection of the obtained contours onto the sagittal plane $y = f(z)$.

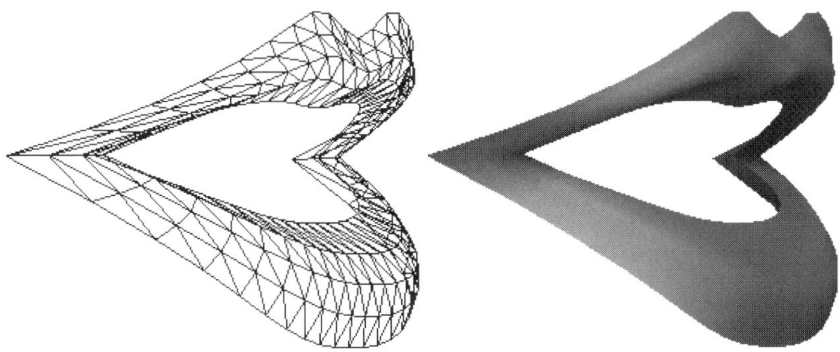

FIGURE 19.6. The 3D lip model displayed through its underlying wire-frame structure and rendered with the Gouraud-shading technique.

corresponding synthetic lips in three extreme cases (open, protruded, and spread lips).

19.4 Animation of the Lip Model

Our 3D model of the lips is implemented on a graphics computer (SGI Indigo-ELAN). The vermilion area is first sampled with 160 rectangles filling in the sur-

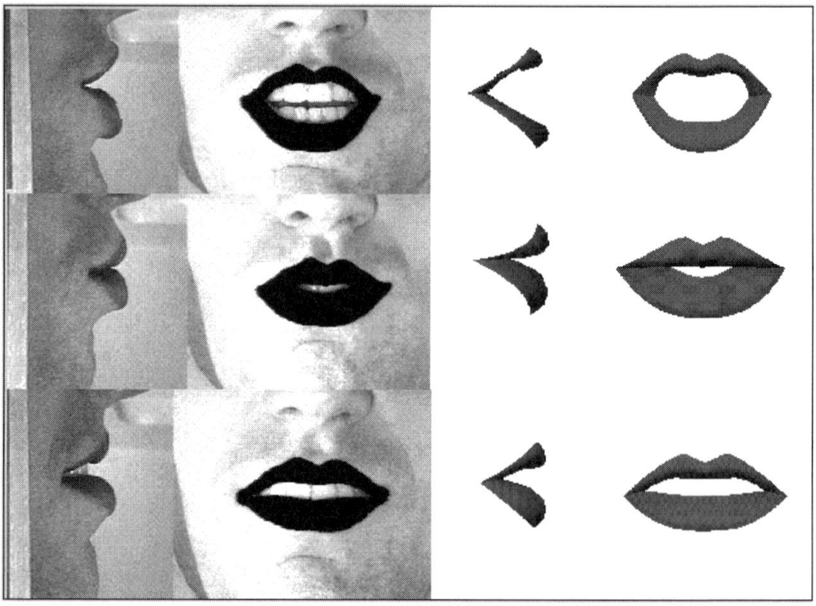

FIGURE 19.7. Comparison between the real lips of the speaker and the lip model.

faces among the five contours. A smooth rendering of the surface is finally obtained using the Gouraud-shading technique. Calculation of the position of each vertex and of the normals to the rectangles, as well as Gouraud shading, are processed at a 50 ips rate on an Indigo-ELAN. All vertices of the mesh can be calculated from the equations of the model. Then the normals to the vertices have to be calculated too, and this is the most time-consuming process because it makes extensive use of vector products. An alternative is to use a differential parametric interpolation method, which speeds up the whole synthesis process. In this approach, a lip shape is considered as the barycenter of a set of extreme lip shapes. Each weight corresponds to a parameter of the model. For a given lip shape to be synthesized, each vertex of the mesh is considered as the barycenter of that of the extreme shapes. The same approach applies for the normals. Because there are five command parameters, the database is made of ten extreme shapes. For a given parameter, the vertices and the normals of a minimum (maximum) shape are stored after the model has been calculated with this parameter at its minimum (maximum), and all other parameters are set at their average value.

The synthesis process makes use only of additions and multiplications. Compared to the original method, the reduction in computation time is dramatic. Figure 19.8, illustrates this parametric interpolation technique along two global parameters, protrusion and opening.

Whatever the synthesis technique used, an easy way to animate the model realistically is to directly measure the command parameters of the model on a real speaker's face. To do so, we can use the software specially designed by Lallouache

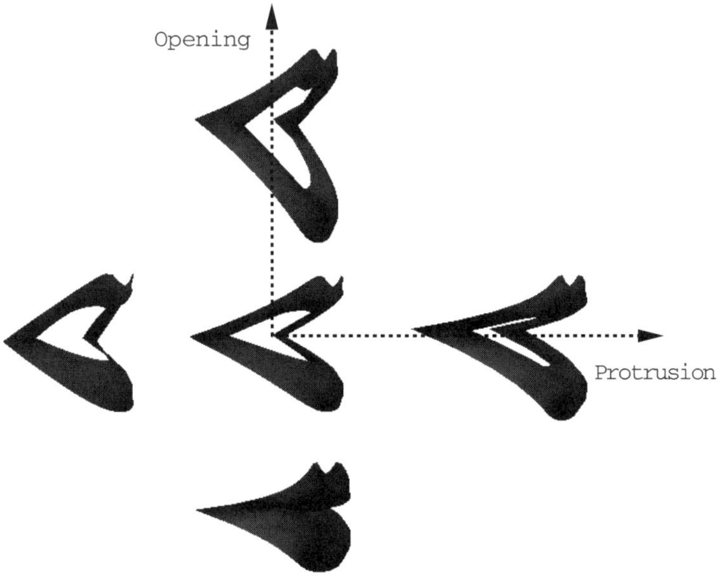

FIGURE 19.8. An illustration of the differential parametric interpolation method. In this example only two global parameters are used: lip protrusion (x-axis) and lip opening (y-axis)

[Lal91] that accurately measures the parameters from a videotape. It generates a file containing the five parameters measured every 20 ms. This file can then be used as a command file to our model. The digitized voice of the natural speaker is synchronized with the display of each frame. Our system allows a highly natural animation at a 50-ips display rate. This is mostly due to the perfect synchronization between the images and the soundtrack.

The visual intelligibility of our lip model is evaluated by Le Goff et al. (chapter 18). It shows that the lip model itself accounts for a third of the visual information carried by the entire image of a natural speaker's face, whatever the degradation of the acoustic message.

19.5 The Jaw Model

After the lips, the most visible articulator is the jaw, to which the chin and the teeth are linked. Because the jaw is made of rigid bone, the animation process is easier than that of the lips. As with any rigid object, jaw motions may be controlled in six degrees of freedom. Its position relative to the skull can thus be defined with three orientation angles (yaw, pitch, roll) and three positions (horizontal, vertical, lateral).

The synthetic jaw we used for our model was first elaborated at McGill University [GOB95] to visualize jaw motion kinematics, during speech or mastication,

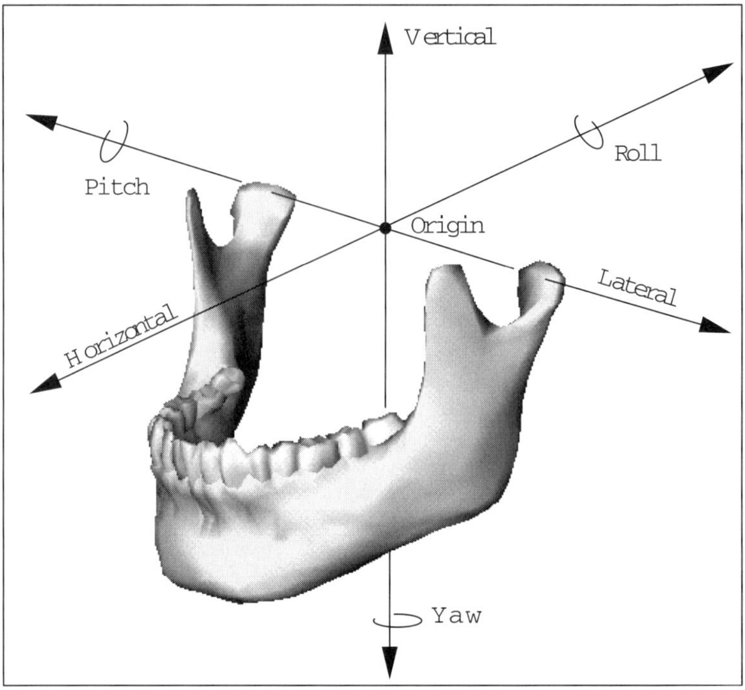

FIGURE 19.9. The coordinate system for three-dimensional jaw motion.

recorded with an optoelectronic measurement system. The visualization uses a 3D digitized upper skull and jaw with their corresponding teeth. Overall the whole facial structure is made of a mesh of 6000 polygons. The coordinate system used to define the jaw motions is represented in figure 19.9. [GOB95] animated this jaw model from three rotations and three translations automatically derived from the motion of a rigid structure attached to the lower teeth of a speaker. The synthetic upper skull and jaw can then be animated in synchrony with the audio part of the natural speech.

19.6 Animation of the Lip and Jaw Models

The second step toward the design of a multiarticulator talking head is the integration of the lip and the jaw models in a single display. To do so, the lip model has been superimposed on the 3D skull with its jaw model, as shown in figure 19.10. Lip gesture was measured as presented above, using a video analysis technique. For the jaw, the best technique would have been to use the optoelectronic measurement system used at McGill. However, this technique requires a device to be plugged onto the speaker's teeth, making lip measurement impossible. Thus, a nonintrusive alternative had to be found. The best alternative we could find was to mark the chin of the speaker with makeup and then to use image processing techniques similar to that developed for lip gesture measurements. Jaw motions in speech are primarily controlled in three degrees of freedom [OB94] namely pitch angle, vertical position, and horizontal position. Two reference points attached to the jaw bone in the sagittal plane are thus necessary for the three motions to be calculated. Practically, the jaw bone is not visible; the skin rolls over it, and those two reference points simply cannot be detected through nonintrusive methods. However, it can be seen from the data obtained by [OM94] that the basic parameters of jaw motion are strongly correlated in running speech. In a first order, those three basic jaw motions can thus be predicted from the displacement of a single point on the jaw. Because the teeth are not always visible, this single point must be detected on the speaker's skin, and discrepancy between the actual jaw motion and that of the reference point on the chin cannot be avoided. A corpus has been widely used at the ICP during the last years to make geometric measurements and to evaluate the contribution of vision to speech intelligibility ([BLMA94]; chapter 18, this volume). Because the speaker's chin was made up with a single dot on the original videotapes, we consider it good enough as a kickoff to animate and then evaluate our lip/jaw synthesizer.

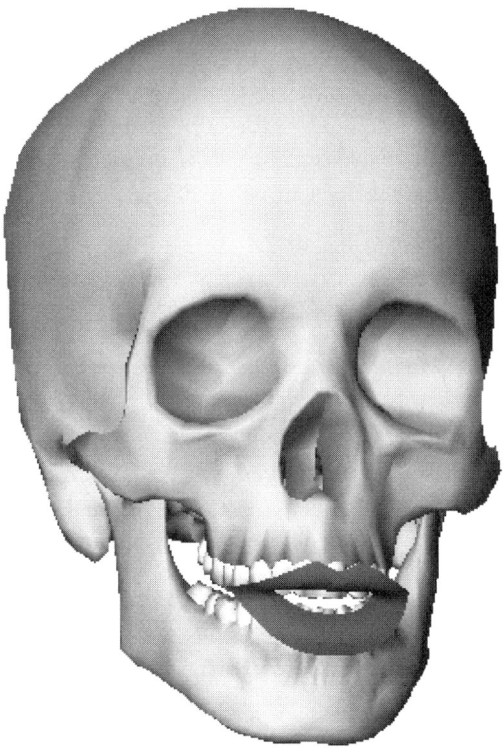

FIGURE 19.10. The 3D digitized skull with our superimposed lip model.

The three jaw motions were finally predicted from the vertical displacement of this dot on the speaker's chin. Animation of both the lips and the jaw was ultimately controlled through six parameters measured every 20 ms. Five parameters were necessary to animate the lip model, and one more for the jaw.

19.7 Evaluation of the Lip/Jaw Model

The measurement method we used to derive the jaw motion kinematics is not optimal because the chin and the jaw motions may differ somewhat. For instance, the jaw lowers to produce /a/ in the word /ababa/ whereas the lower lip raises to close the mouth for the /b/. This makes the chin skin roll up over the jaw bone. Despite this, [GOB95] obtained a noticeable gain in speech intelligibility when the synthetic jaw was added to the synthetic lips, as shown in figure 19.11. They used the same speech material as Le Goff (see chapter 18) in the assessment of the audio-visual intelligibility of various components of a speaker's face, i.e., three French vowels /i, a, y/ and six French consonants /b, v, z, ʒ, R, l/. When synchronized with the lip model, the jaw model enhances its visual intelligibility at several levels. The number of no-responses from subjects were reduced by a factor two. Vowel /i/ is much less confused with vowel /a/, mostly in closing consonantal context (/b/ or /v/). There are fewer confusions between /ʒ/ and /R/, whatever the vocalic context. Finally, /b/ is no longer confused with /v/, especially in a /i/ vocalic context. However, visibility of the jaw also leads to a larger amount of confusion between /i/ and /a/ in a /z/ context. In addition, /l/ and /v/ are more often mixed up with /ʒ/, but this occurs only in rounded vocalic contexts.

19.8 Conclusion

This chapter presents two synthetic articulators that are necessary for the implementation of a 3D audio-visual articulatory speech synthesizer. The 3D model of the lips is a high-resolution model simply controlled by means of five parameters. We have developed a robust procedure to detect lip gestures on a real speaker's face and analysis-synthesis is clearly functional. Animation of the jaw model is easier, but new techniques must be developed so that nonintrusive automatic detection of jaw motion on a speaker's face can be more accurate. Meanwhile, the intelligibility gain observed with the two models clearly shows that our approach is very promising. Parallel work is under progress at the ICP to integrate these two models within an articulatory speech synthesizer as well as within a text-to-audio-visual speech system for French. It is obvious that speech synthesizers will be more and more multimodal in the future. The 3D models we present here are clearly an innovative and promising step in the perspective of its integration into a whole model of a talking head.

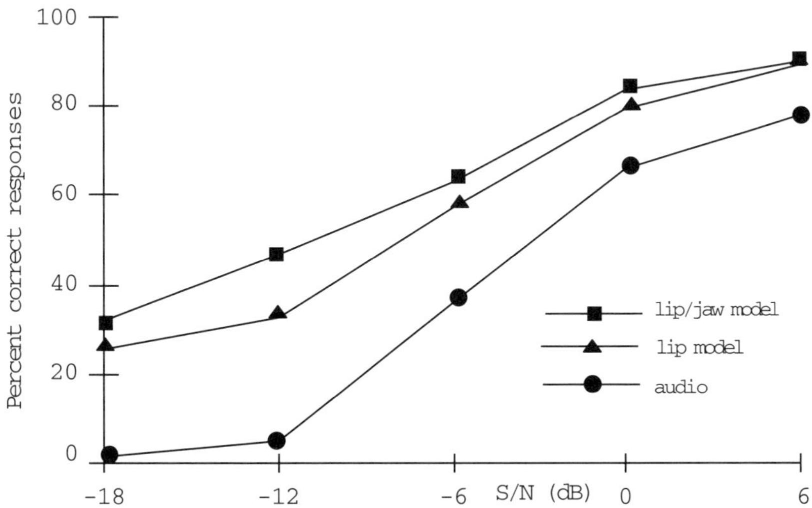

FIGURE 19.11. Audio-visual intelligibility of the lip model and of the lip/jaw model compared to the auditory-alone intelligibility of speech across various levels of degradation by additive noise.

Acknowledgments: This research was supported by the French CNRS and by a grant from the European ESPRIT-BRA programme ("MIAMI" project No. 8579). The lip model could not have been developed without the help of Tahar Lallouache, Tayeb Mohamadi, Omar Angola, Alain Arnal, Christian Abry, and Jean-Luc Schwartz. We want to thank David Ostry for support in the design and control of the jaw model. We are also indebted to Marie-Agnès Cathiard, Sonia Kandel, and Bertrand Le Goff for preparation of the perceptual test.

References

[Adj93] A. Adjoudani. Élaboration d'un modèle de lèvres 3D pour animation en temps réel. In *Mémoire de D.E.A. Signal Image Parole*, l'Institut National Polytechnique, Grenoble, France, 1993.

[BLMA92] C. Beno̧ît, M. T. Lallouache, T. Mohamadi, and C. Abry. A set of French visemes for visual speech synthesis. In *Talking Machines: Theories, Models and Designs*, G. Bailly and C. Benoî̧t, eds. Elsevier Science Publishers B.V., North-Holland, Amsterdam, 485–504, 1992.

[BMJ74] C. A. Binnie, A. A. Montgomery, and P. L. Jackson. Auditory and visual contributions to the perception of consonants. *J. Speech & Hearing Res.* 17:619–630, 1974.

[BMK94] C. Beno̧ît, T. Mohamadi, and S. D. Kandell. Effects of phonetic context on audio-visual intelligibility in French. *J. Speech & Hearing Res.* 37:1195–1203, 1994.

[Erb69] N. P. Erber. Interaction of audition and vision in the recognition of oral speech stimuli. *J. Speech & Hearing Res.* 12:423–425, 1969.

[Erb75] N. P. Erber. Auditory-visual perception of speech. *J. Speech & Hearing Disorders* 40:481–492, 1975.

[GO95] T. Guiard-Marigny and D. J. Ostry. Three-dimensional visualization of human jaw motion in speech. 129th Meeting of the Acoustical Society of America, Washington, DC, May 1995.

[GOB95] T. Guiard-Marigny, D. J. Ostry, and C. Benoît. Speech intelligibility of synthetic lips and jaw. In *Proceedings of the XIII International Congress of Phonetic Sciences*, Stockholm, Sweden, 3, 222–225, August 1995.

[Gui92] T. Guiard-Marigny. Animation en temps réel d'un modèle paramétrisé de lèvres. In *Mémoire de D.E.A. Signal Image Parole*, l'Institut National Polytechnique, Grenoble, France, 1992.

[Lal91] M. T. Lallouache. *Un Poste "Visage-parole" Couleur. Acquisition et Traitement Automatique des Contours des Lèvres.* Doctoral thesis, l'Institut National Polytechnique, Grenoble, France, 1991.

[LGB95] B. Le Goff, T. Guiard-Marigny, and C. Benoît. Read my lips... and my jaw! In *Proceedings of the 4th Eurospeech Conference*, Madrid, Spain, 1, 291–294, Sept. 1995.

[MB92] T. Mohamadi and C. Benoît. Apport de la vision du locuteur á l'intelligibilité de la parole bruitée. *Bulletin de la Communication Parlée*, 2, Cahiers de l'ICP, Grenoble, France, 1992.

[McG85] M. McGrath. *An Examination of Cues for Visual and Audio-Visual Speech Perception Using Natural and Computer-Generated Faces.* Ph.D. thesis, Univ. of Nottingham, UK, 1985.

[Nee56] K. K. Neely. Effect of visual factors on the intelligibility of speech. *J. Acoust. Soc. Amer.* 28:1275–1277, 1956.

[OB94] D. J. Ostry and E. V. Bateson. Jaw motions in speech are controlled in (at least) three degrees of freedom. In *Proceedings of 1994 International Conference on Spoken Language Processing*, 1:41–44, 1994.

[OM94] D. J. Ostry and K. G. Munhall. Control of jaw orientation and position in Mastication and speech. *J. Neurophysiology* 71(4), April 1994.

[RMG87] D. Reisberg, J. McLean, and A. Goldfield. Easy to hear but hard to understand: A lip-reading advantage with intact auditory stimuli. In *Hearing by Eye: The Psychology of Lip-Reading*, B. Dodd and R. Campbell, eds. Lawrence Erlbaum Associates, Hillsdale, NJ, 97–114, 1987.

[SP54] W. H. Sumby and I. Pollack. Visual contribution to speech intelligibility in noise. *J. Acoust. Soc. Amer.* 26:212–215, 1954.

[Sum79] Q. Summerfield. Use of visual information for phonetic perception. *Phonetica*, 36:314–331, 1979.

Appendix: Video Demos

The CD-ROM contains 10 demonstrations of our system.

Section IV

Concatenative Synthesis and Automated Segmentation

20

Section Introduction. Concatenative Synthesis

Joseph P. Olive

A widely used method for converting a string of phonemes into an acoustic signal is the concatenation of segments of naturally spoken utterances. In this method, segments of speech are excised from spoken utterances and are connected to form the desired speech signal.

The concatenative method of synthesis has introduced a variety of research topics, such as:

- the structure of the recorded database (natural sentences, words in carrier phrases or in isolation, or nonsense syllables)
- the size and type of the units stored
- the use of multiple units for different contexts and different prosodic environments
- concatenation algorithms
- the type of parametric representations (formants, LPC, or PSOLA)
- automatization of the inventory collection

The five chapters in this section deal with some of these topics.

Chapter 21 by Portele, Höfer, and Hess argues for a mixed inventory, using both demisyllables and diphones as concatenative units. This chapter shows a definite improvement when the inventory consists of both types of elements. For German, the authors constructed a synthesis scheme with 2,182 elements. The question remains whether a polyphone inventory rather than a demisyllabic plus diphone inventory would produce similar results with the same number of units, or even fewer units.

In contrast to the chapter by Portele et al., Campbell and Black, in chapter 22, suggest the use of prosodically dependent units in their acoustic inventory. Such a scheme will necessitate an inventory that is considerably larger than commonly used inventories. To keep an inventory to a reasonable size, they have devised a method to determine which units are to be used for a given inventory size. The evaluation of their system was based on synthesized speech where the prosody was copied from natural speech; such speech is generally of higher quality than speech synthesized completely by rule. Consequently it is difficult to assess the ultimate quality of a complete rule system.

Whereas Campbell and Black add concatenation units to account for prosodic variability at the expense of continuity of the acoustic parameters at the junctures of the concatenation units, Conkie and Isard describe in chapter 23 a study to insure better continuity of the units. In this study, the authors are concerned about continuity of the acoustic parameters as well as continuity of their derivatives. Tests of their concatenation scheme show an improvement in the quality of the synthesized speech.

The remaining two chapters in this section are concerned with a related topic, namely automatic segmentation of speech into phonemic units. This topic is related to the topic of concatenative inventory elements, because automatic segmentation of the speech signal is a precondition for automatic acoustic inventory creation. However, automatic segmentation has many—and, one might argue, more important—applications besides creating acoustic inventories. For example, progress in getting a better quality for speech synthesis relies on developing better rules for generating segment duration and f0 contours. The development of such rules necessitates analysis of large corpora of natural speech. Because of the magnitude of these corpora, manual segmentation of the speech is prohibitive.

Chapters 24 and 25 on automatized segmentation present similar results. They both demonstrate that their methods work for many types of phonemes. However, for some phoneme combinations, the methods are not reliable enough to perform the task; the phoneme boundaries are not determined with enough accuracy to make these techniques usable at this time without some manual intervention. Whether these limitations can be overcome with larger and better training corpora, or whether fundamentally different methods are required, remains to be seen.

In summary, the chapters in this section demonstrate that considerable research is being devoted to the concatenative method in speech synthesis. So far, this has not produced a consensus, but rather a diversity of approaches.

21

A Mixed Inventory Structure for German Concatenative Synthesis

Thomas Portele
Florian Höfer
Wolfgang J. Hess

ABSTRACT In speech synthesis by unit concatenation a major point is the definition of the unit inventory. Diphone or demisyllable inventories are widely used but both unit types have their drawbacks. This chapter describes a mixed inventory structure that is syllable-oriented but does not demand a definite decision about the position of a syllable boundary. In the definition process of the inventory the results of a comprehensive investigation of coarticulatory phenomena at syllable boundaries were used as well as a machine-readable pronunciation dictionary. An evaluation comparing the mixed inventory with a demisyllable and a diphone inventory confirms that speech generated with the mixed inventory is superior regarding general acceptance. A segmental intelligibility test shows the high intelligibility of the synthetic speech.

21.1 Introduction

Demisyllables [PS60] and diphones [KW56] are the two main paradigms in the design of inventories for concatenative speech synthesis. Each has its advantages and drawbacks. The demisyllable paradigm claims that coarticulation is minimized at syllable boundaries and only simple concatenation rules are necessary. However, this assumption is only partially valid for German. Heavy coarticulation effects occur at syllable boundaries (for instance nasal/lateral plosive releases in *Mitleid* /mɪtl͜aɪt/ or *abmachen* /apmaxən/, or initial devoicing in *Aufsatz* /a͜ʊfzats/ or *Hausbau* /ha͜ʊsba͜ʊ/) [Koh90]. The diphone model, on the other hand, assumes that coarticulative effects occur only between adjacent sounds in a domain limited by the centers of the pertinent sounds. Coarticulation is largely planned [Wha90], and the resulting phenomena (for instance, lip rounding) may extend over several segments. The partial invalidity of the diphone assumption led Olive [Oli90] to augment a diphone inventory with additional units.

Our previous synthesis system HADIFIX [PSPSH92] used a combination of initial demisyllables, diphones (vowel to postvocalic consonant transitions), and suffixes (postvocalic consonant clusters). In our own experience, effects such as the ones described above could not be adequately modeled with HADIFIX, and this difference from natural speech not only degrades the naturalness but also the intelligibility of the synthetic speech. Moreover, we had serious problems with postvocalic sonorants synthesized by two units meeting in the middle of the sound. A formal evaluation of segmental intelligibility [Por93] revealed that unit boundaries inside postvocalic sonorants lead to undesirable spectral jumps and to misperceptions of the pertinent sounds. These experiments convinced us that neither diphones nor demisyllables are sufficient for high-quality synthesis of German utterances.

21.2 Investigating Natural Speech

21.2.1 Material

German phonotactics is governed by syllable-domain phenomena, final devoicing being just one example. Therefore, a syllable-based inventory definition [Fuj75] is appropriate. However, two phenomena must be taken into account:

- The number of syllables in German is very large. Demisyllables [PS60] can be used instead [FML77], but vowel reduction in unstressed syllables might require a number of additional demisyllables.
- Coarticulation at syllable boundaries must be treated.

Regarding the first point, there is some evidence that a complete set of reduced demisyllables is not necessary [DC91, Por94]. In the experiment described here we examined the second point, namely coarticulation at syllable boundaries.

All possible consonant combinations were studied with an in-between syllable border, and also, if not prohibited by German phonotactics, within the syllable in onset or coda. The items were extracted from 48 specifically designed passages read by a male and a female speaker (the latter was also the speaker for the female voice of our synthesis system). About 4,200 items were investigated, 2,100 from each speaker. For each sound combination the range of possible realizations was determined, and a standard realization was established as well as hypercorrect and reduced forms. Temporal aspects were also investigated.

21.2.2 Devoicing

Voiced fricatives and plosives are often devoiced when preceded by an unvoiced obstruent, regardless of their position (see figure 21.1). With the exception of /z/, which becomes /s/, the devoiced sounds are distinguishable from their unvoiced counterparts: devoiced /b,d,g/ are less aspirated and have a weaker burst than /p,t,k/ [Sto71, Kea84]; a devoiced /v/ has less intensity than an /f/. The liquids /l,r/ are also subject to devoicing with /r/ more than /l/ [Koh77]. For liquids, the position of the syllable boundary determines the degree of devoicing. A syllable-initial /l/ preceded by a /k/ (as in *weglaufen* /vɛkla͜ʊfən/) is less likely to be devoiced than an /l/ in the second position of the onset (as in *Wehklage* /ve:kla:gə/). Nasals are devoiced only when preceded by a homorganic stop in reduced position.

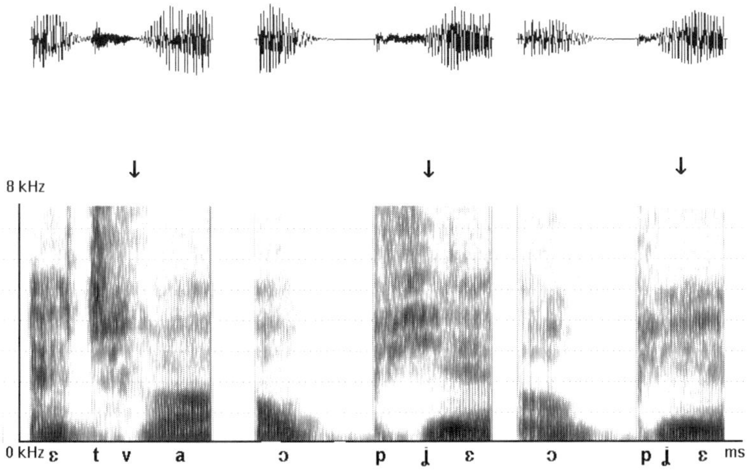

FIGURE 21.1. Different levels of devoicing for fricatives extracted from fluent speech (left, completely devoiced; middle, partially devoiced; right, voiced). The arrows mark the fricatives.

21.2.3 Assimilation

Assimilation usually involves manner or place of articulation. Plosives are often assimilated to a following nasal, lateral, or fricative. The combination /tʃ/ is produced with a retracted /t/ to match the place of articulation of the /ʃ/ [Wan60]. The stop in the combinations /pm/ and /tl/ (and in similar pairs) is released by lowering the velum or lifting the tongue blades, the articulatory gesture to produce the following sound [MM61]. In all these cases the two sounds are effectively articulated together and cannot be separated. They are frequent in German, and a syllable boundary is often present between the two sounds (see figure 21.2).

21.2.4 Position of the Syllable Boundary

Within-syllable combinations are more likely to be reduced. Devoicing is more common, and stops are less aspirated. It is generally held that a /t/ after a syllable-initial /ʃ/ in words such as *Drehstuhl* /dre:ʃtu:l/ is unaspirated, whereas a syllable-initial /t/ in the same context in a word such as *Tischtuch* /tɪʃtu:x/ is aspirated [Fuj79]. This difference is statistically significant (U-test, $p < 0.05$); however, among the data there are aspirated stops in the second position of the onset as well as unaspirated syllable-initial ones. The realization depends on the syllable's prominence [Koh90].

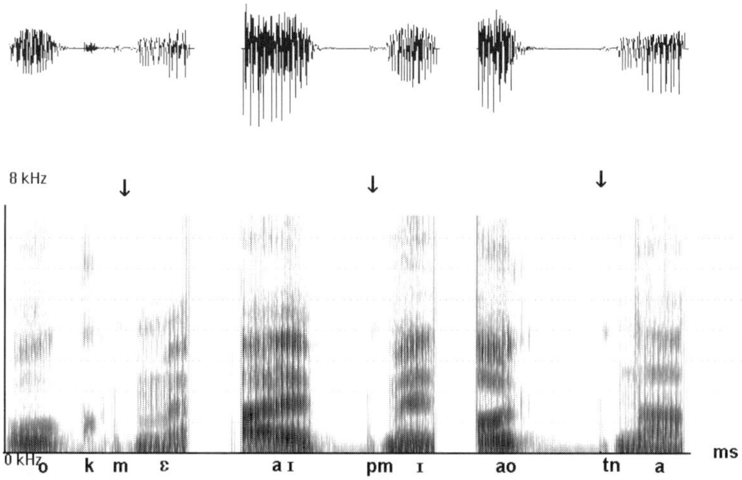

FIGURE 21.2. Stop-nasal combinations extracted from fluent speech. The arrows mark the nasal releases. As with plosive-plosive combinations, the stop is not released when both sounds are homorganic (middle and right).

21.2.5 Pre- and Postvocalic Consonants

Differences between pre- and postvocalic realizations of a phoneme can be observed, especially for liquids; a vowel with a following postvocalic liquid is often produced like a diphthong [Rap36], and sometimes postvocalic liquids are indicated just by a slight change in the formants of the preceding vowel (and an elongated duration) [Hei79]. In general, the preceding vowel has a strong influence on liquids and nasals and also on dorsal obstruents. Intervocalic consonants bear more resemblance to their prevocalic counterparts, at least in the acoustic domain investigated here (see figure 21.3) [SKT84, Bou88].

21.2.6 Conclusions

A unit inventory based on a phonological syllable definition [Twa38] will inevitably lead to unnatural and hypercorrect synthetic speech. On the other hand, there are noticeable differences between phoneme realizations in the coda compared to those in the onset. A syllable-based inventory structure is appropriate for the synthesis of German speech as long as a "syllable" is defined in a phonetical and not phonological way. However, there are doubts whether such a definition can be accomplished [Koh77]. The solution proposed here is designed to avoid the explicit placement of a syllable boundary between two nuclei.

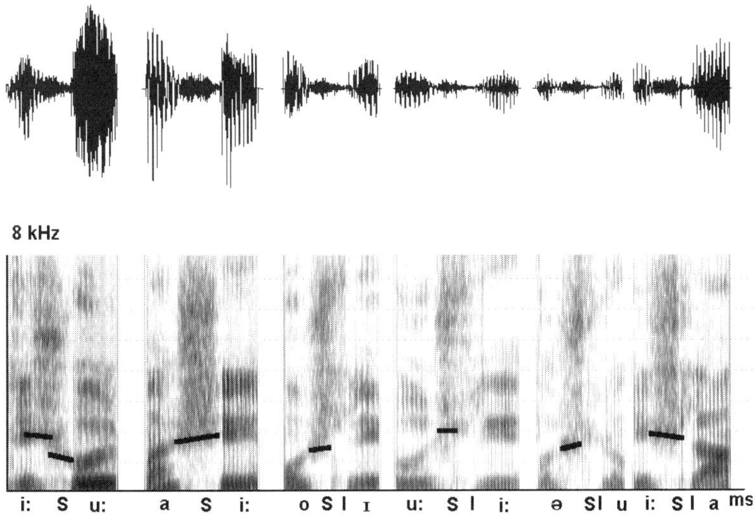

FIGURE 21.3. Context-dependent rounding of the /ʃ/ and its acoustic manifestation as lowered formant-like structure between 1.5 and 2 kHz. The two leftmost examples show an intervocalic /ʃ/ that is articulated compatible to the following vowel. The four rightmost examples demonstrate the influence of the syllable boundary on the combination /ʃl/ (left after /ʃ/; middle left, middle right and right before /ʃ/). The thick line inside the spectrogram denotes the formant-like structure afflicted by lip rounding.

21.3 Inventory Structure and Concatenation Rules

21.3.1 Concatenation Methods

There are three ways to concatenate two units. Their differences are exemplified by the construction of the /n/ in /ana:/:

1. *Diphone concatenation* takes place inside a sound. A sound suitable for diphone concatenation must have some kind of stable part insensitive to contextual influences. The /n/ in /ana:/ is synthesized by combining units /an/ and /na:/ in the middle (or a suitable place [Kra94, CI94]) of the /n/.

2. *Hard concatenation* is the simplest case of putting two sounds together. This happens at each syllable boundary in standard demisyllable systems. In the example /ana:/ the units /a/ and /na:/ are concatenated.

3. *Soft concatenation* also takes place at segment boundaries. However, the concatenation is smoothed by including the transitions that would appear in natural speech. A certain overlap between the two units is necessary. As anticipatory coarticulation is assumed to be more important than persistent coarticulation, the overlap is primarily provided by the first unit. The /n/ in

/ana:/ is generated by the units /an/ and /na:/, but, unlike diphone concatenation, the concatenation takes place at the beginning of the /n/. While the transition to the nasal and the anticipatory lowering of the velum is preserved inside the first /a/, the realization of the intervocalic /n/ is determined by the following vowel. Furthermore, the concatenation point is located at a point of change in the spectrum. Spectral jumps at this point are not as offending as inside a supposedly stable part of speech.

21.3.2 Inventory Structure

The mixed inventory structure (MIS) is designed to avoid hard concatenation whenever possible. Inside a vowel, diphone concatenation is performed, whereas soft concatenation is the method of choice whenever consonants are involved. Seven types of units are used:

1. *Initial demisyllables* (1,086 elements) with voiced and unvoiced versions for the initial consonants that can be devoiced. Here, "initial demisyllable" means "sequence of consonants that share articulatory gestures with a following vowel." Consequently, units such as /pni:/ or /tlu:/ are included.

2. *Final demisyllables* (577 elements) are like classical demisyllables but without voiceless fricatives (except /x/ and /ç/) or combinations of obstruents. Due to the German syllable final devoicing rule, only voiceless obstruents appear in a coda. These sounds are synthesized by soft concatenation of suffixes (see subsection 21.3.4).

3. *Suffixes* (88 elements) consisting of voiceless fricatives and obstruent combinations in rounded and unrounded versions.

4. *Consonant-consonant diphones* (167 elements) to smooth problematic syllable boundaries and to allow synthesizing of words that do not obey the phonotactics of German.

5. *Vowel-vowel diphones* (67 elements) to smooth transitions between vowels whenever the glottal stop is missing or eliminated due to reduction.

6. *Syllables or syllable parts containing syllabic consonants* (122 elements) because context dependency was found to be especially prominent in such syllables. A pilot study with 15 of these units has confirmed the perceptual salience of such units [Por94].

7. *Syllables with /ə/ and voiced initial plosives* (75 elements) are very common prefixes in German, and diphone concatenation inside the /ə/ is especially difficult because of the context dependency of the /ə/.

21.3.3 Inventory Definition

The exact inventory definition was achieved by analyzing a machine-readable pronunciation dictionary with more than 90,000 entries. In this dictionary, inflected forms appear only when irregular. Due to the flexible suffix concept, even the most complicated German inflections such as /rpsts/ in *Herbsts* /hɛrpsts/ or /mpfst/ in *kämpfst* /kɛmpfst/ are supported, whereas, in a genuine demisyllable system, every possibility will have to be covered by a special unit. Many foreign words even with nasal vowels or nonsyllabic vowels are supported; for instance, *Chance* /ʃãsə/, *Szene* /stse:nə/, *Dschungel* /dʒʊŋəl/, or *Skorpion* /skɔrpĭo:n/. The complete inventory consists of 2,182 units, which is in the order of standard inventories for German (about 2,700 diphones in the German version of the CNET synthesizer [BCEW93], and more than 2,000 demisyllables in a demisyllable inventory [KA92]).

21.3.4 Concatenation Rules

The complex structure of the inventory requires a complex set of rules describing the unit selection. Only the major principles are described here (see [Por94] for a complete explanation of the concatenation). A phonemic string such as /ʔapmʊrksən/ is converted into a list of synthesis units in two steps:

- For each syllable nucleus, an environment is determined. If no special nucleus (e.g., a syllabic consonant) is present, the longest matching initial demisyllable and the longest matching final demisyllable are selected. Final obstruent clusters are represented by a list of suffixes that are combined using soft concatenation. The principle is to maximize the environment for each nucleus without respect to any phonologically defined syllable.

  In our example, /ʔapmʊrksən/, the following syllable nucleus environments are obtained:

  1. /ʔap/ with /ʔa/ and /ap/
  2. /pmʊrks/ with /pmʊ/, /ʊr(k)/ (the /k/ is deleted) and /ks/
  3. /ksn̩/ with syllabic [n̩] as one unit (the /ə/ is deleted to simulate its elision due to reduction)

- The nucleus environments are concatenated. If there is an overlap between two adjacent environments, the pertinent sounds of the first environment are deleted. This ad hoc rule assumes the validity of the maximum-onset principle. The transition to the following (deleted) sound is still present in the last sound of the first nucleus environment. The results described in section 21.2, however, indicate that such a simple procedure may sometimes fail. Work is in progress to determine the best places for concatenation by minimizing the spectral distances in the overlapping areas and by appropriate rules describing phenomena related to the position of the syllable boundary, such as lip rounding (figure 21.3) or aspiration. In our experience, in more than 80% of all cases there is an overlap of at least one sound.

If two nucleus environments meet without overlap, two possibilities exist:

1. Hard concatenation is allowed. This holds, for instance, when the second environment begins with a glottal stop.
2. Hard concatenation is prohibited. For instance, concatenating /n/ and /g/ is not permitted because of the partially velarized /n/ before /g/ in natural speech. This phenomenon is instead modeled by the use of an intermediate diphone (diphone concatenation). The appearance of a sound pair in a special diphone table determines whether hard concatenation between these sounds is allowed.

In the unlikely case (less than 1%) that there are sounds not represented by one or the other of two adjacent nucleus environments, a set of "exception diphones" is defined. These units, in part extracted from other units, allow synthesis of every possible combination of German sounds.

In our example, the following will happen:

1. /ʔa(p)/ and /pmʊrks/—the /p/ from the first environment is deleted
2. /pmʊr(ks)/ and /ksn̩/—the /ks/ suffix from the first environment is deleted

The final result is then:

1. initial demisyllable /ʔa/
2. final demisyllable /a(p)/ (/p/ is deleted in this unit)
3. initial demisyllable /pmʊ/
4. final demisyllable /ʊr(k)/ (/k/ is deleted in this unit)
5. syllable with syllabic consonant /ksn̩/

Whenever a vowel is involved, diphone concatenation takes place. Everywhere else soft concatenation is performed.

21.4 Perceptual Evaluation

To check whether the new mixed inventory structure meets our expectations, two perception experiments were carried out: a pair comparison test with standard inventory structures, and a segmental intelligibility test.

21.4.1 *Pair Comparison*

A test set of 22 words was used. It mimics the properties of the German sound architecture to a large extent (i.e., sound frequency, sound number per word, and syllable number per word). Demisyllables, diphones, and mixed units necessary to synthesize these words were defined, spoken by a male speaker, recorded with a 32-kHz sampling rate, and segmented by hand. A complete diphone method was assumed, that is, all sound pairs are represented by their own unit. We decided to treat combinations of a long vowel and a following /ɐ/ as two sounds. This decision is questionable [Koh77] but reduces the number of units considerably. Moreover, as our results indicate, at least the combinations of a vowel and a following /l/ must receive the same treatment. The demisyllable definition was adopted from [KA92]. Altogether, 400 units were generated. Whenever possible, "unit-sharing" was performed insofar as, for instance, the diphone /na/, the demisyllable /na/, and the MIS unit /na/ were extracted from the same utterance. Prosodic manipulations were carried out using the TD-PSOLA-algorithm [MC90]. The original temporal structure was used, and all versions of a word had the same intonation contour. A pair comparison test was used to assess quality differences. Ten subjects participated. The outcome of the test (figure 21.4) is a significant (χ^2-test, $p < 0.0005$) advantage of the new mixed inventory over demisyllables (68% versus 32% preference rate) and over diphones (59% versus 41% preference rate). A result in its own right is the preference of diphones to demisyllables (55% versus 45%), which is also significant (χ^2-test, $p < 0.04$).

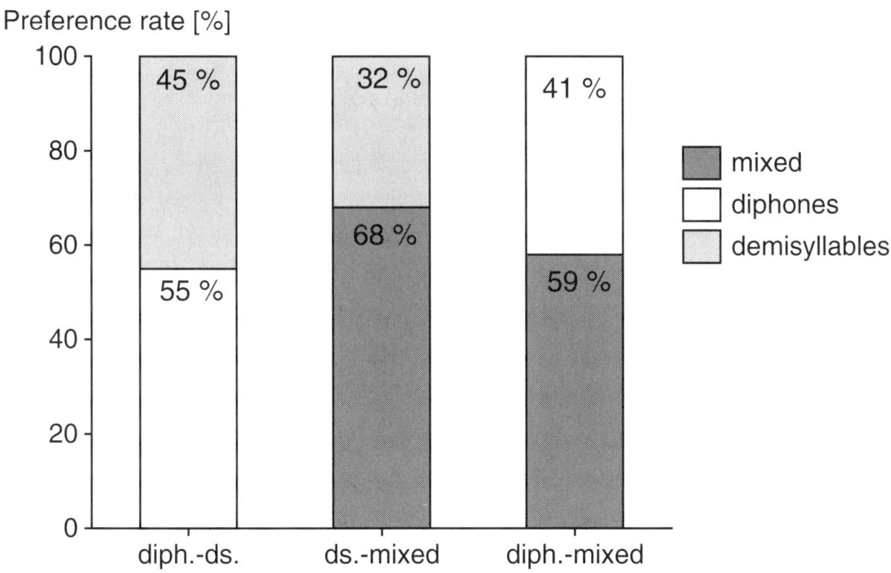

FIGURE 21.4. Results of a pair comparison among the mixed inventory, a demisyllable inventory, and a diphone inventory.

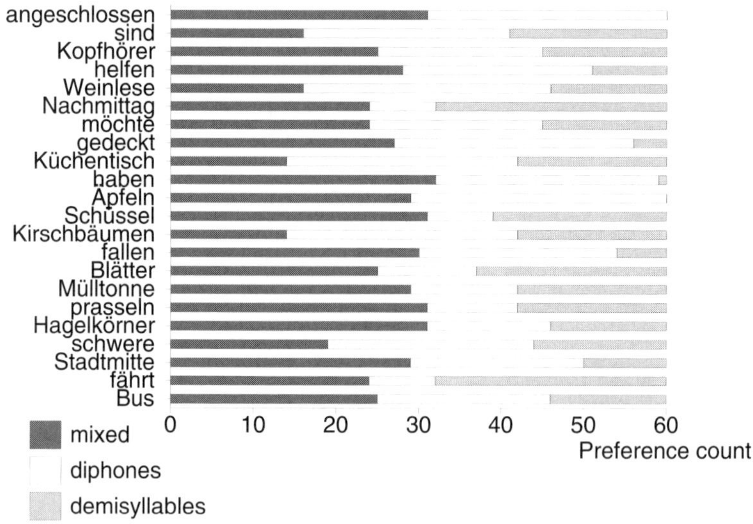

FIGURE 21.5. Results of a comparison among the mixed inventory, a demisyllable inventory, and a diphone inventory. The preference number is displayed for each test word.

A separate analysis for each test word (figure 21.5) shows that the diphone versions of some words performed poorly, especially for words with postvocalic liquids (for instance, *fährt* /fɛːɐt/ (figure 21.6) or *Mülltonne* /mʏltɔnə/), and the demisyllable versions of some other words with large coarticulatory effects across syllable boundaries had very bad preference values (for instance, *angeschlossen* /angəʃlɔsən/ (figure 21.7) or *Stadtmitte* /ʃtatmɪtə/). The mixed inventory structure produced no severe outliers and turned out to avoid the weaknesses of the standard paradigms.

21.4.2 Segmental Intelligibility

The segmental intelligibility was explored by the SAM segmental test [CGN90, Por93]. The test assesses consonant intelligibility in CV, VC, and VCV contexts. All German consonants are combined with the vowels /a,i,u/. In the VC, the CV, and the VCV contexts, 36, 48 and 56 items were included, respectively. The open response form allows analysis of the confusions between the sounds without external constraints. The synthesized stimuli were recorded on a DAT tape and presented over headphones. Eighteen subjects participated.

Figure 21.8 displays the results for the three contexts in comparison with the values obtained in an earlier investigation [Por93] for a human voice and the synthesis system HADIFIX [PSPSH92]. There is a noticeable decrease in the error rates for the new synthetic voice with the mixed inventory (VC: 1.9%, CV: 6.0%, VCV: 6.9%) compared to the HADIFIX voice. Indeed, in the VC and CV contexts the differences between human and synthetic voice are not significant

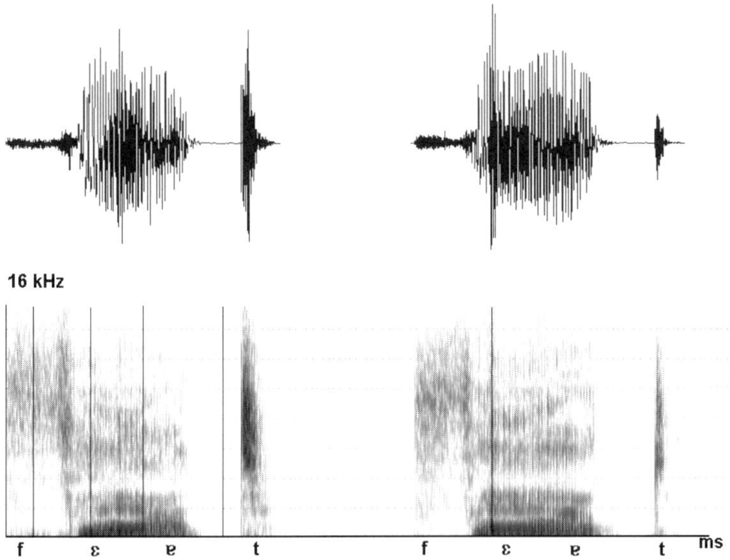

FIGURE 21.6. Comparison of two versions of the word *fährt* /fɛːɐt/ (left, diphone version; right, demisyllable/mixed version). The dashed lines indicate positions where two units meet.

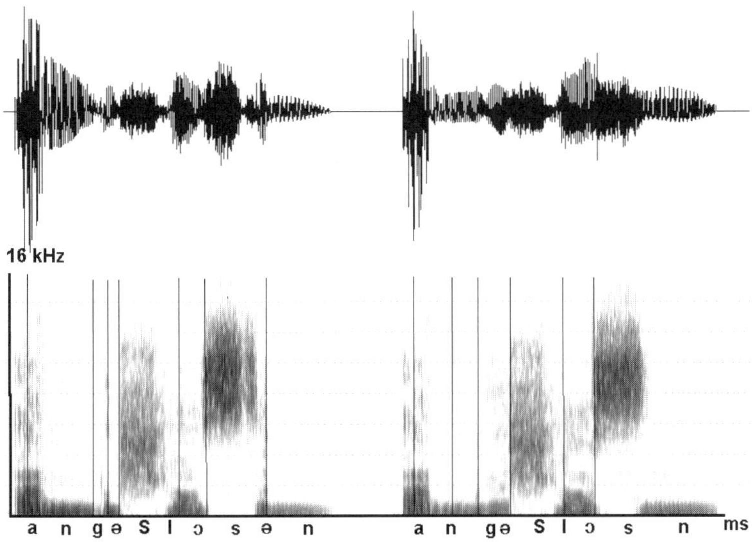

FIGURE 21.7. Comparison of two versions of the word *angeschlossen* /angəʃlɔsən/ (left, demisyllable version; right, mixed version). The dashed lines indicate positions where two units meet. The final syllabic nasal on the right picture could not be modeled with the demisyllable inventory used in this test.

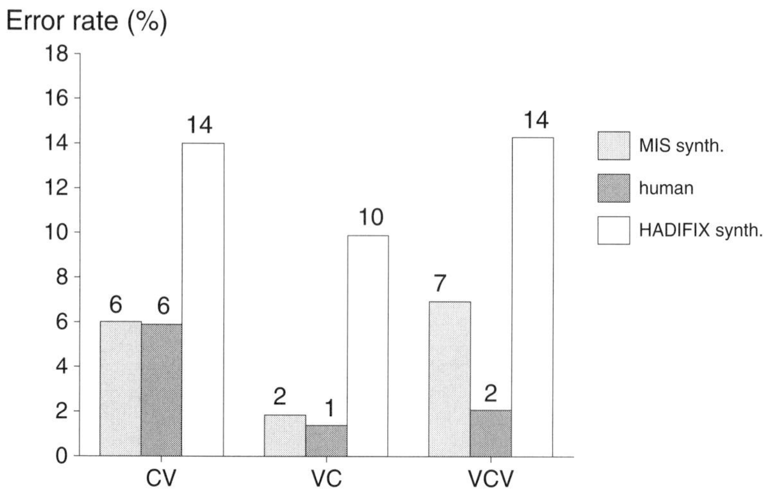

FIGURE 21.8. Segmental intelligibility of VC, CV, and VCV consonants. The mixed inventory (MIS) is compared to the HADIFIX inventory and to a human voice.

(t-test, $p > 0.4$). In the VCV context, however, the error rate for the MIS voice is still three times higher than the one for a human voice. Possible sources of these errors for the MIS voice are:

- *VC context.* There is only one case of systematic misinterpretation. The stimulus /an/ was understood as /aŋ/ in one-third of all responses. A more carefully pronounced unit could solve this problem. Otherwise, the synthetic voice is as intelligible as a human voice.

- *CV context.* More than 50% of all errors were caused by misunderstanding /hi:/ as /ti:/ or /çi:/; and /hu:/ as /bu:/, /tu:/, /ku:/, or /fu:/. Similar results were obtained for the human voice. The high error rates for /h/ seem not to be caused by bad synthesis but rather by difficult discrimination of /h/ with closed vowels; /ha:/ was always recognized correctly. Other error sources were /b/ (9.8% error rate), /ʃ/ (11% error rate), and /ç/ (25% error rate). Most of these errors occured in combination with /i:/. No other systematic problems were detected.

- *VCV context.* In this context, /h/ was also problematic (44.4% error rate). Voiced apical obstruents were perceived as voiceless (6.1% error rate). Nasals were sometimes confused (13.0% error rate), /ana:/ with /aŋa:/; /ini:/ with /imi:/; and, very often, /uŋu:/ with /unu:/. Synthetic intervocalic nasals and /h/ must be modeled in a better way to reach the segmental intelligibility of their natural counterparts. Intervocalic /h/ is, in fact, the only intervocalic consonant for which hard concatenation is performed; this may be the primary source of the increased confusion rate compared to the other consonants.

The segmental intelligibility of the MIS inventory almost meets the standard set by the intelligibility of human voices (at least for this simple test under laboratory conditions).

21.5 Summary

This paper has described an inventory structure based on seven types of units. The definition was based on experiments investigating the relevant acoustic and phonetic facts. Elaborate concatenation rules allow the synthesis of natural-sounding speech as confirmed in two different evaluations: a pair comparison test with diphones and demisyllables, and a segmental intelligibility test.

Acknowledgments: This research was supported within the language and speech project VERBMOBIL by the Bundesministerium für Bildung, Wissenschaft, Forschung und Technologie, and by the Deutsche Forschungsgemeinschaft. We thank two anonymous reviewers for their helpful comments, Dieter Stock for his support, Volker Kraft for valuable discussions, and all the participating subjects for their patience.

References

[BCEW93] O. Boeffard, B. Cherbonnel, F. Emerard, and S. White. Automatic segmentation and quality evaluation of speech unit inventories for concatenation-based, multilingual PSOLA text-to-speech systems. In *Proceedings Eurospeech'93*, Berlin, Germany, 1449–1452, 1993.

[Bou88] V. J. Boucher. A parameter of syllabification for VstopV and relative timing invariance. *J. Phonetics* 16:299–326, 1988.

[CI94] A. Conkie and S. Isard. Optimal coupling of diphones. In *Second ESCA/IEEE-Workshop on Speech Synthesis*, New Paltz, NY, 119–122, 1994.

[CGN90] R. Carlson, B. Granström, and L. Nord. Segmental evaluation using the ESPRIT/SAM test procedures and monosyllablic words. In *First ESCA-Workshop on Speech Synthesis*, Autrans, France, 257–260, 1990.

[DC91] R. Drullman and R. Collier. On the combined use of accented and unaccented diphones in speech synthesis. *J. Acoust. Soc. Amer.* 90:1766–1775, 1991.

[Fuj75] O. Fujimura. Syllable as the unit of speech synthesis. Unpublished paper.

[FML77] O. Fujimura, M. J. Macchi, and J. B. Lovins. Demisyllables and affixes for speech synthesis. In *Ninth ICA*, Madrid, 513, 1977.

[Fuj79] O. Fujimura. An analysis of English syllables as cores and affixes. *Zeitschrift für Phonetik, Sprachwissenschaft und Kommunikationsforschung* 4/5:471–476, 1979.

[Hei79] G. Heike. Prerequisites of speech synthesis on the basis of an articulatory model. *AIPUK* 12:91–99, 1979.

[Kea84] K. A. Keating. Phonetic and phonological representation of stop consonant voicing. *Language* 60:286–319, 1984.

[Koh77] K. Kohler. *Einführung in die Phonetik des Deutschen.* Erich Schmidt, Berlin, 1977.

[Ko90] K. Kohler. Segmental reduction in connected speech in German: Phonological facts and phonetic explanations. In *Speech Production and Speech Modeling*, W. J. Hardcastle and A. Marchal, eds. Kluwer, Dordrecht, 69–92, 1990.

[KA92] V. Kraft and J. Andrews. Design, evaluation, and acquisition of a speech database for German synthesis-by-concatenation. In *Proc. SST-92*, Brisbane, Australia, 724–729, 1992.

[Kra94] V. Kraft. Does the resulting speech quality improvement make a sophisticated concatenation of time-domain synthesis units worthwhile? In *Second ESCA/IEEE-Workshop on Speech Synthesis*, New Paltz, NY, 65–68, 1994.

[KW56] K. Küpfmüller and O. Warns. Sprachsynthese aus Lauten. *Nachrichtentechnische Fachberichte* 3:28–31, 1956.

[MM61] C. Martens and P. Martens. *Phonetik der deutschen Sprache.* Hueber, Munich, 1961.

[MC90] E. Moulines and F. Charpentier. Pitch-synchronous waveform processing techniques for text-to-speech synthesis using diphones. *Speech Comm.* 9:453–467, 1990.

[Oli90] J. Olive. A new algorithm for a concatenative speech synthesis system using an augmented acoustic inventory of speech sounds. In *First ESCA-Workshop on Speech Synthesis*, Autrans, France, 25–30, 1990.

[PS60] G. E. Peterson and E. Sievertsen. Objectives and techniques in speech synthesis. *Language and Speech* 3:84–95, 1960.

[PSPSH92] T. Portele, B. Steffan, R. Preuss, W. F. Sendlmeier, and W. Hess. HADIFIX—A speech synthesis system for German. In *Proceedings ICSLP'92*, Banff, Alberta, Canada, 1227–1230, 1992.

[Por93] T. Portele. Evaluation der segmentalen Verständlichkeit des Sprachsynthesesystems HADIFIX mit der SAM-Testprozedur. In *Fortschritte der Akustik - DAGA'93*, Frankfurt, Germany, 1032–1035, 1993.

[Por94] T. Portele. *Ein phonetisch-akustisch motiviertes Inventar zur Sprachsynthese deutscher Äusserungen.* Dissertation, University of Bonn, 1994.

[Rap36] K. M. Rapp. *Versuch einer Physiologie der Sprache nebst historischer Entwicklung der abendländischen Idiome nach physiologischen Grundsätzen.* Cotta, Stuttgart-Tübingen, 1836.

[SKT84] A. G. Samuel, D. Kat, and V. Tartter. Which syllable does an intervocalic stop belong to? A selective adaptation study. *J. Acoust. Soc. Amer.* 76:1652–1663, 1984.

[Sto71] D. Stock. *Untersuchungen zur Stimmhaftigkeit hochdeutscher Phonemrealisierungen.* Buske, Hamburg, 1971.

[Twa38] W. F. Twadell. A phonological analysis of intervocalic consonant clusters in German. In *Actes du 4e congrès int. des linguistes*, Copenhagen, Denmark, 218–225, 1938.

[Wan60] H.-H. Wängler. *Grundriss einer Phonetik des Deutschen.* Elwert, Marburg, 1960.

[Wha90] D. H. Whalen. Coarticulation is largely planned. *J. Phonetics* 18:3–35, 1990.

Appendix: Audio Demos

The CD-ROM contains sound examples that demonstrate our system.

22

Prosody and the Selection of Source Units for Concatenative Synthesis

Nick Campbell
Alan W. Black

ABSTRACT This chapter describes a procedure for processing a large speech corpus to provide a reduced set of units for concatenative synthesis. Crucial to this reduction is the optimal utilization of prosodic labeling to reduce acoustic distortion in the resulting speech waveform. We present a method for selecting units for synthesis by optimizing a weighting between continuity distortion and unit distortion. The source-unit set is determined statistically from a speech corpus by representing it as the set of sound sequences that occur with equal frequency, i.e., by recursively grouping pairs of segment labels to grow nonuniform-length compound label-strings. Storing multiple units with different prosodic characteristics then ensures that the reduced database will be maximally representative of the natural variation in the original speech. The choice of an appropriate depth to which to prune the database reflects a trade-off between compact size and output voice quality; a larger database is more likely to contain a prosodically appropriate segment that will need less modification to reach a target setting in the concatenated utterance.

22.1 Introduction

As large corpora of natural speech are becoming more widely available, we can consider them not just as source materials for the modeling of language and speech characteristics, but also as a source of units for concatenative synthesis.

Concatenative synthesis systems have traditionally employed only a small number of source units, using one token, or waveform segment, per type of unit in the source-unit inventory. Such systems produce highly intelligible speech quickly and economically, while using only a small amount of computer memory and processing. However, as yet, they have failed to produce really natural-sounding speech.

Part of the reason that even concatenative synthesis still sounds artificial is that the source units for concatenation are typically excised from recordings of carefully read lab speech, which although *phonemically representative*, is constrained to be *prosodically neutral*. The speech tokens used as source units thus encode the relevant static spectral characteristics (or configurations of the vocal tract) for a given sound sequence but fail to adequately model the different dynamic articulatory characteristics of that sequence when it is reproduced in different meaningful contexts.

The prosodic variations in human speech across different speaking styles or between different parts of the same utterance are normally accompanied by corresponding variation in phonation— that is, by changes in voice quality such as the emphasis of higher-frequency energy that comes from the longer closed phase of "pressed voice" or the steeper roll-off of energy that comes with "breathy voice" [Jon95]. Whereas current signal-processing techniques used in concatenative synthesis are adequate to warp the prosody of source units to model coarse variations in fundamental frequency (f_0), duration, and energy, they fail to adapt the spectral characteristics of the concatenated units to encode these fine phonation differences and, as a result, the synthetic speech sounds artificial or hyperarticulated. The simple modification of prosody without an equivalent modeling of its phonation effects is insufficient.

Furthermore, only a limited number of speech tokens are used in the generation of a large variety of speech utterances, and therefore considerable degradation can result from the amount of signal processing required to modify their prosody to suit the wide variety of target contexts. For example, in matching the duration of a unit, it is customary to repeat or delete waveform segments to artificially create a longer or shorter sound. This is inherently damaging to naturalness.

The solution we propose for the above problems requires a larger inventory of speech tokens for source units, allowing several tokens for each type[1] so that the token closest to the target context can be used in synthesis to minimize subsequent signal processing. Because designing such a large inventory can be difficult, and recording such a database can be very time-consuming, we have developed tools to process existing corpora of natural speech to extract suitable units for concatenative synthesis.

Extraction of a maximally varied and representative set of speech tokens from a given speech source requires three stages of processing: (1) segmental and prosodic labeling of the speech corpus, (2) analysis of frequencies and distributions of each segment type, and (3) selection of a reduced-size but optimally representative set of source tokens to cover the variations encountered in each type.

22.2 Segmental and Prosodic Labeling

A basic requirement for a speech corpus to be processed is that it has an accompanying orthographic transcription. This can be aligned segmentally by generating the phone sequences that would be used to synthesize it and using Markov modeling

[1] We henceforth use the term *type* to refer to monophone, diphone, or polyphone *classes*, and the term *token*, to refer to actual *instances* of each type (i.e., waveform segments taken from a speech database).

to perform the segmentation. Manual intervention is still required in the segmental labeling.

Prosodic labeling is automatic, given the sequence of phone labels and the speech waveform. For each segment in the corpus, measures are currently taken of prosodic dimensions that are derived automatically from combined output of the ESPS *get f0* and *fft* programs [ERL93]. With the exception of "duration," the measures are taken at 10-ms intervals throughout the speech signal and then averaged over the duration of the phone-sized segments of speech as delimited by the labels:

- duration
- fundamental frequency
- waveform envelope amplitude
- spectral energy at the fundamental
- harmonic ratio
- degree of spectral tilt

These values are then z-score normalized [Cam92a] for each phone class to express the difference of each segment from the mean for its type in terms of the observed variance for other tokens of that type for each of the above dimensions. To model the position of each token in relation to changes within the respective prosodic contours, first differences of these normalized values are then taken over a window of three phones to the left and right of each segment. The sign of the result indicates whether a segment is part of a rising (*e.g.*, increasing pitch, loudness, or length) or falling contour. The magnitude indicates the rapidity of the change.

From this prosodic information (which is also later used to determine optimal units in the selection process for synthesis), it is possible to discriminate between tokens in otherwise identical segmental contexts.

22.3 Defining Units in the Database

In a labeled speech corpus, the number of *types*, as defined by the monophone labels on the segments, is small (on the order of 20 to 50 for most languages), whereas the number of *tokens* of each type depends on the size of the corpus but also varies considerably between types, from very large (for a few vowels) to extremely few (for some rare consonants). The type and token distributions are defined both by the language and by the contexts from which the speech data were collected; if the corpus is sufficiently large, then all sound types of the language will be represented, but the number of variants for some will be few.

The aim when processing such a corpus to produce synthesis units is to preserve all tokens of the rare types while eliminating duplicate or redundant tokens from the more common types. Because system storage space is often limited, and, even with efficient indexing, searching for an optimal token in a large database can be very time-consuming, efficient pruning of the corpus is necessary in order to control the size of the source database while maximizing the variety of tokens retained from it.

As there are never likely to be two tokens with identical waveform characteristics, "duplicate" is defined here to imply proximity in the segmental and prosodic space, as limited by the requirements of the storage available for units in the synthesis system. That is, subject to system capacity, we can store a number of different tokens of each type to ensure maximally diverse coverage of the acoustic space.

22.3.1 Segmental Types and Tokens

The relation between the prosody of an utterance and variation in its spectral characteristics has long been known [GS89, Tra91]. Lindblom [Lin90] described the continuum of hyper- and hypospeech observed in interactive dialogues, by which speakers tune their production to communicative and situational demands. Sluijter and van Heuven [Sv93, Sv94], also citing work on overall "vocal effort" such as [GS89], showed that stressed sounds in Dutch are produced with greater local vocal effort and hence with differentially increased energy at frequencies well above the fundamental. More recently, Campbell and Beckman [CB95] confirmed that for English too, spectral tilt is affected by linguistic prominence.

It is not yet easy to quantify such phonation-style–related differences in voice quality directly from a speech waveform, but fortunately the differences correlate with grosser prosodic features such as prominence and proximity to a prosodic-phrase boundary [WC95, Cam92b]. To select a subset of tokens that optimally encodes the phonation-style variants of each segment type, we therefore adopt a functional approach and determine instead the contexts in which the prosody is markedly different. Because the weak effects of phonation co-occur with the stronger prosodic correlates, we label the strong to also encode the weak. Thus it is not necessary to be able to detect the changes in phonation style directly in a speech corpus; rather we can capture them from the gross and more easily detectable differences in F_0 and duration to encode the speech segments.

22.3.2 Determining the Units

When reducing the size of a speech corpus to produce source units for concatenative synthesis, we need to store the most representative tokens of each type, ideally several of each, to be sure of enough units to cover prosodic variation, but no more than necessary to model the language. The first step is therefore to determine a set of types that uniformly describes the distribution characteristics of phones in the corpus.

Several methods have already been suggested for the automatic extraction of units for a speech synthesis database [NH88, SKIM92, Nak94, Iwa93, Tak92], but these have concentrated on maximizing the variety of segmental contexts. We emphasise here the importance of also considering the prosodic context.

Under the assumption that listeners may be more sensitive to small variation in common sound sequences (such as in function-words) and more tolerant of variant pronunciation in less common words or in names, we cluster the original monophone labels by frequency to determine the common sequences in the corpus. In this way, the number of "types" is increased by clustering the original phone labels to form nonuniform-length sequences (compound phone-strings) for an optimal representation of the common sound sequences in the corpus.

Because function words are very common in speech, they will tend to emerge more readily from the clustering process. As these words are often produced with special reduction in fluent speech, this clustering allows them to be automatically modeled separately as discrete units, without the need for special linguistic analysis of the corpus.

The algorithm for determining the unit set, given in pseudocode in figure 22.1, is derived from [SKIM92] but uses counts instead of entropy. In a process similar to Huffman coding, the most frequently-occurring label is conjoined with its most frequently co-occurring neighbor to produce a new compound type, and the cycle is repeated using the increased set. At each iteration the number of types grows and the token count of the two most frequent conjoined types correspondingly decreases.

The loop terminates when the threshold that specifies the maximum number of tokens for any type has been reached. This threshold (and by implication, the ultimate number of types) is arbitrarily chosen, according to the initial size of the corpus and the degree of reduction required, to be approximately five times the number of tokens required from each type. The greater the number of tokens per type, the better the prosodic flexibility of the final unit set.

This stage of processing yields a set of new labels, which describe the frequency of co-occurrence of the speech sounds in the corpus. The resulting set of compound types ensures that (with the exception of the very few sparse-token types) each unit is equally likely in the language being modeled. Infrequently occurring types (such as /zh/ in English or /tsa/ in Japanese) will not be clustered, and all tokens of each are preserved.

The remaining tokens of the more common types can now be reduced to prune down the size of the database to meet system limitations. However, rather than simply select the one most typical token of each type, we preserve a number of each, up to the limits of storage space.

22.3.3 Pruning the Database

To ensure best coverage of the prosodic space, we next select the n most prosodically diverse tokens to represent each type of unit. Vector quantization is performed to cluster all tokens of each type in turn according to their prosodic characteristics. For each type, the tokens closest to the centroid of each cluster are then taken as representative.

```
main-loop
set threshold = n
initialize: types = labels

while( number_of_tokens( any type ) > threshold ) do
    current_type = max( count tokens of each type )
    for( most frequent type) do
        max = find_most_frequent_neighbour( current_type )
     return( compound_type ( current, max ))

find_most_frequent_neighbour( current )
    for( tokens in database )
        if( type == current )
        count left_neighbour types
        count right_neighbour types
    return( max( left_neighbour types, right_neighbour types ))

compound_type ( current, most_freq )
    if( most_freq == left_neighbour)
        return( concat( most_freq_neighbour + current_label ))
    else
        return( concat( current_label + most_freq_neighbour ))
```

FIGURE 22.1. Rectangularization algorithm: Subdivide the frequent units and cluster them with their most common neighbors to form longer units.

The number of clusters (n) specifies the depth n to which to prune the database. Thus the ultimate size of the source-unit database will be slightly less than n times the number of types (there usually being less than n tokens of the rarest few types). The size of the quantization codebook is thus determined as a function of the number of tokens to be retained for each type.

The choice of an appropriate depth to which to prune a large source corpus is a trade-off between compact size and synthetic speech quality; a larger source-unit database is more likely to contain a prosodically appropriate segment that will need less modification to reach a target setting in the concatenated utterance. If only one token were to be stored for each type (i.e., to produce the smallest and least flexible unit set that still covers the segmental variety in the corpus), then we would simply choose the token closest to the centroid for each type, without using prosodic vector quantization, to be sure of having the most typical token.

By taking more than one token to represent each type, we are assured not only that all contextual segmental variation in the original corpus will be preserved, but also that we will be better able to match different prosodic environments; the more tokens, the more flexibility of coverage. In this way we no longer have to synthesize using the one most typical token in all prosodic contexts, but can select from several. One of the supposedly redundant units, whose segmental environment is the same but which actually differ in prosodic aspects, can be selected to minimize the amount of waveform distortion needed to match the target prosody for a particular utterance.

22.4 Prosody-Based Unit Selection

Each unit in the database is labeled with a set of features. These features include phonetic and prosodic features (such as duration and pitch power) to which are added acoustic features (such as cepstral frame quantization). The features available are database dependent, although we expect at least phone label, duration pitch, and power. As far as possible, features are specified in terms of z-scores so distances are normalized between features. Other features are used in a database unit description that do not directly affect selection (e.g., position in the waveform file).

For selection of units for synthesis, the target segments (predicted by earlier components of the synthesizer, or for testing purposes taken from natural speech) are specified with a subset of these features to specify the characteristics of the utterance and its prosody.

Because of the richness of this information, we do not (although we could if our databases were so labeled) use all the acoustic measures described in [SKIM92] for selection. We are testing the assumption that appropriate prosodic characterization will capture the acoustic differences in the units, and that we can thereby avoid computationally expensive acoustic measures to achieve faster unit selection.

22.4.1 *The Measures of Distortion*

In this selection model, we define two types of distortion to be minimized to find the best sequence of units.

- *Unit distortion* is defined as the distance $Du(s_i, t_i)$ between a selected unit and a target segment, i.e., the weighted mean distance between non–zero-weighted features of the selected unit feature vector $\{sf_1, sf_2, ..., sf_n\}$ and the target segment vector $\{tf_1, tf_2, ..., tf_n\}$. Distances are normalized between 0 (good) and 1 (bad). Weights too are between 0 and 1, causing Du to also lie between 0 and 1.

- *Continuity distortion* is the distance $Dc(s_i, s_{i-1})$ between a selected unit and its immediately adjoining previous selected unit. This is defined similar to Du as a weighted mean distance between non–zero–weighted features of a selected unit's feature vector and that of the previous unit.

The weights for significant (i.e., non–zero–weighted) features in unit distortion and continuity distortion will be different, as will be the choice of features. Vectors may also include features about a unit's context as well as the unit itself.

Varying the weights allows the relative importance of features to change, for example allowing pitch to play a greater role in selection than duration. The values may also be zero, thus completely eliminating a feature from the selection criteria. The weights for unit distortion will differ from the weights for continuity distortion.

The following diagram illustrates this distinction between distortion measures.

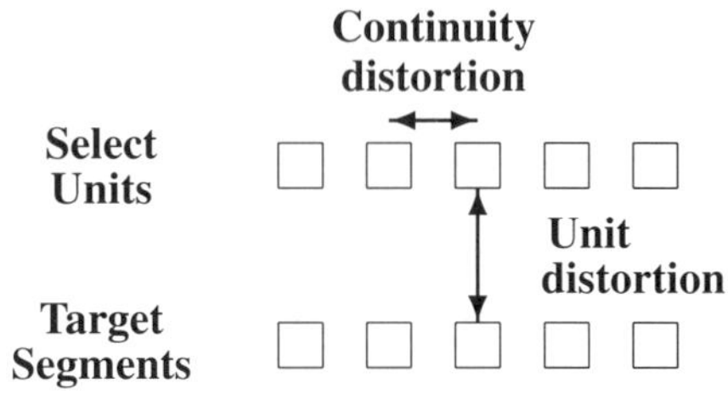

The *best* unit sequence is the path of units from the database that minimizes

$$\sum_{i=1}^{n}(Dc(s_i, s_{i-1}) * WJ + Du(t_i, s_i) * WU)$$

where n is the number of segments in the target, and WJ and WU are further weights. Maximizing WJ with respect to WU minimizes the distortion between selected units at the (possible) expense of distance from the target segments.

Defining the optimal value of the weights so that the *best* selection produces the perceptually best-quality synthesis is nontrivial. Some measure is required to determine if the *best* selection is perceptually better to a listener.

Human perceptual tests constitute one measure, but they are prone to errors and are neither very discriminatory at fine detail nor automatic. Another more objective measure is the mean Euclidean cepstral distance [RJ93, pp 150–171] between (time-aligned) vectors of selected units and target segments. In the special test case of mimicking a natural speech utterance from the speaker of the source database, the target features are completely known, and an objective quantification of the degree of match can be obtained. But it is open to question how closely the cepstral distance measure predicts human perceptual characteristics. This issue is addressed further below.

A beam search algorithm was used to find the best units based on the above minimization criteria. Originally for each database all weights were hand-tuned without any formal measurement (and interestingly they were noticeably different for different databases). The following procedure is now employed to optimize these weights automatically and determine the best combination for each database.

- First an utterance is removed from the database so its segments are no longer available for selection. Its natural segments are used to define the parameters of the test utterance to ensure that testing is independent of any higher levels of the synthesis system.

- The beam search algorithm is used to find the best selection of units that minimize the distance described above, with respect to a set of weights.

- The cepstra of the selected units are time aligned with those of the target segments, and the mean Euclidean cepstral distance between the target (original) segments and the selected (replacement) units is calculated.

- The process is repeated with varying weights until they converge on a minimum mean cepstral distance.

This process is not ideal, as minimizing cepstral distance may not maximize the quality of the synthesized speech, but it is an objective measure which as we will show offers some correlation with human perception. Problems with this measure and some ways it may be improved are discussed next.

22.5 Evaluation

Unit selection is only a small part of the whole synthesis process, and for full synthesis, higher-level modules generate a segment description specifying the appropriate features (e.g., predicting values for f_0, duration, power), and subsequent signal processing is applied to the waveforms of the selected units, modifying them to the desired target values and reducing any discontinuities at the joins between units.

However, for the tests presented here, we were concerned only with unit selection and did not want to confound our results with any errors from higher-level modules. We therefore used as a target the prosody and phone sequences of original natural utterances, and performed no subsequent signal processing. In this way we were able to test the selection of units under controlled conditions without having to take into consideration possible effects of other modules.

22.5.1 Test 1: Full Database

By using a medium-size database of English female radio announcer speech (44,758 phone labels) [OPS95], we varied weightings of four individual features used in measuring unit distortion and overall continuity distortion (WJ). The four features were: local phonetic context, pitch, duration, and power.

To evaluate the relationship between objective cepstral and subjective perceptual distance measures, we asked subjects to score a set of utterances selected using different weights, and compared their scores with the cepstral measures. Six subjects were presented with speech synthesized by concatenation according to the different weightings.

The material consisted of seven combinations of features for selecting two sentences, each repeated three times, in random order (giving 45 utterances in all, including three dummy sentences to allow for acclimatization).

Tests were administered by computer; no records were kept of the time in each waveform during which subjects detected a poorly selected unit, only of the response counts. Subjects were asked to simply indicate perceived discontinuities ("bad segments") in each test utterance by pressing the return key, and to ignore any "clicks" arising from simple abutting of segments. In the ideal case, in which suitable segments were selected, the amount of noise was minimal because like was being joined with like. The typical segment length was about two phones, and in the average case, a discontinuity was noticeable for one in four pairs.

In practice, perceptual scores varied considerably between respondents, with some appearing much more sensitive to abutment noise than others after counts were normalized per speaker, however, analysis of variance showed no significant effect for speaker, nor for utterance, but a clear difference for selection weighting type ($F(6, 231) = 5.865$), confirming that the preferences were in agreement in spite of the differences in individual sensitivity.

The following table compares results from the perceptual test with the cepstral distances for some range of weights. The perception measure represents the average score for each selection type, normalized by subject and target sentence, in standard deviation units (PC is phonetic context).

WJ	$WU = 1.0$				Perceptual Measure	Cepstral Distance
	PC	Power	Dur	Pitch		
0.6	0.6	0.0	0.333	0.333	−0.55	0.2004
0.6	0.2	0.666	1.0	0.0	−0.49	0.2004
0.2	0.2	1.0	0.666	1.0	−0.38	0.1967
0.2	0.2	1.0	0.333	0.333	−0.07	0.1981
0.6	0.4	0.333	1.0	0.333	0.12	0.1981
0.8	0.4	0.0	0.0	0.0	0.24	0.2050
0.8	0.6	0.0	0.0	0.0	0.55	0.2056

The cepstral distance seems to give more importance to unit distortion at the expense of continuity distortion. Human perception favors more weight on WJ (i.e., less continuity distortion). This is because the cepstral measure takes the mean error over each point. Therefore, continuous closeness is favoured over short "burst errors" that occur at bad joins. Humans, however, are upset by burst errors as well as prosodic mismatches, and hence prefer a balance of WJ to WU. Obviously a better automatic distance measure is required that appropriately penalizes burst errors. Although the numeric distances of the cepstral measure are small, the quality of the synthesis varies from very jumpy almost unrecognizable speech to undetectable unit concatenation producing natural-sounding speech.

22.5.2 *Test 2: Reduced Database*

A further test was performed with a reduced database of Japanese. The units for the synthesizer were selected from a corpus of 503 magazine and newspaper sentence readings. In Japanese, which is not a stress-based language, there is not as great a range of prosodic variation as in English, and the damage to naturalness caused by subsequent signal processing is less, but the inclusion of prosodic information in the selection process ensured selection of more appropriate units according to an acoustic measure of spectral characteristics.

The source database consisted of 26,264 phone segments, and included 70 original labels (including segment clusters that could not be reliably separated by hand labelers). After processing, these formed 635 non-uniform units ranging in length from 1 to 7 original labels. It was pruned to a maximum depth of 35 tokens each.

Target segment labels and raw prosodic values of duration, mean pitch, and energy for each phone were extracted from a test set of 100 randomly selected sentences, and each original sentence was removed from the database before resynthesis to ensure that no original units were included. The resynthesized version was then compared with the original, using measures of cepstral similarity. Comparisons were made between the original recording of each sentence and resynthesized versions with and without prosodic selection.

Nonweighted Euclidean measures of the cepstral distance between the original utterance and each resynthesized version were calculated on a phone-by-phone basis using 12 coefficients per 10-ms frame from LPC cepstral coding of the waveforms. Results confirmed that an improvement in spectral match was gained by inclusion of prosodic information in the selection (*seg only vs. seg+pros:* $t = 4.484$, $df = 6474$, $p < 0.001$).

Quartiles of the Euclidean cepstral distance measure

	min	25%	median	75%	max
segmental context alone:	0.0071	0.3965	0.8581	1.7219	8.8546
segmental & prosodic ctxt:	0.0073	0.3167	0.6232	1.4390	10.3748

The improved spectral match in turn confirms a strong connection between prosodic and segmental characteristics of the speech waveform, and shows that the inclusion of prosodic information in the selection of units can result in more natural-sounding synthetic speech.

22.6 Discussion

We have shown that prosodic variation has more than a small effect on the spectral characteristics of speech, and that advantage can be taken of this in the selection of units for concatenative synthesis. We have also shown that a database of nonuniform units can be automatically generated from a labeled corpus and that the prosodic characteristics of contour shape and excursion can be automatically coded. Nothing above will make up for the lack of an appropriate unit in a corpus, and careful choice of this resource is essential; however, a way of making better use of the supposedly redundant duplicate tokens has been suggested.

Most concatenative synthesis methods still employ a relatively small fixed number of source units, under the assumption that any modification of their inherent pitch and duration can be performed independently at a later stage through signal processing. The distortion of the synthesized speech, which is introduced as a result of changing a segment's prosodic characteristics, has until recently been masked by the generally poor, mechanical quality of the generated speech after it has passed through a coding stage. However, as synthesis quality has improved, and as the memory limitations of earlier systems are eased, it now becomes necessary to reconsider the merits of such small unit sets.

Future refinements to objective measurement procedures must include a bias to the cepstral distance measure to increase sensitivity to local concatenation points ("burst errors") and hence better approximate the human preferences. It should also be noted that such acoustic measures, because of the necessary time-alignment, are blind to inappropriate durations, and will not degrade under suboptimal timing patterns, so for this reason, too, they may not correlate well with human perception.

22.7 Conclusion

This chapter has addressed several aspects of the concatenative synthesis process. We have shown that a large corpus of naturally occurring speech can be used as a source of units for concatenative synthesis, and we have described the tools and processes we use for this.

The prosodic labeling is generated automatically from the speech data. The method is thus relatively language independent, and relies only on (a) an adequate size corpus from which to draw the units, and (b) a suitable language interface module by which to generate the transcriptions and predict the prosody for the utterance to be synthesized.

By specifying prosodic variation in terms of variance about a mean, and the slope of prosodic contours in terms of the differential of the normalized measures, we gain the advantage of speaker-independence in our synthesizer processes. The higher-level prosody prediction modules can now specify their targets in normalized terms, and whatever the database, regardless of the prediction values, the retrieved values are constrained to be within the natural range for the speaker's voice. Describing the pitch of a segment as, for example, "moderately high and rising" ensures that the closest unit in the database will be selected, and in many cases the difference between the desired target pitch and retrieved unit's original will be small enough to be perceptually insignificant.

Our unit-generation algorithm produces a unit set that models the collocation frequencies of the input data in terms of its own labels by grouping them into equally likely nonuniform compound units.

At the waveform level, we have questioned the validity of applying signal-processing techniques to warp the prosody of a speech segment, preferring instead to select appropriate units to minimize such postselection distortion. We have shown simple and efficient ways to do this.

Experience with this system encourages us to believe that in the majority of cases it is better to relax our target goals in the direction of the database events rather than to impose an unnatural (and possibly distorting) pitch or duration on the waveform.

The method is currently being tested with several databases from different speakers of both English and Japanese, under different labeling conventions, and appears immune to differences in language or labeling conventions.

Acknowledgments: The authors thank Yasuhiro Yamazaki for supporting this research, and Yoshinori Sagisaka, Norio Higuchi, and two anonymous reviewers for comments on an earlier version of this chapter.

References

[Cam92a] W. N. Campbell. Syllable-based segmental duration. In *Talking Machines: Theories, Models, and Designs*, G. Bailly, C. Benoit, and T. R. Sawallis, eds. Elsevier, North-Holland, Amsterdam, 211–224, 1992.

[Cam92b] W. N. Campbell. Prosodic encoding of English speech. In *Proceedings ICSLP-92*, Banff, Alberta, Canada, 663–666, 1992.

[CB95] W. N. Campbell and M. E. Beckman. Stress, loudness, and spectral tilt. In *Proceedings Acoustical Society Japan*, spring meeting, 1995.

[deJ95] K. de Jong. The supraglottal articulation of prominence in English: Linguistic stress as localised hyperarticulation. *J. Acoust. Soc. Amer.* 97(1):491–504, 1995.

[ERL93] Entropic Research Laboratory, Inc. *ESPS/Waves+Entropic Research Laboratory*, Washington DC, 1993.

[GS89] J. Gauffin and J. Sundberg. Spectral correlates of glottal voice source waveform characteristics. *J. Speech and Hearing Res.* 556–565, 1989.

[IKS93] N. Iwahashi, N. Kaiki, and Y. Sagisaka. Speech segment selection for concatenative synthesis based on spectral distortion minimisation. *Trans. IEICE* E76-A:11, 1993.

[Lin90] B. E. F. Lindblom. Explaining phonetic variation: A sketch of the H&H theory. In *Speech Production and Speech Modelling*, H. J. Hardcastle and A. Marchal, eds. Kluwer, Dordrecht, 403–409, 1990.

[Nak94] S. Nakajima. Automatic synthesis and generation of English speech synthesis based on multi-layered context-oriented clustering. *Speech Comm.* 14:313–324, 1994.

[NH88] S. Nakajima and H. Hamada. Automatic generation of synthesis units based on context-oriented clustering. In *Proc IEEE ICASSP*, New York, 659–662, 1988.

[OPS95] M. Ostendorf, P. Price, and S. Shattuck-Hufnagel. The Boston University Radio News Corpus. Technical Report ECS-95-001, Electrical Computer, and System Engineering Department, Boston University, Boston, MA, 1995.

[RJ93] L. Rabiner and B. Juang. *Fundamentals of speech recognition.* Prentice Hall, Englewood Cliffs, NJ, 1993.

[SKIM92] Y. Sagisaka, N. Kaiki, N. Iwahashi, and K. Mimura. ATR – ν-TALK speech synthesis system. In *Proceedings of ICSLP 92*, vol. 1, pages 483–486, 1992.

[Sv93] A. M. C. Sluijter and V. J. van Heuven. Perceptual cues of linguistic stress: Intensity revisited. In *Proceedings ESCA Prosody W/S*, Lund, Sweden, 246–249, 1993.

[Sv94] A. M. C. Sluijter and V. J. van Heuven, Spectral tilt as a clue for linguistic stress. Presented at 127th ASA, Cambridge, MA, 1994.

[TAS92] K. Takeda, K. Abe, and Y. Sagisaka. On the basic scheme and algorithms in non-uniform unit speech synthesis. In *Talking Machines: Theories, Models, and Designs*, G. Bailly, C. Benoit, and T. R. Sawallis, eds. Elsevier, North-Holland, Amsterdam, 93–106, 1992.

[Tra91] H. Traunmüller. Functions and limits of the F1:F0 covariation in speech. In *PERILUS XIV*, Stockholm University, 125–130, 1991

[WC95] C. W. Wightman and W. N. Campbell. Improved labelling of prosodic structures. ATR Technical Report No. TR IT-0061, July 1994.

Appendix: Audio Demos

There are four short sample speech files that accompany this chapter to illustrate the unit selection procedures.

The source database is section f2b from the BU-RADIO-NEWS CORPUS, consisting of 123 utterances (41,252 phone units) from a female American news announcer, recorded in a studio. We are grateful to Mari Ostendorf of Boston University for providing access to this corpus.

These synthesis samples do *not* represent full text-to-speech, but only illustrate the unit selection process described in the chapter. The prosody is taken from analysis of the natural speech, and no subsequent signal processing is performed. Synthesis is by simple concatenation of raw speech samples taken directly from the corpus. In a complete synthesis system, subsequent signal processing would be required, but here we want to focus on the quality of the selected units without modification.

First is the original target utterance before it was removed from the source database.

Next are three resynthesised versions of the same utterance, using segments selected in various ways from the units remaining in the source database.

The first version puts priority on closeness to the prosodic target but pays no attention to how well the selected units may join (i.e., it just minimizes unit distortion and ignores continuity distortion). The prosody of the selected units is very close to the target, but the unit boundaries are noisy.

The next sample minimizes continuity distortion at the expense of unit distortion. The result is a smoother flow between units, but the prosody is at times very far from the desired target, and significant loss of naturalness would occur if a process such as PSOLA were applied to force them to fit.

Obviously, we need a balance of weighting between minimizing unit distortion and continuity distortion. Tuning the weights as described in the chapter allows us to optimize the weights between the various distances. The resulting selection with tuned weights has both close prosodic distance and reasonable joins.

The number of units selected for each example varies. There are almost twice as many units in the example which minimizes unit distortion as there are in the others (because minimizing joins is not important). Most units are one or two phones in length with a few examples of demisyllable selection.

23

Optimal Coupling of Diphones

Alistair D. Conkie
Stephen Isard

ABSTRACT We describe several ways of measuring spectral mismatch at diphone joins. We report on the effect for synthetic speech of choosing diphone boundaries so as to minimize the various measures.

23.1 Introduction

Time domain PSOLA [CM89] has proven to be a high-quality resynthesis technique for concatenative synthesis, and specifically for diphone synthesis. However, it is not particularly well suited to spectral manipulation for the smoothing of joins at unit boundaries, so it is especially important to choose units in a way that minimizes spectral mismatch. For improving the quality of diphone speech, one approach is to enlarge the set of stored units, either incorporating units larger than the diphone to cut down on the number of boundaries across which mismatch can occur, and/or to provide a redundant set of units from which ones with minimal mismatch can be chosen [Sagi88]. The optimal coupling technique [Ver90, TI91] is a variant on the latter approach, in which the boundaries of a given diphone are not fixed in advance but chosen in order to provide the best fit with the neighboring diphones in the particular utterance being synthesized.

In practice, "best fit" has to amount to minimizing some physical measure of mismatch. Previous attempts at this have not given an entirely satisfactory correlation with subjective assessments of quality [Boe92]. In this chapter, we report our experience using four different measures of mismatch, and the different styles of join that result either from minimizing one of the measures or using existing sets of fixed diphone boundaries.

23.2 The Unoptimized Diphone Sets

The fixed-boundary diphones that served as the point of departure for this work come from two diphone sets, a large British English set with many "allodiphones" and a minimal Spanish set. The English set is part of the synthesizer described in [CIMV90], which was used for the evaluation reported in [Syd92]. The Spanish set originated from an unpublished student project [Lop92] and was further developed by Alistair Conkie. In both cases, nonsense words were generated to cover all of the required diphones, using a scheme similar to that described in [IM86], and diphone boundaries were originally chosen by hand in stable regions of spectrogram and waveform displays. Some boundaries were later adjusted on the basis of experience with the synthesizer.

23.3 Measures of Mismatch

We have calculated the measures below for both the English and Spanish diphone sets.

23.3.1 Simple Frame Mismatch

This simple frame mismatch measure is based on a mel-cepstral analysis of the nonsense words from which our diphones are extracted, with a 25.6-ms analysis window and 5-ms frame shift. The nonsense words were automatically aligned with their phonetic transcriptions using an HMM aligner, as in [TI91], and the alignments were hand-corrected in a small number of cases so that HMM alignment errors would not confuse the comparison between fixed and optimized join styles. This yielded, for our English diphone set, 2,283 phone pairs of the form *P1-P2* from which to extract the diphones, and 661 for our Spanish set.

For this case, we take the mismatch between two frames to be the Euclidean distance between vectors of 13 mel-scale cepstral coefficients, including a 0th coefficient representing energy. Optimized join based on this measure consists in joining the left-hand diphone up to a frame x and the right-hand one from a frame y onwards such that the mismatch between the frames x and y is minimal (see figure 23.1).

Table 23.1 displays statistics for this measure on three types of data:

- Natural speech: For all diphones, the mismatch across the hand-selected diphone boundary in the nonsense word from which the diphone is taken. This gives a baseline of mismatch to be found in relatively stable regions of natural speech.

- Fixed diphones: For all diphone pairs of the form *P1-P2, P2-P3*, the mismatch at the boundary when the hand-selected diphone *P1-P2* is joined to the hand-selected *P2-P3*.

- Optimized diphones: The mismatch at the boundary when mismatch is minimized in joining *P1-P2* and *P2-P3*.

The means from table 23.1 are plotted in figure 23.2.

For both languages, the results are broadly what one might expect, with the optimized figures falling in between the natural speech and the fixed boundary diphones. The absolute values for the two languages are also similar, although

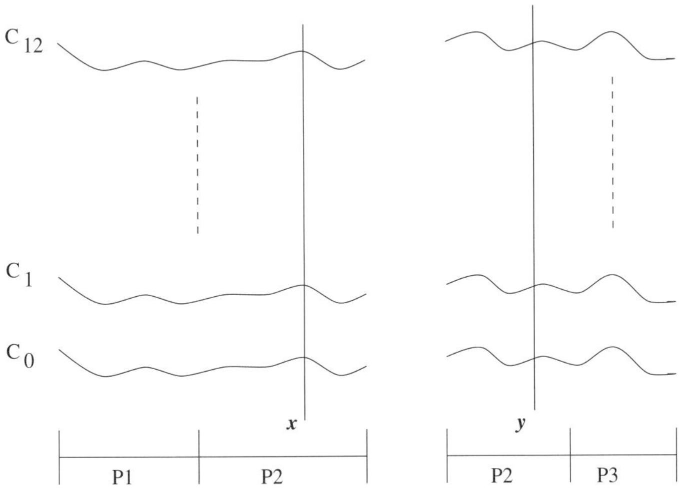

FIGURE 23.1. Minimizing cepstral mismatch in joining the diphones *P1-P2* and *P2-P3*.

TABLE 23.1. Simple frame mismatch over all diphones.

	English			Spanish		
	max	mean	SD	max	mean	SD
Natural speech	6.08	0.62	0.52	4.42	0.60	0.45
Fixed diphones	9.59	2.26	1.14	12.31	2.66	1.47
Optimized diphones	5.66	1.05	0.41	4.67	1.16	0.48

mismatch is slightly greater for the Spanish set in all conditions. The biggest difference is in the mismatch at fixed diphone boundaries, where optimization gives a correspondingly greater improvement. The fact that for English the maximum optimized value is actually smaller than the maximum natural speech value suggests that there may be cases for which minimizing this measure will not necessarily have the effect of making the joined diphones more similar to natural speech.

Experience with our synthesizer has given us the impression that mismatch at diphone joins has its greatest effect on synthesis quality within vowels, where abrupt spectral changes will be easiest to hear. We therefore made a separate calculation of the mismatch at joins within vowels, that is for diphone pairs (*P1-V, V-P2*), where *V* ranges over all vowels. The results are shown in table 23.2 and plotted in figure 23.3.

The pattern of means is similar to that for the whole diphone sets but there is a greater reduction in the natural speech mean and the spread around it—as indicated by the maximum and standard deviation (SD)—than for the optimized diphones. The threat of "over-optimization" raised with regard to the maximum values of table 23.1 is not present here. These results probably give a more realistic picture of the advantage to be gained by minimizing simple cepstral mismatch. The mismatch at optimized joins is greater relative to the other two categories.

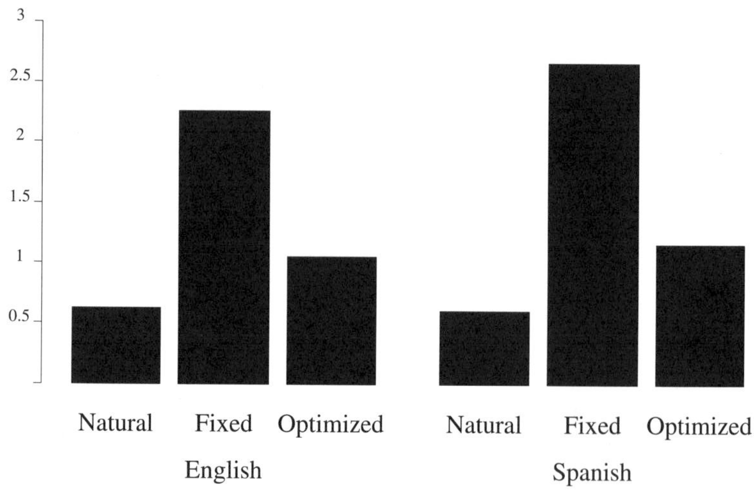

FIGURE 23.2. Means for simple frame mismatch over all diphones.

TABLE 23.2. Simple frame mismatch within vowels.

	English			Spanish		
	max	mean	SD	max	mean	SD
Natural speech	1.43	0.44	0.18	1.21	0.35	0.17
Fixed diphones	6.15	1.91	0.66	6.23	2.23	0.82
Optimized diphones	3.35	1.00	0.32	3.08	1.20	0.38

Some of the apparent gain from optimization with regard to the whole diphone set represents the reduction of mismatch present in natural speech for some classes of phonemes. In fact, the figures for vocalic consonants—liquids, glides, and nasals—are almost identical to those for vowels. It is these classes for which optimization is consistent in making the joined diphones more like natural speech with regard to the simple mismatch measure.

We have found the same relationship between statistics computed over all diphone joins and those computed over vowel joins for all of the measures that we have considered. We will report only on the vowel joins in the rest of the paper.

23.3.2 Mismatch of Frames Plus Regression Coefficients

It is possible that the measure above, while reducing spectral jumps at diphone boundaries, will still permit abrupt changes of direction in the contours of spectral parameters. Figure 23.4 illustrates a hypothetical example.

One way of avoiding such effects is to extend the vector of cepstral coefficients by including for each cepstral coefficient a linear regression coefficient calculated over a window of a chosen size, as is done in some speech recognition systems (e.g., [HL89]). Keeping the difference in regression coefficients small across a boundary

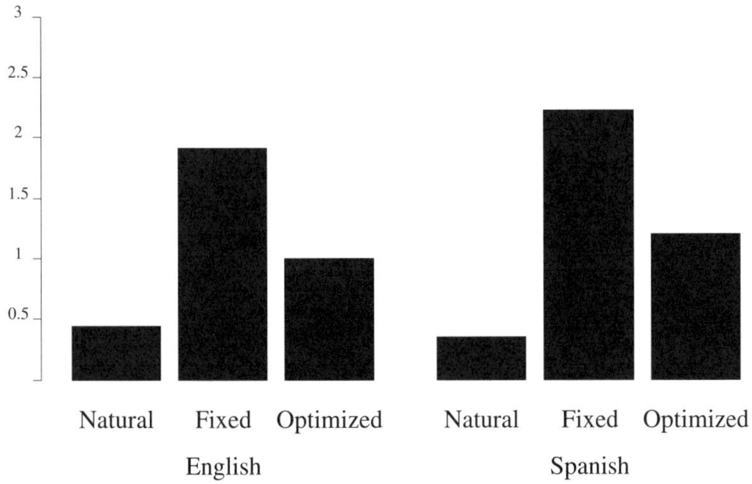

FIGURE 23.3. Means for simple frame mismatch within vowels.

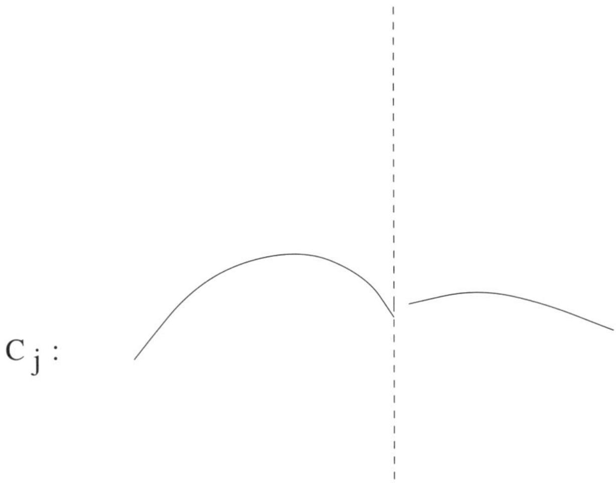

FIGURE 23.4. Hypothetical cepstral track showing abrupt change of direction.

should enforce a certain amount of smoothness in the contour. The calculation of the measure is illustrated in figure 23.5. To compensate for differences in absolute magnitude, coefficients are weighted by their variances across all instances of the phone under consideration.

This measure is, of course, dependent on the size of the window over which regression coefficients are calculated. We have done computations with window sizes ranging from two to five frames on either side of the boundary. In fact, the measure turns out to vary remarkably little with window size. Table 23.3 reports

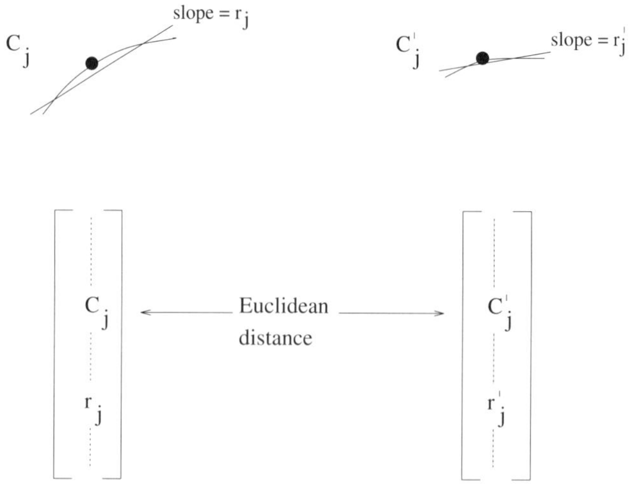

FIGURE 23.5. Distance measure on a vector of cepstral coefficients and regression coefficients.

TABLE 23.3. Mismatch of cepstral plus regression coefficients over a three-frame window for vowel diphones.

	English			Spanish		
	max	mean	SD	max	mean	SD
Natural speech	9.86	2.34	0.84	4.72	1.70	0.83
Fixed diphones	8.68	3.48	0.91	8.24	3.51	0.94
Optimized diphones	5.53	2.26	0.56	4.96	2.42	0.58

the measure across vowels for a window of three frames, and figure 23.6 plots the means.

The difference between natural speech and fixed diphones is smaller for this measure than for the previous one. For English, optimization gives less mismatch than in natural speech. Taken together with the fact that the measure behaves differently for the English and Spanish diphone sets, these points give reason to doubt that minimizing the measure will necessarily make diphone speech more like real speech.

23.3.3 Linear Fit Over a Window of Frames

An alternative to including regression coefficients in the frames is to compute not the simple Euclidean distance between frames, but the deviation of coefficients from their best fit line over the chosen window size, as in figure 23.7.

Again, we have calculated the measure over a range of window sizes. The variability of the measure on natural speech tends to increase with window size.

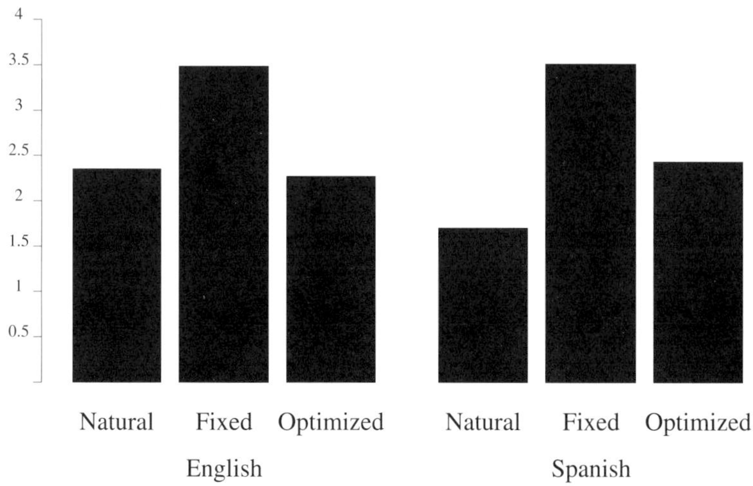

FIGURE 23.6. Means for mismatch of cepstral plus regression coefficients over a three-frame window measured on vowel diphones.

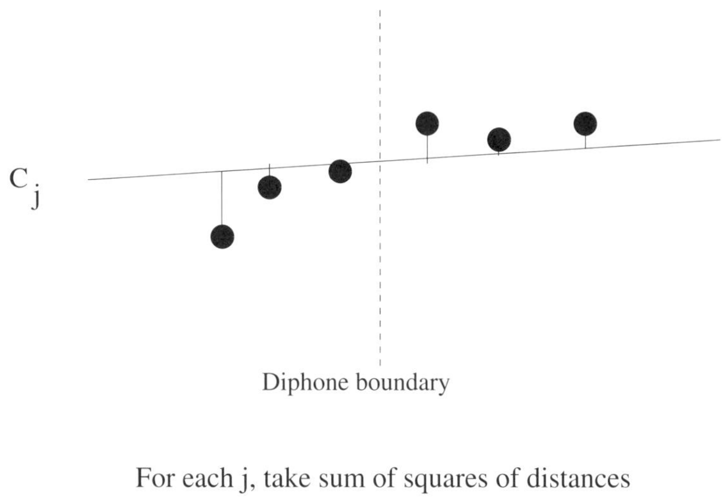

FIGURE 23.7. The linear fit measure.

Table 23.4 gives the results for a window size of two frames on either side of the join, with means plotted in figure 23.8.

For both languages, the natural speech shows substantially less mismatch than the fixed diphones, and optimization eliminates a large proportion of the diphone mismatch.

TABLE 23.4. Linear fit for vowels with a window size of two.

	English			Spanish		
	max	mean	SD	max	mean	SD
Natural speech	1.00	0.17	0.12	0.58	0.10	0.09
Fixed diphones	7.76	0.98	0.65	7.83	1.22	0.88
Optimized diphones	2.46	0.32	0.18	2.01	0.41	0.25

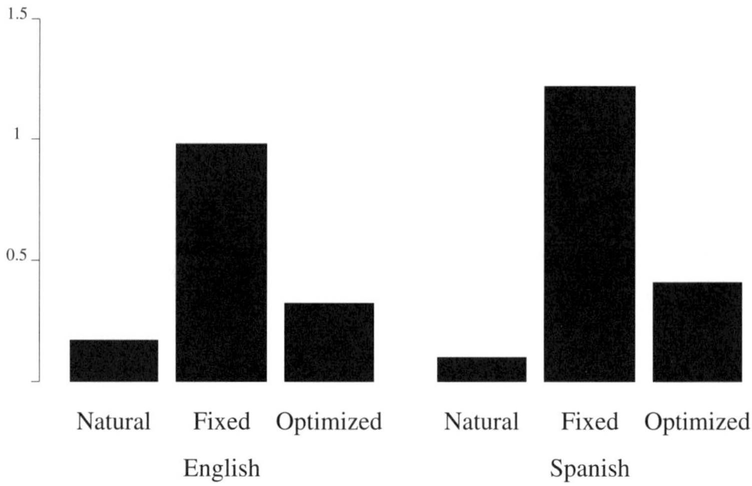

FIGURE 23.8. Means for linear fit for vowels with a window size of two.

23.3.4 Least Mismatch Giving a Chosen Duration

In most applications of synthetic speech, the durations of phones are controlled so as to approximate natural speech rhythm and phrasing. [CM89] explains a method for manipulating phone durations in PSOLA diphone speech by the duplication or elimination of pitch periods. If the smoothest coupling of an adjacent pair of diphones turns out to yield a phone whose duration has to be substantially altered to fit the desired timing of an utterance, the distortion introduced by changing the duration may give a worse result than a slightly suboptimal coupling whose duration is correct in the first place and does not need to be changed. With this consideration in mind, we tried minimizing linear fit as above while constraining the length of the resultant phone to the value dictated by the duration module of our synthesis system. There are no strictly comparable statistics to present for this measure because mismatch varies with phone duration. For a sample of typical phone durations, we found mismatch values generally approximating those for fixed diphones, but sometimes appreciably larger.

23.4 Assessment

23.4.1 General Considerations

Although minimizing the various measures above can guarantee less spectral mismatch in our synthetic speech, we still need to know whether the quality of the speech improves as a result. Informal listening tests brought us to the opinions that:

1. Minimizing mismatch with a fixed duration is unsatisfactory on its own. The quality of synthesis is degraded below that for fixed diphone boundaries in a substantial number of cases: A more sophisticated method that weighted duration and mismatch might perform better. However, this is also the computationally most expensive method because the dependence of diphone cutpoints on duration rules out precomputation of optimal values. We have therefore not proceeded further with it.
2. For phrase length materials, it is usually difficult to distinguish between simple mismatch and the linear fit method of section 23.3.3 without very careful listening. There is no obvious class of cases for which they differ, although cases do occur for which each is preferable to the other.
3. When linear fit does go wrong, it is often by leaving out too much of the middle of the vowel. We have not investigated individual cepstral tracks to see how these cases occur, but there were enough to bring us to the opinion that the method should not be used on its own without some further form of constraint.
4. The three variable-duration optimization methods of sections 23.3.1–23.3.3 often sounded better than synthesis with fixed cutpoints, especially in cases where vowel formants move monotonically from onset to offset, with no clear steady state in between, or where formant steady states are not synchronized. These are cases where formant jumps are likely to occur with fixed cutpoints.

23.4.2 Perceptual Test

Only opinion 4 above yields a clear hypothesis to test in a perceptual experiment. We have compared fixed cutpoints with ones for which simple mismatch is minimized in an English word identification task. In order that subjects' answers on such a task should not simply reflect their knowledge of the lexicon, the words used must have near neighbors with which they can in principle be confused.

Materials

We chose 20 CVC monosyllabic words whose spectral properties at vowel onset and offset could be expected to differ substantially and which had near neighbors. Examples were lace (lease, lice, less), caught (cut, cot, curt), and wait (white, wheat, wet). The neighbors themselves were not included in the set so that direct contrasts would not influence subjects' judgments. Each word was synthesized with fixed cutpoints and optimized cutpoints. Only vowels were optimized. The words were given the duration of a stressed syllable in a nonfinal polysyllabic word, as determined by our duration algorithm. The same durations were used for both synthesis types. Initial consonants were synthesized as if following the same consonant at the end of a previous word, rather than following a silence, and analogously for final consonants. The intention was to create words similar to ones excised from running speech, rather than spoken in isolation. The task needed to be made sufficiently difficult for the subjects to produce enough errors for a comparison of synthesis types to be possible, and the identification of excised words from natural speech is known to be difficult for experimental subjects [PP64]. Our fixed cutpoint diphone synthesis is generally quite comprehensible, and ceiling effects might have been encountered if we had synthesized the equivalent of isolated words.

The words were randomized in two different orders and for each order half the words were arbitrarily designated as set A and the other half as set B. For each order, two tapes were made: one with the A words synthesized with fixed cutpoints and the B words with optimized cutpoints, and the other with synthesis types reversed. Words were spaced at 2 second intervals on the tapes.

Method

Sixteen speakers of British English with no reported hearing problems were divided into four groups, each hearing a different tape. They were asked to write down the words they heard on a score sheet.

Results

We report results for vowels correct, since the stimuli for the two synthesis methods differed only in their vowel diphones, and the stimuli were such that errors, especially involving voicing, were to be expected in final consonants. Overall, subjects scored 84% vowels correct on the optimized stimuli as opposed to 64% correct for the fixed cutpoint stimuli (see figure 23.9). The advantage for the optimized stimuli both for subjects and for words was significant at the 0.05 level. Of the 16 subjects, 13 scored better on the optimized words. Six of the words were heard correctly by all subjects in both synthesis methods. Of the remaining 14, 4 showed a small advantage for the fixed cutpoint version and 10 showed an advantage for the optimized version. Two words were identified correctly by all subjects in the optimized version and by none in the fixed version. There were no words whose optimized version was misheard by all subjects.

Furthermore, for those words where there was no advantage for the optimal version, the average spectral mismatch for the fixed version was 0.443, which is close to the value found at vowel midpoints in natural speech, whereas for those words for which an advantage did appear for the optimal version, average mismatch in the fixed version was significantly higher at 2.17. It therefore seems likely that it is the minimizing of spectral mismatch, rather than some other, accidental, factor that gives the optimized version its advantage.

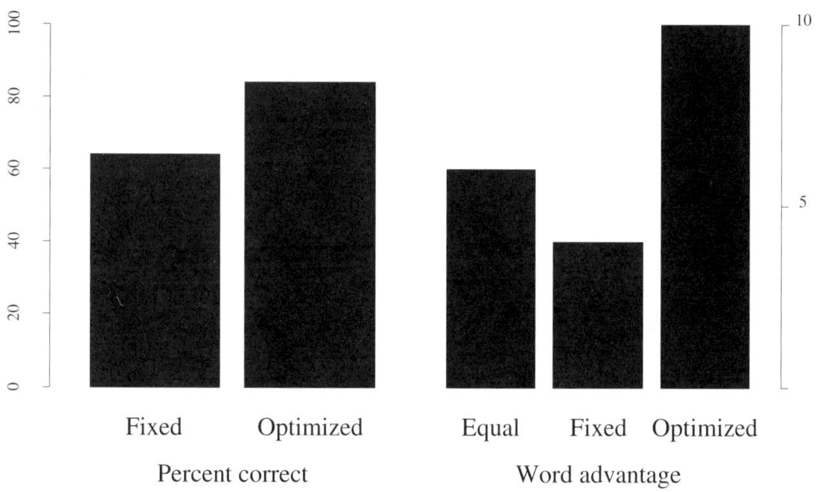

FIGURE 23.9. Results of perceptual test.

23.5 Conclusion

Optimizing cutpoints by simple frame mismatch does improve the quality of our synthetic speech. If the cutpoints are precomputed and a table stored, there is little extra run-time computation, and the table is small by comparison to the diphones themselves, allowing real-time synthesis on a wide variety of machines with no special arithmetic or signal-processing hardware. It is possible that a combination of mismatch and durational constraints could give further improvement.

References

[Boe92] O. Boeffard et al. Automatic generation of optimized unit dictionaries for text to speech synthesis. In *Proceedings ICSLP 92*, Banff, Alberta, Canada, 1211–1214, 1992.

[CIMV90] W. N. Campbell, S. D. Isard, A. I. C. Monaghan, and J. Verhoeven. Duration, pitch and diphones in the CSTR TTS system. In *Proceedings ICSLP 90*, Kobe, Japan, 825–828, 1990.

[CM89] F. Charpentier and E. Moulines. Pitch-synchronous waveform processing techniques for text-to-speech synthesis using diphones. In *Proceedings Eurospeech '89*, Paris, 2:13–19, 1989.

[HL89] M. J. Hunt and C. Lefèvbre. A comparison of several acoustic representations for speech recognition with degraded and undegraded speech. *Proceedings IEEE ICASSP '89*, Glasgow, Scotland, 262–265, 1989.

[IM86] S. D. Isard and D. A. Miller. Diphone synthesis techniques. In *Proceedings International Conference on Speech Input/Output*, IEE Conference Publication No. 258, London, 77–82, 1986.

[Lop92] E. Lopez. *A Diphone Synthesiser for Spanish.* Unpublished MSc dissertation, Department of Artificial Intelligence, University of Edinburgh, 1992.

[PP64] I. Pollack and J. M. Picket. Intelligibility of excerpts from fluent speech: Auditory vs. structural context. *J. Verbal Learning and Verbal Behaviour* 3:79–84, 1964.

[Sagi88] Y. Sagisaka. Speech synthesis by rule using an optimal selection of non-uniform synthesis units. In *Proceedings IEEE ICASSP 1988*, 449–452, 1988.

[Syd92] H. Sydeserff et al. Evaluation of speech synthesis techniques in a comprehension task. *Speech Comm.* 11(2-3):189–194, 1992.

[TI91] P. Taylor and S. Isard. Automatic diphone segmentation. In *Proceedings Eurospeech '91*, Genova, Italy, 1:341–344, 1991.

[Ver90] J. Verhoeven. Context-sensitive diphones as units in speech synthesis. In *Proceedings I.O.A.*, Windermere, England, 12(10):275–281, 1990.

24

Automatic Speech Segmentation for Concatenative Inventory Selection

Andrej Ljolje
Julia Hirschberg
Jan P. H. van Santen

ABSTRACT Development of multiple synthesis systems requires multiple transcribed speech databases. Here we explore an automatic technique for speech segmentation into phonemic segments applied to an Italian single-speaker database. The output segmentation is compared to manual segmentations by two human transcribers. The performance is very good on voiced stop to vowel boundaries and unvoiced fricative-to-vowel boundaries, whereas vowel-to-vowel and voiced fricative-to-vowel boundaries are estimated less accurately.

24.1 Introduction

Construction of a concatenative text-to-speech synthesis system usually requires the segmentation and labeling of speech recorded by a single speaker, and this segmentation and labeling must be done only once. However, every time a new language or voice is required, the process of segmentation and labeling has to be completely redone. An automatic procedure for transcribing speech would alleviate much of the time-consuming effort that goes into building a TTS system. It could take advantage of the fact that training and testing involve a single speaker [LR93], which presents a much easier problem than the processing of speaker-independent speech databases [LR91]. In addition to enhanced efficiency, an automatic procedure ensures consistency in the placement of phoneme boundaries within the constraints of its knowledge of the speech signal, as specified by the speech model. However, due to the limited amount of speech used to train the algorithm, plus the inherent limits in parametrization of the speech signal and the speech model structure, the accuracy of the transcription is inferior to that achieved by human transcribers.

24.2 Automatic Transcription Algorithm

The transcription system used in this work is based on a single ergodic continuously variable duration hidden Markov model (CVDHMM) in which each state is a different phoneme. Each state of the CVDHMM is modeled as a three-state left-to-right conventional hidden Markov model (HMM) using a separate continuous probability density function (pdf) for each of the three states. Each pdf consists of a parameter space rotation and a weighted mixture of Gaussian pdfs. This type of hidden Markov model also allows the use of explicit duration models. We use two-parameter gamma distributions to model phoneme durations. The transcription system acts as a phoneme recognition system constrained by a phonotactic model.

24.2.1 The Phonotactic Models

Training of the acoustic model and the speech segmentation algorithm both require a separate phonotactic model for each utterance. This way we ensure that the path taken through the states of the CVDHMM exactly corresponds to the expected phonemic transcription of the utterance. The phonotactic model is in a form of a finite-state network and it can accept a network of possible phonetic realizations if desired. It also allows us to assign probabilities to different phonetic realizations when we do not have detailed transcriptions of the speech data, but have a priori knowledge about different ways of realizing speech utterances. This way the segmentation algorithm, acting as a speech recognizer, selects the most likely phonetic realization and performs segmentation at the same time. In the experiments described here such alternative transcriptions were not available for either the training or testing of the models. Only one phoneme sequence was available for every utterance, and the algorithm was used just for segmentation of those segments.

24.2.2 The Duration Models

Each acoustic phone model has a duration model associated with it. The model is a two-parameter gamma distribution, which is based on parameters estimated from the training data. Those are the mean observed duration and the variance of duration. An additional parameter is used to constrain the recognition search algorithm. It is the 99th percentile of the duration distribution, which gives the longest hypothesized duration for that phoneme.

24.2.3 The Phoneme Acoustic Models

It has been shown that the best segmentation performance is achieved by using context-independent phoneme models [LR91]. Speech recognition results always improve when context-dependent models are used. Those phoneme models have many different versions, which are chosen depending on the left and right phonemic context of the phoneme we are trying to model. However, a sequence of such context-dependent models can lose its correspondence to the underlying phoneme sequence during the iterative training process. Because such acoustic models occur only in fixed ellic environments (phonemic context is fixed), their alignments are driven by the acoustic effects. For example, this often means that such models represent just the body of a particular phoneme; the next model, which always occurs in this model combination, represents the following transition and the body of the following phoneme.

Each phoneme model can be visualized as a three-state left-to-right HMM. Because we use an explicit duration model paradigm, we do not use transition probabilities for the phoneme models. They serve only as a crude duration model, which we have replaced with a more accurate explicit duration model. Each state of the phoneme model is represented by a probability density function. Although we generate the model from speech by a single speaker, we use Gaussian mixtures. In addition, it has been shown that the speech parameters are highly correlated [Ljo94] and that distribution-specific decorrelation significantly improves the accuracy of the phoneme models. Thus each distribution consists of a decorrelating matrix and a conventional weighted Gaussian mixture. The decorrelation matrix is the matrix of eigenvectors obtained from the sample covariance matrix of the training data for that distribution.

24.2.4 The Segmentation Algorithm

The segmentation algorithm is a modified version of the Viterbi algorithm [Vit67], which allows for the explicit duration models.

We calculate the likelihood, P_j, that a CVDHMM state/phone q_j corresponds to the observations over the time ρ from $t-\tau+1$ to t, where t is the current position in the whole utterance in the modified Viterbi algorithm, τ is the hypothesized length of the hypothesized phone, and ρ is a time index used in the phone internal recursion.

$$P_{j_{\rho+1}}(m,\tau) = \sum_{l\in L_m} P_{j_\rho}(l,\tau) b_{j_l}(O_{\rho+1}) \qquad t-\tau+1 \le \rho < t \tag{24.1}$$

$b_{j_l}(O_{\rho+1})$ is the likelihood that the observation O at time $\rho+1$ was generated by the lth distribution (state of the phoneme HMM) of the jth phone model.

In other words, $P_{j_t}(M,\tau)$ is the likelihood that the observation vectors $O_{t-\tau+1}, \ldots, O_t$ were generated from distributions $1, \ldots, M$.

We now define the conventional recursion for a second-order CVDHMM, equation (24.1), where $\alpha_t(j,k)$ provides the forward likelihood of the best scoring path from the beginning of the utterance to time t, ending in phone j, assuming that the right context is the phone k, which starts at time $t+1$. The likelihood of observing a phone j during the time period from $t-\tau+1$ to t is $P_{j_t}(M,\tau)$.

$$\alpha_t(k) = \max_\tau \alpha_{t-\tau}(j) a_{jk} d_j(\tau) P_{j_t}(M,\tau) \qquad 1 \le t \le T. \tag{24.2}$$

We can recover the most likely state sequence by remembering which j and τ maximized the expression in equation (24.2) and backtracking when we reach the end of the recursion.

24.2.5 *The Training Algorithm*

The input speech is parametrized into cepstra and energy with their first and second time derivatives, using a window of 20 ms and a window shift of 2.5 ms. We use 9 cepstra, 9 $\Delta-$cepstra, 6 Δ^2-cepstra, and 3 energy parameters, giving a total of 27 parameters extracted every 2.5 ms. Given the speech parameters and a phoneme sequence, the system returns the location of phoneme boundaries. The model is initialized using uniform segmentation of the training utterances. It is then used to segment those same utterances. We then repeat this procedure iteratively until there is virtually no change in the boundaries and the acoustic models. Because all of the utterances were embedded in carrier phrases, we use the whole utterance for a few iterations, but then the carrier phrase is ignored and the models are based only on the target phrases for the final few iterations.

24.3 Segmentation Experiments

The segmentation experiments were conducted on a database of Italian utterances, all of which were embedded in carrier phrases. We used a subset of 100 utterances for testing, with four different transcriptions. One was derived from concatenation of dictionary-preferred pronunciations, the same as all of the training data. Two other transcriptions were done separately by two different human transcribers, and the fourth represented the consensus of the two human transcribers. The phonemic segments were clustered into seven categories, as shown in table 24.1.

Table 24.2 explains the various metrics used to describe the results in the experiments.

We use the consensus transcription as the reference transcription, except when we compare the human transcribers. However, the test phoneme sequences can

TABLE 24.1. Phonemic categories used to evaluate the accuracy of the automatic segmentation algorithm.

Symbol	Phonetic Category
V	Vowel
P	Unvoiced stop
B	Voiced stop
S	Unvoiced fricative
Z	Voiced fricative
L	Liquid, glide
N	Nasal

TABLE 24.2. Definitions of quantities used in subsequent tables.

Symbol	Description
> 10 ms	Percentage of cases where the absolute difference exceeded 10 ms, 20 ms, and so forth for 30 ms, 40 ms and 50 ms
M	Average difference in the boundary placement (indicates biases)
MDA	Median absolute difference
SD	Standard deviation of differences
N	Number of such boundaries in the test data

TABLE 24.3. The segmentation results for the automatic segmentation algorithm when compared to the consensus boundaries.

BND	M	MDA	SD	> 10ms	> 20ms	> 30ms	> 40ms	> 50ms	N
P-V	−1.6	7.5	19.4	38.6	22.8	14.0	8.8	1.8	57
V-V	−4.5	12.8	35.7	59.2	38.8	28.6	14.3	12.2	49
V-N	−4.8	10.0	22.8	50.0	20.6	8.8	4.4	2.9	68
V-B	−13.9	12.5	20.3	60.0	30.0	13.3	10.0	10.0	30
V-L	−23.2	20.5	19.6	75.5	51.0	32.7	20.4	6.1	49
V-P	2.2	5.9	11.6	20.0	7.5	2.5	2.5	0.0	40
V-Z	−15.8	19.5	14.1	76.9	48.7	17.9	0.0	0.0	39
N-V	0	16.2	23.6	77.8	29.6	16.7	9.3	3.7	54
B-V	0	3.7	8.8	20.5	2.3	2.3	0.0	0.0	44
L-V	11.1	13.2	21.1	63.9	29.5	14.8	4.9	3.3	61
Z-V	15.4	21.3	24.8	78.6	52.4	40.5	16.7	7.1	42
S-V	2.7	5.2	20.8	18.8	9.4	9.4	3.1	3.1	32

differ from the reference phoneme sequences because those are obtained in several different ways. Only the boundaries that appear in both the reference and the test transcriptions are used in the performance evaluation. At most a few percent of the boundaries were excluded from the evaluation.

The first experiment compared the automatic segmentation based on the automatically obtained transcription from the dictionary with the consensus boundaries. The results can be seen in table 24.3.

Table 24.4 shows the results after the removal of the bias.

When the extra effort is made to provide the manually obtained transcription for training the model, the automatic segmentation performance is not changed, as can be seen by comparing the results in table 24.5 to the results in table 24.4.

The differences between the automatic segmentation output and the manual segmentations can be very large. The differences between two different human transcribers remain much smaller, as can be seen in table 24.6, in which their segmentation is compared with the bias removed.

TABLE 24.4. The segmentation results for the automatic segmentation algorithm when compared to the consensus boundaries with the bias removed.

BND	M	MDA	SD	> 10ms	> 20ms	> 30ms	> 40ms	> 50ms	N
P-V	0	7.1	19.4	36.8	22.8	14.0	8.8	3.5	57
V-V	0	12.3	35.7	65.3	36.7	26.5	18.4	10.2	49
V-N	0	6.8	22.8	29.4	14.7	5.9	2.9	2.9	68
V-B	0	6.4	20.3	33.3	13.3	10.0	10.0	6.7	30
V-L	0	13.5	19.6	57.1	30.6	10.2	6.1	2.0	49
V-P	0	5.8	11.6	27.5	7.5	2.5	2.5	0.0	40
V-Z	0	9.0	14.1	41.0	10.3	5.1	2.6	0.0	39
N-V	0	16.2	23.6	77.8	29.6	16.7	9.3	3.7	54
B-V	0	3.7	8.8	20.5	2.3	2.3	0.0	0.0	44
L-V	0	11.4	21.1	52.5	27.9	6.6	4.9	3.3	61
Z-V	0	18.8	24.8	81.0	47.6	16.7	7.1	4.8	42
S-V	0	7.8	20.8	25.0	9.4	9.4	3.1	3.1	32

TABLE 24.5. The segmentation results for the automatic segmentation algorithm applied to the manually obtained transcriptions when compared to the consensus boundaries with the bias removed.

BND	M	MDA	SD	> 10ms	> 20ms	> 30ms	> 40ms	> 50ms	N
P-V	0	7.1	19.4	36.8	22.8	14.0	8.8	3.5	57
V-V	0	12.3	35.7	65.3	36.7	26.5	18.4	10.2	49
V-N	0	6.8	22.8	29.4	14.7	5.9	2.9	2.9	68
V-B	0	6.4	20.3	33.3	13.3	10.0	10.0	6.7	30
V-L	0	13.5	19.6	57.1	30.6	10.2	6.1	2.0	49
V-P	0	5.8	11.6	27.5	7.5	2.5	2.5	0.0	40
V-Z	0	9.0	14.1	41.0	10.3	5.1	2.6	0.0	39
N-V	0	16.2	23.6	77.8	29.6	16.7	9.3	3.7	54
B-V	0	3.7	8.8	20.5	2.3	2.3	0.0	0.0	44
L-V	0	11.4	21.1	52.5	27.9	6.6	4.9	3.3	61
Z-V	0	18.8	24.8	81.0	47.6	16.7	7.1	4.8	42
S-V	0	7.8	20.8	25.0	9.4	9.4	3.1	3.1	32

24.4 Results

After the mean bias was removed, the automatic segmentation achieved very good results for transitions between voiced stops and vowels (97.7% of the boundaries were within 20 ms of the boundary produced by the human transcribers consensus), unvoiced fricatives and vowels (90.6% were within 20 ms of the consensus boundary), and vowels and unvoiced stops (92.2% were within 20 ms of the consensus boundary). For these types of boundaries the human transcribers never disagreed by more than 20 ms.

TABLE 24.6. The segmentation differences between the two human transcribers with the bias removed.

BND	M	MDA	SD	> 10ms	> 20ms	> 30ms	> 40ms	> 50ms	N
P-V	0	1.3	2.5	0.0	0.0	0.0	0.0	0.0	57
V-V	0	14.1	22.6	65.3	26.5	14.3	10.2	4.1	49
V-N	0	3.6	13.1	7.2	2.9	1.4	1.4	1.4	69
V-B	0	1.9	4.4	6.7	0.0	0.0	0.0	0.0	30
V-L	0	6.6	14.8	32.7	8.2	4.1	2.0	2.0	49
V-P	0	8.0	9.6	25.0	0.0	0.0	0.0	0.0	40
V-Z	0	5.9	8.4	23.1	0.0	0.0	0.0	0.0	39
N-V	0	3.8	12.1	18.5	9.3	3.7	1.9	1.9	54
B-V	0	1.0	3.5	0.0	0.0	0.0	0.0	0.0	44
L-V	0	8.8	18.9	45.9	26.2	9.8	3.3	3.3	61
Z-V	0	1.4	4.5	2.4	0.0	0.0	0.0	0.0	42
S-V	0	0.7	1.4	0.0	0.0	0.0	0.0	0.0	32

Other types of boundaries were determined less accurately. Vowel-to-vowel transitions, for example, were within 30 ms of the consensus boundary 73.5% of the time and voiced fricatives-to-vowel boundaries, the same as nasal-to-vowel boundaries, were within 30 ms of the reference boundary in 83.3% of the cases. Human transcribers also had difficulties with the vowel-to-vowel boundaries, and had a comparable performance for the liquid-to-vowel boundaries as the automatic technique.

References

[Ljo94] A. Ljolje. The importance of cepstral parameter correlations in speech recognition. *Computer, Speech and Language* 8:223–232, 1994.

[LR93] A. Ljolje and M. D. Riley. Automatic segmentation of speech for TTS. In *Proceedings, 3rd European Conference on Speech Communication and Technology, Eurospeech*, Berlin, 1445–1448, 1993.

[LR91] A. Ljolje and M. D. Riley. Automatic segmentation and labeling of speech. In *Proceedings, IEEE International Conference on Acoustical Speech and Signal Processing*, Toronto, 473–476, 1991.

[Vit67] A. J. Viterbi. Error bound for convolutional codes and an asymptotically optimal decoding algorithm. *IEEE Trans. on Information Theory* IT-13:260–69, 1967.

25

The Aligner: Text-to-Speech Alignment Using Markov Models

Colin W. Wightman
David T. Talkin

ABSTRACT Development of high-quality synthesizers is typically dependent on having a large corpus of speech that has an accurate, time-aligned, phonetic transcription. Producing such transcriptions has been difficult, slow, and expensive. Here we describe the operation and performance of a new software tool that automates much of the transcription process and requires far less training and expertise to be used successfully.

25.1 Introduction

The past several years have seen fundamental changes in the methods of speech communication research and in the approaches to speech technology development. We can now test theories on a scale that was impractical only a few years ago. Moreover, where theory fails, we can now contemplate a purely empirical approach to describing speech phenomena. The use of corpus-based methods, in which knowledge is automatically extracted from the patterns found in large collections of speech data, has become the primary methodology today. This change has been partly driven by several factors related to computational technology, including: (1) enhanced communication across disciplines (e.g., engineering, linguistics, computer science, psychology); (2) the availability of large, standardized speech databases; (3) mathematically and computationally tractable models of speech acoustic realizations (e.g., hidden Markov models); (4) more powerful computers; (5) the development of standardized, portable speech research and development (R&D) tools; and (6) demands from potential applications.

The practical manifestations of these changes are emerging in the form of new speech products and greatly improved performance of laboratory speech recognition systems. Speech synthesis technology is also beginning to reap the benefits of these changes. For example, striking improvements have been made in text-to-speech (TTS) synthesis systems as a result of empirical studies of natural productions at the prosodic and segmental levels. Large-scale text analyses have led to more robust methods for part-of-speech determination, homograph disambiguation [SHY92], and stressing of complex nominals [Spr90]. We can anticipate that this trend will continue.

To make use of corpus-based methods it is necessary to have access to a large speech database in which the phenomena of interest have been accurately and consistently transcribed. In particular, developers of advanced TTS systems generally work with subword units. Development of convincing models of segment duration variation as a function of context is important to the implementation of high-quality TTS systems, and speech corpora with accurate, time-aligned, phonetic transcriptions are essential to the development of such models. The development of concatenative synthesizers is also predicated on the availability of large, transcribed corpora. Although there are already several such corpora available (e.g., [LKS87]), developers often want to base their systems on speech from a particular speaker or in a particular domain. Consequently, there is a great need for a fast, inexpensive, and accurate means of obtaining time-aligned phonetic labeling of arbitrary speech.

In addition to TTS systems, several other technical areas can benefit from the availability of large, time-aligned, phonetically labeled databases. Basic speech and language studies concerned with timing of events, dialectical variation, coarticulation and segmental-prosodic interactions usually require some annotation of the speech signal at the segmental level. In the speech recognition domain, acoustic models used as the basis for the recognizers can be trained more quickly if good initial estimates of the phone and word boundaries can be obtained. Finally, synchronization of cross-modal stimuli with speech is of interest to those studying perception as well as in the entertainment industry where realistic facial animations can be achieved through such synchronization.

Traditionally, phonetic transcription of speech has been done by hand and has required considerable expertise. Consequently, time-aligned phonetic transcriptions were obtained slowly and at considerable expense as graduate-level linguists laboriously marked and labeled each segment by hand. To reduce the time and cost associated with this task and the development of speech systems that require such transcriptions, some sort of automated processing is clearly required. Ideally, such an automated system would provide for

1. Accessibility: The system should be readily available and run on most UNIX workstations.
2. Ease of use: The requirements for human intervention in the process should be reduced to a minimum and should not require extensive training (or a detailed understanding of phonetics), and convenient graphical interfaces should be provided for all operator tasks.
3. Good accuracy of alignment, and speaker-independent operation.

Perhaps the ultimate answer to this need would be an *ideal* speech recognizer. Speech recognizers based on subword units such as phonemes actually recognize the phone sequence of an utterance but generally report only the word-level results. It requires only a trivial change to report the phone sequence. Moreover, most recognizers also identify the optimal time-alignment between the speech signal and the recognized phone sequence so reporting of the time-aligned phone sequence is again merely a question of including the appropriate information in the output. However, speech recognizers are fallible, and the need to transcribe arbitrary utterances by many speakers requires the use of recognizers, which will make significant errors. For many applications, especially TTS systems, the errors made by current recognizers are too serious and too frequent to permit the use of their output for the transcription of speech corpora.

Most of the errors made by speech recognizers, however, are due to the fact that the content of the speech utterance is unknown. That is, the uncertainty of the word/phone sequence results in an extremely large search space with sufficient ambiguity that the probability of error is quite large. However, the task of transcribing a corpus of speech can be considerably less complex than this: *A human transcriber can specify the correct word sequence for the utterance.* Given the correct word sequence, the phone sequence is almost completely determined (except for a few possible variations in word pronunciation). This means that the only significant uncertainty is the alignment of the transcription with the speech waveform. Thus, although speech recognizers cannot be used directly to transcribe a corpus, they can be used to align a transcription with the utterance. We refer to this process as *aligning* the speech utterance and its transcription.

The idea of automatic text-speech alignment is an old one. Several implementation approaches have been taken with varying degrees of success. At the 1993 EuroSpeech Conference example, no less than thirteen papers addressing various aspects of automatic alignment were presented. Much of the previous work in this area has focussed on the automatic segmentation of speech based solely on the acoustic signal without making use the transcription (e.g., [HH93, SN93, RR93, GZ88]). Of the reported systems that make use of the word-level transcription, many are quite limited in their applicability (e.g., are speaker dependent, label only broad classes rather than individual phones, only align at word-level). Moreover, the broadly applicable systems that have been reported (e.g., [ABFGG93, LR91]) are often inconvenient to use and/or are not easily accessible to many who could benefit from them. In this chapter, we describe an automatic alignment tool, the Aligner, that meets the criteria listed above. In particular, the Aligner determines the phone sequence and boundaries automatically from the given word sequence and the speech signal while running in near real-time on common workstations.

In the following sections, we begin by examining the operation of the Aligner. We describe the basic sequence of operations that a typical user would follow, delving into the internal algorithms where needed. We then report on an extensive series of evaluations done to determine the accuracy of the results produced by the Aligner.

25.2 Aligner Operation

As described above, the basic division of labor between the user and the software requires the user to specify the correct word sequence for the speech utterance. Consequently, the software must provide a convenient method for entering the word sequence and a meaningful display of the final results so that they can be reviewed and, if desired, modified. To achieve this, the Aligner we describe here is the result of integrating technologies and databases from several sources under a uniform graphical user interface (GUI). Recognizing that speech data is often accompanied by word-level transcriptions, the Aligner includes a "batch mode" of operation in which the user interaction is done in advance for a whole set of files, which are then aligned without requiring user input. The basic sequence of operations for this batch mode is the same as when the Aligner is used interactively, however, so we have restricted our discussion here to the latter case.

25.2.1 Generating the Phone Sequence

The interactive mode of the Aligner begins by using the *waves+* software [ERL95] to display the speech waveform. The user can then delimit a section of the utterance, listen to it, and type the word sequence for that section into a transcription entry window. When this has been done, the Aligner automatically generates a pronunciation network that contains the most common phone sequences for the given word sequence. The interactive mode works only on the section delimited by the user rather than the entire utterance. This is done because many corpora contain long utterances, which are much easier to transcribe in smaller sections. The batch mode, however, aligns entire utterances regardless of length.

The pronunciation network is generated by looking up each word in a dictionary. The Aligner uses an English-pronouncing dictionary, which contains more than 113,000 entries, including multiple pronunciations for some words and many proper and place names. The pronunciations are represented using the DARPA symbol set, and stress and syllable boundaries are included. The dictionary access algorithms further expand the effective lexicon size by handling all regular inflections, plurals, possessives, and so forth. Thus, for example, if a word is not found in the dictionary but concludes with the suffix *ing*, then a set of rules are applied to remove the suffix and the dictionary is searched again to locate the root. If the root is found, its pronunciation(s) is(are) modified to reflect the suffix.

When building the pronunciation network, optional silences are included between each word. That is, the network includes a link directly from the last phone in one word to the first phone of the next, as well as a link to a silence, which is then linked to the next word. Likewise, if multiple pronunciations are found for a word, the pronunciation will include an optional branch for each additional pronunciation. The silences and pronunciations that best represent the speaker's production of the word sequence will then be determined automatically as part of the recognition process.

If the Aligner encounters a word that is not in the dictionary and that cannot be generated automatically (either due to a typographic error by the user or because a correctly entered word is not in the dictionary), additional input must be solicited from the user. When a word is not found, the user is presented with a display showing the transcription that was entered, the word that was not found, and a list of the 50 dictionary entries on either side of the place the missing word should have been. At this stage, the user can either correct the transcription or create a new entry in the dictionary. This process can be repeated as often as needed until the Aligner is able to construct the complete pronunciation network.

It is also possible for the user to include specific phonemes in the transcription. This is important for transcribing disfluencies, as it allows the user to describe word fragments that are not (and should not be) included in the dictionary. It also permits limited coverage of nonspeech sounds provided these can be reasonably expressed as a sequence of the available acoustic models. Direct specification of a phoneme sequence does require some linguistic knowledge on the part of the user. However, simply specifying the phonemes from the relatively limited DARPA set, and not having to mark their boundaries, is a much more approachable task, and most users find this relatively easy when provided with a tabulation of the phones and their sounds.

25.2.2 Aligning the Transcription

Once the pronunciation network for a section of speech has been generated, a search is performed to determine the best path through the network and the best time alignment between that path and the original speech waveform. The Viterbi decoding algorithm used to perform this search requires three inputs: (1) the pronunciation network, which describes the possible transitions from one phone to the next; (2) acoustic models, which statistically describe the acoustic features of each phoneme; and (3) a sequence of observations, which, in the case of the Aligner, are feature vectors derived from the speech waveform.

Rather than work directly from the speech signal itself (which provides simply amplitude as function of time), a richer representation is used in which information about frequency content and energy can be represented. This is done by segmenting the speech signal into a sequence of overlapping *frames* and extracting a set of features based on the signal contained in each frame. In this way, a sequence of feature vectors are obtained from the original signal. In the Aligner, each frame contains 25.6 ms of signal, and successive frames are offset by 10 ms. The features for each frame are obtained by applying a Hamming window, computing the first 12 mel-frequency cepstral coefficients and the signal energy, and estimating their derivatives. It is important to note that, due to the granularity of the feature extraction (100 frames per second), it is not possible to locate a specific event in the speech waveform more precisely than the 10 ms spacing of the feature vectors.

The acoustic models themselves are three-state, hidden Markov models, with states arranged left-to-right with no skips. The models were trained as context-independent monophones. Each state distribution is represented as a mixture of five Gaussians. Models for each symbol in the standard DARPA set were trained. In addition, models for /cl/ and /vcl/ (unvoiced and voiced stop consonant closures) were trained. No explicit models were trained for breaths, lip smacks, and so forth. These are all collapsed into the *silence* model.

The actual alignment of the feature vectors with the pronunciation network is done by a Viterbi decoder [YRT89]. In essence, the Viterbi decoder finds the sequence of states in the pronunciation network that is most likely to have produced the sequence of feature vectors observed (maximum likelihood criterion). In so doing, the optimal alignment between the network states and the feature vectors (which are tied to specific times in the original waveform) is determined. Thus, the ending time for each phone can be determined by looking up the time associated with the last feature vector assigned to the last network state for that phone. Likewise, the ending time for a word is simply the ending time for the last phone in that word. Syllable boundaries, as marked in the lexicon, are inserted appropriately in the time-aligned phone string.

Adjustment of the Viterbi *beam width* (or search-tree pruning threshold) permits one to trade speed of operation for increased robustness. With a conservatively broad beam, which yields very robust performance, the Aligner requires 201 s to align a 222-word utterance of 87 s duration sampled at 16 kHz when run on an SGI Indigo R3000 computer. This time includes feature computations and dictionary lookup. The same computation with a narrower beam requires 119 s, but yields identical alignments for the speech sample in question (fluent reading casually recorded using a lapel microphone in an office environment).

The acoustic model training and Viterbi decoder used by the Aligner are based on the *HTK* HMM Toolkit developed by Young and Woodland of Cambridge University's Engineering Department and distributed by Entropic Research Laboratory. The TIMIT speech database [LKS87] was used for training and evaluating the Aligner.

25.3 Evaluation

The Aligner has been tested informally on many different classes of speech from noisy casual telephone recordings sampled at 8 kHz to carefully recorded laboratory speech. Although the results obtained from these exercises appear quite reasonable, a more rigorous evaluation is required.

The TIMIT corpus used for training the Aligner was chosen for its availability, size, and reasonable coverage of co-occurrence phenomena, talkers and dialects. Its hand-corrected segment labels not only provide a convenient means for bootstrapping the acoustic models but, by setting aside part of the corpus, also provide a means for evaluating the alignment results. Of the 630 talkers in TIMIT, each producing 10 sentences, we reserved 16 for evaluation purposes (one male and one female from each of the eight dialect regions). Note that the female-to-male ratio in all of TIMIT is about 2 to 3.

Thus, to obtain some objective assessment of its performance, we have evaluated the Aligner formally on the reserved portion of TIMIT not included in the training set as well as on all of TIMIT (6300 sentences; 630 talkers). This was done for both 8 kHz and 16 kHz phone models. The results for the two sampling rates are very similar, and we report only on the performance obtained with the 16 kHz models.

Because the Aligner uses the DARPA symbol set and the sequences all correspond to more or less canonical pronunciations, it is necessary to algorithmically affiliate symbol boundaries from the Aligner with those in the TIMIT transcriptions. (The TIMIT transcriptions are quasi-phonetic and often do not correspond to canonical pronunciations). The Aligner and TIMIT agree exactly on the symbol for a given segment about 67% of the time. The remaining instances require some heuristic to affiliate Aligner symbols with TIMIT symbols so that boundary times may be compared. We used dynamic programming with a distinctive phonetic-feature–based distance metric to determine the best symbol-string–to–symbol-string mapping over the whole utterance. When one-to-many or many-to-one mappings were required, only the endpoints of the corresponding segments were scored. It should be emphasized that the disagreements over a segment label (which occurred on roughly one-third of all segments) are not "errors" in the sense that the Aligner got the wrong phone. Because the Aligner's output is derived from the phone strings generated from the transcription, the only way in which the Aligner can produce a "wrong" label is if the word sequence is not transcribed correctly or if the dictionary contains an erroneous pronunciation. The mismatches that occurred in our evaluation are due largely to differences between the TIMIT transcriptions and the standard DARPA symbol set.

The Aligner cannot mislabel a segment, given correct inputs, so the only real errors the algorithm can produce are in the time alignment of the boundaries. To evaluate these, we used the Aligner to re-align the reserved test set and compared the boundary times produced by the Aligner with those that were determined by hand during the production of the TIMIT corpus. The boundary differences are the result of subtracting the Aligner time from the TIMIT reference time. Thus, positive mean values indicate that the Aligner is placing the boundary earlier in time than the TIMIT boundary. No outliers were rejected.

To provide some insight into the nature of the Aligner-TIMIT boundary differences, and to permit some comparison with previous alignment work, we divided the acoustic symbols into the four classes described in table 25.1. The results in tables 25.2, 25.3, 25.4 characterize the time differences between the Aligner phone boundaries and those marked in TIMIT for each of the possible class-to-class transitions. Thus, referring to tables 25.1 and 25.2, we observed 328 occurrences of transitions from sonorants to fricatives (class 0 to class 1). These occurrences had a mean error of 1.6 ms, an RMS-error of 13.6 ms, and 90% of all such transitions had errors of less than 18 ms. By recalling that the granularity of our feature vectors is only 10 ms, we can see that the performance is quite good in this case.

Table 25.2 reports results only for the boundaries of segments for which there was an exact match between the Aligner and TIMIT segment labels.

TABLE 25.1. The four classes into which the phone symbols used in the Aligner were grouped for evaluation purposes.

Class ID	Symbols in Class
0	r,er,l,el,w,y,aa,ae,ah,ao,aw,ax,axr,ay, eh,ey,ih,ix,iy,ow,oy,uh,uw,ux
1	zh,ch,jh,v,z,s,sh,f,th,dh,hh
2	d,g,b,p,t,k,dx,en,m,n,ng
3	silence

In table 25.3, we present similar data but include all boundaries, not just those with exact matches. Because some of these boundaries are for segments for which there is not a direct mapping between the TIMIT labels and the DARPA symbol set, it is not surprising that the magnitude of the errors increases somewhat. Table 25.3 also reports the results of comparisons on the *entire* TIMIT corpus. Because this includes the training data as well as the test data, one would normally expect the performance to be better than that obtained only on the test data. However, as can be seen from the table, that is not the case for most boundary types. This suggests that the errors introduced by the inconsistent label inventory of TIMIT is greater than the performance difference resulting from the use of a resubstitution estimate. The data obtained using all of the TIMIT corpus was also used to evaluate percentiles for the alignment error, as plotted in figure 25.1. This shows that, although a small number of boundaries have errors in excess of 100 ms, more than 90% of all boundaries have errors of less than 35 ms. Finally, in table 25.4, we compare the overall statistics for male and female speakers.

TABLE 25.2. Occurrence counts (N), mean, rms deviation from TIMIT locations, and the absolute deviation within which 90% of all occurrences fell (all in seconds) for all symbol class sequences (Seq.). Data are presented for boundaries following exact symbol matches in the reserved test set.

	Exact Matches Test			
Seq.	N	mean	rms	90%
00	876	−.0012	.0278	.038
01	328	.0016	.0136	.018
02	567	.0006	.0176	.023
03	40	−.0015	.0667	.046
10	572	.0011	.0117	.018
11	45	−.0001	.0226	.032
12	135	−.0006	.0143	.020
13	68	−.0031	.0213	.024
20	748	.0028	.0131	.018
21	147	.0047	.0150	.024
22	90	−.0017	.0155	.028
23	55	.0067	.0440	.059

TABLE 25.3. Occurrence counts (N), mean, rms deviation from TIMIT locations, and the absolute deviation within which 90% of all occurrences fell (all in seconds) for all symbol class sequences (Seq.). Data are presented for all scoreable boundaries in the reserved test set (All Bounds. Test), and for all scoreable boundaries in all of TIMIT (All Bounds. TIMIT).

	All Bounds. Test				All Bounds. TIMIT			
Seq.	N	mean	rms	90%	N	mean	rms	90%
00	1020	−.0026	.0289	.041	38341	−.0040	.0277	.042
01	539	.0027	.0143	.020	20770	.0030	.0167	.023
02	1021	.0013	.0207	.023	41025	.0021	.0196	.023
03	79	−.0002	.0501	.051	2829	−.0043	.0317	.045
10	624	.0010	.0130	.018	23500	−.0012	.0140	.018
11	49	.0005	.0220	.032	1790	.0019	.0219	.033
12	139	−.0009	.0143	.021	5927	−.0010	.0154	.023
13	75	−.0040	.0208	.023	2781	−.0026	.0233	.029
20	929	.0012	.0183	.020	37999	.0012	.0187	.020
21	223	.0036	.0208	.027	8384	.0016	.0192	.028
22	174	.0072	.0254	.050	8396	.0068	.0276	.048
23	102	.0117	.0481	.07	3568	.0080	.0448	.07
30	103	.0181	.0436	.045	3612	.0122	.0253	.041
31	78	.0175	.0317	.049	3164	.0178	.0313	.044
32	74	.0173	.0291	.045	2604	.0214	.0344	.049

TABLE 25.4. Overall statistics computed using all scoreable boundaries over all of TIMIT (All); all TIMIT females (Females); all TIMIT males (Males); and on the reserved test set (Test).

Data	N	mean	rms
All	210998	.0016	.0227
Females	64726	.0013	.0228
Males	146272	.0017	.0227
Test	5389	.0021	.0234

25.4 Discussion and Conclusions

In this chapter, we have described the Aligner, a new, commercial software tool that enables most users to quickly produce accurate, time-aligned phonetic transcriptions of arbitrary speech. Such transcriptions are crucial for the development of many types of speech systems as well as for conducting speech research. The Aligner allows the user to generate the transcriptions simply by specifying the word sequence contained in the utterance and thus does not require an extensive knowledge of phonetics.

Essentially a graphical interface wrapped around a constrained, HMM-based recognizer, the Aligner produces accurate alignments in near real time on most workstations. For most "well defined" boundaries such as those between sonorants and fricatives, the Aligner can achieve rms deviations of less than 20 ms from

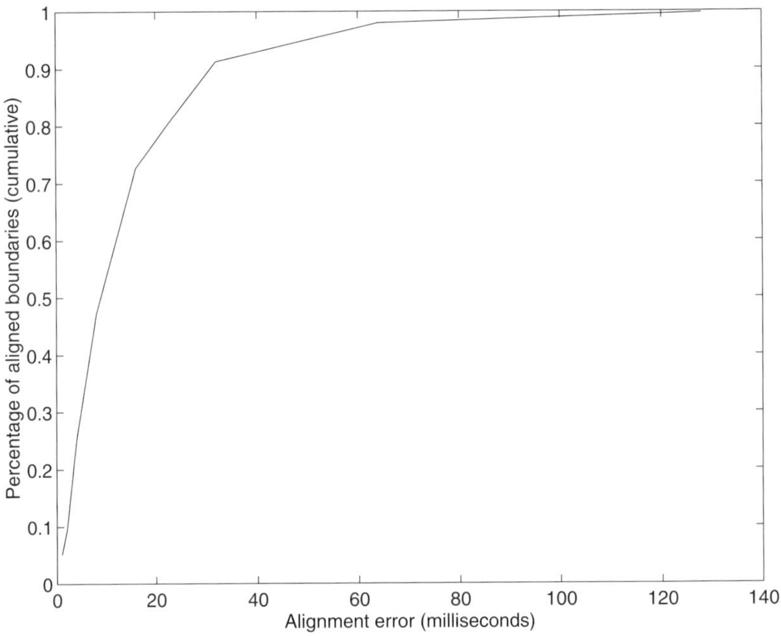

FIGURE 25.1. Accuracy of aligned boundaries over all of TIMIT.

the hand-corrected TIMIT boundaries. The performance appears stable across male/female speaker populations and within and outside of the training set. Most of the silence-to-phone and phone-to-silence boundaries (which have larger errors) occur in utterance-initial or utterance-final locations, where the judgment of hand labelers tends to be least reliable. Furthermore, the boundary between silence and a stop consonant is often equivocal. Many sonorant-to-sonorant "boundaries" are very difficult to define objectively. Our evaluation of the aligner shows its performance to be consistent with these observations: errors are larger in situations in which human labelers are more likely to differ.

The Aligner described here provides sufficiently accurate and consistent segment identification and location for many types of corpus-based studies. Should more precise boundary locations be required, a speaker-dependent aligner with more detailed acoustic models can be easily bootstrapped from this Aligner. The graphical interface provided with the Aligner permits rapid verification of segment labels and their boundaries.

Acknowledgments: The authors would like to acknowledge the tremendous contribution of Mari Ostendorf and several of her students at Boston University in correcting and expanding the dictionary used in the Aligner as well in formulating many of the rules used to expand its coverage.

Acknowledgments: The authors would like to acknowledge the tremendous contribution of Mari Ostendorf and several of her students at Boston University in correcting and expanding the dictionary used in the Aligner as well in formulating many of the rules used to expand its coverage.

REFERENCES

[ABFGG93] B. Angelini, F. Brugnara, D. Falavigna, D. Giuliani, R. Gretter, and M. Omologo. Automatic segmentation and labeling of English and Italian speech databases. In *Proceedings of Eurospeech*, Berlin, 653–656, 1993.

[ERL95] Entropic Research Laboratory, Inc. *Waves+ Reference Manual*. Entropic Research Laboratory, Washington, DC, 1995.

[FZBP87] W. Fisher, V. Zue, J. Bernstein, and D. Pallett. An acoustic-phonetic data base. In *Proceedings of the 113th Meeting of the Acoustical Society of America*, Indianapolis, IN, 8(S1):S92–S93, 1987.

[GZ88] J. Glass and V. Zue. Multi-level acoustic segmentation of continuous speech. In *IEEE International Conference on Acoustics, Speech, and Signal Processing*, New York, 429–432, 1988.

[HH93] K. Hubener and A. Hauenstein. Controlling search in segmentation lattices of speech signals. In *Proceedings of Eurospeech*, Berlin, 1763–1766, 1993.

[LR91] A. Ljolje and M. Riley. Automatic segmentation and labeling for speech. In *IEEE International Conference on Acoustics, Speech, and Signal Processing*, Toronto, Canada, 473–476, 1991.

[LKS87] L. F. Lamel, R. H. Kasel, and S. Seneff. Design and analysis of the acoustic-phonetic corpus. In *Proceedings of the DARPA Speech Recognition Workshop*, 26–32, 1987.

[PFBP88] P. Price, W. M. Fisher, J. Bernstein, and D. S. Pallett. The DARPA 1000-word resource management database for continuous speech recognition. In *IEEE International Conference on Acoustics, Speech, and Signal Processing*, New York, 651–654, 1988.

[RR93] W. Reichl and G. Ruske. Syllable segmentation of continuous speech with artificial neural networks. In *Proceedings of Eurospeech*, Berlin, 1771–1774, 1993.

[SN93] H. Shimodaira and M. Nakai. Accent phrase segmentation using transition probabilities between pitch pattern templates. In *Proceedings of Eurospeech*, Berlin, 1767–1770, 1993.

[Spr90] R. Sproat. Stress assignment in complex nominals for English text-to-speech. In *Proceedings of European Speech Communication Association Workshop on Speech Synthesis*, Autrans, France, 129–132, 1990.

[SHY92] R. Sproat, J. Hirschberg, and D. Yarowsky. A corpus-based synthesizer. In *Proceedings of the International Conference on Spoken Language Processing*, Banff, Alberta, Canada, 1992.

[YRT89] S. Young, N. Russell, and J. Thornton. *Token Passing: A Simple Conceptual Model for Connected Speech Recognition Systems*. Cambridge University Engineering Department Technical Memorandum CUED-TR38, Cambridge, 1989.

Section V

Prosodic Analysis of Natural Speech

26

Section Introduction. Prosodic Analysis: A Dual Track?

René Collier

The last decade or so has witnessed a lively interest in the study of prosodic phenomena. A major impetus for this new trend has been coming from speech technology, in particular from speech synthesis. In this field, a strong need has arisen to provide systems for text-to-speech conversion with more pleasant and natural-sounding variations in melody, rhythm, voice quality, and other prosodic features. Perhaps, surprisingly, perhaps, little or no knowledge could be derived from the existing literature on the prosody of many of the languages for which the technology was being developed. Indeed, more often than not, existing descriptions appeared to be too qualitative in nature and provided an insufficient basis for the definition of synthesis rules. Facing this discrepancy between phonetic supply and technological demand, speech researchers have attempted to unravel regularities in intonation, accentuation, temporal variation, and other descriptions, by using new procedures for automatic rule extraction. These rules may lead to acceptable synthesis to the extent that they adequately model recurring patterns in the acoustic structure of speech. However, they are less likely to provide insight into the production and perception of prosody and its role in speech communication. This technology-driven approach, to be referred to as *analysis-for-synthesis*, may nevertheless be an efficient short-cut in an applications-oriented line of work. It may even be an indispensible support for the development of speech-based products and services in the near future. Yet, whenever possible, preference should be given to a research methodology that combines basic insight through experimentation with practical testing through application.

Interestingly, the first attempts to analyze prosody in a controllable way were based on synthesis. The well-known *analysis-by-synthesis* procedure was particularly useful in the study of intonation at a time when reliable pitch meters were still lacking. Thus, for instance, in the early years of intonation research at the Institute for Perception Research, or Instituut voor Perceptie Onderzoek (IPO) (i.e., in the 1960s), use was made of the Intonator, a vocoder that allowed the investigator to impose an artificial pitch contour on the resynthesized segmental carrier of an original utterance. In this way, a simplified course of F_0 could be generated, which was indistinguishable from its natural counterpart and therefore contained all the perceptually essential features.

The level of abstraction, attained through the stylization process, quickly leads to generalizations and predictions about the course of F_0 in not yet observed utterances. A small number of carefully analyzed utterances, often not more than a few dozen, can be a sufficient basis on which to develop a prototypical *grammar of intonation*. This explicit prediction of regularities can then be verified and refined in confrontation with independent observations. These can be of two sorts: a set of additional data from natural speech or (re)synthesized utterances with rule-based prosody. The latter are then evaluated for naturaleness in confrontation with the intuitions of a panel of listeners.

In earlier years, synthesis-by-rule of prosodic features was often performed outside the context of text-to-speech conversion, that is, without a view on *speech technology* as an application domain. In fact, the same was true of segmental synthesis, which often served the sole purpose of testing hypotheses about the perceptual importance of acoustic cues. However, today's speech-based applications confront the developer with problems that require immediate solutions. The emphasis is on the result—the quality of the output speech (in relation to the cost)—and not on the process—the way in which such result is obtained. In other words, speech *synthesis* technology need not embody whatever we happen to know about the production and perception of speech. Incidentally, the much favored statistical approach to automatic speech *recognition* shows even more clearly that for machine performance to be successful, it need not mimic the process of human speech processing. In the case of prosodic *analysis-for-synthesis*, exemplified in this section, the preferred research strategy appears to be a data-driven approach, too. An overview of prosodic analysis of natural speech is offered in chapter 27 by Gronnum.

Chapter 28 by Hirai and colleagues presents a procedure for the automatic extraction of prosodic rules from a set of data. In particular, an approach is developed that discovers regularities in the melodic structure of Japanese utterances and relates them to elements of their linguistic structure (such as the location of boundaries between syntactic constituents). The approach takes as its starting point the quantitative model of (Japanese) intonation developed by H. Fujisaki. The development of intonation rules then amounts to specifying the location and the strength of the two sorts of commands that are the backbone of this model. The same model is applied to the analysis of a database of German utterances, described in chapter 32 by Möbius.

Rules for intonation need not predict F_0 variations in every detail that can be observed in natural speech. The synthetic pitch contours may be simplified approximations to the natural ones, as long as they sound acceptable. Therefore, rule extraction can be facilitated if the input to the learning algorithm is preprocessed by some sort of data-reduction procedure. Chapter 29 by Mertens and colleagues illustrates this point: an automatic stylization approach reduces the input F_0 curve to only the perceptually relevant variations. In turn, this simplification may be the starting point for the automatic labeling of pitch contours in terms of phonological categories.

In the time domain, the same search for regularities is pursued in an attempt to predict the duration of speech sounds or silence (in particular pause). Shih and Ao show in chapter 31 how a greedy algorithm discovers the (combination of) factors that determine the nature and the extent of duration adjustments in the segmental sounds of Mandarin Chinese. A similar statistical approach is used by Barbosa and Bailly (chapter 30) to predict the emergence of pauses in French as a function of decreasing speech rate.

Finally, certain applications may require the use of different speaking styles, such as the expression of attitude or emotion. Chapter 33 by Higuchi and associates describes how stylistic changes affect the F_0 parameters in the Fujisaki model.

In conclusion, the analysis of prosodic structure that is reported upon in this section, mainly reflects the endeavors in a technology-driven line of work. Contributions to better theoretical models concerning the production or perception of prosody were clearly a minority at the Mohonk workshop. As stated above, this situation is clearly the consequence of an increasing demand for high-quality synthesis that cannot always await the solutions based on the slow progress of basic research. The well-known dichotomy of "knowledge-based" versus "data-driven" solutions also applies to prosody-by-rule. Evidently, the new generation of prosodists has to develop a dual-track approach to the subject matter of their research: the fast track of innovation and the slow track of invention. Both approaches seem equally necessary, although they may be favored by different types of investigators or funded by different sorts of organizations.

27

Section Introduction. Prosodic Analysis of Natural Speech

Nina Gronnum

No description of a language is complete without an account of its prosody: those aspects of speech that pertain not to individual sounds but to strings of segments, from the smallest units, morae, through syllables, words, and phrases, to utterances and paragraphs. Prosody, or suprasegmentals, encompasses syllable tones, word accents and stress, and the nonlexical phenomena rhythm (including pausing), prominence gradation and intonation (including sentence accents and juncture cues)—features that are all encoded in (the perceptual correlates of) durational relations, intensity relations, and the time-varying course of fundamental frequency, within and across the relevant temporal and structural domains.

Prosody as an integral part of language description has always been recognized in linguistics and phonetics, but has not always been in the foreground of empirical research. However, prosodic analysis experienced a considerable boost about 20 years ago, and the past two decades have seen an increasing number of publications on prosodic research emerge, both phonological and phonetic (be it physiological, acoustic, or perceptual). Indeed, every respectable conference on speech has separate sessions dedicated to prosody. The reason lies, I believe, in two simultaneous developments: in a heightened awareness on the part of some linguists of the need for better language teaching materials and methods—where prosody was to play a more major role—and in a growing frustration among other linguists and speech technologists over the rather poor (monotonous and unnatural) quality of the output from speech synthesizers. Together with the development of more sophisticated technological tools, which justified the demand for improvement in both camps, empirical and theoretical studies of prosody got off the ground.

The range of languages analyzed also increased. We now have prosodic descriptions of a variety of typologically different languages and, what is more, the emphasis is often on modeling human production and perception processes. From the massive amounts of data, we can now begin to sift out, with some confidence, what appear to be universal regularities of behavior, which must be grounded in general human capacities for speech production and perception—an obvious advantage for rule-governed synthesis. We are simultaneously forced to appreciate how enormously different languages are, also prosodically.

Granted that good prosody is paramount for the acceptability and naturalness of synthetic speech, the demand in text-to-speech (as opposed to close-copy) synthesis is for models of prosody with very strong predictive power. This translates into a need for analysis of very comprehensive speech materials. But as if that were not enough, it is perhaps also to the technological applications that we owe the impatient interest in spontaneous, natural speech, that is, speech produced under less restrained circumstances than what has come to be known as laboratory speech. It is easy to motivate the necessity for experimental control over all the variables, orthogonal as well as mutually dependent, which converge in the temporal and melodic patterns of an utterance. And while "lab speech" therefore has its undisputed justification as a means to expose the fundamental prosodic structure(s) of a given language. It is equally certain, however, that the style of delivery appropriate for reading aloud strongly manipulated speech materials under laboratory conditions is not appropriate for every occasion, for example, not for some of the applications to which text-to-speech systems are put. However, analyses based on spontaneous speech corpora have hitherto been prohibitively time-consuming because the mathematical and statistical know-how and the appropriate software tools have not been available to tease out the influence from each and all of those factors that contribute to the complex temporal and fundamental frequency patterns of utterances and subsequently formulate the corresponding production rules.

Where not so many years ago text-to-speech synthesis could not really do justice to the theoretical knowledge about speech acoustics and perception, the situation seems to be reversed today. Technology is not the limit and it is apparent that we are short of data and models. The following chapters in this section go some way toward remedying this deficit. They cover typologically different languages: Mandarin Chinese, Japanese, German, French, and Mexican Spanish. The concern is mainly with prosody in natural speech, right down to modeling speaker attitudes (in Japanese) and pause location and duration (in French), and in several instances the analysis procedures are founded on models (or hypotheses) of human production and perception. The authors certainly employ the latest in highly sophisticated acoustic, digital, and statistical analysis techniques.

I shall close with memento: The responsibility for successful text-to-speech synthesis cannot be exclusively conferred on phonetics and speech technology. Theoretical linguistics must step in somewhere soon, because without a comprehensive description and understanding of the prosody/pragmatics interface we will not reach the final goal: natural-sounding synthetic speech from text.

28

Automatic Extraction of F_0 Control Rules Using Statistical Analysis

Toshio Hirai
Naoto Iwahashi
Norio Higuchi
Yoshinori Sagisaka

ABSTRACT This chapter describes an automatic derivation of F_0 control rules using a superpositional F_0 control model and a tree-generation–type statistical model. In this derivation, the superpositional model was used to parameterize the F_0 contours of speech data. The F_0 control rules, which predict the parameters from linguistic information, were formed statistically by analyzing the relationship between the parameter value and the linguistic information. An experiment using 200 Japanese-read sentences showed the effectiveness of this derivation algorithm. Throughout this experiment, two clear F_0-control characteristics of the superpositional model were derived: (1) the dominant factor controlling amplitude of the accent command was accent type, and (2) for the phrase command the dominant factor was the number of morae in the previous phrase.

28.1 Introduction

Appropriate fundamental frequency (F_0) control is one of the most important factors in improving the naturalness of synthetic speech. The optimization of F_0 control has been studied using a computational model with large speech corpora [Sag90, AS92, Tra92]. In these nonparametric modelings, superpositional decomposition of F_0 patterns was not carried out in order to be free from parametric estimation of superpositional models, although such decomposition is implicitly embedded in some of the models [Sag90, SK92]. However, a parameter expression using superpositional models does have an advantage over nonparametric modeling. First, a superpositional expression can reduce the number of degrees of freedom, which gives natural constraints on parameter optimization. Second, and more important, as a superpositional model has the potential to separately express structurally different F_0-controls, the results of a statistical analysis of these control parameters can be interpreted in a more direct fashion.

The superpositional model is applied to many languages [Ohm67, FS71, Mae76, Tho79, Gar79, Vai83]. For Japanese intonation, Fujisaki's model [FS71, FH84] is widely used. Fujisaki was inspired by the research of Öhman [Ohm67], which he extended to his model. Fujisaki's model has been applied to many languages (e.g., English [FHS81, Fuh92], German [MDP91, MF94]). A linear tonal sequence model for Japanese was proposed by Pierrehumbert and Beckman [PB88]. However, the study described in this chapter adopts Fujisaki's model because it is suitable for the quantitative analysis described here.

After parametrization of the F_0 contours of many sentences, statistical analysis is applied to them to elucidate the relationship between these parameters and the factors that affect them. Linear regression and tree regression have been previously used as statistical analysis methods in prosodic modeling [KTS92, Ril92]. This study uses a new statistical optimization scheme [IS93] to integrate traditional linear regression and tree regression.

This chapter proposes automatic derivation of F_0 control rules by using the parametrization of F_0 contours with Fujisaki's model and the statistical analysis shown above in brief. The following sections introduce this new algorithm and present experimental results using 200 Japanese-read sentences.

28.2 Algorithm for Automatic Derivation of F_0 Control Rules

28.2.1 *Overview of the Rule Derivation Procedure*

Figure 28.1 shows an outline of the procedure for the derivation of F_0 control rules. There are two steps: an F_0 contour decomposition step using a superpositional model and a rule derivation step.

In the F_0 contour decomposition step, a parametric representation of the F_0 contour was obtained using Fujisaki's superpositional model [FH84]. This model is described in more detail below.

In the rule derivation step, statistical analysis is carried out on the parameters of the F_0 contours by factoring out linguistic parameters obtained from corresponding input text. The following subsections detail these steps.

28.2.2 *F_0 Contour Decomposition*

Reduction of the number of parameters for F_0 control is important to construct a statistical computational model of F_0 control. The superpositional model is an effective parametric model to reduce the degrees of freedom. The outline of Fujisaki's model is described next [FH84]. F_0 contours are described by the following

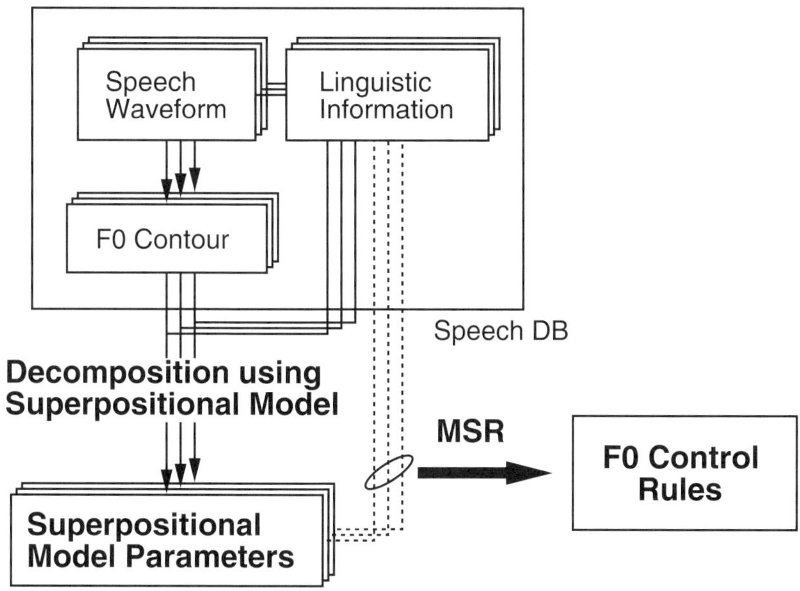

FIGURE 28.1. Outline of F_0 control rule derivation procedure.

expression in the model:

$$\ln F_0(t) = \ln F_{\min} + \sum_{i=1}^{I} A_{pi} G_{pi}(t - T_{0i}) + \sum_{j=1}^{J} A_{aj}\{G_{aj}(t - T_{1j}) - G_{aj}(t - T_{2j})\}$$

where

$$G_{pi}(t) = \begin{cases} \alpha_i^2 t \exp(-\alpha_i t), & \text{for } t \geq 0, \\ 0, & \text{for } t < 0, \end{cases}$$

and

$$G_{aj}(t) = \begin{cases} \mathrm{Min}[1 - (1 + \beta_j t)\exp(-\beta_j t), \theta_j], & \text{for } t \geq 0, \\ 0, & \text{for } t < 0. \end{cases}$$

The symbols indicate:

$F_{\min}$: base level upon which all the phrase and accent components are superposed to form an F_0 contour
I: number of phrase commands
J: number of accent commands
A_{pi}: amplitude of the ith phrase command
A_{aj}: amplitude of the jth accent command
T_{0i}: time of occurrence of the ith phrase command
T_{1j}: onset of the jth accent command

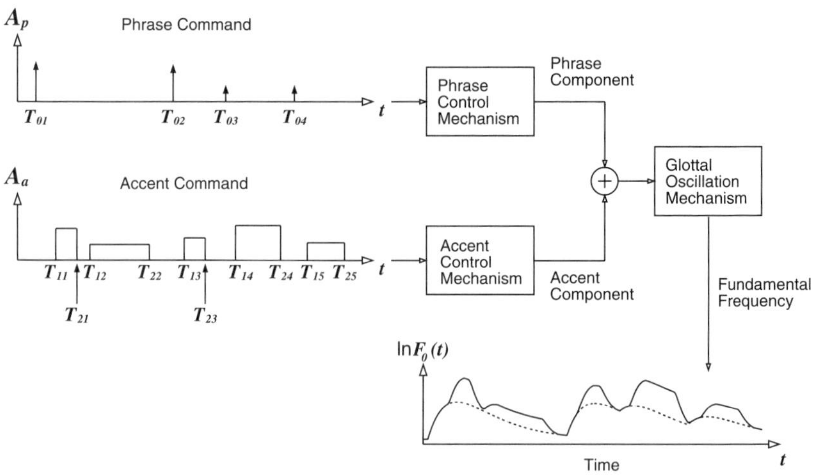

FIGURE 28.2. A functional model for the process of generating F_0 contours of sentence [FH84].

T_{2j}: end of the jth accent command
α_i: natural angular value of the phrase control mechanism to the ith phrase command
β_j: natural angular value of the accent control mechanism to the jth accent command
θ_j: ceiling level of the accent component for the jth accent command

The generation process for F_0 control of a sentence is illustrated schematically in figure 28.2. The portions controlled by a phrase command and an accent command are referred to with the terms major phrase and minor phrase, respectively.

These parameters are derived using an analysis-by-synthesis (ABS) approach, in which many parameter combination sets are tested against the original F_0 contour. If there exist F_0 peaks and the foot of those peaks in the F_0 contour, those parameters are derived stably by ABS owing to the computational model's constraints on the form of the function. In the parametrization of F_0 contours, three parameters (α_i, T_{0i}, and A_{pi}) for a phrase command and four parameters (β_j, T_{1j}, T_{2j}, and A_{aj}) for an accent command are used in ABS. Therefore, the search dimension of a sentence which has I phrase commands and J accent commands is $(3I + 4J)$.

However, these parameters are not independent in the viewpoint of least squares for F_0 estimation error. For example, the amplitude of each accent command (A_{aj}) will be derived according to the other parameters by minimizing the difference between the two assumed time series of the accent component: one is calculated from assumed A_{aj}, and the other is the phrase component and the F_0 baseline $F_{\min}$ subtracted from F_0 contour. The partial differential equation of summation of squared F_0 estimation error by A_{aj} is a linear equation of A_{aj} for $j = 1, 2, \cdots, J$. By solving these equations, each A_{aj} is derived from the other parameters analytically. With this approach, the dimension decreases by J, as no accent amplitude needs to be found in ABS, and the calculation time is expected to be shorter. Therefore, this approach was used in the present work for effective decomposition of F_0 contours. The algorithm to derive the amplitude of each accent command is described in more detail in Appendix A.

28.2.3 Statistical Rule Derivation

By analyzing the relationship between linguistic information and the features of each command generated in the previous step, a model can be statistically trained on a large corpus to predict the commands from the linguistic information.

As the features of command, (α_i and β_j), timing information (T_{0i}, T_{1j}, T_{2j}), and amplitude (A_{pi}, A_{aj}) of each command can be considered. The natural angular values are treated as constants [FH84], and timing information is predicted with a limited number of rules [OFH84, HFK85]. In contrast, the amplitude of each command is controlled by many factors closely related to linguistic properties that are difficult to fully quantify. This study investigates how precisely the amplitude of command is predicted by using common linguistic information closely related to the linguistic properties of the utterance. The amplitude estimation models for phrase command and accent command were generated separately. By using both models, each command amplitude is predicted from provided linguistic information for the derivation of the F_0 contour of a sentence. These models are F_0 control rules.

Linear regression and tree regression have been used previously in prosodic modeling [KTS92, Ril92]. Linear regression is based on the assumption of linear independence between the factors, and it cannot easily represent the dependency of multiple factors. In tree regression, no specific functional form is assumed for prediction. A binary decision tree is formed by splitting data sets by specifying the control factors, but as each split has a model parameter, the degrees of freedom of the model have a tendency to be high.

Multiple split regression (MSR) has been proposed [IS93] to overcome the inadequacies of the linear model and the tree-regression model. MSR can be regarded as a superset of the linear-regression and tree-regression models where the number of free parameters is optimized by allowing parameter sharing in a tree-regression. As with conventional tree-regression models, MSR generates a binary decision tree that can be interpreted as a set of tree-structured F_0 control rules.

In this study, we adopted MSR as the statistical computational model. The resulting binary tree can be regarded as a set of rules to predict the command amplitude by adding the weights assigned to nodes in the same manner as tree regression; and the associated parameters are optimized to generate the appropriate amplitude of each command.

By interpreting the binary tree structure, major control factors and their effect on the amplitude can be determined. For example, if a factor is frequently used in many classifications, or if the absolute value of a parameter is larger than other parameters, the factor can be considered dominant.

28.3 Experiments of F_0 Control Rule Derivation

28.3.1 Speech Data and Conditions of Parameter Extraction

Modeling experiments were performed using 200 Japanese-read sentences. The speech data from one male professional narrator was used for this experiment [STAKU90] using sentences collected from newspapers and magazines. To characterize an F_0 contour by phonetic parameters, accentual phrase boundaries and accented position were marked from listening to the speech. The F_0 contour of each sentence was obtained using the algorithm described by Secrest and Doddington [SD83]. In the extraction of F_0, the window width was 49 ms, window shift was 5 ms, window type was $\cos^4$, and the LPC order was 12. Three-point median smoothing was applied to the F_0 patterns.

Among the parameters of the superpositional model, the natural angular values of the phrase control mechanism and the accent control mechanism were fixed ($\alpha = 3.0 \text{ sec}^{-1}$, $\beta = 20.0 \text{ sec}^{-1}$) [FH84]. The minimum F_0 (= $F_{\min}$) was set to 66.7 Hz, according to the mean $F_{\min}$ of the manually obtained analysis results of 10 sentences. The onset of each phrase command and the onset and offset of each accent command were based on phonetically labeled data and accent type[1] data determined by listening.

The mean estimation error for F_0 using the superpositional model was 1.28 semitones[2] (10.7 Hz at 150 Hz). The number of phrase commands and accent commands were 739 and 1,193, respectively.

28.3.2 Linguistic Factors for the Modeling

Five factors were selected for the control of phrase command amplitude as shown in table 28.1. The length of a major phrase was used as a factor to predict the amplitude of each phrase command because the longer the major phrase is, the larger the amplitude becomes. Preliminary experiments showed a positive correlation between the number of morae in a major phrase and the amplitude of the corresponding phrase command. The punctuation comma also affects the amplitude of

---

[1]In standard Japanese, there are three basic rules for accentuating a word [Kin58]. (1) The subjective tone of each mora (a unit of (consonant+) vowel) is distributed into only two levels, high or low. (2) The high tone sequence appears only once in a word. (3) The tone changes at the end of the initial mora. According to these rules, there exist $n + 1$ *accent types* in words of n morae. For example, all the possible accent types of 4 morae words are: type 0 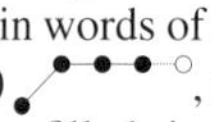, type 1 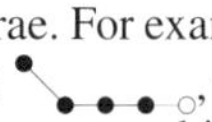, type 2 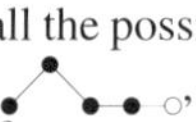, type 3 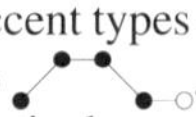, and type 4 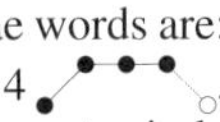, where a filled circle represents subjective frequency of a particular mora and an empty circle shows that of an auxiliary whose existence serves to discriminate two different types that are otherwise indistinguishable (type 0 and type n) . The last mora of high tone sequence is called the *accent nucleus*, except in the case of type 0. The accent type describes the presence or absence of accent nucleus, and if there is a nucleus the type number indicates its position.

[2]One *semitone* is equal to $\frac{1}{12}$ octave.

TABLE 28.1. Factors used for the control of phrase command amplitude.

Length of syntactic unit	(Number of morae)	Pre. maj. phr. Cur. maj. phr. Fol. maj. phr.
Syntactic structure indicated by punctuation mark	(Presence/absence of comma at the neighbor boundary of min. phr.)	Pre. boundary Fol. boundary

the phrase command. The speaker decided on breath positions from the positions of commas, and the correlation between pause duration and the amplitude of a phrase command is high [FHTK85].

Table 28.2 summarizes 16 factors for estimating the accent command amplitude. The amplitude of a nonzero accent command is larger than the amplitude of a type zero accent command. Previous research has reported that the amplitude of the accent command depends on whether the accent type is type zero [FK88]. The accent type of the previous minor phrase also affects the amplitude of the accent command of the current minor phrase. The modifying information has been used to analyze the F_0 pattern in previous research [HS80, SK92]. Abe and Sato [AS92] noted that the part of speech of a word in a minor phrase also affects the F_0 pattern. The position of the minor phrase in the major phrase was considered as a factor in order to check the gradual decreasing of accent command amplitude, that is, downstep. Similar to the phrase command, the mora count of a minor phrase and existence of commas were used to predict the amplitude of an accent command. Although it is not clear that the attributes of the neighbor minor phrase affect the amplitude, these are included in the factors.

TABLE 28.2. Factors used for the control of accent command amplitude.

Lexical information	Accent type	Pre. min. phr. Cur. min. phr. Fol. min. phr.
	Part of speech	Pre. min. phr. Cur. min. phr. Fol. min. phr.
Length of syntactic unit	Number of mora	Pre. min. phr. Cur. min. phr. Fol. min. phr.
	Number of min. phr. to the modifier	Pre. min. phr. Cur. min. phr.
Position in syntactic unit	Serial position in maj. phr. Relative position in maj.phr.	
	Number of preceding accented minor phrase	
Syntactic structure indicated by punctuation mark	Presence/absence of comma at the neighbor boundary of min. phr.	Pre. boundary Fol. boundary

The number of free parameters used in MSR was set to 17, which showed the convergence of estimation accuracy for each command analysis.

28.3.3 F_0 Control Rule Interpretation

The F_0 control rules were studied to determine which factors affected command amplitude. The F_0 estimation error was also calculated to determine the predictive accuracy.

Estimation Model for Phrase Command

By examining the binary tree for the model for phrase command amplitude, 9 out of 17 parameters are shared over the entire feature vector space. In interpreting the output of MSR, an extensive sharing of parameters is advantageous. The features of the initial five splits are shown in table 28.3. All these split features indicate splitting of the entire feature vector space. Each weight value whose corresponding condition is satisfied is added to the initial value, 0.79, to derive the amplitude of phrase command.

The absolute value of the second split was the largest (0.21) of all splits. The second and the third split were both based on the mora count of the previous major phrase. This means that the mora count of the previous major phrase was the dominant factor in estimating the amplitude of the phrase command. Table 28.4 shows the summary of the relationship between mora count of preceding major phrase and its weight. If the previous phrase contains six or fewer morae, it comes under the second and third classification of split description in table 28.3. As a result, the weight for the phrase command whose previous phrase is less than six morae becomes the summation of the weights of these classifications; that is, $(-0.21) + (-0.05) = -0.26$ in table 28.4. The absence of the previous major phrase means that there is no effect of the previous phrase component. Therefore, the mora count of the previous major phrase is assumed to be infinity. In the case of table 28.4, it corresponds to the range "Mora count of preceding major phrase is equal to or greater than 13."

According to these tables, we can conclude:

If there is no comma near the onset of the phrase command,
then the amplitude of the phrase command decreases.

If the previous major phrase is short,
then the amplitude of the phrase command decreases.

If the number of morae in the major phrase is less than or equal to 12,
then the amplitude of the phrase command decreases.

Estimation Model for Accent Command

In the binary tree for estimating accent command amplitude, the number of model parameters assigned for accent type was 4 out of the total 17. This indicates

TABLE 28.3. Initial five splits for derivation of phrase command amplitude.

Split	Factor	Weight†
1	no comma	−0.10
2	the mora count of previous major phrase ≤ 6	−0.21
3	the mora count of previous major phrase ≤ 12	−0.05
4	the mora count of succeeding major phrase ≤ 7	−0.06
5	the mora count of current major phrase ≤ 12	−0.05

†initial amplitude was 0.79

TABLE 28.4. Relationship between mora count of preceding major phrase and its weight.

Mora count of preceding major phrase	≤ 6	$7 \sim 12$	≥ 13
Weight†	−0.26	−0.05	0

†initial amplitude was 0.79

that type of accent was the dominant factor in predicting the amplitude of accent command. The relationship between accent type and its weight is shown in table 28.5. Excluding accent types 0, 5, and 6, there is a negative correlation between the length of a high tone period and the amplitude of an accent command. Considering that the period between command onset and offset in the zero-accent case is long (it is as long as the word, minus the first mora), it is reasonable that the weight for this type should be the smallest in the table, and the zero-accent thus does not constitute an exception to the negative correlation observed between accent type and weight that applies within the nonzero accent type.

Estimation of F_0 Contour

F_0 estimation errors for closed and open data were calculated in order to evaluate the estimation accuracy of F_0 patterns. A closed test was executed by estimating F_0 patterns of 200 sentences. In an open test, the speech data set was separated into two parts, and the F_0 patterns in one group (50 sentences) were predicted by the F_0 control rules derived from the other part (150 sentences).

The mean estimation error of the F_0 contour was used for the objective evaluation of the naturalness of the F_0 pattern in this study. These errors are shown in table 28.6. In figure 28.3, the predicted F_0 patterns of a sentence by the models trained using open and closed data are shown (b), (c) with the best approximation results from the superpositional model (a). This sentence is se-

TABLE 28.5. Relationship between accent type and its weight.

Accent Type	1, 2	3, 4, 6	$5, 7 \sim$	0
Weight†	±0	−0.10	−0.25	−0.35

†initial amplitude was 0.88

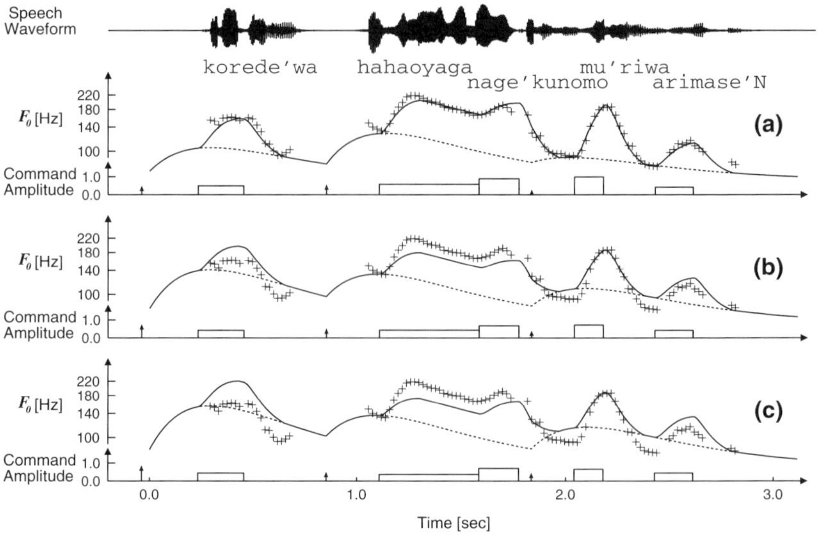

FIGURE 28.3. Estimation of F_0 patterns using F_0 control rules. The waveform, (a) best approximation result of F_0 estimation using the superpositional model, (b) closed-test estimation result, and (c) open-test estimation result of "koredewa hahaoyaga nagekunomo muriwa arimasen" (it is reasonable for his mother to heave a sigh). The + signs and solid curves are observed F_0 and predicted F_0 contour, respectively. The dotted lines indicate the phrase component on the basis of ln $F_{\min}$. Phrase command is shown as an arrow, and accent command is shown as a box.

lected according to each estimation error, which approximates the corresponding mean estimation error. The estimation errors in this figure are 1.02 semitones (8.6 Hz at 150 Hz), 2.33 semitones (18.9 Hz), and 3.15 (25.0 Hz). The timings of each command were taken over the best approximation result of F_0 estimation: (a) in figure 28.3. It is understandable that the estimation error of (a) is smallest because there is no constraint for the parametrization of the F_0 contour using the Fujisaki intonation model in its best approximation. Although there were many discrepancies, like the contours (b) and (c) derived by the models, almost all of the synthesized speech using the contours were acceptable due to the constraints provided by the formulation of Fujisaki's model.

TABLE 28.6. F_0 estimation error.

	Estimation error, Units: semitone (Hz†)
Closed test	2.27 (18.4)
Open test	2.91 (23.2)

†error at 150 Hz

28.4 Summary

We have proposed a method for automatic derivation of F_0 control rules using MSR and the superpositional F_0 model. The experimental results show that the dominant factors for the phrase command and accent command prediction are the number of morae of the previous major phrase and the type of accent, respectively. The negative correlation between the length of a high tone period and the amplitude of an accent was derived automatically, which is an extension of former rule. This result supports the effectiveness of the approach proposed in this chapter.

For future research, we are planning to use the probability of voicing for each sample of the observed F_0 contour as a weighting of the F_0 estimation error. Further investigation will also concern the analysis of various speaking styles using the method shown here and an examination of the difference between the dominant factors of those speaking styles.

References

[AS92] M. Abe and H. Sato. Two-stage F0 control model using syllable based F0 units. In *Proceedings ICASSP '92*, San Francisco, CA, 53–56, 1992.

[FH84] H. Fujisaki and K. Hirose. Analysis of voice fundamental frequency contours for declarative sentence of Japanese. *J. Acoust. Soc. Jpn. (E)* 5(4):233–242, 1984.

[FHS81] H. Fujisaki, K. Hirose, and M. Sugito. Fundamental frequency contours of English sentences. In *Trans. Committee on Speech Research, Acoust. Soc. Jpn.*, S81-03, 17–24, 1981 (in Japanese).

[FHTK85] H. Fujisaki, K. Hirose, S. Tada, and H. Kawai. Generation of prosodic symbols for synthesis of spoken sentences of Japanese. In *Rec. Spring Meeting, Acoust. Soc. Jpn.*, 141–142, 1985 (in Japanese).

[FK88] H. Fujisaki and H. Kawai. Realization of linguistic information in the voice fundamental frequency contour. In *Proceedings ICASSP '88*, New York, 663–666, 1988.

[FS71] H. Fujisaki and H. Sudo. Synthesis by rule of prosodic features of connected Japanese. In *Proceedings of 7th International Congress on Acoustics*, vol. 3, 133–136, 1971.

[Fuj92] H. Fujisaki. The role of quantitative modeling in the study of intonation. In *Proceedings International Symposium on Japanese Prosody*, Nara, Japan, 163–174, 1992.

[Gar79] E. Gårding. Sentence intonation in Swedish. *Phonetica* 36:207–215, 1979.

[HFK85] K. Hirose, H. Fujisaki, and H. Kawai. A system for synthesis of connected speech—Special emphasis on the prosodic features. In *Trans. of the Committee on Speech Research, Acoust. Soc. Jpn.*, S85–43, 1985 (in Japanese).

[HS80] K. Hakoda and H. Sato. Prosodic rules in connected speech synthesis. *Sys. Comp. Cont.* 11(5):28–37 1980. (Translated from *Trans. of IEICE Jpn.* J63–D(9):715–722, 1980.)

[IS93] N. Iwahashi and Y. Sagisaka. Duration modelling with multiple split regression. In *Proceedings Eurospeech '93*, Berlin, 329–332, 1993.

[Kin58] H. Kindaichi. *Meikai Nihongo Akusento Jiten.* Sanseido, Tokyo, 1958 (in Japanese).

[KTS92] N. Kaiki, K. Takeda, and Y. Sagisaka. Linguistic properties in the control of segmental duration for speech synthesis. In *Talking Machines: Theories, Models, and Designs*, G. Bailly and C. Benoit, eds. North-Holland, Amsterdam, 1992.

[Mae76] S. Maeda. A characterization of American English intonation. PhD thesis, Massachusetts Institute of Technology, Cambridge, MA, 1976.

[MDP91] B. Möbius, G. Demenko, and M. Pätzold. Parametric description of German fundamental frequency contours. In *Proceedings ICPhS '91*, Aix-en-Provence, France, 222–225, 1991.

[MF94] H. Mixdorff and H. Fujisaki. Analysis of F_0 contours of German utterances using a model of the production process. In *Rec. Spring Meeting, Acoust. Soc. Jpn.*, Tokyo, Japan, 287–288, 1994 (in Japanese).

[OFH84] E. Ohira, H. Fujisaki, and K. Hirose. Relationship between articulatory and phonatory controls in the sentence context. In *Rec. Spring Meeting, Acoust. Soc. Jpn.*, Tokyo, Japan, 111–112, 1984 (in Japanese).

[Ohm67] S. Öhman. *Word and Sentence Intonation: A Quantitative Model.* Speech Transmission Laboratory Quarterly Progress and Status Report, KTH, STL-QPSR 2-3/1967, 20–54, 1967.

[PB88] J. Pierrehumbert and M. Beckman. *Japanese Tone Structure.* Linguistic Inquiry Monograph Series, The MIT Press, Cambridge, MA, 1988.

[Ril92] M. Riley. Tree-based modelling of segmental durations. In *Talking Machines: Theories, Models, and Designs*, G. Bailly and C. Benoit, eds. North-Holland, Amsterdam, 1992.

[Sag90] Y. Sagisaka. On the prediction of global F0 shape for Japanese text-to-speech. In *Proceedings ICASSP '90*, Albuquerque, NM, 325–328, 1990.

[SD83] B. G. Secrest and G. R. Doddington. An integrated pitch tracking algorithm for speech system. In *Proceedings ICASSP '83*, Boston, MA, 1352–1355, 1983.

[SK92] Y. Sagisaka and N. Kaiki. Optimization of intonation control using statistical F0 resetting characteristics. In *Proceedings ICASSP '92*, San Francisco, CA, 49–52, 1992.

[STAKU90] Y. Sagisaka, K. Takeda, M. Abe, S. Katagiri, T. Umeda, and H. Kuwabara. A large-scale Japanese speech database. In *Proceedings ICSLP '90*, Kobe, Japan, 1089–1092, 1990.

[Tho79] N. Thorsen. Interpreting raw fundamental frequency tracings of Danish. *Phonetica* 36:57–58, 1979.

[Tra92] C. Traber. F_0 generation with a database of natural F_0 patterns and with a neural network. In *Talking Machines: Theories, Models, and Designs*, G. Bailly and C. Benoit, eds. North-Holland, Amsterdam, 1992.

[Vai83] J. Vaissière. Language-independent prosodic features. In *Prosody: Models and Measurements*, A. Cutler and D. R. Ladd, eds. Springer-Verlag, Heidelberg, 53–66, 1983.

Appendix A: Algorithm for the Extraction of Accent Command Amplitude

The amplitude of an accent command of the superpositional model will be extracted from other model parameters without the ABS approach. By minimizing the accumulated square error of predicted F_0, the amplitude is derived analytically using least squares.

The vector of estimation error of F_0 $\boldsymbol{e}$ is

$$\boldsymbol{e} = \boldsymbol{f}_0 - (\boldsymbol{f}_{\min} + \sum_{i=1}^{I} A_{pi}\boldsymbol{g}_{pi} + \sum_{j=1}^{J} A_{aj}\boldsymbol{g}_{aj})$$

where

$$\boldsymbol{f}_0 = (\ln F_0(\Delta t), \ln F_0(2\Delta t), \cdots, \ln F_0(N\Delta t))^T,$$

$$\boldsymbol{f}_{\min} = (\underbrace{\ln F_{\min}, \ln F_{\min}, \cdots, \ln F_{\min}}_{N})^T,$$

$$\boldsymbol{g}_{pi} = \begin{pmatrix} G_p(\Delta t - T_{0i}) \\ G_p(2\,\Delta t - T_{0i}) \\ \vdots \\ G_p(N\,\Delta t - T_{0i}) \end{pmatrix}, \text{ and}$$

$$\boldsymbol{g}_{aj} = \begin{pmatrix} G_a(\Delta t - T_{1j}) - G_a(\Delta t - T_{2j}) \\ G_a(2\,\Delta t - T_{1j}) - G_a(2\,\Delta t - T_{2j}) \\ \vdots \\ G_a(N\,\Delta t - T_{1j}) - G_a(N\,\Delta t - T_{2j}) \end{pmatrix}.$$

Δt and N are sampling rate of F_0 and number of F_0 data, respectively. Error E, the squared norm of $\boldsymbol{e}$, is

$$E = \boldsymbol{e}^T \cdot \boldsymbol{e}.$$

In order to simplify the following explanation, $\boldsymbol{e}$ is separated into two parts:

$$\boldsymbol{e} = \boldsymbol{d} - \hat{\boldsymbol{d}}$$

where

$$\boldsymbol{d} = \boldsymbol{f}_0 - (\boldsymbol{f}_{\min} + \sum_{i=1}^{I} A_{pi}\boldsymbol{g}_{pi}) \quad \text{and} \quad \hat{\boldsymbol{d}} = \sum_{j=1}^{J} A_{aj}\boldsymbol{g}_{aj}.$$

The problem of extracting the amplitude of each accent command from the other parameters is resolved by least squares of E. This is because E is desired to minimize variation of the value of each amplitude of an accent command. The partial deviation of E by A_{aj} is equal to zero:

$$\frac{\partial E}{\partial A_{aj}} = 2\,((\frac{\partial(\boldsymbol{d} - \hat{\boldsymbol{d}})}{\partial A_{aj}})^T (\boldsymbol{d} - \hat{\boldsymbol{d}})) = 2\,(-\boldsymbol{g}_{aj}{}^T (\boldsymbol{d} - \hat{\boldsymbol{d}})) = 0 \qquad (j = 1, 2, \cdots, J),$$

$$\boldsymbol{g}_{aj}{}^T\boldsymbol{d} = \boldsymbol{g}_{aj}{}^T\hat{\boldsymbol{d}} = \begin{pmatrix} \boldsymbol{g}_{aj}{}^T\boldsymbol{g}_{a\Delta} & \boldsymbol{g}_{aj}{}^T\boldsymbol{g}_{a\Theta} & \cdots & \boldsymbol{g}_{aj}{}^T\boldsymbol{g}_{aJ} \end{pmatrix} \begin{pmatrix} A_{a1} \\ A_{a2} \\ \vdots \\ A_{aJ} \end{pmatrix}. \quad (28.1)$$

By considering the j is $1, 2, \cdots, J$ in equation (28.1),

$$\begin{pmatrix} \boldsymbol{g}_{a\Delta}{}^T\boldsymbol{d} \\ \boldsymbol{g}_{a\Theta}{}^T\boldsymbol{d} \\ \vdots \\ \boldsymbol{g}_{aJ}{}^T\boldsymbol{d} \end{pmatrix} = \begin{pmatrix} \boldsymbol{g}_{a\Delta}{}^T\boldsymbol{g}_{a\Delta} & \boldsymbol{g}_{a\Delta}{}^T\boldsymbol{g}_{a\Theta} & \cdots & \boldsymbol{g}_{a\Delta}{}^T\boldsymbol{g}_{aJ} \\ \boldsymbol{g}_{a\Theta}{}^T\boldsymbol{g}_{a\Delta} & \boldsymbol{g}_{a\Theta}{}^T\boldsymbol{g}_{a\Theta} & \cdots & \boldsymbol{g}_{a\Theta}{}^T\boldsymbol{g}_{aJ} \\ \vdots & \vdots & \ddots & \vdots \\ \boldsymbol{g}_{aJ}{}^T\boldsymbol{g}_{a\Delta} & \boldsymbol{g}_{aJ}{}^T\boldsymbol{g}_{a\Theta} & \cdots & \boldsymbol{g}_{aJ}{}^T\boldsymbol{g}_{aJ} \end{pmatrix} \begin{pmatrix} A_{a1} \\ A_{a2} \\ \vdots \\ A_{aJ} \end{pmatrix}.$$

From this result, each amplitude of an accent command is extracted by a matrix calculation:

$$\begin{pmatrix} A_{a1} & A_{a2} & \cdots & A_{aJ} \end{pmatrix}^T = \mathbf{H}^{-1} \begin{pmatrix} \boldsymbol{g}_{a\Delta} & \boldsymbol{g}_{a\Theta} & \cdots & \boldsymbol{g}_{aJ} \end{pmatrix}^T \boldsymbol{d}$$
$$(\mathbf{H} = (h_{jk}), \quad h_{jk} = \boldsymbol{g}_{aj}{}^T\boldsymbol{g}_{ak}).$$

The squared norm E is calculated from the matrices and vectors defined above:

$$\begin{aligned} E &= (\boldsymbol{d} - \hat{\boldsymbol{d}})^T(\boldsymbol{d} - \hat{\boldsymbol{d}}) \\ &= \boldsymbol{d}^T\boldsymbol{d} - 2\,\boldsymbol{d}^T\hat{\boldsymbol{d}} + \hat{\boldsymbol{d}}^T\hat{\boldsymbol{d}} \\ &= \boldsymbol{d}^T\boldsymbol{d} \\ &\quad - 2\,\boldsymbol{d}^T \begin{pmatrix} \boldsymbol{g}_{a\Delta} & \boldsymbol{g}_{a\Theta} & \cdots & \boldsymbol{g}_{aJ} \end{pmatrix} \begin{pmatrix} A_{a1} & A_{a2} & \cdots & A_{aJ} \end{pmatrix}^T \\ &\quad + \begin{pmatrix} A_{a1} & A_{a2} & \cdots & A_{aJ} \end{pmatrix} \mathbf{H} \begin{pmatrix} A_{a1} & A_{a2} & \cdots & A_{aJ} \end{pmatrix}^T. \end{aligned}$$

□

Appendix B: Audio Demos

(1) Using best approximation F_0 contour by the superpositional model,
(2) using F_0 contour estimated from closed test, and
(3) using F_0 contour estimated from open test.

29

Comparing Approaches to Pitch Contour Stylization for Speech Synthesis

Piet Mertens
Frédéric Beaugendre
Christophe R. d'Alessandro

ABSTRACT This chapter describes two approaches to pitch contour stylization as well as a perception experiment to evaluate and compare both methods. The first approach uses an automatic stylization procedure, based on perceptual criteria. It outputs the sequence of audible pitch events (static tones, dynamic tones, complex dynamic tones) in the utterance. Both the tonal perception model and the algorithm are described in some detail. The second approach, known as close-copy stylization, is a manual procedure in which a straight-line approximation of the pitch contour is obtained interactively, by resynthesis of the stylized contour and auditory comparison with the original. A perception experiment using synthetic stimuli with stylized contours was run in order to compare and evaluate both approaches. The stylized contours can hardly be distinguished from the natural contours. Tonal perception stylization gives slightly better results than straight-line stylization.

29.1 Introduction

Several approaches have been proposed for the generation of intonation in speech synthesis systems. The prosodic models used in these systems are quite different; they are defined in terms of: (1) pitch target values derived directly from a phonological representation [Pie81, Hir92]; (2) voice source commands [Ohm67, FK88]; (3) standardized pitch movements, obtained using some stylization of F_0 contours [tHa90]. This chapter focuses on the latter approach. Both in synthesis and analysis (or recognition), the stylized pitch contour is a meaningful level of representation. For synthesis, this is obvious: the stylized pitch controls the synthesizer. But also in the case of analysis, the stylization based on perceptual criteria is a meaningful representation because, as will be shown, it provides a transcription of the prosodic auditory events in the utterance.

The purpose of this chapter is to compare two approaches to pitch contour stylization. To this effect a perception experiment was run in which the subjects had to decide whether the pitch contours of a pair of utterances were identical. These stimulus pairs contained resynthesized utterances using either the original pitch contour, or a stylized contour obtained by one of the procedures under analysis. The proportion of answers for which the subjects hear no difference between the original and the modified contours provides an estimation of the quality of the stylizations. The subjects were native speakers of French; they were judging French stimuli.

The structure of the chapter is as follows. The remainder of this introduction is concerned with stylization in general: what is stylization, what are the components of a stylization procedure, and how can stylizations be classified and compared? The next two sections each describe one particular stylization strategy. The first is an automatic procedure that simulates tonal perception; it was developed recently [tHa95], and will be described in detail. We also give a concise overview of tonal perception. The second approach is the well-known close-copy stylization, which approximates the pitch contour by a sequence of straight lines [tHa90]. The terms "straight-line stylization" and "pitch movements approach" are used also to refer to the second type of stylization. For each of these approaches, we also mention some already available experimental data. Section 29.4 describes a new experiment aimed at a comparison of the two stylization approaches. Finally, the last section provides a general conclusion.

Stylization is often viewed as a way to reduce the amount of information contained in the fundamental frequency tracing in such a way as to retain only those parts of the pitch curve that have a linguistic function in speech communication, and hence are necessary for the synthesis of prosody. Because still too little is known about which parts of F_0 contours are relevant, and how to determine them, there are several approaches to intonation stylization.

When comparing stylization systems, it is useful to decompose the overall process into three successive components. The first is F_0 determination, as F_0 is a major input to the stylization algorithm. The second component is the actual stylization. The result is a simplified pitch curve, whatever procedure is used to obtain this curve. This component can be followed by a classification step, in which parts of the pitch curve are recognized as instances of discrete units within a particular intonation model. In some systems the last two components are merged within a single step, in particular when the intonation model is seen as the set of (normalized) pitch movements.

Stylization procedures can be classified on the basis of the underlying model. The stylization procedure can be purely mathematical, without any reference to the way the speech signal is processed by the human listener. Most current approaches are of this type. However, when for a given utterance one compares the physical F_0 curve with what one hears, it is obvious that many variations go unnoticed, whether these variations correspond to parts of sounds or to parts of syllables. So what can be heard is only a subset of what is measured. As a result, the stylization can be based on the way in which humans perceive pitch changes in speech signals; that is, it can be based on tonal perception, and in its strongest form stylization is a computer simulation of tonal perception. Although the mathematical and perceptual approaches can be equally successful for speech synthesis, the latter allows one to gain insight into the process of human perception of prosody.

Another dimension for classification is the criterion used in assessment. When evaluating stylization through comparison of the stylized utterances with their original counterparts, one can use a strong or a weaker criterion; either one verifies whether both versions are indistinguishable, or whether they are functionally equivalent. But these are two completely different questions: in the former case the optimal stylization is the auditory image, in the latter it can be something else, such as the units in the intonation model.

29.2 Automatic Stylization Based on Tonal Perception

This section deals with the stylization procedure based on perceptual criteria. It will be referred to as the automatic tonal perception stylization (ATS). Here we explain the rationale of the approach, quickly introduce concepts of tonal perception, describe the algorithm, and give some of the results obtained with this method.

29.2.1 Rationale

To start, there is an obvious question that needs to be asked: *Why would one make the effort to simulate tonal perception?* The first motivation is due to its theoretical scientific interest. A simulation is a kind of verification procedure. Via listening tasks with stylized stimuli, we can test the accuracy of the perceptual model and modify it where necessary. As the stylization is controlled by two parameters, which are perceptual thresholds (the glissando threshold and the differential glissando threshold; see next subsection), it can be used to measure these thresholds. The stylization thus becomes a tool for basic research on tonal perception. The second motivation is that the stylized contour is an *estimation* of the pitch pattern as perceived by the average human listener, rather than the sequence of the (recognized) language-specific and theory-specific intonation units.[1] The stylized contour thus reflects a representation after low-level perceptual processing, prior to any categorization involving a language-specific intonation grammar. As a result, this auditory representation can be defined and investigated (i.e., measured) on its own, without reference to some particular intonation model, or even without reference to the communicative function of pitch in speech! Third, the stylization is independent of any particular linguistic intonation model and can in fact be used to construct such a model in an unbiased way.

---

[1] It should be noted, however, that the segmentation into syllabic nuclei is to some extent determined by language-specific factors, such as the existence of syllabic consonants in the language. In the current implementation for French, syllables are formed around vocalic nuclei, and the part of the F_0 contour that is analyzed corresponds to the voiced part of the syllable.

29.2.2 Tonal Perception and Prosodic Analysis

Both the peripheral auditory system and the perceptual system shape the speech signal into a mental auditory signal that is quite different from the acoustic signal, to say the least. To illustrate sensory and perceptual processing, one could mention the frequency analysis in the cochlea, the role of frequency response nonlinearity and critical bands for human pitch determination, and so forth. Given these phenomena, it will be clear that (1) the pitch perceived by the human listener does not closely match the fundamental frequency measured in the acoustic signal and (2) the communicative function of prosody can be conveyed only by those pitch events that are preserved in the auditory signal, rather than by any measurable F_0 variation. Consequently, it is useful to obtain the auditory representation. We do not claim that a perceptual model should mimic the peripheral auditory system (although this would, of course, provide the most accurate simulation), but rather that it should take into account the major perceptual effects observed in psychoacoustics.

We briefly describe three perceptual effects related to frequency variations. The first is known as the *glissando threshold*. A fundamental frequency variation that takes place during a given time interval will be perceived as a pitch movement if the rate of change exceeds some minimal amount; this amount depends on the duration of the transition: the shorter the stimulus, the larger the required frequency change. Frequency variations below this threshold are perceived without pitch change (i.e., they are perceived as static tones). The glissando threshold has been measured for pure tones and synthetic vowels generated with a linear frequency change. It should be pointed out that the stimuli used for the determination of the glissando had a constant amplitude. If a glissando threshold for continuous speech could be accurately determined, we would be able to determine which frequency changes are heard as dynamic pitch changes and which as static pitch events.

Of course, the frequency variations observed in actual speech usually exhibit more complex patterns. One observes changes in slope, for instance when a rise is followed by a fall, or when a slow rise changes into a steep rise. Now, let us assume that small slope changes go unnoticed: in this case we can consider the entire frequency variation as a single movement, measure its frequency change, and confront it with the glissando threshold. However, if a given change in slope is audible as such, the variation should be divided into two parts at the point of change in direction, and the frequency change in each part should be evaluated with respect to the glissando threshold. For this reason it is important to know under which conditions a change in slope is perceived. The critical slope change is called the *differential glissando threshold*. There has been very little research on this effect. Note that the proposed procedure still assumes that amplitude changes (as observed in speech) have no effect on tonal perception.

As shown by House [Hou90], the perception of pitch variations is influenced by changes in amplitude and in spectral composition. For instance, a signal with constant fundamental frequency that shows rapid and substantial amplitude dips of some minimal duration will be perceived as a sequence of tones, starting at the amplitude dips. The same holds for signals containing unvoiced parts. Speech signals are indeed characterized by rapid amplitude changes (e.g., for plosives) and by unvoiced intervals (e.g., unvoiced fricatives). But speech signals are also characterized by slow amplitude changes (e.g., nasal consonants) and progressive transitions from quasi-periodic to aperiodic sounds. A complete quantitative model of this *segmentation effect* is required to deal with those common cases in an appropriate way. Its effect would be to transform the pitch contour at the auditory level into a sequence of short duration tones corresponding to the syllabic nuclei. However, the lack of a quantitative model for this effect makes it difficult to say which parts of the signal make up the syllabic nuclei. As a first approximation, our tonal perception model uses the voiced parts of syllables as the intervals corresponding to the tones.

Automatic perceptual stylization simulates these three perceptual effects. The segmentation effect results in a segmentation of the speech signal into a sequence of short pitch variations. The differential glissando is used to decompose such a pitch variation into uniform pitch movements (rise, fall, level); they are called *tonal segments*. Finally, the glissando threshold determines which tonal segments correspond to audible pitch changes and which are static.

29.2.3 *Description of the Algorithm*

The stylization procedure consists of several processing steps, some of which are purely acoustic (pitch determination, voicing determination) whereas others are related to perception. We first give an overview of the main processing steps and later describe them in more detail. Figure 29.1 gives a schematic description of the algorithm.

The perceptual model evaluates fundamental frequency variations for syllable-sized fragments of the speech signal. This requires the determination of fundamental frequency and a segmentation of the signal, which, in the current implementation, provides the sequence of voiced portions, one for each syllable in the speech signal.

In the next stage, a short-term perceptual integration (see below, weighted time-average model) is applied to the F0 of each voiced fragment, resulting in a somewhat smoothed pitch contour.

For each voiced portion, the obtained pitch curve is divided into uniform parts (*tonal segments*) on the basis of two perceptual parameters: the glissando threshold and the differential glissando threshold. Ideally, a tonal segment will correspond to a single audible pitch event (rising, falling, or level). Each syllable contains one or more tonal segments, each of which is either static or dynamic (rise or fall). The tonal segment is characterized by the time and pitch of its starting and ending points. The actual stylization is trivial: it consists of a linear interpolation between the start and end points. It will be viewed as an estimation of the perceived pitch (and of the audible pitch movements).

This representation can be further processed within the context of a language-specific intonation grammar in order to go from the level of auditory events to that

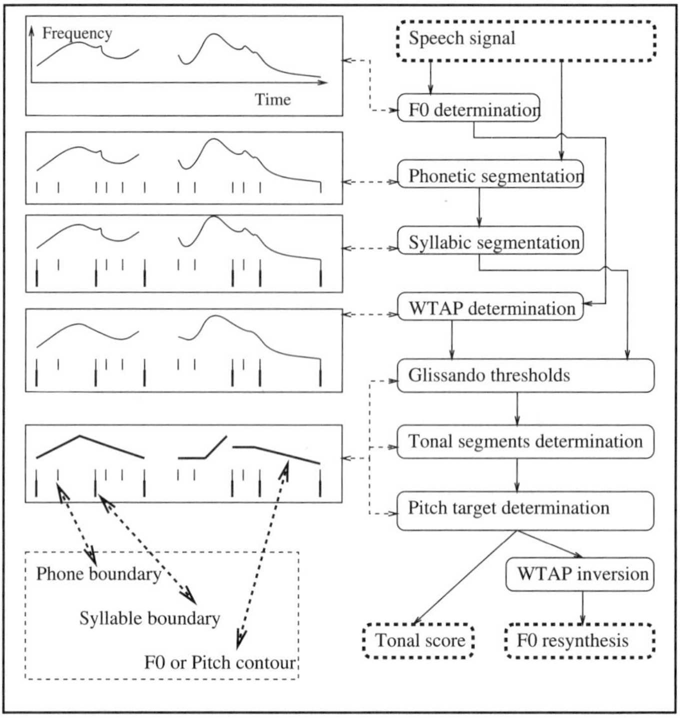

FIGURE 29.1. Automatic intonation analysis algorithm. The left side of the illustration gives a schematic representation of the shape of the pitch contour, in relation to the processing steps, shown on the right side. WTAP stands for weighted time-average pitch.

of the linguistic units [Mer87a]. However, the latter aspect will not be dealt with here.

In what follows, each processing step is described in more detail (see also [dM95]).

1. *Fundamental frequency measurement.* In principle, any kind of pitch determination algorithm can be used, provided its precision is as good as that of human listeners. Most current pitch extractors meet this requirement. Their average accuracy is sufficient. However, results can vary substantially from one algorithm to the other, especially for transitions (unvoiced to voiced, plosive to vowel, glottal stops) and vocal fry. In our implementation the spectral comb method was used as a basic extractor, combined with a postprocessor, which traps many octave shifts.

2. *Voicing determination.* The current implementation uses a simple voiced/unvoiced detection based on energy and zero-crossing rate. Of course, more sophisticated approaches could be used.

3. *Syllabic segmentation.* As said above, the perceptual segmentation effect decomposes the speech signal into a sequence of short tones corresponding to the syllabic nuclei. In the absence of a quantitative model for this effect, the syllabic nucleus is obtained as the voiced portion of the syllable. To avoid artifacts in the resynthesized pitch due to segmentation errors rather than to the stylization itself, an accurate segmentation is required, and for this reason the phonetic labeling provided by the LIMSI speech recognizer [GLAA93] was used. An additional algorithm groups the phonetic segments into syllables. This segmentation is then aligned with the voicing decision in such a way that the sequence of voiced parts is obtained, one per syllable. Another type of segmentation into syllabic nuclei is proposed in [Mer87b].

4. *Short-term pitch integration.* The auditory system seems unable to follow rapid short-term changes in fundamental frequency. There is evidence that an integration process takes place in pitch perception. This phenomenon was observed in a study on vibrato perception [dC94], which proposes a weighted time-average (WTA) model for the perception of short tones. When this WTA model is applied to the F_0 data of each voiced part, a smoother pitch curve is obtained.

5. *Stylization.* Stylization will depend on the settings of two parameters, each of which corresponds to a perceptual threshold: the glissando threshold and the differential glissando threshold.

   The following procedure is applied to each syllable in the utterance, more specifically to the pitch values in the voiced region of those syllables. The syllabic pitch contour is divided into parts with uniform slope, called "tonal segments," in such a way that pitch changes below threshold are normalized to static tonal segments, and that slope changes between two successive tonal segments must be audible (otherwise they should be merged in a single tonal segment). While weighted time average pitch values are used for contour segmentation, F_0 is used for evaluating the frequency changes in relation to the thresholds. The algorithm imposes no limitation whatsoever on the number of tonal segments within one and the same syllable; consequently, any number of pitch movements per syllable are accepted: there can be none (static), one (rise or fall), two (rise-fall, etc.) or more (e.g., rise-fall-rise). The stylized contour is given by the linear interpolation between the WTA pitch values at the boundaries of the tonal segment(s) in the syllable.[2] For static tonal segments the pitch value of the end point is extrapolated throughout the entire segment.

[2]Interpolation is done on a linear frequency scale (Hz), whereas a logarithmic (semitone) scale could have been used (as is the case for close-copy stylizations). There is no evidence in the literature about the perceptual relevance of the differences between the two types of interpolation. A comparison between straight-line stylization and other types of interpolations based on pitch targets [tHa91] shows that these methods were perceptually equivalent.

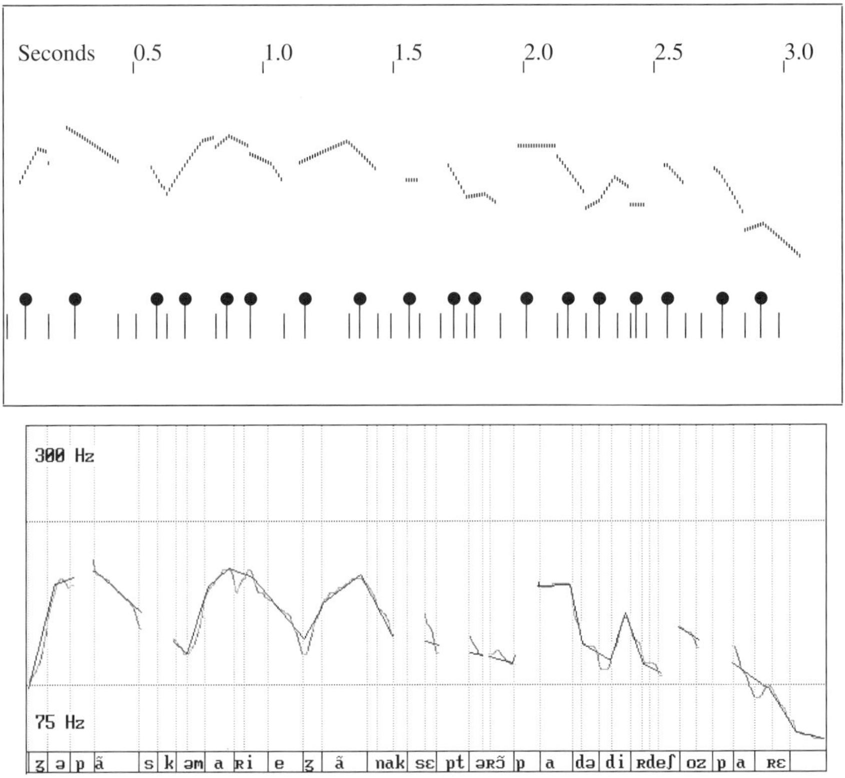

FIGURE 29.2. Automatic tonal perception stylization (top) and straight-line stylization (bottom) for the utterance "Je pense que Marie et Jean n'accepteront pas de dire des choses pareilles." The vertical markers in the upper part indicate boundaries between phonetic segments, and the bullets indicate vowel onsets.

The result of this stylization is referred to as the *tonal score*.

As can be seen in figure 29.2, the tonal score sometimes contains blanks between successive syllables, even though these blanks correspond to voiced parts in the speech signal. This is due to the simple voicing detection algorithm, which also takes into account the energy level.

6. *Resynthesis*. It is possible to reconstruct a synthetic F_0 contour, starting from the stylized pitch contour of the tonal score. In principle this reconstruction is needed because the tonal score is a perceptual representation based on the integrated pitch data, whereas the synthetic speech is an acoustic signal. If the stylized pitch contour were used directly as the pitch for synthetic signal, the perceptual integration would be applied twice: first during the stylization and second by the auditory system of the subject listening to the synthetic signal. The reconstruction of the synthetic F_0 is obtained by passing the tonal score through the inverse of the weighted time-average model.

Figure 29.3 illustrates the pitch contour after different processing steps of the algorithm. The first curve represents F_0, i.e., the output of the pitch determination algorithm. The second curve represents pitch at the output of the weighted time-average model. One can notice that small variations are smoothed. The third curve is the stylized pitch, i.e., the tonal description of the intonation contour. Finally, the last curve is the stylized contour passed through the inverse WTA model; this is the pitch control used in resynthesis.

29.2.4 Discussion

When the model parameters are set to the thresholds as observed in psychoacoustics, the resulting stylization very closely matches the measured pitch (i.e., WTA-pitch) contour.[3] These "standard" thresholds were measured for acoustically simple stimuli (pure tones, synthetic vowels), which are presented in isolation and repeated several times. The thresholds for continuous speech will undoubtedly be higher, because of the acoustic (spectral) complexity and the absence of stimulus repetitions. An important asset of the stylization based on tonal perception is that the stylization itself can be used to measure glissando and differential glissando thresholds for continuous speech. Examples of stylized contours obtained with ATS using different values for the thresholds are presented in *sound example 4* (see [AM95]).

There are, however, some problems that need to be solved first. A major problem is that the approximation of the syllabic nucleus, as the voiced part of the syllable, is inaccurate. New psychoacoustic research is needed to provide a solution. Another problem is the errors introduced by large microprosodic excursions due to unvoiced-voiced coarticulation, in combination with the smearing effect of the WTA model. In order to avoid these errors, a simple microprosody preprocessor (such as described in [Mer87a, Mer89]) can be used. However, it would be preferable to study the perceptual processing of typical microprosodic patterns and to adapt the model for short-term perceptual integration of pitch accordingly.

29.3 Manual Straight-Line Stylization

Manual straight-line stylization (MSLS) is a procedure by which the observed pitch contour is replaced by a less complex contour, having the form of a concatenation of straight lines. It is based on the hypothesis that unnecessary details of the natural melodic curves can be ruled out without any perceptual change. No structural assumption has been made up to now about the nature of such details. For instance, some of the pitch variations related to micromelody (defined here as segmental

[3]The analysis procedure described above was tested in a same-different task using synthetic speech stimuli, based on the original or the stylized pitch contour. This experiment is described in [AM95], and is not repeated here.

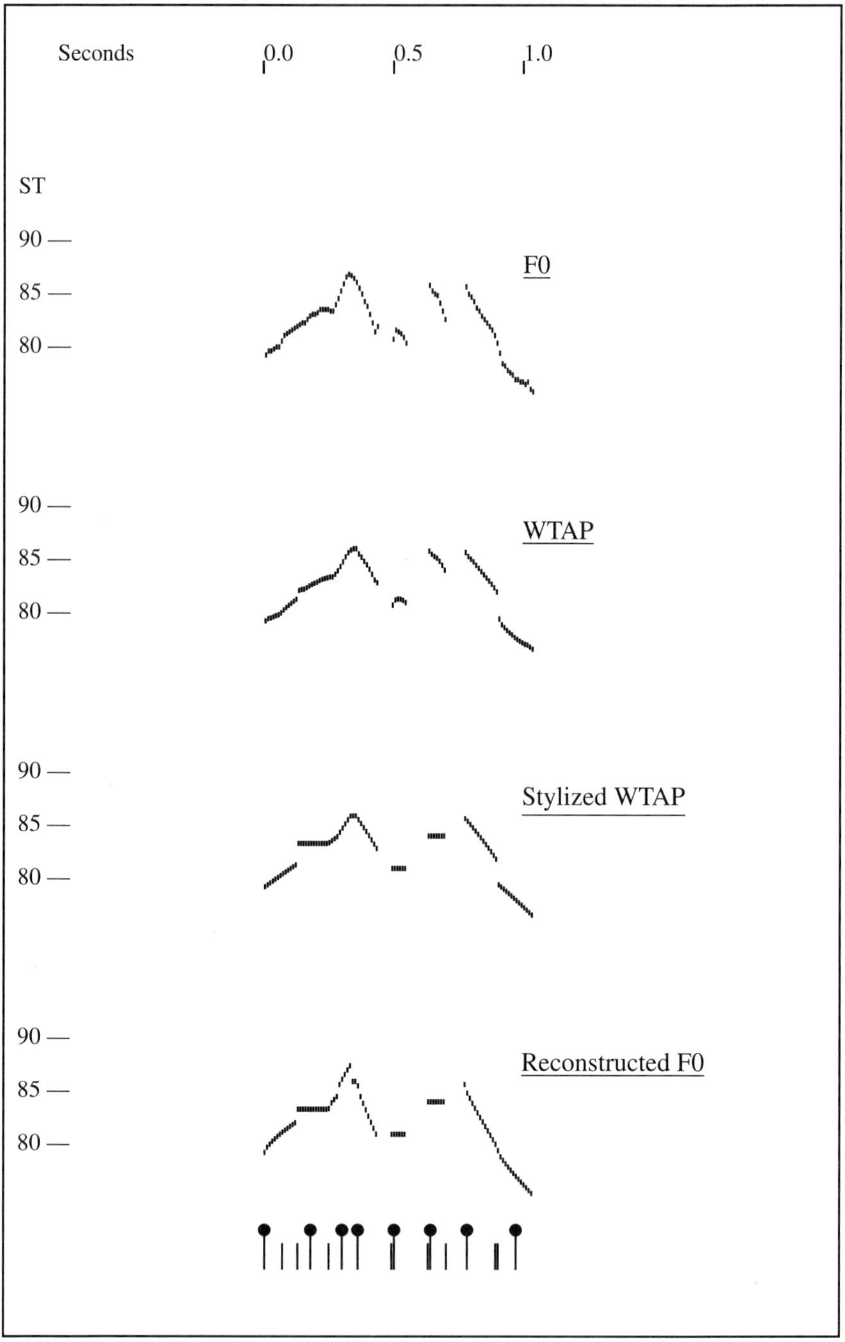

FIGURE 29.3. Pitch curves between processing steps in the stylization algorithm for the utterance "Anne-Marie était effondrée." WTAP is the weighted time-average pitch. Stylized WTAP is the stylized pitch contour. The lower tracing is the pitch contour used for resynthesis.

influences on melody) may be deleted, if they are not perceived, but other details must be preserved. The only aim of stylization is to obtain reduced contours, which must be perceptually identical to the original ones.

An approach to performing such a task was proposed in [tHa90]. In the process of stylization, natural melodic contours are reduced to a concatenation of straight lines on a logarithmic frequency scale (semi-tones/second).

The stylization is obtained interactively by means of an analysis-through-resynthesis technique. The process of stylization is a loop containing three steps: (1) a piece of the pitch contour is replaced by a straight line, according to the (manual) selections by the phonetician; (2) a speech signal with the modified pitch contour is synthesized; and (3) the phonetician listens to the synthetic signal and compares it to the original one. This procedure is repeated for that part of the contour until the original utterance and the modified utterance (i.e., with the stylized contour) are judged equivalent. The same procedure is applied to all parts of the contour.

As a general principle, a minimum number of straight lines is searched for a given contour. These straight lines are called *pitch movements*. The concatenation of pitch movements is called the *stylized pitch contour*. The resynthesized sentences using stylized pitch contours are called *close-copy stylizations*. Again, stylization is only a way to perform data analysis, and no special meaning is associated with pitch movements, so far. It is clear that the data reduction performed provides a better basis for further analyses of pitch contours.[4] Linear predictive coding (LPC) is often used as an analysis/resynthesis technique to obtain close-copy stylizations; but, of course, other techniques can be used as well. *sound example 1* (see [dM95]) presents LPC close-copy stylizations.

This methodology was applied to French in [BdLT92], the aim was to develop a melodic model of French intonation for use in a text-to-speech synthesizer.

While stylization enables a simplified description of intonation contours, without a loss of prosodic information, the degree of abstraction achieved is insufficient to serve as a description of the melodic properties of the language. On the one hand, it is clear that the human auditory system imposes some limits below which two pitch contours cannot be distinguished, and these limits can be determined for each acoustic dimension of prosody (e.g., pitch slope, frequency range, duration of the variation, direction). On the other hand, even if two pitch contours can be differentiated from a perceptual point of view, the prosodic information conveyed by them may still be identical. Taking these constraints into account requires a next stage of data reduction, based on the perceptual equivalence of pitch contours. The data reduction is obtained by a classification of the stylized pitch movements into normalized elementary units. This stage is essential for the development of

[4] A comparison between straight-line stylization and other types of interpolations based on melodic target values is reported in [tHa91]. It appeared that these methods were perceptually equivalent. In particular, the angular points created at transitions between two straight lines do not have any special effect on the resulting melody, as compared to smooth transitions.

a melodic model for TTS synthesis. *Sound example 2* (see [dM95]) illustrates the differences between stylized contours and contours consisting of standardized movements.

As this classification was to be used as the phonetic specification of prosody in a text-to-speech system for French, a set of rules was developed to automatically generate intonation contours from written text, tagged with syntactic information [BdLT92]. Examples of the output of the LIMSI text-To-speech synthesizer using these rules are recorded in *sound example 3* (see [dM95]).

29.4 Comparing Perceptual and Straight-Line Stylizations

This section describes a perception experiment to compare the two types of stylization described above. First some general observations will be made about the underlying assumptions and the results of the two approaches.

29.4.1 Differences Between the Two Approaches

Straight-line stylization is a manual procedure, whereas tonal perception stylization is automatic. For the sake of comparison, we will assume that close-copy stylization can be obtained automatically; indeed such an algorithm has already been proposed [SDHG93].

Both stylization algorithms have some characteristics in common. They both take into account amplitude variation and voicing, although in a different way. In the case of ATS this is done explicitly in the phonetic segmentation step. In the case of automatic close-copy stylization it is implicit in the weighting of F0 data points. Both approaches need a way of handling microprosodic perturbations. This aspect can be integrated in ATS because the phonetic segmentation provides the identification of the phonetic segments. Automatic straight-line stylization uses vowel onset detection.

A major difference between ATS and MSLS is the scope of pitch movements. In ATS, the syllable was chosen as the basic intonation unit (syllabic tones) at the linguistic level, although at the auditory level a syllable can contain multiple tonal segments, of course. This was motivated by the perceptual relevance of syllables for the segmentation of pitch contours. By contrast, straight-line stylization is based on units (pitch movements) that may encompass several syllables, or only a part of a syllable. This effect is visible in figure 29.2. Generally, MSLS is more global: it represents larger intonation units (containing several syllables). The same pitch movement can group a series of static or dynamic tones. The three tones at the end of *n'accepterons* have clearly no individual linguistic significance, and it is simpler in this case to group them in a single pitch movement.

It should be pointed out, however, that the linguistic intonation model [Mer87a] underlying the ATS approach also defines units ranging over several (unstressed) syllables as well as prosodic groups comprising both stressed syllables and sequences of (one or more) unstressed syllables. These larger units are supposed to be formed at a higher level of perceptual processing and are language-specific. A description of the great variety of pitch contours observed in spontaneous speech would require a very large amount of basic pitch patterns (corresponding to prosodic groups); it would require a smaller number of units if pitch movements (defined in terms of the pitch variation) were used; however, it can be described most economically as the combination of syllable-sized components.

The pitch movement approach seems particularly efficient in terms of simplicity of intonation rule design. But a drawback of the approach is that the pitch movement inventory was designed to model a given speech corpus, and it is not clear whether the set of pitch movements obtained can actually be used to synthesize pitch contours in any speaking style.

For these reasons, we think that the grouping of syllabic tones within the same pitch movement should be done at a higher level. It is not a matter of stylization, because it is dependent on several factors such as stress and the intonation grammar of the particular language.

29.4.2 Perception Experiment

An experiment was conducted in which the two types of stylization were presented to the same group of subjects for comparison.

Stimuli

For this experiment, 20 sentences were selected from a speech database of 60 sentences read by one male speaker of Parisian French, taking into account some syntactic, phonotactic and lexical constraints. All sentences were relatively short (between two and eight words) in order to avoid problems of short-term auditory memory when comparing the two versions (original and stylized) of the sentence. All stimuli were generated using LPC resynthesis. For each sentence, the stimulus groups are labeled V1, V2A, V2B, V3, and V4. V1 is the resynthesized original signal, V2A corresponds to the ATS stylization, V2B is the close-copy stylization (MSLS), and V3 and V4 are two alternative versions of the MSLS, with pitch contours that are increasingly different from the original contour. They were included in the test material in order to obtain a range of pitch contours going from identical, over almost identical, to clearly different.

For category V1, LPC-resynthesized sentences (with the original pitch contour) were used rather than the original sentences because of the quality degradation introduced by LPC: the quality difference between an original sentence and the corresponding LPC stylized version would have been easier to detect than the difference in intonation.

In the V3 and V4 categories, the alternative versions were derived from manually stylized contours (i.e., from V2B) in which either the slope, the timing, or the frequency level of a movement (of the overall close-copy contour) had been modified. A change in slope will affect the duration of the pitch movement; the timing of the start of the movement was modified such that the end time of the movement remained unchanged. For these modifications, we referred to perceptual experiments on the differential thresholds of pitch and pitch change [IS70, tHa81]. The modifications of slopes and levels chosen for categories V3 and V4 were of the order of 1.5 and 3 times the thresholds of "just noticeable difference," respectively. (For more details, see [Bea94], p. 65, table 2.1.)

Procedure

Subjects were asked to concentrate on intonation alone, and not on other aspects of the signal. For each pair, they had to indicate whether the two stimuli in the pair were identical with respect to intonation.

The subject sat in front of a computer with a mouse device. Each stimulus pair was presented once, after which the subject had to enter his response ("same" or "different") by clicking in the appropriate box on the screen.

Stimuli of the five conditions (V1-V1, V1-V2A, V1-V2B, V1-V3, V1-V4) were presented in random order. The order of the two stimuli within a pair was also varied randomly between X-Y and Y-X (e.g., V1-V3 and V3-V1).

Subjects

There were 10 subjects. None of them had known hearing loss. Tonal audiograms were made before the experiment to verify this. One of the authors also participated in the experiment as a subject (d'Alessandro).

Results

Table 29.1 summarizes the results of this experiment. About 93% of pairs in category V1-V1 were perceived as identical (100% were identical), and about 90% of pairs in category V1-V2A and 88% of pairs in category V1-V2B were also perceived as identical. The scores were only about 50% and 24% for categories V1-V3 and V1-V4, where the difference between the two parts of the stimulus pair becomes progressively larger. The scores thus follow the expected trend: the larger the difference between the two parts of the stimulus pair, the lower the proportion of "same" answers. This indicates that the subjects were able to perform the task, to perceive modifications in the pitch contours.

The fact that we did not obtain 100% "same" ratings for the V1-V1 category is normal in perceptual tests and can be ascribed to unsystematic errors of judgment and variation in the subjects' levels of attention.

The mean difference between the ratings for identical stimuli (V1-V1) and those for pairs containing the automatic stylization (V1-V2A) is only about 3%. The mean difference between the V1-V1 category and the manual stylization (V1-V2B) is only about 5%. This might indicate that even if a slight difference between original and stylized contours exists, it can be ignored, and the stylized contours can thus be considered perceptually equal to the original ones.

We can conclude that the two types of stylization (MSLS and ATS) give fairly similar results in terms of the perceptual equivalence between stylized and natural contours.

On the average, tonal perception stylization gives slightly better results than straight-line stylization. However, the results are very close: the mean difference

TABLE 29.1. Perceptual ratings for manual straight-line stylization and automatic tonal perception stylization. The columns are the subject identification, the total number of stimulus pairs in the test set (NSP), followed by the proportion of stimulus pairs judged identical for the five stimulus types. V1 is the resynthesized original utterance, V2A is the synthetic utterance with the pitch contour obtained with the automatic tonal perception stylization, V2B is the synthetic utterance with the pitch contour obtained using the manual straight-line stylization, and V3 and V4 are alternative versions of the manual stylization.

Subject	NSP	V1V1	V1V2A	V1V2B	V1V3	V1V4
JKas	348	82.8	82.3	67.6	38.1	19.1
XLap	449	89.4	90.4	86.9	71.4	52.8
CdAl	567	92.9	82.3	86.7	40.1	3.1
BDov	449	94.3	92.7	84.5	29.9	8.4
SRos	453	94.1	95.4	85.2	47.7	22.2
ABra	456	100	90.9	97.7	53.9	13.2
TLeb	467	95.3	97.0	97.9	53.0	27.8
LLac	452	89.7	86.5	90.3	58.6	24.8
SBen	575	98.8	88.7	95.2	42.3	28.9
MJar	455	94.1	93.6	85.0	63.3	41.0
ALL	4681	93.1	90.0	87.7	49.8	24.1

between the V1-V2A and V1-V2B categories is only about 2%. It should be pointed out that 4 subjects out of 10 rated straight-line stylization higher than tonal perception stylization. The fact that ATS gives better overall results than MSLS is somewhat surprising because the latter uses an interactive procedure in which audible differences will be eliminated as much as possible during the stylization procedure, thanks to the auditory feedback. A possible explanation for the good results for ATS is that it contains more line segments than straight-line stylization, resulting in a better match with the original pitch contour. Examples of stylized contours obtained with ATS using different values for the thresholds are presented in *sound example 4*.[5]

29.5 Conclusion

The experiment described in this chapter demonstrates that the automatic stylization of pitch contours based on tonal perception produces a simplified contour that is hardly distinguishable from the original, and is as good as, or even better than, the stylization obtained with the manual, interactive procedure known as close-copy stylization.

---

[5]These examples correspond to stimuli V1, V2, V3, V4 in [dM95].

Pitch contour stylization in general is a powerful tool for designing prosodic models in speech synthesis. Such a model was built for French, according to the close-copy stylization (or pitch movement, or straight-line) methodology [Bea94]. It is integrated within the LIMSI text-to-speech system. Two types of problems with this approach to stylization were encountered. On the one hand, the stylization process is time consuming, and the experimenter has to make ad hoc decisions regarding the relevant features and movements that are needed. On the other hand, the pitch movements obtained could be dependent on the specific characteristics of the speech corpus used and are not strongly linguistically motivated.

Therefore, it was decided to design another type of intonation stylization to overcome the above-mentioned problems. This stylization procedure is automatic and grounded on perception. Because it is automatic, the procedure is fast and efficient. Because it is grounded on perception, the procedure separates the linguistic and perceptual-acoustic aspects of F_0 contours.

As for the perceptual equivalence between stylized and natural F_0 contours, the two types of stylization processes seem almost comparable. This means that F_0 contour stylization is not unique, but is dependent on the underlying perceptual and linguistic assumptions.

The pitch movements approach is an efficient representation for designing intonation synthesis rules, with the above-mentioned limitations in mind. Automatic tonal stylization represents intonation at a lower level of description. Therefore, it should make a better framework for intonation rule writing, but more rules would be needed. ATS makes no assumptions on what is relevant and what is not in the processing of prosodic features by the human listener; it merely applies the findings of psychoacoustics. It could also be used for further automatic processing of intonation (such as automatic transcription of prosody) and for computer assisted teaching of intonation.

Acknowledgments: The authors would like to thank the subjects for their kind help in the course of this research, as well as the two anonymous reviewers for their valuable comments.

References

[BdLT92] F. Beaugendre, C. d'Alessandro, A. Lacheret-Dujour, and J. Terken. A perceptual study of French intonation. In *Proceedings of ICSLP'92*, Banff, Alberta, Canada, 739–742, 1992.

[Bea94] F. Beaugendre. *Une étude Perceptive de l'Intonation du Français.* Doctoral dissertation, LIMSI report NDL 94-25, Université de Paris-sud, Orsay, 1994.

[dC94] C. d'Alessandro and M. Castellengo. The pitch of short-duration vibrato tones. *J. Acoust. Soc. Amer.* 95(3):1617–1630, 1994.

[dM95] C. d'Alessandro and P. Mertens. Automatic pitch contour stylization using a model of tonal perception. *Comp. Speech and Lang.* 9:257–288, 1995.

[FK88] H. Fujisaki and H. Kawai. Realization of linguistic information in the voice fundamental frequency contour of the spoken Japanese. In *Proceedings of IEEE-ICASSP*, New York, 663–666, 1988.

[GLAA93] J. L. Gauvain, L. F. Lamel, G. Adda, and M. Adda-Decker. Speaker independent continuous speech dictation. In *Proceedings of Eurospeech'93*, Berlin, 125–128, 1993.

[Hir92] D. J. Hirst. Prediction of prosody: An overview. In *Talking Machines: Theories, Models, and Designs*, G. Bailly and C. Benoit, eds. North-Holland, Amsterdam, 1992.

[Hou90] D. House. *Tonal Perception in Speech*, Lund University Press, Lund, Sweden, 1990.

[IS70] A.V. Issachenko and H.-J. Schädlich. *A Model of Standard German Intonation.* Mouton, The Hague, Paris, 1970.

[Mer87a] P. Mertens. L'intonation du Francais. De la description linguistique à la reconnaissance automatique. Unpublished doctoral dissertation, Catholic University of Leuven, 1987.

[Mer87b] P. Mertens. Automatic segmentation of speech into syllables. In *Proceedings of the European Conference on Speech Technology*, Edinburgh, UK, 9–12, 1987.

[Mer89] P. Mertens. Automatic recognition of intonation in French and Dutch. In *Proceedings of Eurospeech 89*, Paris, France, 1:46–50, 1989.

[Ohm67] S. Ohman. Word and sentence intonation: A quantitative model. *K.T.H. Quarterly Progress and Status Report*, 2:20–54, 1967.

[Pie81] J. Pierrehumbert. Synthesizing intonation. *J. Acoust. Soc. Amer.* 70:985–995, 1981.

[Spa93] G. W. G. Spaai, A. Storm, A. S. Derksen, D. J. Hermes, and E. F. Gigi. *An Intonation Meter for Teaching Intonation to Profoundly Deaf Persons.* IPO Manuscript no. 968, Inst. for Percept. Res., Eindhoven, 1993.

[tHa81] J. 't Hart. Differential sensitivity to pitch distance, particularly in speech. *J. Acoust. Soc. Amer.* 69(3):811–821, 1981.

[tHa90] J. 't Hart, R. Collier, and A. Cohen. *A Perceptual Study of Intonation.* Cambridge University Press, Cambridge, 1990.

[tHa91] J. 't Hart. F_0 stylization in speech: Straight lines versus parabolas. *J. Acoust. Soc. Amer.* 90(6):3368–3370, 1991.

Appendix: Audio Demos

A sound demonstration is provided.

30

Generation of Pauses Within the z-score Model

Plínio Almeida Barbosa
Gérard Bailly

ABSTRACT We have previously proposed [BB94] a model for the generation of segmental durations that proceeds in two steps: (1) prediction of the timing of a salient acoustic event per syllable according to phonotactic and syntactic information, and (2) application of a repartition model that determines the duration of each individual segment between these events. This chapter focusses on the repartition model and describes how the initial model has been enriched to account for the emergence of pauses as speech rate is decreased. It describes a perceptual evaluation of the whole model. This evaluation shows that, for the same distribution of prediction errors, a precise timing of these events is perceptually more relevant than a segment-based method aiming at predicting precisely each individual segmental duration.

30.1 Introduction

Many generation models for segmental duration are based on statistical analysis of ad hoc corpora. In these models, a large set of coefficients accounts for different factors influencing the duration of segments. Several types of factors are used: intrinsic factors such as mode and place of articulation of the phoneme, contextual factors such as identities of the surrounding segments, phonotactic constraints such as the position of the segment within the syllable, phonological factors such as the word prominence, or linguistic factors such as the word class [Kla82, BS87, OSh81, van94]. These models can be referred to as phoneme-based models of prediction. In the most recent models, parameters are tuned automatically on large corpora in order to get the least squares minimum of segmental prediction errors. All segments have a priori the same weight in the optimization problem.

Other models use intermediate higher-level units, such as the syllable [MG93, Cam92] or the foot [Wit77, KOH86]. These models focus on macroscopic rhythms and isochronies. The choice of a *rhythmic unit* is expected to reveal prosodic regularities or units [GJM94]. Compared to phoneme-based models, these models are expected to be more robust to variation in speech rate [MG93] and segmental components [Cam92]. These models rely on the fact that rhythm can be defined as a perceptual structure in which units are grouped according to recurrent patterns. The tentative definition of rhythm proposed by Woodrow [Woo51, p. 1232]:

By rhythm, in the psychological sense, is meant the perception of a series of stimuli as a series of groups. The successive groups are ordinarily of similar pattern and experienced as repetitive. Each group is perceived as a whole and therefore has a length lying within the psychological present

is confirmed by Fraisse [Fra74, p. 75]:

Sans perception globale, il n'y a pas de structure rythmique[1]

Regularity of these patterns is thus considered as the prime object of the perception of rhythm. The exact durations of the segmental constituents of each unit are thus less important than the functions assumed by the rhythmical patterns of the groups. The nature of these units and the language-specific constraints acting on the regularity of the patterns have been largely discussed (see the debate on isochrony/isosyllabism in [Lea74, Leh77, Dau83, Noo91, Wh94]).

Our generation model is based on an original rhythmic unit called the inter-perceptual center group (IPCG). This IPCG is delimited by *salient acoustic events*. That is, the *perceptual centers* (see section 30.2) and the succession of IPCGs evidence rhythmical groups that are characterized for French as a gradual lengthening of IPCGs. The perceptual experiment described in section 30.5 demonstrates that listeners are quite sensitive to the distribution of "prediction errors" along the utterance: The relative timing of salient acoustic events such as the vocalic onset is more relevant for the perception of momentary tempo than the distribution of segmental durations between these events.

Relative timing of these events is conditioned by both linguistic and phonotactic factors and is sensitive to variation in speech rate. Our model is based on the assumption that momentary tempo is encoded by the speaker and decoded by the listener according to a reference clock.[2] Such a hypothesis is clearly stated by other authors [All75, Fra74]. This is also corroborated by data on speech acquisition [Kon91, Smi78], which show that babbling departs from an initial isochrony near 16 months of age to incorporate language-specific rythmic structure. Systematic deviations of IPCG durations from this reference clock can be thus interpreted as rhythmical patterns assuming various linguistic or paralinguistic functions. We use this paradigm to control speech rate: The duration of each IPCG is expressed in terms of clock units.[3] Then the repartition model divides the duration of each IPCG among its segmental constituents. We show that this repartition model can easily be in charge with pause emergence: durations of pauses are thus intimately tied with rhythmical patterns because pauses as other segments contribute to "fill" each IPCG when the stretch of segments is too large. This is a partial answer to

---

[1] Without global perception there is no rhythmical structure.

[2] Existence of such internal clocks has been clearly hypothesized by neurophysiologists to coordinate biological movements [TSR90, Lli89]. Coordination of movements is controlled by two cooperative structures: a timekeeper, which delivers periodic signals, and a motor function, which uses these landmarks to deliver muscular activations.

[3] The actual value of the clock is computed for each utterance using the average duration of unstressed IPCGs.

Fant's rhythmical coherence of pauses and tempo observed for English, French, and Swedish [Fan91, p. 248]:

the average inter-stress interval within a short time memory span of about 4 seconds preceding a pause ... synchronises an internal beat generating clock which sets a preferred pause duration.

This tendency to restart phonation at a time in relation with prior tempo—eventually with intervening "silent" beats—has been evidenced also by Couper-Kuhlen [Cou91] for turn-taking.

It is important to notice that in all duration models, pause is generally considered separately. In Klatt's model, for example, pause durations are given a priori and not integrated in the same mechanism responsible for generating segmental durations [Kla82, p. 761]

Rule 1: Insert a brief pause before each sentence-internal main clause and at other boundaries delimited by an orthographic comma.

The models proposed by Bartkova and Sorin for French and by O'Shaughnessy for Canadian French modify also an inherent (or intrinsic) segment duration as a function of the phonemic environment. Their models also take a priori durations for pauses that are placed according to phonotactic and syntactic criteria. In this kind of generation, corpus-dependent results may be easily obtained due to exhaustive computation of specific context effects.

The choice of a higher-level unit such as the syllable or the foot avoids the tuning of a large set of coefficients (about 80 in Bartkova-Sorin's model!). It also should enable an easier integration of duration in the prosodic structure via the characterization of the durational contours: these contours can be predicted in parallel with fundamental frequency contours by automatic learning procedures recently developed using dynamical predictors [Tra92] or stored prototypes [Aub92, AS92]. Nevertheless, none of the models using rhythmic units integrates the pause phenomenon as a by-product of *performance*: In Campbell's model, pause is not considered for generation. In Kohler's model ([Koh86], chapter 37 this volume), pause is also attributed at the beginning of the application of rules. Monnin and Grosjean [MG93] predict the duration of the last vowel of each word and the optional following pause from an index of linguistic complexity. An average correlation of 0.86 between the normal and slow rates is found. Although the concept of performance structures was introduced using very slow rates [GG83] in order to obtain a pause following each word, the performance structure is now built using the normal rate. No quantitative model is proposed to describe how the duration of each final group is shared by the vowel and the optional pause.

We next describe a model that integrates the automatic placing and calculation of pauses (silent—if present—and lengthened segments) as a continuous function of speech rate. Our model is guided by rhythmic constraints in a coherent and homogeneous framework.

30.2 Rhythm and the Perceptual Center

30.2.1 Arguments for a Reference Point per Syllable

Strong arguments for the characterization of momentary tempo by only one event per syllable come mainly from synchronization of speech with other acoustic stimuli or gestural activities. Research developed by Marcus and his colleagues [MMF76, Mar76] on perceptual centers (PCs) show that listeners perceive regularity in syllable sequences despite important differences among absolute durations [MMF76, p. 405]:

we were forced to ask ourselves what it was that was regular in a rhythmic list. To simplify our discussions we defined this as the P-centre of each item.

The PC does not seem to correspond to any simple articulatory or acoustic correlate, but there is a clear interaction between speaker production and perception systems in order to produce speech sequences that the listener perceives as isochronous [All75, Leh77]. Subjects must be capable of relating to some points of reference located in the speech flow in order to perceive such sequences as isochronous. This point of reference (or beat) was coined by Marcus [Mar75] as the perceptual center of the word (for monosyllabic ones). Similar experiments involving alternation or synchronization of speech and tapping or musical performance [Fra74, GA92] shows the consistency of this paradigm.

30.2.2 The Cyclic Attractor

These experiments do not concern connected speech and force the subjects to focus on a precise synchronization. However, more ecological experiments involving speech and tapping [All75, BAL91] report converging results. Allen shows that subjects place their down-beats near the onsets of stressed syllables. For French, Berthier et al. show that there is a temporal relationship between the percussion of the knife on the whistle and the vocal onset that follows (from 0 to 100 ms) but never with the end of the vowels preceding it. One of the interesting questions to answer here is *what* is synchronized with *what*. By hypothesis, the performance of such tasks requires an internal time-measuring device or clock that provides target time points at which the tapping movements or syllabic gestures must produce their effect. The concept of an internal clock put forward by neurophysiologists [HM87, Sha82, SSV92] is central to biological movements. The natural frequency of the opening/closing of the vocal tract driven by the jaw (equal to 6 Hz $\pm$ 1 [SGE80]) could be easily adjusted to the frequency of regular beats of other gestural activies such as the leg/foot or arm/hand movements, which can reach 4-6 Hz. However, in most experiments the beat is not strictly synchronous with the metronome (clicks, taps), but rather precedes it. What could be the explanation of this? A partial answer is given by Fraisse [Fra80]: The subjects were asked to perform the beat with their hands or their feet. What is basically observed is that in the foot condition the lead of the beat over the click was larger than in the hand condition. The difference is roughly equivalent to the difference in nerve conduction time due to the foot/brain versus the hand/brain distance. This suggests that synchrony is maintained at the level of the two perceived events and that the motor commands can be issued well ahead to secure synchrony of the two percepts [Pri92]. We have thus to keep in mind that momentary tempo is a perceptual construct [All75, Leh77].

30.2.3 Acoustic Correlate of P-Centers

Perception experiments carried out by various authors [Mar81, Pom89, Sco93] confirm that despite the nature of experimental conditions (perception or production experiments) and the nature of phonemes, the PC seems to be located at the vicinity of vocalic onset. Although their experiments do not concern connected speech, we have set the location of the PC at the vocalic onset when the syllable is not preceded by a silence. If there is a silence, the PC is usually placed earlier in the syllable:[4] in this case, only we have taken the beginning of the left-most voiced consonant of the syllable onset as the PC, that is, at the major increase in low-frequency energy within the syllable.[5]

Two consecutive PCs define the boundaries of the inter-perceptual-center interval (IPCI). The segments it contains form the IPCG. In the next sections, two prediction models for duration that use a rhythmic programming unit are confronted: Campbell's model, which uses the syllable, and our model, which uses the IPCG.

30.2.4 Importance of P-Centers as Sensory Regulators

Our experimental and theoretical work is strongly based on the fact that speech is produced by biological movements that have to be coordinated with other activities as important as breathing. Section 30.2.2 showed that such a sensorimotor synchronization can be explained by the coupling between internal and such "external" clocks. Such claim is not beyond the scope of speech synthesis: Speech is multimodal, and multimedia applications will need to synchronize speech with other movements (artificial or synthetic). Our two-stage model of control of segmental durations, which first computes the timing of one event per syllable, satisfies these requirements. We next demonstrate that this approach offers an efficient way to control speech rate and the emergence of pauses that are not constrained to appear in ad hoc and predefined positions.

[4]Scott has implemented a model for PC location based on the first derivative of the intensity function of the acoustic signal. According to her model, significant increases of intensity in the band frequency from 195 to 1638 Hz strongly determine the PC location.

[5]The exact PC location could be easily associated with each synthesis unit in concatenative synthesis systems.

TABLE 30.1. Parameters of the z-score model for two speakers, EV and FB, computed over logatoms. Means ± standard deviations of log-transformed segmental durations are given.

phon	EV	FB	phon	EV	FB	phon	EV	FB
a	4.58 ± .30	4.74 ± .34						
ϵ	4.43 ± .32	4.68 ± .27	œ	4.61 ± .34	4.68 ± .34	ɔ	4.62 ± .38	4.69 ± .30
e	4.49 ± .33	4.73 ± .33	Φ	4.65 ± .31	4.75 ± .27	o	4.67 ± .32	4.81 ± .27
i	4.49 ± .38	4.55 ± .33	y	4.59 ± .40	4.61 ± .29	u	4.64 ± .40	4.69 ± .31
˜ɑ	4.80 ± .30	4.89 ± .29	˜ɔ	4.81 ± .29	4.89 ± .31	˜ε	4.75 ± .28	4.93 ± .31
w	4.36 ± .21	4.51 ± .25	j	4.28 ± .23	4.47 ± .29	ɥ	4.33 ± .26	4.48 ± .23
l	4.16 ± .19	4.48 ± .26	∇	4.25 ± .19	4.48 ± .32	ə	4.16 ± .28	4.86 ± .43
p	4.60 ± .19	4.58 ± .53	t	4.51 ± .26	4.54 ± .55	k	4.60 ± .23	4.56 ± .43
b	4.34 ± .19	4.48 ± .24	d	4.28 ± .21	4.48 ± .23	g	4.32 ± .18	4.42 ± .24
v	4.34 ± .24	4.59 ± .28	z	4.39 ± .23	4.71 ± .20	ʒ	4.43 ± .23	4.62 ± .21
f	4.73 ± .25	5.04 ± .38	s	4.87 ± .24	5.26 ± .37	ʃ	4.81 ± .23	4.99 ± .25
m	4.46 ± .24	4.61 ± .23	n	4.35 ± .26	4.59 ± .24	ŋ	4.75 ± .24	4.68 ± .23

30.3 The Campbell Model

The elasticity principle [Cam92] in its strongest version says that all segmental durations in a syllable frame are obtained by a same and single factor z—the so-called z-score or normalized duration [WSOP92]—as follows:

$$Dur_i = \exp(\mu_i + z\sigma_i), \tag{30.1}$$

where μ_i and σ_i are the mean and standard deviation of the log-transformed durations (in ms) of the realizations of the phoneme i. These z-scores are computed over the syllable by:

$$\sum_i Dur_i = syllable\ duration. \tag{30.2}$$

The successive use of equations (30.1) and (30.2) is referenced as the repartition algorithm. Campbell's model proceeds in two steps:

- prediction of the syllable duration from phonotactic and phonological information by a statistical model (a multilayer perceptron)
- use of the repartition algorithm [Cam92]

As input to the perceptron, Campbell describes the syllable by six factors: (1) number of phonemes; (2) nature of the vowel (reduced, lax, and tense vowels or syllabic consonants or diphthongs); (3) position in intonational group; (4) type of foot (used to describe the rhythmic context of the syllable); (5) stress; and (6) word class to which the syllable belongs (lexical or functional words). In the output, the syllable duration is expressed in natural logarithmic form.

We have already demonstrated the optimal consistency of the rhythmic behavior of the segmental units inside an IPCG [BB92]: the z-scores of the syllable onset are positively correlated with the preceding rime whereas they are negatively correlated with the following nucleus. If the z-scores of all IPCGs of a sentence are represented graphically, temporal organization of these units exhibits monotonous ascending movements toward each phrase accent with a reset just after accent realization. For French, these patterns delimit rhythmic units characterized by a gradual lengthening toward the accented syllable [Pas92]. Such elementary macrorhythmic "contours" can be easily captured by dynamical nonlinear predictors.

30.4 The Barbosa-Bailly Model

30.4.1 Prediction of IPCG Durations

As in Campbell's model, our duration predictor proceeds in two stages. Nevertheless, significant differences have to be noticed:

- A sequential network [Jor89] constrained by an internal clock (a measure of the speaking rate for each utterance computed as the mean of the IPCIs in the utterance that do not correspond to a prosodic marker [BB94]) generates the timing of PC locations in terms of internal clock units (see figure 30.1).
- The IPCIs are then distributed among the IPCG constituents according to the repartition model, which presently includes the emergence of pauses.

In the network input, prosodic and phonological information that seems to be relevant for the prediction of the duration of the current IPCI is sequentially de-

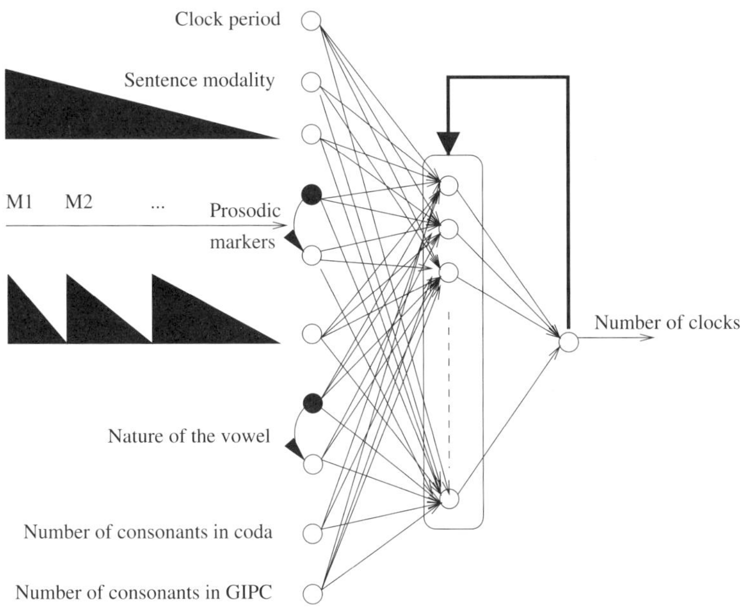

FIGURE 30.1. The sequential network for generating timing of PC locations in terms of internal clock units. The network transforms simple ramps indicating the length and the function of each linguistic unit of the utterance into rhythmic contours according to speech rate and phonotactic constraints. The input cells shown with filled circles store the values of connected cells with delay -1, i.e., following prosodic marker and next nucleus.

scribed. Two cells remain constant: (1) the frequency of the internal clock and (2) the sentence modality. Two other cells depict possible basic rhythmic patterns. Ramps indicate the extent of two basic prosodic units: (3) the sentence and (4) the prosodic group.[6] These linear ramps are set to a value equal to the number of IPCGs of the unit and reach zero value at the end of the unit. The six remaining input cells are fed with: (5) current prosodic marker, (6) next prosodic marker, (7) nature of current vowel, (8) nature of next vowel, (9) number of consonants in the IPCG, and (10) the coda (to be able to correct the PC estimate in case of pause emergence).

The parameters of the repartition model for French are given in table 30.1 for two speakers of our text-to-speech system. These parameters have been obtained by statistical analysis of the segmental durations of logatoms produced in isolation (speaker FB) or with a carrier sentence (speaker EV). Only data of the di- and trisyllabic logatoms are used. Although FB is globally faster than EV, well-known characteristics of intrinsic phonemic durations are captured: nasal vowels longer than oral ones, voiced consonants shorter than unvoiced ones, and so forth.

The repartition algorithm was modified to include pause emergence. For this, an original procedure for recording versions of the same sentence at different speech rates was tested.

30.4.2 *The Corpus*

A corpus with 20 sentences uttered at five different speaking rates was recorded to study the influence of pause emergence on overall rhythmic structure. In order to simplify the problem of locating PCs, all sentences of the corpus are CV sequences (all pauses were thus preceded by a vowel). Effective separated rates were obtained by controlling the speaker's speaking rate with synthetic questions. The speaker was instructed to answer these questions with sentences presented on a screen at the same speaking rate as that of the questions. The speaking rate of the synthetic questions was controlled by multiplying the coefficients μ_i and σ_i of each phoneme by a scaling factor, as suggested by Wightman and colleagues [WSOP92]. This procedure has been particularly effective in differentiating five speaking rates from very slow to very fast in a rather continuous way. The average values of the five speech rates are given in table 30. 3.

[6]The prosodic group is defined on a linguistic basis: It consists of each content word and any depending function words. The prosodic groups are linked by prosodic markers [Bai89]. These markers are indexed by the degree of cohesion between the adjacent prosodic groups.

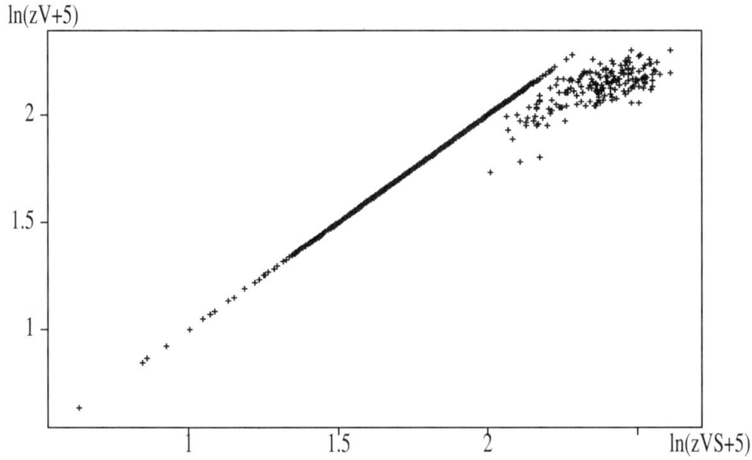

FIGURE 30.2. Scatterplot of log-transformed actual z-scores z_V of each vowel of the corpus versus log-transformed virtual z-scores z_{VS} for all speech rates.

30.4.3 *Incorporating Pause Phenomenon to the Repartition Algorithm*

In order to compute the silence duration that must be assigned to each IPCI in the generation stage, we have studied the relation between the actual z-scores of vowels $z_V = (\ln(Dur_V) - \mu_V)/\sigma_V$ and virtual z-scores of the same vowels. These virtual z-scores are computed by adding to the actual duration of the vowel Dur_V the duration of the optional adjacent silence as $z_{VS} = (\ln(Dur_{\text{vowel}} + Dur_{\text{silence}}) - \mu_V)/\sigma_V$.

As presented in figure 30.2, $z_V = z_{VS}$ up to the critical point $z_{VS} \simeq 2.4$. Then, between $z_{VS} = 2.4$ and $z_{VS} \simeq 4.5$, the speaker has two strategies:

(1) The speaker carries on lengthening the vowel with no pause, that is, a subjective pause is produced; this corresponds to the end of the segment $z_V = z_{VS}$ already mentioned.

(2) The speaker introduces a silent pause that corresponds to the scatterplot, which becomes more and more dense as z_{VS} increases.

Finally, for large z_{VS}, only the strategy (2) is retained, and a pause is always inserted.

Figure 30.3 presents a zoom on the scatterplot presented in figure 30.2. Figure 30.3 shows that despite the emergence of a pause, the speaker still lengthens the vocalic part of the IPCG. Our model of pause emergence uses the regression line (see equation (30.3) below) for $z_V \neq z_{VS}$ as the computational model of pause duration.

$$(z_V + 5) = (z_{VS} + 5)^{0.59} \exp(0.72) \tag{30.3}$$

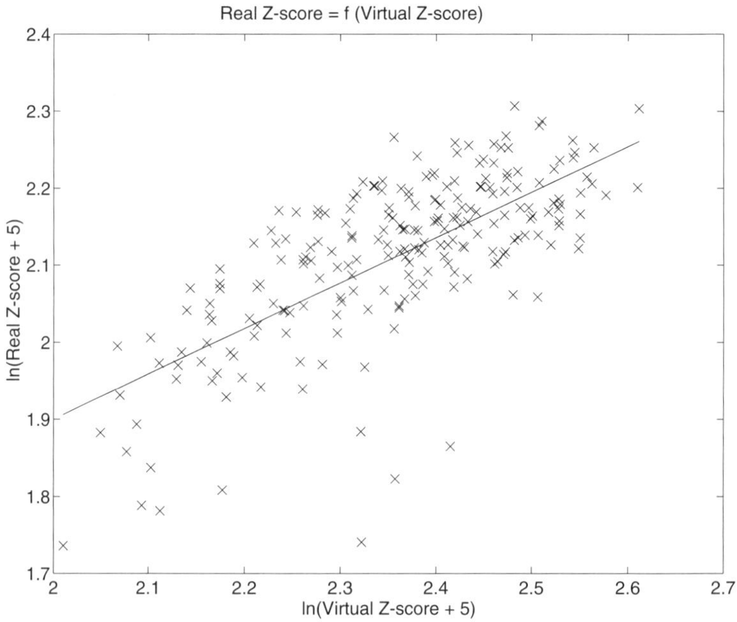

FIGURE 30.3. Scatterplot of log-transformed actual z-scores versus log-transformed virtual z-scores for $z_{VS} \neq z_V$.

A pause can thus theoretically "emerge" when this regression line crosses the straight line $z_V = z_{VS}$. This critical point is thus $z_{VSc} = 0.79$. No silence per se is generated between z_{VSc} and the minimal $z_{VS} \simeq 2.4$ because a minimum silence duration is required for relaxing speech articulators. A minimum silence duration was set for each speech rate as a result of analysis carried out on the corpus (see table 30.2). Surprisingly, this minimum duration does not vary very much as a function of speech rate and remains around $Dur_{\text{silence}} \simeq 60$ ms.

The modified repartition algorithm proceeds as follows:

- Computation of the z-score for a given IPCG (z_g).
- If z_g is smaller or equal to the critical z_{VSc}, the procedure is over: segmental durations are obtained by using equation (30.1).
- If z_g is greater than z_{VSc}, the z-score of the vowel z_V is obtained by regression equation (30.1), by setting $z_{VS} = z_g$.
- The segmental durations are computed with the repartition algorithm with $z_g = z_V$ and added up. The difference between this result and the original IPCG duration gives the duration of the silence.
- If the silence duration is greater than the minimum the procedure is over.
- If not, no silence is inserted and the z-score of the IPCG is kept equal to z_g.

TABLE 30.2. Averages ± standard deviations of absolute prediction errors for three sound classes using the modified repartition algorithm on original data. The distribution of errors is compared for three different rhythmic units: the syllable (SY), the inter-vocalic onset group (VO), and the IPCG (PC).

rate	*Vowels*			*Consonants*			*Silences*		
	SY	VO	PC	SY	VO	PC	SY	VO	PC
v. slow	34 ± 32	35 ± 32	34 ± 29	43 ± 46	42 ± 43	37 ± 46	67 ± 58	76 ± 62	55 ± 53
slow	38 ± 40	42 ± 41	36 ± 32	32 ± 32	34 ± 31	28 ± 26	73 ± 52	73 ± 52	73 ± 56
normal	26 ± 26	30 ± 38	29 ± 23	27 ± 26	28 ± 26	25 ± 22	54 ± 37	64 ± 36	56 ± 35
fast	17 ± 19	18 ± 19	19 ± 20	21 ± 25	21 ± 27	19 ± 21	75 ± 77	68 ± 99	49 ± 78
v. fast	12 ± 12	11 ± 13	12 ± 14	13 ± 13	12 ± 17	11 ± 12	59 ± 33	93 ± 28	44 ± 20

TABLE 30.3. Number of errors obtained by the repartition algorithm in placing the silences. The number of actual silences by speech rate in the natural utterances is also given (there are 285 IPCGs by rate in the corpus).

rate	clock (ms)	minimum pause (ms)	location errors	actual silences
very slow	360	75	15	99
slow	325	67	13	64
normal	270	57	11	32
fast	210	51	—	26
very fast	165	59	1	5

It is important to note that no constraint on location—such as accentable position—is imposed on the silences; they are placed as a result of the modified repartition algorithm described above (see comments on the last audio example, see section 30.6). We applied this modified repartition algorithm to the original data for the five speech rates; the distribution of errors is given in table 30.2. This table shows that the IPCG predicts the durations of consonants and pauses with more accuracy than the syllable or the inter-vocalic onset group.

Using original IPCG durations, only a few silences were placed in positions not actually chosen by the speaker (see table 30.3); but all positions were assigned to a latent location for accent realization (a prosodic marker).

30.4.4 Automatic Learning

As described above, a sequential network was trained in order to generate the successive IPCIs for each sentence of the corpus. Fifty sentences (ten sentences in five rates) from the corpus (learning set) were chosen for the learning phase. The network generalizes quite successfully the corresponding IPCI sequences for all sentences in the corpus (compare standard deviations of errors between learning and test sentences in table 30.4). Synthetic segmental and silent durations were obtained by applying the modified repartition algorithm described above.

The errors between the original segmental durations and the ones obtained by our model were computed for each speech rate and for two types of unit: segments

TABLE 30.4. Means (and standard deviations) in milliseconds from the histograms of errors between the segmented and generated durations for learning-set and test-set sentences.

	learning set		*test set*	
rate	silences	segments	silences	segments
very slow	−41(137)	−1(49)	3(207)	4(56)
slow	70(116)	−2(40)	−82(147)	6(46)
normal	67(122)	0(37)	−105(113)	5(43)
fast	−189(84)	3(30)	64(144)	0(28)
very fast	−4(170)	2(25)	−49(193)	−1(23)

and silences (see table 30.4). The histograms of these errors for all learning-set sentences were strongly correlated with the normal distribution (minimum 95%).

We conducted an experiment to test the perceptual relevance of IPCG's organization. The stimuli were obtained by a PSOLA [CM90, BBW92] analysis-resynthesis technique, in which only the segmental durations were modified.[7] Two versions of each utterance were compared: the first one (tagged as *model*) is paced by our model and the second (tagged as *random*) is obtained by adding a Gaussian noise to the original segmental durations. The noise distribution is equal to the distribution of prediction errors produced by our model. Three distinct distributions are used for vowels, consonants, and silences (see table 30.2). The two versions differ only in the time structure of errors: The *model* version is expected to predict with more accuracy the sequence of PCs.

30.5 Perception Test

Fifteen subjects studying at ICP participated in this perception experiment. Each session lasted between 7 and 10 minutes. In an ABBA test the subjects were asked to select the utterance with the most adequate prosody. Listening may be repeated once. A question mark could be used if both utterances seemed similar to the subject. All subjects considered, 89% of *model* utterances were chosen as being the most natural. In 15% of the answers there was doubt between the utterances. Individual scores can be seen in figure 30.4.

Comments are quite instructive to evaluate what perceptual cues were used by listeners. Some of them are listed below:

- "When the utterances were continuous without interruption they seemed natural to me" (HL).

[7]The insertion/deletion of short-term signals is ruled by emergence functions computed by temporal decomposition [BMA89]. Fundamental frequency values are recoded with only three values by segment: beginning, middle, and final, and energy is recorded by an average value.

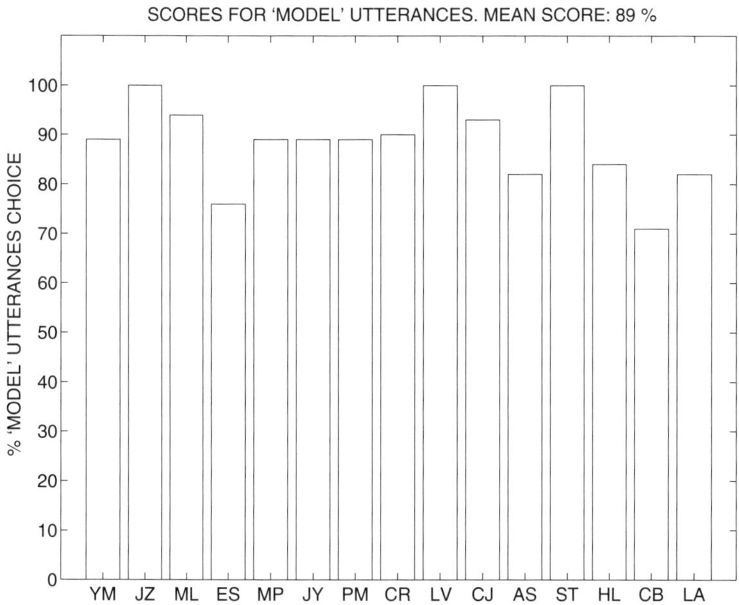

FIGURE 30.4. Individual scores for the choice of *model* utterances. Please note that inter-subject results are similar.

- "My decisions were guided by word sequencing: if the sequence was interrupted or rhythm was not continuous, these utterance did not seem natural to me." (ML).
- "Utterances with a rhythmic change at the middle of a word did not seem natural to me!" (LV).

Subjects' preference for one utterance in the pair involves judgments of global aspects of rhythm (see last two comments above). In both stimuli the nature of the segments does not seem unnatural to them. Results are very similar among subjects. These results thus indicate the reliability of rhythmic judgments.

Results show a clear preference for the utterances obtained with our generation procedure. They support a model that maintains the basic rhythmic structure of natural speech by using a new higher-level programming unit: the IPCG.

30.6 Conclusions

We have presented a complete model for the rhythmic control of synthetic utterances. This model is based on the assumption that the rhythmic structure of speech is perceived by extracting salient acoustic events from the speech signal. This extraction is thus parallel to the acoustic-to-phonetic decoding of utterances and can explain why prosody and segments can be processed separately [CN88, PMK93].

As the synthetic rhythm is controlled by anchoring each syllable to its perceptual center—syllables expand or retract according to the repartition algorithm to fill in the IPCGs—the rhythmic trajectory produced can be further synchronized with other parametric trajectories such as melody or loudness to characterize in a unified way the prosodic structure of the synthetic utterances. We are currently working on the prediction of melody using a global approach [MAB95].

Acknowledgements

The first author was supported by CNPq/Brazil when he was working at ICP. Now he works as a post-doctoral student at the Instituto de Estudos da Linguagem, Unicamp, Campinas, Brazil. We thank Frédéric Bimbot for his voice and our listeners who offered their ears to the perception task.

REFERENCES

[All75] G. Allen. Speech rhythm: its relation to performance universals and articulatory timing. *J. Phonetics* 3:75–86, 1975.

[AS92] M. Abe and H. Sato. Two-stage F_0 control model using syllable based F_0 units. In *Proceedings IEEE International Conference on Acoustics, Speech and Signal Processing*, vol. 2, 53–56, 1992.

[Aub92] V. Aubergé. Developing a structured lexicon for synthesis of prosody. In *Talking Machines: Theories, Models and Designs*, G. Bailly and C. Benoit, eds. Elsevier B.V., North-Holland, Amsterdam, 307–321, 1992.

[Bai89] G. Bailly, Integration of rhythmic and syntactic constraints in a model of generation of French prosody. *Speech Comm.* 8:137–146, 1989.

[BAL91] V. Berthier, C. Abry, and T. Lallouache. Coordination du geste et de la parole dans la production d'un instrument traditionnel. In *Proceedings, Twelfth XIIe International Congress of Phonetic Sciences*, vol. 4, Aix-en-Provence, France, 34–37, 1991.

[BB92] P. Barbosa and G. Bailly. Generating segmental duration by p-centres. In *Fourth Rhythm Workshop: Rhythm Perception and Production*, C. Auxiette, C. Drake, and C. Gérard, eds. Ville de Bourges, Bourges - France, 163–168, 1992.

[BB94] P. Barbosa and G. Bailly. Characterisation of rhythmic patterns for text-to-speech synthesis. *Speech Comm.* 15:127–137, 1994.

[BBW92] G. Bailly, T. Barbe, and H. Wang. Automatic labelling of large prosodic databases: tools, methodology and links with a text-to-speech system. In *Talking Machines: Theories, Models and Designs*, G. Bailly and C. Benoit, eds. Elsevier B.V., North-Holland, Amsterdam, 323–333, 1992.

[BMA89] G. Bailly, P. F. Marteau, and C. Abry. A new algorithm for temporal decomposition of speech. application to a numerical model of coarticulation. In *Proceedings IEEE International Conference on Acoustics, Speech and Signal Processing*, 508–511, 1989.

[BS87] K. Bartkova and C. Sorin. A model of segmental duration for speech synthesis in French. *Speech Comm.* 6:245–260, 1987.

[Cam92] W. Campbell. *Multi-level Timing in Speech*. Ph.D. thesis, University of Sussex, Sussex, UK, 1992.

[CM90] F. Charpentier and E. Moulines. Pitch-synchronous waveform processing techniques for text-to-speech using diphones. *Speech Comm.* 9(5-6):453–467, 1990.

[CN88] A. Cutler and D. Norris. The role of strong syllables in segmentation for lexical access. *J. Experimental Psychology: Human Perception and Performance* 14:113–121, 1988.

[Cou91] E. Couper-Kuhlen. A rhythm-based metric for turn-taking. In *Proceedings, Twelfth International Congress of Phonetic Sciences*, vol. 1, Aix-en-Provence, France, 275–278, 1991.

[Dau83] R. M. Dauer. Stress-timing and syllable-timing re-analyzed. *J. Phonetics* 11:51–62, 1983.

[Fan91] G. Fant. Units of temporal organization. Stress groups versus syllables and words. In *Proceedings, Twelfth International Congress of Phonetic Sciences*, vol. 1, Aix-en-Provence, 247–250, France, 1991.

[Fra74] P. Fraisse. *La psychologie du rythme*. Presses Universitaires de France, Paris, 1974.

[Fra80] P. Fraisse. Des synchronisations sensori-motrices aux rythmes. In *Anticipation et Comportement*, J. Requin, ed. Editions du CNRS, Paris, 233–257, 1980.

[GA94] C. Gérard and C. Auxiette. The processing of musical prosody by musical and nonmusical children. *Music Perception* 9:471–503, 1992.

[GG83] J. P. Gee and F. Grosjean. Performance structures: A psycholinguistic and linguistic appraisal. *Cognitive Psychology* 15:418–458, 1983.

[GJM94] L. A. Gerken, P. W. Jusczyk, and D. R. Mandel. When prosody fails to cue syntactic structure: 9-months-olds' sensitivity to phonological versus syntactic phrases. *Cognition*, 20:237–265, 1994.

[HM87] D. Hary and G. P. Moore. Synchronizing human movement with an external clock source. *Biological Cybernetics* 56:305–311, 1987.

[Jor89] M. I. Jordan. Serial order: A parallel, distributed processing approach. In *Advances in Connectionist Theory: Speech*, J. L. Elman and D. E. Rumelhart, eds. Lawrence Erlbaum, Hillsdale, NJ, 1989.

[Kla82] D. H. Klatt. The KLATTalk text-to-speech conversion system. In *Proceedings, IEEE International Conference on Acoustics, Speech and Signal Processing*, Paris, France, 1589–1592, 1982.

[Koh86] K. J. Kohler. Invariability and variability in speech timing: from utterance to segment in German. In *Invariance and Variability in Speech Processes*, J. Perkell and D. H. Klatt, eds. Lawrence Erlbaum, Hillsdale, NJ, 268–298, 1986.

[Kon91] G. Konopczynski. Acquisition de la proéminence dans le langage émergent. In *Proceedings, International Congress of Phonetic Sciences*, vol. 1, Aix-en-Provence, France, 333–337, 1991.

[Lea74] W. A. Lea. *Prosodic Aids to Speech Recognition: IV. A General Strategy for Prosodically-guided Speech Understanding*. Univac Report PX10791, Sperry Univac, DSD, St. Paul, MN, 1974.

[Leh77] I. Lehiste. Isochrony reconsidered. *J. Phonetics* 5:253–263, 1977.

[Lli89] R. R. Llinás. The role of the intrinsic electrophysiological properties of central neurons in oscillation and resonance. In *Cell to Cell Signalling: From Experiments to Theoretical Models*, A. Goldbeter, ed. Academic Press, New York, 3–16, 1989.

[MAB95] Y. Morlec, V. Aubergé, and G. Bailly. Evaluation of automatic generation of prosody with a superposition model. *International Congress of Phonetic Sciences*, Stockholm, Sweden, 1995.

[Mar75] S. M. Marcus. Perceptual centres. Unpublished fellowship dissertation, King's College, Cambridge, UK, 1975.

[Mar76] S. M. Marcus. *Perceptual centres*. PhD thesis, Cambridge University, Cambridge, 1976.

[Mar81] S. M. Marcus. Acoustic determinants of Perceptual center (p-center) location. *Perception and Psychophysics* 30(3):247–256, 1981.

[MG93] P. Monnin and F. Grosjean. Les structures de performance en français: Caractérisation et prédiction. *L'Année Psychologique* 93:9–30, 1993.

[MMF76] J. Morton, S. Marcus, and C. Frankish. Perceptual centers (p-centers). *Psychological Revue* 83(5):405–408, 1976.

[Noo91] S. G. Nooteboom. Some observations on the temporal organisation and rhythm of speech. In *Proceedings, International Congress of Phonetic Sciences*, vol. 1, Aix-en-Provence, France, 228–237, 1991.

[OSh81] D. O'Shaughnessy. A study of French vowel and consonant durations. *J. Phonetics* 9:385–406, 1981.

[Pas92] V. Pasdeloup. Durée inter-syllabique dans le groupe accentuel en français. In *XIXe Journées d'Etudes sur la Parole*, 531–536, 1992.

[PMK93] V. Pasdeloup, J. Morais, and R. Kolinsky. Are stress and phonemic string processed separately? Evidence from speech illusions. In *Proceedings of the European Conference on Speech Communication and Technology*, vol. 2, Berlin, 775–778, 1993.

[Pom89] B. Pompino-Marschall. On the psychoacoustic nature of the p-center phenomenon. *J. Phonetics* 17:175–192, 1989.

[Pri92] W. Prinz. Distal focussing in action control. In *Fourth Rhythm Workshop: Rhythm Perception and Production*, C. Auxiette, C. Drake, and C. Gérard, eds. Bourges, 65–71, 1992.

[Sco93] S. Scott. *Perceptual Centres in Speech—An Acoustic Analysis*. Ph.D. thesis, University College, London, 1993.

[SGE80] V. N. Sorokin, T. Gay, and W. Ewan. Some biomechanical correlates of the jaw movements. *J. Acoust. Soc. Amer.* 68:S32, 1980.

[Sha82] L. H. Shaffer. Rhythm and timing in skill. *Psychological Review* 89:109–122, 1982.

[Smi78] B. L. Smith. Temporal aspects of English speech production: A developmental perspective. *J. Phonetics* 6:37–67, 1978.

[SSV92] A. Semjen, H. H. Schulze, and D. Vorberg. Temporal control in the coordination between repetitive tapping and periodic external stimuli. In *Fourth Rhythm Workshop : Rhythm Perception and Production*, C. Auxiette, C. Drake, and C. Gérard, eds. Bourges, 73–78, 1992.

[Tra92] C. Traber. F_0 generation with a database of natural F_0 patterns and with a neural network. In *Talking Machines: Theories, Models and Designs*, G. Bailly and C. Benoit, eds. Elsevier B.V., North-Holland, Amsterdam, 287–304, 1992.

[TSR90] M. Turvey, R. Schmidt, and L. Rosenblum. Clock and motor components in absolute coordination of rhythmic movements. *Haskins Laboratories Status Report on Speech Research*, New Haven, CT, 231–242, 1990.

[van94] J. P. H. van Santen. Assignment of segmental duration in text-to-speech synthesis. *Computer, Speech and Language* 8:95–128, 1994.

[WH94] B. Williams and S. M. Hiller. The question of randomness in English foot timing: A control experiment. *J. Phonetics* 22:423–439, 1994.

[Wit77] I. H. Witten. A flexible scheme for assigning timing and pitch to synthetic speech. *Language and Speech* 20:240–260, 1977.

[Woo51] H. Woodrow. Time perception. In *Handbook of Experimental Psychology*, S. Stevens, ed. Wiley, New York, 1224–1236, 1951.

[WSOP92] C. Wightman, S. Shattuck-Hufnagel, M. Ostendorf, and P. Price. Segmental durations in the vicinity of prosodic boundaries. *J. Acoust. Soc. Amer.* 91(3):1707–1717, 1992.

Audio Demos

Original and synthetic signals are given in the CD-ROM to assess the quality of our generation. Three versions of three sentences of the corpus at different speech rates are given: (a) the original signal, (b) the *random* version, and (c) the *model* version. These last two synthetic versions are obtained by TD-PSOLA resynthesis applied to (a). The three sentences are:

1: Very fast: "Tu peins les sommets du Mont Karimo!"

2: Normal: "Tôt dans la matinée, Thibaud peint les sommets du Mont Karimo!"

3: Very slow: "Ta main, posée sous le manteau, visait les sommets du Mont Karimo, tôt dans la matinée!"

Then three purely synthetic versions produced by our current text-to-speech system driven by the original prosodic contours of the sentences are given.

Finally sentence 2 is synthesized using the original F_0 and energy contours of the *very slow* version and the durations computed with a clock at 450 ms. Six pauses have been inserted, including one after the unaccented word "dans."

31

Duration Study for the Bell Laboratories Mandarin Text-to-Speech System

Chilin Shih
Benjamin Ao

ABSTRACT We present in this chapter the methodology and results of a duration study designed for the Mandarin Chinese text-to-speech system of Bell Laboratories. A greedy algorithm is used to select text from on-line corpora to maximize the coverage of factors that are important to the study of duration. The duration model and some interesting results are discussed.

31.1 Introduction

This chapter reports the design and results of a study of Mandarin Chinese segmental durations. The project is part of a Mandarin text-to-speech system, and our primary goal is to model the duration pattern of natural speech to improve the naturalness of the text-to-speech system. An ideal duration study for a text-to-speech system should investigate all Mandarin speech sounds in all contexts and capture all the factors that affect duration. Such a goal, of course, is impossible to reach. Therefore the initial step of our task is to decide how to scale down the scope of our study so the task will be manageable without losing crucial information.

Previous duration studies on Mandarin took the form of controlled experiments, in which a limited number of contextual factors were examined in a fixed sentence frame [Fen85, Ren85]. Controlled experiments provide excellent contrast from minimal pairs, which are useful in identifying factors that lead to durational variations. A practical concern with this approach is that it is not possible to cover all factors and their interactions. Furthermore, experiments are better suited to answer predefined, well-focused questions, while offering very little about factors that are not included in the experimental design. Speech databases [Kla73, Ume77, CH82] are more versatile and better suited for exploratory studies. However, small databases are limited in scope, and the construction of large databases is extremely time consuming. Moreover, the lack of controlled environments makes it more difficult to ascertain the effects of factors.

To circumvent the problems of both experimental studies and speech databases, we follow the methodology proposed by van Santen [van93]. In the methodology, a large on-line text corpus is coded in terms of all factors that are either known or suspected to affect duration. Subsequently, a greedy algorithm is used to select text from this corpus. The effect of this is that the size of the database is minimized without sacrificing the richness of contextual factors. Once the database has been recorded and segmented, statistical methods [van92a, van94] are used that allow prediction of durations for contextual combinations not present in the database. This methodology makes it possible to construct a durational model for the text-to-speech system with satisfactory performance within a relatively short time frame.

Due to the nature of a text-to-speech system, we can further control the size of the database by limiting our investigation to one speaker, and primarily to those factors that can be predicted from text.

31.2 Database

The source of our database is the ROCLING Chinese text corpus, which consists of more than 9 million characters of newspaper text collected during a six-month period from October 1979 to March 1980 in Taiwan. Aside from news articles, there are also mixed genre texts in the corpus, such as essays, short stories, and kung-fu fiction.

We first extracted from the ROCLING corpus 15,620 sentences and short paragraphs that were 25 to 50 characters long, each sentence (or paragraph) containing several phrases. The character strings were segmented into words and transcribed into phonetic representation using an automatic segmenter [SSGC94]. We represent Mandarin sounds with a system that is very similar to *pinyin*, the official transliteration system used in China. However, when a pinyin symbol is ambiguous or uses two letters, we assign a unique, one-letter symbol. Table 31.1 gives the correspondence between pinyin and our notation where there is a difference. If a pinyin symbol is ambiguous, we provide a disambiguating environment in parentheses.

All Mandarin stops and affricates are voiceless; for example, *b, d*, and *g* represent voiceless unaspirated stops. *S, C*, and *Z* represent the retroflex fricative, retroflex unaspirated affricate, and retroflex aspirated affricate, respectively. *J* is a retroflex vowel, which occurs only with retroflex consonants. *Q* is a central apical vowel, which occurs only with the dental affricates *z* and *c* and the dental fricative *s*. *U* is a high front-rounded vowel. *R* is a heavily retroflexed vowel with the unique

TABLE 31.1. Conversion chart of symbols.

Pinyin	(sh)i	(d)e	j(u)	(s)i	er	ou	ei	ai	a(n)
Our Symbol	J	E	U	Q	R	O	A	I	F
Pinyin	ao	(d)i(e)	(d)u(o)	yu(e)	sh	ch	zh	(i)n	ng
Our Symbol	W	y	w	Y	S	C	Z	N	G

is a high front-rounded vowel. *R* is a heavily retroflexed vowel with the unique property that it must be the only sound in a syllable; it does not co-occur with any initial or coda consonants. *F* is an allophone of *a*, which is fronted and raised in the context of a following alveolar nasal *N*. *E* is our symbol for schwa. Even though Mandarin destressed vowels are often reduced to schwa, this vowel, unlike the schwa in English, can be fully stressed (i.e., carry full tone). Diphthongs are treated as single units. Our symbols *A, I, O*, and *W* represent pinyin *ei, ai, ou*, and *ao*, respectively. Syllable final consonants (i.e., codas) in Mandarin are very restricted. Only alveolar nasal *N* and velar nasal *G* are allowed in that position.

Every segment in the on-line text was coded with a set of factor values before the search began. Based on previous reports on Mandarin duration [Fen85, Ren85] and literature on other languages [Noo72, Leh72, Kla73, Oll73, HU74, Por81, CH82, AHK87, CH88, van92a, WSOP92, FM93] we choose the following factors as the focus of our investigation, with the number of values indicated in parentheses.

1. Identity of the current segment (46)
2. Identity of the current tone (6)
3. Identity of the previous segment (10)
4. Identity of the previous tone (6)
5. Identity of the next segment (10)
6. Identity of the next tone (6)
7. Degree of discourse prominence (3)
8. Number of preceding syllables in the word (3)
9. Number of following syllables in the word (3)
10. Number of preceding syllables in the phrase (3)
11. Number of following syllables in the phrase (3)
12. Number of preceding syllables in the utterance (2)
13. Number of following syllables in the utterance (2)
14. Syllable type (9)

Factor 1 has 46 values that correspond to 46 segments, including 15 vowels, 4 diphthongs, 3 glides, 21 consonants, and 3 coda consonants. Factors 2, 4, and 6 have each 6 values that correspond to the 4 full tones, the neutral tone (0), and a sandhi tone (5). Each of factors 3 and 5 each groups sounds into 10 categories. Factor 7 has three values: normal reading, some prominence, and strong prominence. Factors 8 through 11 have three values each, 0, 1 and 2, where 0 means that the segment in question lies at the boundary, 1 means that it is one syllable away, and 2 means that it is 2 or more syllables away from the boundary. Factors 12 and 13 have two values each, 0 and 1, where 0 means that the segment lies at the boundary and 1 means that it is 1 syllable or more away from the boundary. Factor 14 has 9 values, corresponding to 9 syllable types.

Factor 7 (discourse prominence) cannot be calculated from text information alone; this factor was not included in the input coding for text selection. The values of this factor were obtained later by transcribing the recorded database.

During the text selection phase, phrasing was coded solely on the basis of punctuation. After the text was selected and the database recorded, phrasing was recoded to correspond to pauses. Each paragraph-sized unit chosen from the corpus was considered an utterance, and each utterance in our database contains at least 2 phrases. There were 3,845 phrases in total, so on average each utterance contained 9 phrases.

There is no factor coding word stress directly, because word stress in Chinese is not as clearly defined acoustically or perceptually as in a stress language such as English. Destressing, however, is clear and is traditionally described as a process of tonal reduction. So in effect the factor on current tone also coded two levels of stress: full-toned syllables (1-5) are stressed, whereas neutral-tone syllables (0) are destressed.

The factor values of each segment are grouped into factor-triplets, a unit we judge to be an acceptable compromise between controlling the number of possible factor combinations and preserving interesting factor interaction. Each factor-triplet consists of the current segment, the current tone, and one of the other 11 factors (factor 7 excluded). Each segment in the on-line text is now represented by 11 factor-triplets. These factor-triplets represent types of interaction in which we are interested and for which we are deliberately searching. The following example illustrates how a sentence in the on-line text is transcribed and coded into factor-triplets. The first factor triplet *x 2 b** means that this is a segment *x*, occurring in a tone 2 syllable, and is preceded by silence.

```
Sample Text:

刑事組幹員認為幕後可能有販毒集團，乃喬裝購毒品，串通被補的
廖清裕打電話給林，李兩人，約好交貨時間地點.

Word Segmentation and Transcription:

x2iGS4J z3u g4FNY2eN r4ENw2A m4uh40 k3En2EGy30 f4FNd2u
j2itw2FN } n3I qy2WZw1aG g40d2up3iN } Cw4FNt1oG b4Ab3ud0E
ly4Wq1iG4U d3ady4eNhw4a g3A l2iN } l3i ly3aGr2EN } Y1eh3W
jy1Whw4o S2Jjy1eN d4idy3eN }

Factor Triplets:

x_2_b* x_2_B* x_2_f3 x_2_F1 x_2_w0 x_2_x1 x_2_p0 ...
i_2_b2 i_2_B* i_2_f9 i_2_F1 i_2_w0 i_2_x1 i_2_p0 ...
G_2_b6 G_2_B* G_2_f1 G_2_F1 G_2_w0 G_2_x1 G_2_p0 ...
S_4_b9 S_4_B1 S_4_f3 S_4_F1 S_4_w1 S_4_x0 S_4_p1 ...
J_4_b2 J_4_B1 J_4_f0 J_4_F1 J_4_w1 J_4_x0 J_4_p1 ...
...
```

There were a total of 1,385,451 segments, or 556,353 syllables, in the input text, with 8,233 unique types of factor-triplet. To ensure that as many types of factor-triplet were covered with the smallest number of sentences, we use a greedy algorithm [van92b] to search through the 15,620 sentences. During each search a covered. In other words, every sentence is chosen for some unique factor-triplets contained therein, at least at the time it is chosen. Redundant sentences in the sense of factor coverage are effectively eliminated, therefore drastically reducing the size of the recorded database without sacrificing factor coverage.

In our case, the search terminated after 427 sentences were chosen when 100% of the input factor-triplets were covered. These sentences are long, each comprising several phrases and are for all practical purpose similar to short paragraphs.[1] The 427 chosen sentences/paragraphs contain 38,881 segments, 19,150 syllables, and each of the 8,233 factor-triplets occurs at least once.

Figure 31.1 compares the performance of the greedy algorithm to random selection of text. The effectiveness of a greedy algorithm is apparent. Whereas 427 sentences selected by the greedy algorithm cover 100% of the factor-triplets present in the input, 427 randomly selected sentences cover only 74%. If we accept 74% coverage, 42 sentences selected by the greedy algorithm will be sufficient. As more sentences are accumulated, most of the frequently occurring factor-triplets are covered and it becomes increasingly difficult to find a new one. The last 129 sentences chosen by the greedy algorithm each added just one new factor. In comparison, the slow increase is still much better than random selection, in which many sentences merely repeat the frequent factors that have already been covered. From the 427th to the 1000th randomly selected sentences, there was only a 3% increase in the coverage of factor-triplets.

After manual correction of transcription and word segmentation errors, the selected sentences were recorded by a male native Beijing Mandarin speaker in a soundproof room, using a Brüel and Kjær microphone 2231. The transcription was edited once again to match the recorded speech. Phrasing and prominence levels were also transcribed to match the reading. The recorded speech was then manually segmented using Waves (Entropic Inc.) on an SGI Indigo workstation.

The segmentation of the speech data followed a set of rigid rules [FGO93, OGC93]. We used both the spectrogram and the waveform to determine segment boundaries, and listened to the speech to confirm the placement of boundaries. Typically, segment boundaries were placed when a sudden change in the formant structure was visible. When such a location could not be found, as in the middle of adjacent vowels, the segment boundary was placed at the energy minimum in the transitional region. When no acoustic cues could be found, as sometimes happens between two identical vowels, the boundary was placed at the midpoint between the two vowels. The boundaries of obstruent consonants were usually easy to identify. The closure and release portions of all plosives were measured separately. The closure portion of an utterance- or phrase-initial plosive, which coincides with silence, was always marked as having no duration of its own and the data were

[1]Three of the 427 sentences turn out to be incomplete. They do not make sense in isolation and are eliminated. The recorded database therefore contains 424 sentences. Furthermore, a few awkward phrases were edited to facilitate fluent reading.

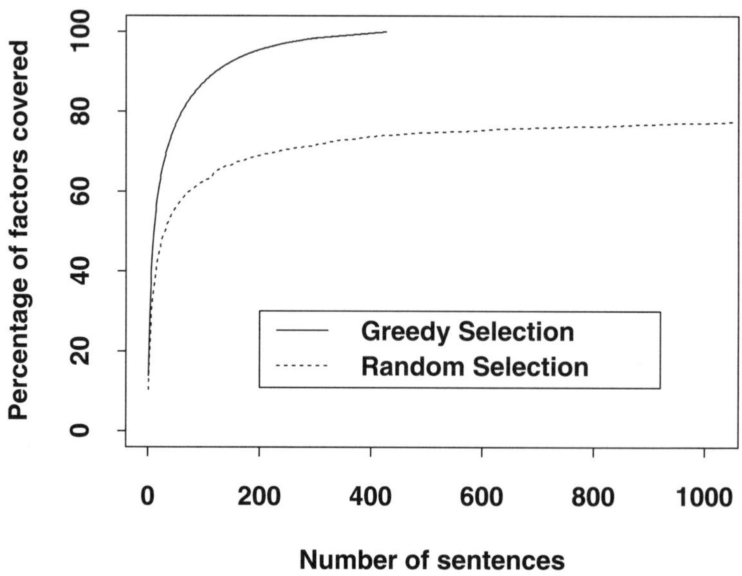

FIGURE 31.1. Comparison between random and greedy sentence selection.

discarded in later analyses. The burst-aspiration duration was measured from the onset of the burst to the onset of the vowel.

After segmentation, the factor values and duration of each segment were coded again into a matrix form suitable for statistical analyses. Excluding pauses and phrase initial closure duration, the final database we used for statistical analyses had 46,265 lines, each line containing the factor values and the duration of a segment. We divided the database into six groups for the purpose of analysis because each group of sounds may respond to a given factor in different ways: (1) vowels, (2) fricatives, (3) burst and aspiration of plosive, (4) closure of plosive, (5) sonorants, and (6) syllable coda consonants.

31.3 Duration Model

In most durational studies, results are analyzed using the raw means of duration measurements. This is fine as long as the experiments are carefully controlled and the factors involved are balanced. The raw mean could be quite misleading in a natural speech database such as ours in which the frequency distribution is unbalanced. A segment may occur in some environments more often then in others and bias the result. For example, the retroflex vowel *R* (pinyin *er*) is the longest vowel if we look at the raw mean. However, the mean duration of *R* turns out to be artificially long because *R* can occur only in the syllable structure *V* due to a phonotactic constraint, and *V* is the syllable type that yields the longest vowel duration in Mandarin (see section 31.3.1). The absence of short samples from other syllable types is responsible for the long *R* in the raw mean. When we isolate the *V* syllable type and calculate its raw mean, *R* is shorter than low vowels and diphthongs.

To avoid the problems described above, we follow van Santen and use corrected means, for which estimated durations are corrected for the effect of various factors. Segment durations are compared to others that occur in the same coded conditions. The idea is similar to comparing the duration of segments in controlled experiments. See [van92a] for mathematical proof and calculation.

When there are too many unevenly distributed gaps in the data matrix, it will not be possible to estimate corrected means. These gaps could be the result of either phonotactic constraints in Mandarin or accidental gaps in the database. In order to reduce the empty cells, we need to collapse some levels of certain factors. This procedure was carried out very carefully, consulting information from raw mean, number of samples, standard deviation, and sometimes t-tests. Only levels that are phonetically similar and affect durational variation in similar ways are collapsed. For example, when we coded our data for statistical analysis, we revived the identity of the preceding and following segments in order to investigate possible effect from individual sounds. After carefully examining the effect of each preceding sound on the following vowel, we collapsed the values of the *preceding segment* for the vowel category from 46 to 11. The resulting values are high vowel, mid-vowel, low vowel and diphthong, coda consonant, sonorant, glide, aspirated stop, nonaspirated stop, aspirated affricate, nonaspirated affricate, and fricative.

We perform a number of analyses after combining factor levels, such as computing corrected means by each factor, computing two-way corrected means to investigate the pattern of interaction of any two factors of interest to us, and building additive and multiplicative models by computing the estimated intrinsic durations of segments and the coefficient of contextual factors. The multiplicative model in general performs better than the additive model, so in the following we report only the result from the multiplicative model, where $Dur_{i(f2,\ldots,fn)}$ is the predicted duration of a given vowel i with factor levels $f2, \ldots, fn$ for factors $F2, \ldots, Fn$ respectively. $IDur_i$ is the intrinsic duration of the vowel, or $F1$, and $F2_{f2}, \ldots, Fn_{fn}$ are the coefficients of the other factor levels.

$$Dur_{i(f2,\ldots,fn)} = IDur_i \times F2_{f2} \times F3_{f3} \times \ldots \times Fn_{fn}$$

Our results agree with well-known durational phenomena reported in the literature in general terms, but often with refinement on details. For example, Oller [Oll73] found that vowels are lengthened in final position and consonants are lengthened in initial position. In our data, consonant-lengthening is found in all initial positions, and most strongly word-initially, but vowel lengthening is found only in the phrase final position and not in the word final and utterance final positions.

Among the 14 factors that we investigated, the identity of the following tone is the only one that shows nearly no effect on all six classes of sound. The factors that consistently have a strong effect on all classes of sounds are the identity of the

TABLE 31.2. Corrected means of vowels in ms.

o	J	E	Q	i	U	u	e
99	109	113	116	120	121	121	128
R	A	O	I	F	W	a	
134	135	138	147	149	155	160	

current segment and prominence. All the other factors have some effect on some classes but not on others.

31.3.1 Vowels

Our data show very clear patterns of intrinsic duration of vowels [Hou61, AHK87, van92a]. For example, we observed the same scale of vowel duration under various degree of prominence and in different positions of a phrase. The shortest vowel is *o*, followed by the two apical vowels *J* and *Q* and the schwa *E*; all of them are shorter then the high vowels *i*, *u*, and *U*. Diphthongs *A*, *O*, *I*, and *W* are longer than high and mid-vowel, and the longest vowel is the low vowel *a*. The best estimates of corrected means of vowels for the entire dataset is given in table 31.2.

Aside from vowel identity, the following are the most important factors that affect vowel duration in the multiplicative model. We rank the importance of the factors by an index number, which is the ratio of the two extreme levels of the factor. For example, the index number 1.82 for the factor *prominence* is obtained by dividing the highest coefficient 1.29 *(level 2)* by the lowest coefficient 0.71 *(normal)*. Given the multiplicative model, a vowel with prominence level 2 will be 1.82 times longer than a normal vowel.

1. *Syllable type* (1.89): open syllable without glide > open syllable with glide > closed syllable without glide > closed syllable with glide
2. *Prominence* (1.82): level 2 > level 1 > normal
3. *Previous phone* (1.73/1.27): across syllable boundary > within syllable boundary; among across syllable: non-low vowel > nasal coda > low vowel and diphthong; among within syllable: unaspirated plosive and sonorant > fricative and glide > aspirated plosive

The following factors have some effect on vowel duration:

1. *Identity of tone* (1.49/1.11): full tone > neutral tone; among full tones, 3 > 2 > 4 > 1
2. *Utterance position* (1.39): nonfinal > final
3. *Following phone* (1.33): diphthong > monophthong > plosive and fricative > sonorant
4. *Phrasal position* (1.31): final > nonfinal

Previous tone, following tone, and within-word position have very little effect on the duration of vowels.

Two index numbers were given for the factors *previous phone* and *identity of tone*; the first numbers, 1.73 and 1.49, are derived in the usual way. These numbers are high primarily because the two factors in question incorporate complex conditions. The *previous phone* of a vowel can be an initial consonant within the same syllable, or it can be the last phone of the previous syllable. The second index number (1.27) for previous phone excludes the across-syllable conditions, and therefore reflects the effect of the initial consonants on vowels. In the factor *identity of tone*, the high index number is caused by the level *tone 0*, which, as explained earlier, refers to the absence of a full tone and most closely resembles the phenomenon of destressing. The second index number (1.11) excludes *tone 0* and reflects the range of the effect of full tones.

The intrinsic scale of vowels from our study is slightly different from [Fen85], in which the duration of the mid-vowel *o* is similar to that of *e*. The discrepancy in *o* is due to different segmentation strategies: we segmented syllables such as *mo* as having three segments *mwo*, whereas [Fen85] treats it as having two segments, combining the *w* portion into the duration of *o*, causing the *o* duration to be artificially long. There is another environment in which *o* occurs in Mandarin: before a nasal coda and without a glide, as in the syllable *gong*. That is an environment in which the segmentation issue wouldn't be a problem, but [Fen85] didn't study *o* in this environment. We have a complete set of data on vowels occurring before a nasal coda. The result of that subset of data also confirms that *o* is the shortest vowel.

The fact that the utterance-final vowels have considerably lower coefficients than nonfinal vowels does not necessarily mean that there is a utterance-final shortening effect in Mandarin. Because the end of an utterance is by definition the end of a phrase, we coded utterance-final vowels as being phrase-final as well. As a result, utterance-final vowels would be lengthened in the model due to their phrase-final status. When in reality there is no utterance-final lengthening effect (see section 31.4.2), a comparable level of shortening for this position needs to be built into the model to offset the lengthening effect associated with the phrase-final position.

31.3.2 Fricatives

Among fricatives, *h* and *f* are short, whereas *s* and *x* are long. Table 31.3 gives the best estimates of corrected means of fricatives.

The important factors affecting fricative duration are:

1. *Following phone* (1.37): high vowel > mid-vowel > low vowel

TABLE 31.3. Corrected means of fricatives in ms.

h	f	S	x	s
98	100	113	119	122

2. *Prominence level* (1.36): with prominence > normal
3. *Position in the word* (1.25) : initial > noninitial
4. *Tone* (1.23): full tone > neutral tone
5. *Syllable type* (1.21) : syllable without glide > syllable with glide

All other factors have index numbers smaller than 1.15. The factors that have nearly zero effect include the following tone, the number of following syllables in the utterance, and the previous phone.

31.3.3 Burst and Aspiration

Intrinsic burst-aspiration duration is given in table 31.4. Not surprisingly, manner of articulation is the most important factor determining the length of the burst and aspiration: Unaspirated stops and affricates have shorter burst-aspiration duration than aspirated ones, and in either the aspirated or the unaspirated category, stops have shorter bursts/aspiration duration than affricates. Place of articulation has a consistent effect, but the effect is small in comparison to the variations caused by manner of articulation. Among stops, bilabials have shorter burst-aspiration duration than alveolars, which in turn have shorter burst-aspiration duration than velars. Among affricates, the retroflex affricates have shorter burst-aspiration duration than dentals, which in turn are shorter than palatals.

Phone identity, with an index number of 10.06, is undisputably the most important factor controlling the duration of the burst and the following aspiration or frication. The next important factor is *the following phone*, with an index number of 1.84.

1. *Following phone* (1.84) : apical vowel > high vowel > low vowel

Other factors that have some effect on burst-aspiration duration include:

1. *Position in word* (1.23) : initial > noninitial
2. *Tone* (1.20): tone 2 > others
3. *Prominence level* (1.19) : level 2 > level 1 > normal
4. *Syllable type* (1.16): without glide > with glide
5. *Preceding phone* (1.15): high vowel > diphthong > apical vowel > nasal coda > low vowel

TABLE 31.4. Corrected means of burst-aspiration duration in ms.

b	d	g	Z	z	j	p	t	k	C	c	q
11	13	21	29	43	46	80	80	86	95	99	113

TABLE 31.5. Corrected means of closure duration in ms.

Ccl	ccl	qcl	zcl	tcl	jcl	Zcl	kcl	dcl	gcl	pcl	bcl
13	13	15	15	16	16	17	18	18	19	20	21

The preceding and the following tone; the number of preceding syllables in the phrase; and the number of following syllables in the word, the phrase, and the utterance have little effect. It is unclear why tone 2 lengthens the burst-aspiration duration. It could be a matter of personal style.

31.3.4 Closure

The intrinsic closure duration is given in table 31.5. The manner of articulation, again, plays a major role: affricates have shorter closure duration than stops, and aspirated ones have shorter duration than unaspirated ones.

Factors affecting the closure duration include:

1. *Position in word* (1.37): initial > noninitial
2. *Tone* (1.31): full tone > reduced tone
3. *Prominence level* (1.17): with prominence > normal
4. *Preceding phone* (1.14): vowel > nasal
5. *Following phone* (1.10): high vowel > low vowel

Preceding and following tones, the position in the utterance, and syllable type have little effect.

31.3.5 Fitted Duration

Even though our speaker read the database in a dramatic style, with frequent shift of speaking rate and liberal usage of exaggeration, the performance of our predictive model is very good. The root mean square of the difference between the observed and the fitted values by the multiplicative model is 25 ms. Figure 31.2 compares the natural segmental duration (solid line) and the fitted segmental duration (dotted line) of the shortest sentence in our database: *Nian2-ji4 da4 le0, zuo5-er3 ou5-er3 bu4 shu1-fu0*, "Due to (my) old age, my left ear sometimes feels uncomfortable" (see audio files on CD-ROM).

The duration of the first vowel nucleus *e* (spelled as *a* in *nian*) is derived as follows:

$$Dur_e(65msec) =$$
$$I\,Dur_e(127.94msec)$$
$$\times F2_{tone[2]}(1.075) \times F3_{previous-phone[y]}(0.875)$$

FIGURE 31.2. Comparison of observed and fitted duration.

$$
\begin{aligned}
&\times F4_{previous-tone[null]}(0.996) \times F5_{next-phone[N]}(1.102)\\
&\times F6_{next-tone[4]}(1.027) \times F7_{stress[0]}(0.709)\\
&\times F8_{preceding-syllable-in-word[0]}(1.012)\\
&\times F9_{following-syllable-in-word[1]}(0.975)\\
&\times F10_{preceding-syllable-in-phrase[0]}(1.016)\\
&\times F11_{following-syllable-in-phrase[2]}(0.911)\\
&\times F12_{preceding-syllable-in-utterance[0]}(0.963)\\
&\times F13_{following-syllable-in-utterance[2]}(1.12)\\
&\times F14_{syll-type[cgvc]}(0.688)
\end{aligned}
$$

31.4 Discussion

Our result confirms previous findings on duration in general areas. The duration scales of vowel categories and consonant categories are similar to those from previous Mandarin studies [Fen85, Ren85], even though the database and the methodology are quite different. Because our database is much more extensive, we are able to explore areas that have not been studied before. We discuss two interesting cases below: (incomplete) compensatory effect and the lack of discourse final lengthening.

31.4.1 Compensatory Effect

We use two examples to illustrate compensatory effects: vowels and other segments in a syllable, and vowels and coda consonants. The vowel duration in a syllable is affected by the structure of a syllable. The duration of a vowel in the simplest syllable structure V is on average 3.5 times the duration of the same vowel in the most complicated syllable structure CGVC. We plot the raw mean duration of vowels by syllable type in the top panel of figure 31.3 in ascending order. With the exception of CVC and CGVV, the differences between all adjacent pairs are significant at the $p < 0.001$ level. The presence of an initial consonant, a glide, or

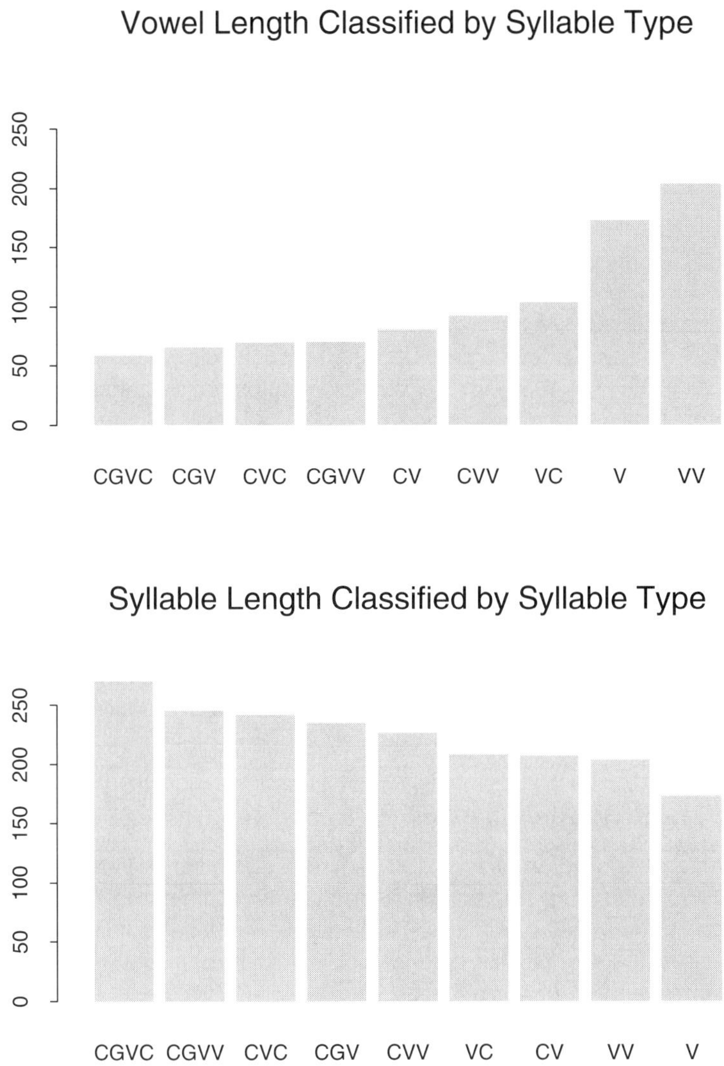

FIGURE 31.3. Vowel and syllable length classified by syllable type.

a coda consonant in a syllable shortens the main vowel. Everything being equal, a diphthong (VV) is longer than a simple vowel.

However, the compensatory effect is incomplete. There are still considerable differences in syllable length. The more phonemes there are in a syllable, the longer the syllable duration is. The raw mean duration of syllable length by type is plotted in the bottom panel of figure 31.3 in descending order. The duration of the

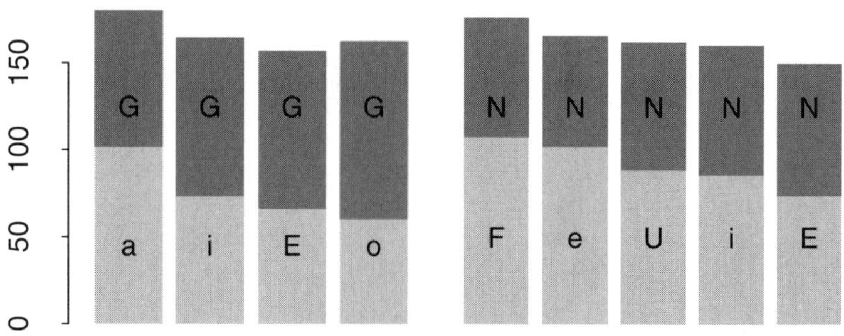

FIGURE 31.4. Compensatory effect between vowel and coda.

longest syllable type, CGVC, is 1.5 times the duration of the shortest one, V. The differences among the two-segment group VC, CV and VV are not significant. Also, the difference between CGVV and CVC is not significant. The differences between all other pairs are significant at the $p < 0.001$ level.

Vowel and coda consonants also exhibit compensatory effects (see figure 31.4). The velar nasal coda G (91 ms) is longer than the alveolar nasal coda N (71 ms), and the vowel before the velar coda is shorter. The compensatory effect is also observed within each class. Given the same coda, a longer vowel tends to be accompanied by a shorter coda. Again, the compensatory effect is not complete. The longest vowel and coda combination comes from the longest coda G and the longest vowel *a*; the shortest combination comes from the shortest coda N and E, the shortest vowel that co-occurs with N.

31.4.2 Lack of Utterance Final Effect

One clear finding from our study is that there is no utterance-final lengthening effect. Table 13.6 compares the raw mean duration of utterance final syllables with some other conditions. The mean duration of utterance-final syllables is 207 ms, which is shorter than the mean of the utterance-penultimate syllables (221 ms). In contrast, the mean duration of phrase-final syllables (utterance-final excluded) is 254 ms, which is considerably longer than the mean duration of phrase-penultimate syllables (216 ms), and the mean duration of all the nonfinal, nonpenultimate syllables (214 ms). We found the same pattern looking at vowel durations. Figure 31.5 plots the vowel durations broken down by position and syllable type. Consistently, utterance-final vowels are comparable to utterance-penultimate, phrase-penultimate, and the nonfinal, nonpenultimate vowels (the *other* condition), whereas phrase-final vowels are longer. There is only one sample

TABLE 31.6. Final and nonfinal syllable duration in ms.

Utt final	Utt penul	Phr final	Phr penul	Others
207	221	254	216	214

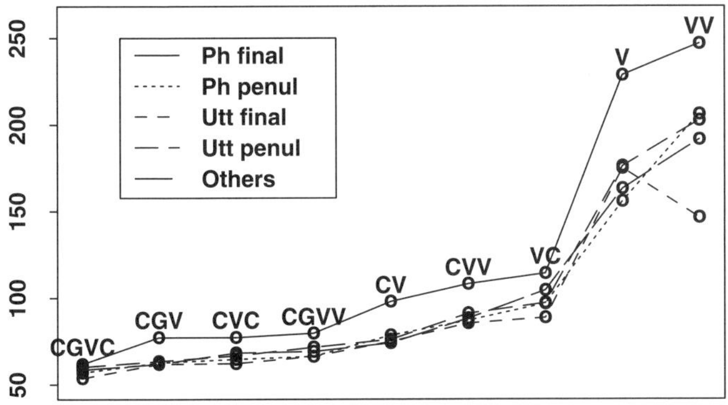

FIGURE 31.5. Vowel duration classified by position and syllable type.

of an utterance-final VV syllable in our database, so its low value may be spurious. The fact that the phrase-final vowels are longer while the phrase-penultimate vowels are similar to other vowels suggests that the domain of final-lengthening is confined to the last syllable of the phrase. If there were an utterance-final effect, we would expect to see a difference between the duration of the utterance-final and the utterance-penultimate vowels.

Our result is most similar to the sentence-final shortening effect of Japanese [TSK89]. Interestingly, Takeda et al. also found a mixture of effects. In conversational speech there are lengthening effects in both the prepausal position (similar to our phrases-final position) and sentence-final position (similar to our utterance-final position). However, in read speech there are lengthening effects in phrase-final position but shortening effects in sentence-final position. Our database contains read speech only; therefore, the finding is entirely consistent with [TSK89].

On the surface, our finding seems to contradict many previous reports on the final-lengthening effect [Kla75, EB88, Ber93]. However, because there is at the same time a considerable amount of phrase-final lengthening in our database, we interpret our data to be consistent with previous findings. Sentences used in previous experimental studies were comparable in size to our phrases. The final-lengthening effects reported in some discourse studies [Kla75, CH88] were actually phrase-final effects; there were very few samples of discourse final syllables in those two studies. Our sentences are more comparable to short paragraphs, consisting of several phrases, and exhibit full discourse structure, with dramatic discourse-final lowering toward the end. We suspect that the lack of discourse-final lengthening is linked to the dramatic drop in F_0 and amplitude.

31.5 Conclusion

One of the most important differences between this study and previous studies on Chinese duration is the design of the database. The major advantage of our methodology is the efficiency of the database. Using a greedy algorithm to select text produced a small database rich in factors that are relevant to duration studies. Moreover, our database is not limited to the chosen factors and turns out to be an excellent source for exploratory study. More factors are collected as a by-product of collecting the specified factors; some others may be created as the result of the reader's rendition of the text. One example is the variation in prominence.

Another encouraging aspect is the degree to which our result confirms previous findings in well-known areas, suggesting that no discrepancies are introduced by the differences in materials and in statistical methods. Against that background, we are confident in interpolating our result to previously untapped areas.

The major findings from this study include the intrinsic scales of all categories of Mandarin sounds, and the major factors affecting their durations. We reported the scales of vowel, fricative, burst-aspiration duration, and closure duration, and ranked the effects of 14 factors on them. We also find incomplete compensatory effects, and the lack of utterance-final lengthening.

Acknowledgments: We acknowledge ROCLING for providing us with the text database. We also wish to thank Jan van Santen. It would be impossible to do this project without his extensive advice and duration analysis tools.

References

[AHK87] J. Allen, S. Hunnicut, and D. H. Klatt. *From text to speech: The MITalk system.* Cambridge University Press, Cambridge, UK, 1987.

[Ber93] R. Berkovits. Utterance-final lengthening and the duration of final-stop closures. *J. Phonetics* 21(4):479–489, 1993.

[CG86] R. Carlson and B. Cranström. A search for durational rules in a real-speech data base. *Phonetica* 43:140–154, 1986.

[CH82] T. H. Crystal and A. S. House. Segmental durations in connected speech signals: Preliminary results. *JASA* 72:705–716, 1982.

[CH88] T. H. Crystal and A. S. House. Segmental durations in connected-speech signals: Current results. *JASA* 83:1553–1573, 1988.

[EB88] J. Edwards and M. E. Beckman. Articulatory timing and the prosodic interpretation of syllable duration. *Phonetica* 45(2):156–174, 1988.

[Fen85] L. Feng Beijinghua yuliu zhong sheng yun diao de shichang (Duration of consonants, vowels, and tones in Beijing Mandarin speech). In *Beijinghua Yuyin Shiyanlu (Acoustics Experiments in Beijing Mandarin)*, Beijing University Press, Beijing, 131–195, 1985.

[FGO93] R. M. French, A. Greenwood, and J. P. Olive. *Speech Segmentation Criteria.* Technical report, AT&T Bell Laboratories, 1993.

[FM93] J. Fletcher and A. McVeigh. Segment and syllable duration in Australian English. *Speech Comm.* 13:355–365, 1993.

[Hou61] A. S. House. On vowel duration in English. *JASA* 33:1174–1178, 1961.

[HU74] M. S. Harris and N. Umeda. Effect of speaking mode on temporal factors in speech: Vowel duration. *JASA* 56:1016–1018, 1974.

[Kla73] D. H. Klatt. Interaction between two factors that influence vowel duration. *JASA* 54:1102–1104, 1973.

[Kla75] D. H. Klatt. Vowel lengthening is syntactically determined in a connected discourse. *J. Phonetics* 3:129–140, 1975.

[Leh72] I. Lehiste The timing of utterances and linguistic boundaries. *JASA* 51(6.2): 2018–2024, 1972.

[LR73] D. Lindblom and K. Rapp. Some temporal regularities of spoken Swedish. *Publication of the Institute of Linguistics, University of Stockholm*, 21:1–59, 1973.

[Noo72] S. G. Nooteboom. *Production and Perception of Vowel Duration*. University of Utrecht, Utrecht, 1972.

[OGC93] J. P. Olive, A. Greenwood, and J. Coleman. *Acoustics of American English Speech: A Dynamic Approach*. Springer-Verlag, New York, 1993.

[Oll73] D. K. Oller. The effect of position in utterance on speech segment duration in English. *JASA* 54:1235–1247, 1973.

[Por81] R. F. Port. Linguistic timing factors in combination. *JASA* 69:262–274, 1981.

[Ren85] H. Ren. Linguistically conditioned duration rules in a timing model for Chinese. In *UCLA Working Papers in Phonetics 62*, I. Maddieson, ed. UCLA, Los Angeles, 1985.

[SSGC94] R. W. Sproat, C. Shih, W. Gale, and N. Chang. A stochastic finite-state word-segmentation algorithm for Chinese. In *Proceedings of the 32nd Annual Meeting of the Association for Computational Linguistics*, New Mexico State University, 66–73, 1994.

[TSK89] K. Takeda, Y. Sagisaka, and H. Kuwabara. On sentence-level factors governing segmental duration in Japanese. *JASA* 86:2081–2087, 1989.

[Ume77] N. Umeda. Consonant duration in American English. *JASA* 61:846–858, 1977.

[van92a] J. P. H. van Santen. Contextual effects on vowel duration. *Speech Comm.* 11(6):513–546, 1992.

[van92b] J. P. H. van Santen. Diagnostic perceptual experiments for text-to-speech system evaluation. In *Proceedings of ICSLP*, Barff, Alberta, Canada, 555–558, 1992.

[van93] J. P. H. van Santen. Perceptual experiments for diagnostic testing of text-to-speech system. *Computer Speech and Language* 7(1):49–100, 1993.

[van94] J. P. H. van Santen. Assignment of segmental duration in text-to-speech synthesis. *Computer Speech and Language* 8(2):95–128, 1994.

[WSOP92] C. W. Wightman, S. Shattuck-Hufnagel, M. Ostendorf, and P. J. Price. Segmental durations in the vicinity of prosodic phrase boundaries. *JASA* 91:1707–1717, 1992.

Appendix: Audio Demos

The CD-ROM contains eight speech files.

32

Synthesizing German Intonation Contours

Bernd Möbius

ABSTRACT This chapter the adaptation of Fujisaki's quantitative model to the analysis of German intonation and its application to the synthesis of fundamental frequency contours by rule. We investigated the sources of variation of the model's parameter values, which were obtained by analyzing utterances in a database. On the basis of this analysis, a set of rules was constructed that captures linguistic as well as speaker-dependent features and generates artificial intonation contours for target utterances. Acceptability of the rule-generated intonation patterns was tested in a series of perceptual experiments. The rules have been implemented in the HADIFIX speech synthesis system for German.

32.1 Introduction

Because adequately modeling prosodic properties enhances intelligibility and naturalness of artificially produced speech, the generation of intonation by rule is an important component of any text-to-speech or speech synthesis system. This statement holds even if the speech synthesizer is capable of producing synthetic speech of high segmental quality [Kla87, Kra94]. The temporal course of the fundamental frequency F_0 is considered the most outstanding acoustic correlate of intonation.

In this chapter I present the application of the quantitative intonation model presented by Fujisaki [Fuj83] as a framework for synthesis-by-rule of German intonation. This model aims at a functional representation of the production of F_0 contours by a human speaker. By using only a small number of control parameters, the model approximates measured F_0 contours very accurately. I demonstrate that intonation contours can be efficiently analyzed, and predicted, by interpreting the components and parameters of Fujisaki's model in terms of linguistic features and categories. I argue that a superpositionally organized model is particularly suitable for a quantitative description of intonation and for the analytical separation of the various factors that determine the shape of F_0 contours.

Two major classes of intonation models have evolved during the last two decades. There are, on the one hand, hierarchically organized models that interpret F_0 contours as a complex pattern resulting from the superposition of several components (e.g., [Fuj83, Gro92]). Their counterparts are usually seen in the models that claim that F_0 contours are generated from a sequence of phonologically distinctive tones by applying phonetic realization rules (e.g., [Pie80, Lad83]).

The main difference between these types of models can be seen in the way they define the relation between local movements and global trends in the intonation contour, or, in other words, in the view of the relation between word accent and sentence intonation. The underlying problem is that word- and utterance- (or phrase-) prosodic aspects all express themselves by the same acoustic variable: the variation of fundamental frequency as a function of time. There is no way of deciding, either by acoustic measurements or by perceptual criteria, whether F_0 movements are caused by accentuation or by intonation. A separation of these effects, however, can be done on a linguistic, more abstract level of description. Here rules can be formulated that predict accent- or intonation-related patterns independent of, as well as in interaction with, each other.

Autosegmental theory allows for the independence of various levels of suprasegmental description and their respective effects on the intonation contour by an appropriate phonological representation. Thus, according to [EB88], the most promising principle of intonation models ought to be seen in the capability to determine the effects of each individual level and their interactions. Although probably not intended by the authors, this is one of the most striking arguments in favor of a hierarchical approach and superpositional models of intonation.

Superpositionally organized models lend themselves to a quantitative approach: Contours generated by such a model result from an additive superposition of components that are, in principle, orthogonal to, or independent of, each other. The components, in turn, can be related to certain linguistic categories or nonlinguistic properties. Thus, the factors contributing to the variability of F_0 contours can be investigated separately. In addition, the temporal course pertinent to each individual component can be computed independently. A production-oriented model providing components for accentuation, on the one hand, and sentence or phrase intonation, on the other hand, and generating the pertinent patterns by means of parametric commands appears to be particularly promising.

The only approach that exploits the principle of superposition in a strictly mathematical sense is the model proposed by Fujisaki and co-workers (e.g., [FHO79, Fuj83, Fuj88]). This particular model has several advantageous properties. Because it satisfies the principle of superposition, the effect of a given factor can be determined for a predefined temporal segment or for a given linguistically or prosodically defined unit, such as a phrase or a stress group. For every desired point in time in the course of an utterance, the resulting F_0 value can be computed. The values of the model parameters (see section 32.2) are constant at least within one stress group. This data reduction is an important aspect for certain applications, such as speech synthesis. The smooth contour resulting from the superposition of the model's components is appropriate for the approximation of naturally produced F_0 contours.

Generally speaking, adequate models are expected to provide both predictive and explanatory elements [Coo83]. In terms of prediction, models have to be as precise and quantitative as possible, ideally being mathematically formulated. A model provides explanations if it is capable of analyzing a complex system in such a way that both the effects of individual components and their combined results become apparent. Fujisaki's model meets both requirements, and all effects can be described uniquely by their causes.

The model does not, however, explain by itself why a given component behaves the way it does. The particular approach and the application presented in this chapter aim at providing these explanations, especially by applying a linguistic interpretation of the model's components.

Another explanatory approach can be seen in the potential physiological foundation, in terms of laryngeal structures and interactions of laryngeal muscles, as discussed by Fujisaki [Fuj83]. To my knowledge, his model is the only one that explicitly includes a quantitative simulation of the F_0 production and control mechanisms inherent in a human speaker; the approach is based on work by Öhman and Lindqvist [OL66].

Several authors provide physiological evidence for decomposing the F_0 contour. Results of studies by Atkinson [Atk78] and Collier [Col75, Col87] suggest that the slowly falling slope of the phrase contour (see figure 32.1) can be explained by the gradually sinking subglottal air pressure in the course of the utterance, whereas the local F_0 rises and falls are caused by active laryngeal gestures. However, Fujisaki [Fuj88] provides a different explanation; he suggests that the phrase and accent components of his model are related to two independent control mechanisms in the laryngeal structures, each with its own muscular reaction time. According to this hypothesis, both local accent patterns and the longer-term phrase patterns are the results of active laryngeal gestures.

Fujisaki's model represents each partial glottal mechanism of fundamental frequency control by a separate component. Although it does not include a component that models intrinsic or coarticulatory F_0 variations, such a mechanism could easily be added in case it is considered essential for natural-sounding synthesis (see, e.g., [Kla87]). In the studies presented here, coarticulatory F_0 perturbations were manually corrected and smoothed. As for intrinsic F_0 variations, it was assumed that the effect is evened out across the speech data under investigation. It is worth mentioning, however, that the vowel-intrinsic effect on the F_0 contour can have the same order of magnitude as the effect of linguistic factors (see [Fis90] for an overview).

32.2 Intonation Model

Fujisaki's model additively superposes a basic F_0 value ($F_{\min}$), a phrase component, and an accent component on a logarithmic scale (figure 32.2). The control mechanisms of the two components are realized as critically damped second-order systems responding to impulse functions in the case of the phrase component and rectangular functions in the case of the accent component. These functions are generated by two different sets of parameters: (1) the timing and amplitudes of

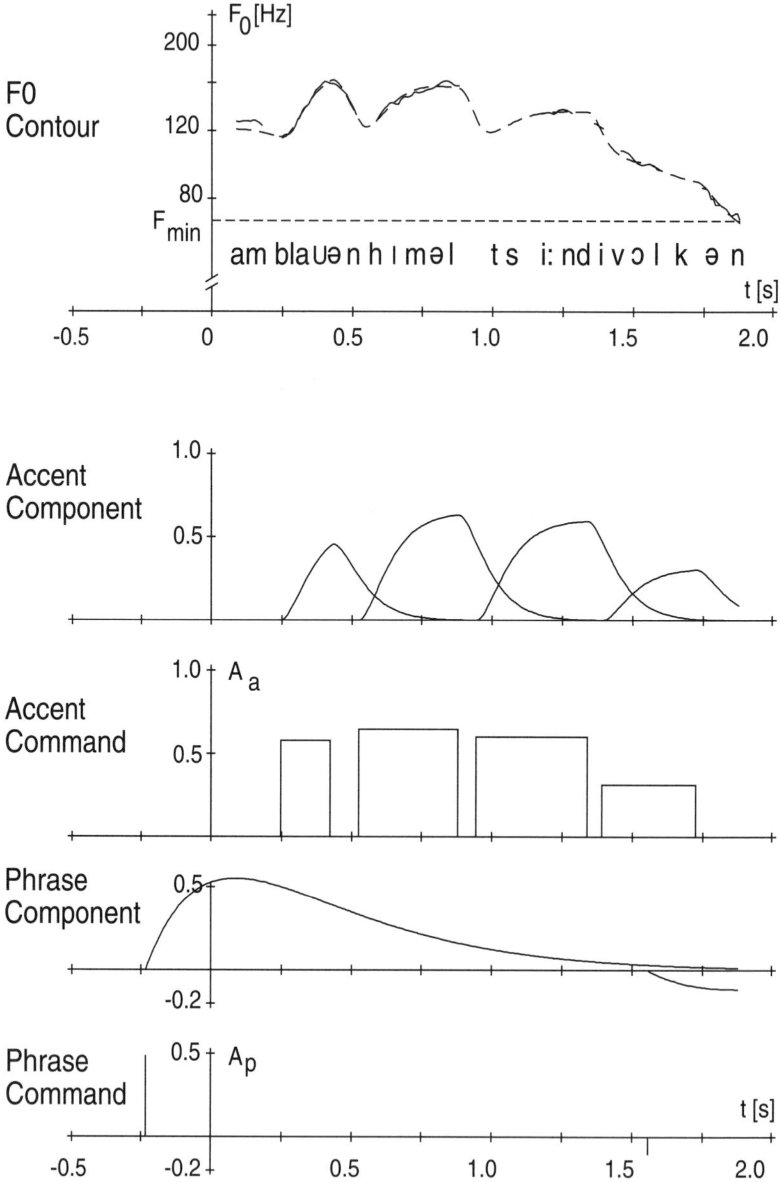

FIGURE 32.1. Top: Close approximation (dashed line) of the F_0 contour of the declarative utterance *Am blauen Himmel ziehen die Wolken* (male voice). Below: Optimal decomposition (but subjected to linguistic constraints) of the F_0 contour into the phrase and accent components and underlying commands.

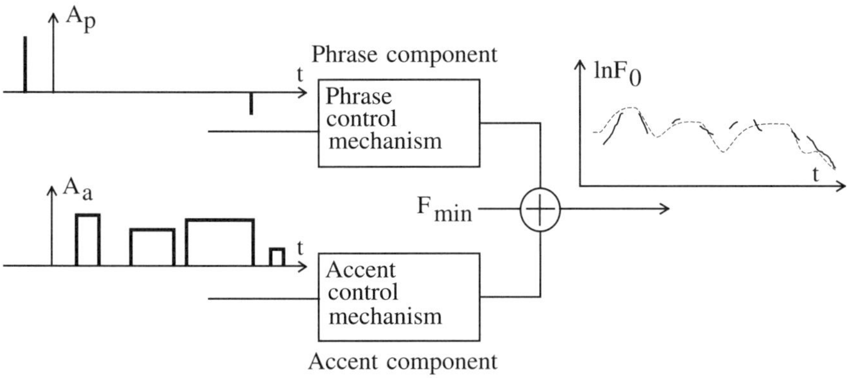

FIGURE 32.2. Block diagram of Fujisaki's quantitative model that additively superposes a basic F_0 value (F_{min}), a phrase component, and an accent component on a logarithmic scale (ln F_0). The control mechanisms of the components respond to impulse commands (phrase component) and rectangular commands (accent component), respectively (A_p = amplitude of phrase commands; A_a = amplitude of accent commands; t = time).

the phrase commands as well as the damping factors of the phrase control mechanism, and (2) the amplitudes and the timing of the onsets and offsets of the accent commands as well as the damping factors of the accent control mechanism. All these parameter values are constant for a defined time interval: the parameters of the phrase component within one prosodic phrase, the parameters of the accent component within one accent group, and the basic value F_{min} within the whole utterance.

The F_0 contour of a given utterance can be decomposed into the components of the model by applying an analysis-by-synthesis procedure. This is achieved by successively optimizing the parameter values, eventually leading to a close approximation of the original F_0 course. Thus, the model provides a parametric representation of intonation contours (see figure 32.1).

As I have argued elsewhere [Mob93a, MPH93], the quantitative description of intonation can be more efficient if modeling a given F_0 contour and extracting the pertinent parameters are subjected to the constraints given by a linguistic and prosodic interpretation in the first place and by the criterion of optimal approximation in a mathematical sense only in the second place.

The key elements of our interpretation of the model are as follows:

- The phrase component of the model represents the global slope and the slow variations of the F_0 contour in the utterance. Obviously, the phrase component is very suitable to describe F_0 declination since the phrase contour reaches its maximum rather early and descends monotonically along the major part of the utterance. Therefore, the contour that results from adding the phrase component to the basic value F_{min} serves as a baseline of the intonation contour, the magnitude of the phrase command amplitude being a direct measure for F_0 declination in the utterance.

- Beside the obligatory utterance-initial phrase command, we provide additional phrase commands only at major syntactic boundaries, for example, between main and subordinate clauses, thereby resetting the declination line. Inserting phrase commands wherever the criterion of optimal approximation seems to demand it [FHO79] is rejected.

- The conspicuous final lowering of F_0, which is regularly observed in declarative utterances and often in wh-questions, is modeled by a negative phrase command. Likewise, positive utterance–final phrase commands are provided for other sentence modes, such as yes/no and echo questions. Thus, the phrase component of the model can be related to the linguistic category *sentence mode*, via the shape of the phrase contour and the underlying commands and parameter values. There are both global (the overall slope) and local (final rise or lowering) cues that contribute to differentiating between sentence modes.

- Local F_0 movements that are asscociated with accented syllables are represented by the accent component and superposed onto the global contour. By closely following Thorsen's definition of *stress groups* [Tho79] I apply an accent group concept, an accent group being defined as a prosodic unit that consists of an accented syllable optionally followed by any number of unaccented syllables. Accent groups are independent of word boundaries but sensitive to major syntactic boundaries, as will be shown below.

The concept of accent groups fits in the hierarchical structure of the model. Whereas the linguistic category *sentence mode* is reflected in the phrase component, the linguistic feature *word accent* is manifested in the locations and the shapes of the accent commands. Consequently, the F_0 course of a given accent group should be modeled by the contour generated by exactly one accent command. Thus, the parameter configurations of the accent component can be interpreted as correlates of the linguistic feature word accent.

The speech data under investigation were manually segmented. In addition, they were labeled for accented syllables, accent group boundaries, phrase boundaries, and sentence mode; a rudimentary manual part-of-speech tagging provided word-class information.

The method of determining and standardizing the parameter values of the model will be described in the following sections.

32.3 Parameter Estimation

In principle, the parameter values that approximate the F_0 contour of a given utterance can be determined automatically or by hand. Nevertheless, only an automatic procedure guarantees that the optimal values are extracted in an objective and reproducible way. Preliminary experiments showed that there are considerable intra-

and interindividual divergencies when an interactive (i.e., partly manual method) is used. Therefore, the parameter values of the model are determined by means of a computer program [Pat91] that automatically approximates measured F_0 contours by successively optimizing the parameters within the framework of the linguistic interpretation of the model. Input information comprise the F_0 data for the given utterance and the locations of accent group boundaries.

Based on the principle of superposition, the step of determining the phrase command parameters and the basic value F_{min}, which is the first step in the algorithm, can be separated from the subsequent determination of the accent command parameters. The contour resulting from F_{min} and the phrase parameters is approximated to the measured F_0 curve. Once the parameters of the phrase component are optimized, the resulting difference signal is interpreted by the accent component of the model.

The accent component is made up of partial contours that are in turn generated by accent commands. Each accent group is modeled by the contour resulting from exactly one accent command. The individual accent groups are processed from left to right. There is no global optimization of the whole accent component, but there is a local approximation to the F_0 curve for each accent group.

Two restrictions are applied that support left-to-right processing of the F_0 contour. The first one prevents the contour optimized so far from being a posteriori affected by a succeeding accent command; the second restriction warrants that the approximation of an accent group is not made impossible by inadequate parameter values of the previous command.

The speech material used in this study can be characterized as typical "laboratory speech." It covers declarative sentences with one or two prosodic phrases as well as three types of interrogative sentences, namely echo questions, yes/no questions, and wh-questions. The two-phrase sentences each contain a main and a subordinate clause, which were intended to be realized as two prosodic phrases. Six male and three female speakers were asked to read the orthographically presented sentences aloud.

The top portion of figure 32.2 illustrates the close approximation of a measured F_0 contour.

32.4 F_0 Synthesis by Rule

The potential sources of variation of the parameter values were explored by means of statistical procedures, taking into account both linguistic and speaker-dependent factors. The results have been presented at full length in [Meb93a], so only the major trends and findings will be presented here.

Damping factors. The damping factors of the phrase and accent components are treated as damping time constants measured in units of the number of oscillations per second (i.e., Hertz). Our experiments confirm the claim [Fuj83] that the approximation of F_0 contours is not impaired by keeping them constant across speakers and utterances. For the phrase component, a standard value of 3.1/s proved to be both appropriate for the purpose of approximation and reasonable as far as the physiological foundation of the model is concerned. A constant value of 16/s corresponding to the arithmetic mean for all speakers and all accent groups proved to be suitable for the damping factor of the accent component.

Basic value F_{min}. For all speakers, the dispersion of the basic value F_{min} is relatively small, yielding 50% of the observed values within the range of about 3.0 Hz around the arithmetic mean for the respective speaker. This finding suggests that it is reasonable to keep F_{min} constant for a given speaker. Typical values are 75–80 Hz for male speakers and 145–150 Hz for female speakers.

Phrase command timing and amplitude. Because the phrase component serves as a baseline to the intonation contour, with the peak of each phrase contour, coinciding with the beginning of the utterance, or the prosodic phrase, the exact timing of a phrase command directly depends on the value of the damping factor (3.1/s). Therefore, the first phrase command is set at 323 ms before the onset of the utterance. This also reflects findings from studies on F_0 production and control, which reveal prephonatory activities of the laryngeal muscles [JWBK89].

The phrase command amplitude is, mathematically speaking, a multiplicative factor that determines the excursion of the phrase contour in the frequency domain (see figure 32.1). The amplitude values are largely speaker, or rather speaker-type, dependent. Sentence mode is the most important linguistic factor; it is globally signaled by the contour of the phrase component. Whereas the phrase contours of wh-questions are very similar to those of declaratives, yes/no questions and the syntactically unmarked echo questions show a much less steep declination (see also [Tho79] for Danish). Typical phrase contours for these three interrogative modes are shown in figure 32.3.

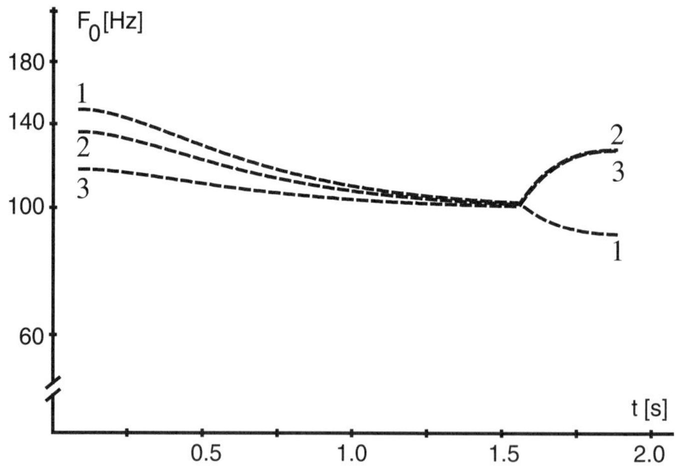

FIGURE 32.3. Rule-generated (dashed line) and original F_0 contours of the declarative utterance *Am blauen Himmel ziehen die Wolken* (male voice). (Sound sample 1 for original contour)

Accent command amplitude. Analogous to the phrase component, the accent command amplitude is a multiplicative factor that modifies the excursion of an accent contour in the frequency domain and thus the height of the pitch peak. The location of the accent group in the utterance turned out to be the most important linguistic factor explaining the variability of accent command amplitudes. Utterance-final accent commands show significantly smaller amplitudes than accent commands in any other position in the utterance. Other important factors are the part-of-speech class for the word carrying the accent, with nouns requiring higher amplitudes than other classes, and the presence of a phrase boundary. Amplitudes of accent commands preceding a phrase boundary tend to be about 25% higher than in other positions.

Accent command duration. Duration of an accent command can be reliably predicted from the duration of the respective accent group. There is a high correlation ($r = 0.84$) between these two variables. That is, more than 70% of the variance observed in durations of accent commands can be explained by accent group duration. An effect of phrase-final lengthening is observed for several speakers.

Accent command position. The most important factor controlling the relative temporal position of an accent command within a given accent group is the location of the accent group in the utterance. Whereas in non-final positions the temporal distance between the beginning of the accent group and the command onset is about 10% of the accent group duration, this distance tends toward zero in utterance-final accent groups.

Based on the analysis described above, a set of rules was constructed that controls the adjustment of the parameters. The rules capture speaker-dependent as well as linguistic features, such as sentence mode, sentence accent, phrase boundary signals, or word accent, and they generate an artificial intonation contour for a given target utterance. The input information needed is the location of accented syllables in the utterance, the durations of accent groups, and, although less important, part-of-speech information for the words that carry accent.

32.5 Perceptual Experiments

The adequacy of the rules was tested in a series of perceptual experiments that aimed at evaluating the acceptability of the rule-generated intonation contours and the adequate realization of linguistic categories.

Listeners were asked to judge the stimuli by their melodic and stress features. Three kinds of stimuli were presented in these experiments: (1) utterances with original F_0 information, (2) utterances with close approximations of the original F_0 courses by the intonation model, (3) utterances with rule-generated intonation contours. Original F_0 data were replaced by applying the PSOLA algorithm [MC90]. One female and one male "voice" were used in the study.

In the first experiment, listeners were not consistently able to discriminate between original utterances and those with closely approximated F_0 data based on differences in the contours. There were, however, perceivable differences in the sound quality of the stimuli due to two problems with the F_0 manipulations that were not satisfactorily resolved. First, although the PSOLA algorithm generally performed well, the quality of its output depends on the correctness of pitch marker positions. Even though the markers were set manually, they were not always in optimal positions, especially in laryngealized sections of the waveform. Second, the smooth contour generated by the model lacks the vocal jitter that is observed in natural speech [BO88], and may therefore have evoked an impression of unnaturalness in those parts of the utterance for which F_0 stays nearly constant.

A comparison test between close approximations and rule-generated contours yielded an overall difference of about one point on a five-point scale in favor of the close approximations. Splitting stimuli by sentence mode, however, showed that this difference is significantly smaller for declaratives than for interrogatives, declaratives being nearly as acceptable to the listeners as the close approximations of original contours. Details on these experiments are presented in [MP92].

In a follow-up experiment [Mob93b] the rules for three interrogative sentence modes (echo, yes/no, and wh-questions) were improved by extending and enriching the training data and by including into the analysis linguistic factors that were omitted in the previous study. For instance, accentuation and simultaneous signaling of interrogative sentence mode superposed and compressed on the utterance-final syllable was not adequately modeled in the earlier version of the rules. In addition, rules for deaccentuation in compounds and in the case of two adjacent accented syllables were incorporated into the new set of rules.

Two versions of each test sentence were presented to the listeners, the first with a rule-generated intonation contour, and the second with the original F_0 data as extracted by a pitch determination algorithm. In order to avoid, as far as possible, any differences in the sound quality of the stimuli (see above) that may affect listeners' judgments of prosodic properties, both kinds of stimuli were processed by means of PSOLA. Again one female and one male "voice" were used for this experiment.

The results yielded an overall difference of about one point on a seven-point scale in favor of the original utterances. The stimulus version turned out to be the only statistically significant factor, stable across sentence modes and "voices"; the factors of sentence mode and "voice" were not significant. While higher ratings for utterances with original F_0 data as opposed to rule-generated contours meet the expectations, the difference is sufficiently small. A more detailed analysis is presented in [Mob93b].

Illustrations of F_0 contours generated by rule are given in figures 32.4 through 32.7. Note that the original F_0 contours are presented for reference only. The degree of similarity or discrepancy between the two types of contours is not meant to be regarded as an example of good approximation nor as an indication of the acceptability of a given rule-generated contour. In figures 32.4 and 32.6, for example, the prominence-lending pitch movement on the last accented syllable (*WOLken* and *SCHWEden*, respectively) is generated by the model as a rising one, whereas in the original F_0 contour a falling movement is observed. Both accent types seem to

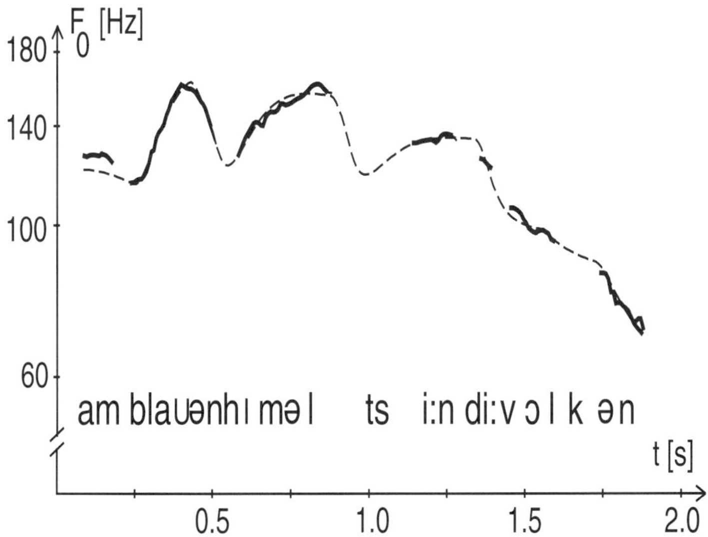

FIGURE 32.4. Typical phrase contours for the interrogative modes wh-question (1), yes/no question (2), and echo question(3), for $F_{\text{min}} = 100$ Hz.

be functionally equivalent, as listeners rated the rule-generated contours as highly acceptable.

Audio renditions of the pertinent speech files are available on the CD-ROM.

32.6 Conclusions

In this chapter, the application of a quantitative intonation model to the synthesis-by-rule of German intonation was presented. In concluding, I discuss the limitations of the approach and of the current rule set as well as point out open questions that require additional studies.

Subjecting the approximation algorithm to linguistic constraints, as opposed to a purely mathematically based optimization, enhanced the interpretation of the model parameters and their variability. Yet, this interpretation is not always straightforward. The effect of phrase boundaries on the amplitude values of accent commands, for instance, is somewhat contradictory. On the one hand, accent commands immediately preceding a nonfinal phrase boundary tend to be about 25% stronger than in other positions. On the other hand, however, utterance-final accent commands show significantly smaller amplitude values (see chapter 4). While this effect seems to confirm Grønnum's finding [Gro90] that German does not have obligatory sentence accents, the fact that utterance-final accent commands are even weaker suggests that the magnitude of the utterance-final phrase command (neg-

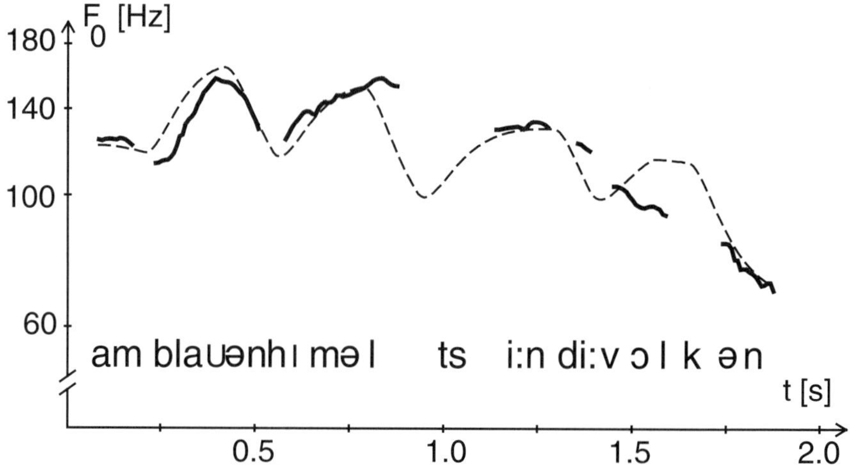

FIGURE 32.5. Rule-generated (dashed line) and original F_0 contours of the yes/no question *Sind die Rinder noch auf der Weide?* (male voice). (Sound samples 2 for original intonation and 3 for rule-generated intonation)

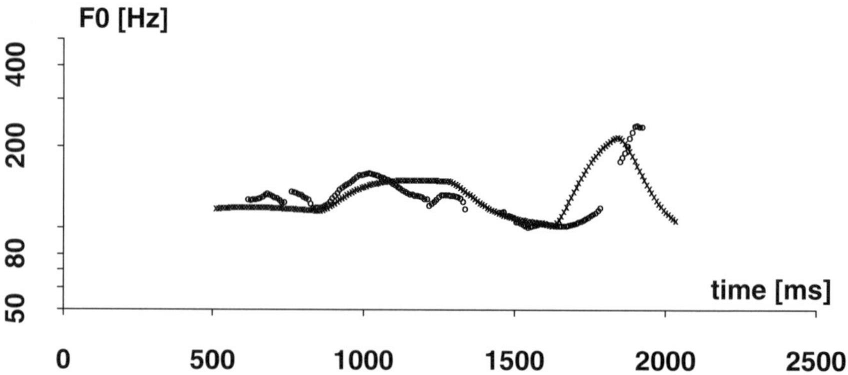

FIGURE 32.6. Rule-generated (dashed line) and original F_0 contours of the wh-question *Wie heißt die Hauptstadt von Schweden?* (female voice). (Sound samples 4 for original intonation and 5 for rule-generated intonation)

ative in declaratives and wh-questions, and positive in other interrogative modes) may have been underestimated [Gro95].

Another area for improvement is the synchronization of the intonation contour with the underlying segmental structure. Currently, there is no precise temporal alignment of rises and falls depending on the syllable structure.

An obvious limitation of the current rule set lies in the amount and structure of the speech data. Whereas the number of subjects (six male and three female speakers) seems to be appropriate to get a reasonable estimation of speaker-dependent features, the coverage of syntactic structures should be extended in future investigations, including multiphrasal utterances and additional sentence modes (e.g.,

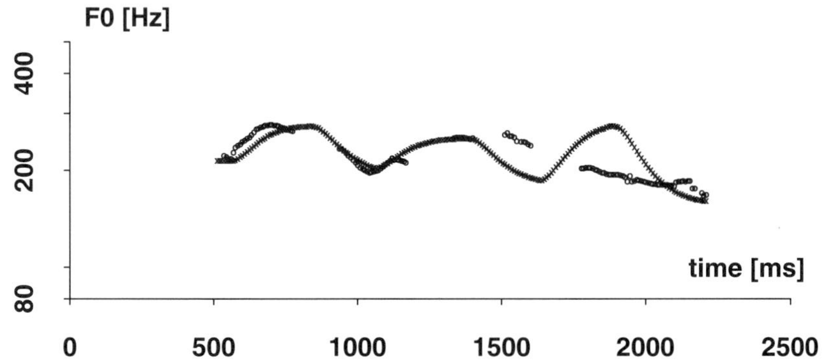

FIGURE 32.7. Rule-generated (dashed line) and original F_0 contours of the echo question *Hans hat dir nichts davon gesagt?* (female voice). (Sound samples 6 for original intonation and 7 for rule-generated intonation)

exclamations). Covering speaking styles other than "laboratory" speech is also desirable.

Furthermore, because the rules are based on the results of statistical analyses, the parameter values they provide are averages, producing contours that were not actually observed for any real speaker. On the other hand, they were shown to capture speaker-dependent features; they produce intonation patterns that to a fair degree correspond to what the modeled speaker could have produced. Thus, one should expect a mixture of frequently very good predictions with occasionally rather poor ones, the latter being due to either insufficient data or inadequate predictive power for a particular context. These expectations were generally confirmed by the results of acceptability tests.

The rules have been implemented in the German speech synthesis system HADIFIX developed at IKP Bonn [PSRSH92]. HADIFIX generates synthetic speech by concatenating three types of nonparametric acoustic inventory units, namely initial demisyllables, diphones for the transitions from vowels to postvocalic consonants, and suffixes for postvocalic consonant clusters (HADIFIX = *HAlbsilben, DIphone, SufFIXe*).

HADIFIX is not a full-fledged text-to-speech system; it takes a phoneme string and accent markers as input. The system includes rules for the concatenation of the acoustic inventory units as well as for the adjustment of syllable and segment durations. Intonation contours are generated using the model and the rules presented in this chapter.

Acknowledgments: The experiments presented in this chapter were carried out at IKP, University of Bonn, supported by grants from the German Research Council (DFG) and the German Federal Ministry of Research and Technology (BMFT). The author wishes to thank Grażyna Demenko, Wolfgang Hess, Julia Hirschberg,

Matthias Pätzold, Thomas Portele, and Jan van Santen, for support and valuable discussions; and the two reviewers, René Collier and Nina Grønnum, for helpful suggestions.

References

[Atk78] J. E. Atkinson. Correlation analysis of the physiological factors controlling fundamental voice frequency. *J. Acoust. Soc. Amer.* 63:211–222, 1978.

[BO88] R. J. Baken and R. F. Orlikoff. Changes in vocal fundamental frequency at the segmental level: control during voiced fricatives. *J. Speech Hearing Research* 31:207–211, 1988.

[Col75] R. Collier. Physiological correlates of intonation patterns. *J. Acoust. Soc. Amer.* 58:249–255, 1975.

[Col87] R. Collier. F_0 declination: The control of its setting, resetting, and slope. In *Laryngeal Function in Phonation and Respiration*, T. Baer, C. Sasaki, and K. Harris, eds. College Hill Press, Boston, 403–421, 1987.

[Coo83] F.S. Cooper. Some reflections on speech research. In *The Production of Speech*, P. F. MacNeilage, ed. Springer, New York, 275–290, 1983.

[EB88] J. Edwards and M. E. Beckman. Articulatory timing and the prosodic interpretation of syllable duration. *Phonetica* 45:156–174, 1988.

[FHO9] H. Fujisaki, K. Hirose, and K. Ohta. Acoustic features of the fundamental frequency contours of declarative sentences in Japanese. *Annual Bulletin of the Research Institute for Logopedics and Phoniatrics (Tokyo)* 13:163–172, 1979.

[Fis90] E. Fischer-Jørgensen. Intrinsic F_0 in tense and lax vowels with special reference to German. *Phonetica* 47:99–140, 1990.

[Fuj83] H. Fujisaki. Dynamic characteristics of voice fundamental frequency in speech and singing. In *The Production of Speech*, P. F. MacNeilage, ed. Springer, New York, 39–55, 1983.

[Fuj88] H. Fujisaki. A note on the physiological and physical basis for the phrase and accent components in the voice fundamental frequency contour. In *Vocal Physiology: Voice Production, Mechanisms and Functions*, O. Fujimura, ed. Raven, New York, 347–355, 1988.

[Gro90] N. Grønnum. Prosodic parameters in a variety of Danish standard languages, with a view towards Swedish and German. *Phonetica* 47:182–214, 1990.

[Gro92] N. Grønnum. *The Groundworks of Danish Intonation: An Introduction*. Museum Tusculanum Press, Copenhagen, 1992.

[Gro95] N. Grønnum. Personal communication. 1995.

[JWBK89] M. Jafari, K. H. Wong, K. Behbehani, and G. V. Kondraske. Performance characterization of human pitch control system: an acoustic approach. *J. Acoust. Soc. Amer.* 85:1322–1328, 1989.

[Kla87] D. H. Klatt. Review of text-to-speech conversion for English. *J. Acoust. Soc. Amer.* 82:737–793, 1987.

[Kra94] V. Kraft. Does the resulting speech quality improvement make a sophisticated concatenation of time-domain synthesis units worthwile? In *Proceedings, ESCA Workshop on Speech Synthesis*, New Paltz, New York, 65–68, 1994.

[Lad83] D. R. Ladd. Phonological features of intonational peaks. *Language* 59:721–759, 1983.

[MC90] E. Moulines and F. Charpentier. Pitch-synchronous waveform processing techniques for text-to-speech synthesis using diphones. *Speech Comm.* 9:453–467, 1990.

[Mob93a] B. Möbius. *Ein quantitatives Modell der deutschen Intonation – Analyse und Synthese von Grundfrequenzverläufen.* Niemeyer, Tübingen, Germany, 1993.

[Mob93b] B. Möbius. Perceptual evaluation of rule-generated intonation contours for German interrogatives. In *Proceedings, ESCA Workshop on Prosody*, Lund, Sweden, 216–219, 1993.

[MP92] B. Möbius and M. Pätzold. F_0 synthesis based on a quantitative model of German intonation. In *Proceedings ICSLP '92*, Banff, Alberta, Canada, 361–364, 1992.

[MPH93] B. Möbius, M. Pätzold, and W. Hess. Analysis and synthesis of German F_0 contours by means of Fujisaki's model. *Speech Comm.* 13:53–61, 1993.

[OL66] S.E.G. Öhman and J. Lindqvist. Analysis-by-synthesis of prosodic pitch contours. *Royal Inst. of Technology (Stockholm), STL-QPSR 4/1965*, 4:1–6, 1966.

[Pat91] M. Pätzold. Nachbildung von Intonationskonturen mit dem Modell von Fujisaki – Implementierung des Algorithmus und erste Experimente mit ein- und zweiphrasigen Aussagesätzen. Unpublished M.A. thesis, University of Bonn, 1991.

[Pie80] J. Pierrehumbert. *The Phonology and Phonetics of English Intonation.* MIT Press, Cambridge, MA, 1980.

[PSRSH92] T. Portele, B. Steffan, R. Preuss, W. F. Sendlmeier, and W. Hess. HADIFIX—A speech synthesis system for German. In *Proceedings ICSLP '92*, Banff, Alberta, Canada, 1227–1230, 1992.

[Tho79] N. Thorsen. Lexical stress, emphasis for contrast, and sentence intonation in Advanced Standard Copenhagen Danish. *Annual Report of the Institute of Phonetics*, Vol. 13, University of Copenhagen, 59–85, 1979.

Appendix: Audio Demos

The CD-ROM contains sound samples that correspond to the figures in the chapter as follows:

Figure 32.4, original contour — sound sample 1
Figure 32.5, original contour — sound sample 2
Figure 32.5, rule-generated contour — sound sample 3
Figure 32.6, original contour — sound sample 4
Figure 32.6, rule-generated contour — sound sample 5
Figure 32.7, original contour — sound sample 6
Figure 32.7, rule-generated contour — sound sample 7

33

Effect of Speaking Style on Parameters of Fundamental Frequency Contour

Norio Higuchi
Toshio Hirai
Yoshinori Sagisaka

ABSTRACT The authors have analyzed the fundamental frequency (F_0) contours of Japanese utterances in simulated conversations with four different speaking conditions—unmarked, hurried, angry, and gentle—for the conversion of speaking style. The parameters of the F_0 generation model proposed by Fujisaki—the minimum value of F_0 (F_{min}), the amplitude of the phrase commands (A_{p}), and the amplitude of the accent commands (A_{a})—were extracted and compared among speaking styles. Based on the results of the analysis of F_0 parameters, simple F_0 conversion rules from unmarked style to the other three speaking styles were made. By applying these conversion rules to unmarked-style speech, both with and without modification of segmental duration, speech samples converted to the other three speaking styles were synthesized and used for the subjective evaluation of speaking style. Conversion rules from unmarked style to hurried and gentle styles by the modification of F_0 and segmental duration was feasible, and in the case of hurried style conversion by the modification of F_0 alone was also valid. It was found that modifying the F_0 parameters and segmental duration were effective to express the difference due to speaking style except for the case of angry style. It was suggested that the difference in spectrum plays an important role, especially in the case of speech spoken in an angry style.

33.1 Introduction

The quality of synthetic speech has been improved by recent research [SKIM92, KHSY94], but most modifications were designed to simulate natural speech, such as that uttered by a professional narrator/announcer when he/she reads just one type of text. As a result, synthetic speech is still too monotonous to use in many man-machine applications, and control of speaking style to mimic conversation has not yet been achieved. The control of speaking style is one of the most interesting and attractive targets in speech synthesis for the next generation [KS93, VCM93, LCRC94, OVWCR94] and chapter 39 of this volume). This control will make synthetic speech more suitable for such practical applications as voice Q-A (question-answer) systems and speech translation systems.

Speaking style influences a range of acoustic features of speech including prosodics, and to a lesser extent, spectral features. The influence is, however, language dependent. This chapter characterizes the prosody characteristics of Japanese sentences by quantitatively analyzing the fundamental frequency contours and modifying them for the conversion from unmarked style to the other three speaking styles based on the F_0 generation model proposed by Fujisaki [FH84]. This model can reduce the number of parameters naturally, which makes the parameter conversion easier. More important, this model has the potential to express structurally different F_0-controls separately, permitting the results of the parameter analysis to be interpreted in a much simpler fashion. Results of an initial study on segmental duration are also presented.

33.2 Speech Material

Speech samples uttered by a male professional narrator in simulated conversation were recorded and analyzed. The speaking styles analyzed here are of the following four varieties, which were selected because they could be produced reliably and naturally by the narrator and because they were expected to have substantially different effects upon the acoustics.

- unmarked style (instruction-free style)
- hurried style
- angry style
- gentle style

The simulated conversations between the narrator and an interlocutor were performed twice in each speaking style, with speakers exchanging roles. These simulated conversations were carried out by assuming adequate situations for subjects to naturally utter in these different styles. The scenarios were as follows:

Customs A conversation between a tourist and a customs official at an international airport

Railroad station A conversation between a passenger and an employee at a railroad station

Post office A conversation between a customer and an official at a post office

There were 18 utterances in the first conversation, 12 in the second, and 19 in the third. From these, we excluded 6 from the first, 3 from the second, and 4 from the third because they were too short for F_0 parameter estimation (being interjections such as "*hai,*" meaning "yes"). Consequently, 35 utterances in each speaking style

TABLE 33.1. Sample sentences in subjective evaluation tests for conversion of speaking style.

No.	Pronunciation	Meaning	Length (Mora count)
1-1	nimotsuwa koredakedesuka	Is this all you have	11
1-2	shoHhiNmihoNwa arimaseNne	Do you have a sample of the commercial product	14
1-3	beQsoHno nimotsuwa arimasuka	Did you send anything separately	14
2-1	anoH, yotsuyamade ikuradesuka	Hmm, how much does it cost for Yotsuya	14
2-2	yotsuyaekimade ichimai kudasai	Please sell me a ticket for Yotsuya.	15
2-3	sokono ryoHgaekide seNeNsatsuwo hyakueNdamani kuzushitekara kaQte kudasai	Please buy a ticket after changing a one-thousand-yen bill into one-hundred-yen coins.	36

were analyzed, and 14 were excluded. The total number of utterances in the four different speaking styles analyzed here is 140.

Three sentences each of the scenarios *customs* and *railroad station* were used for the subjective evaluation tests, which are described in section 33.4.2. Speech sample number, pronunciation, meaning, and length of each are shown in table 33.1.

33.3 Analysis of Parameters of Fundamental Frequency Contour

33.3.1 Extraction of Parameters of Fundamental Frequency Contour

A quantitative analysis was performed using parameters of the F_0 generation model proposed by Fujisaki [FH84]. The observed F_0 contour can be decomposed into three components: the minimum value of F_0, the phrase component, and the accent component.

The minimum value of F_0 depends mainly on the size of the vocal folds, and in previous studies it was not clear whether the minimum value of F_0 was controlled consciously. The following analysis suggests that it should be adjusted to speaking style.

The phrase component is the response of the phrase command through a smoothing function, which can be approximated by an impulse-response of a critically damped system. The phrase command indicates the initial position of each phrase, and it is closely correlated to phrase structure.

The accent component is the response of the accent command through another smoothing function, which can be approximated by a step-response of a critically damped system. It carries information about pitch accent patterns, which contrast words in Japanese (see chapter 28, this volume), and also about whether each word is prominent in its sentence.

Parameters used here for the comparison of speaking styles are the following, called F_0 parameters in this chapter:

F_{min} : The minimum value of F_0

A_{p} : The amplitude of the phrase commands

A_{a} : The amplitude of the accent commands

Each F_0 parameter is determined by a semi-automatic analysis-by-synthesis method that consists of two stages, similar to the technique described by Hirai et al. in chapter 28. In the first stage, the timing parameters are decided based on labeled data, and the amplitude of the phrase commands are chosen so that the phrase component does not exceed the observed value of F_0. The difference between the phrase component and the observed F_0 value is calculated. Then, the amplitudes of the accent commands are determined by solving simultaneous linear equations so as to minimize the estimation error between the contour generated by the model and the above-mentioned difference. In the second stage, each parameter value is optimized using a hill-climbing method. The approximation of F_0 contours using the Fujisaki model was good in most of the sentences, as shown in figure 33.1.

33.3.2 Comparison of F_0 Parameters Among Speaking Styles

Figures 33.2, 33.3, and 33.4 show the parameter values of F_{min}, A_{p} and A_{a}, respectively. The vertical axis indicates the F_{min}, A_{p} or A_{a} values of the utterances spoken in three different speaking styles, and the horizontal axis indicates those of the same part of the same sentences in unmarked style.

As shown in figure 33.2, the F_{min} value is concentrated in a narrow range for each speaking style except for two utterances in unmarked style. The ranking of the average value of F_{min} (from highest to lowest) is angry, hurried, gentle, and unmarked.

The A_{p} value covers a wider range from the line of equivalence than that of A_{a} in all speaking styles, as shown in figures 33.3 and 33.4. The differences between these two figures suggest that the constraints on the accent command are stronger than those on the phrase command because the accent command is more strongly governed by lexical content than is the phrase command. Furthermore, the points

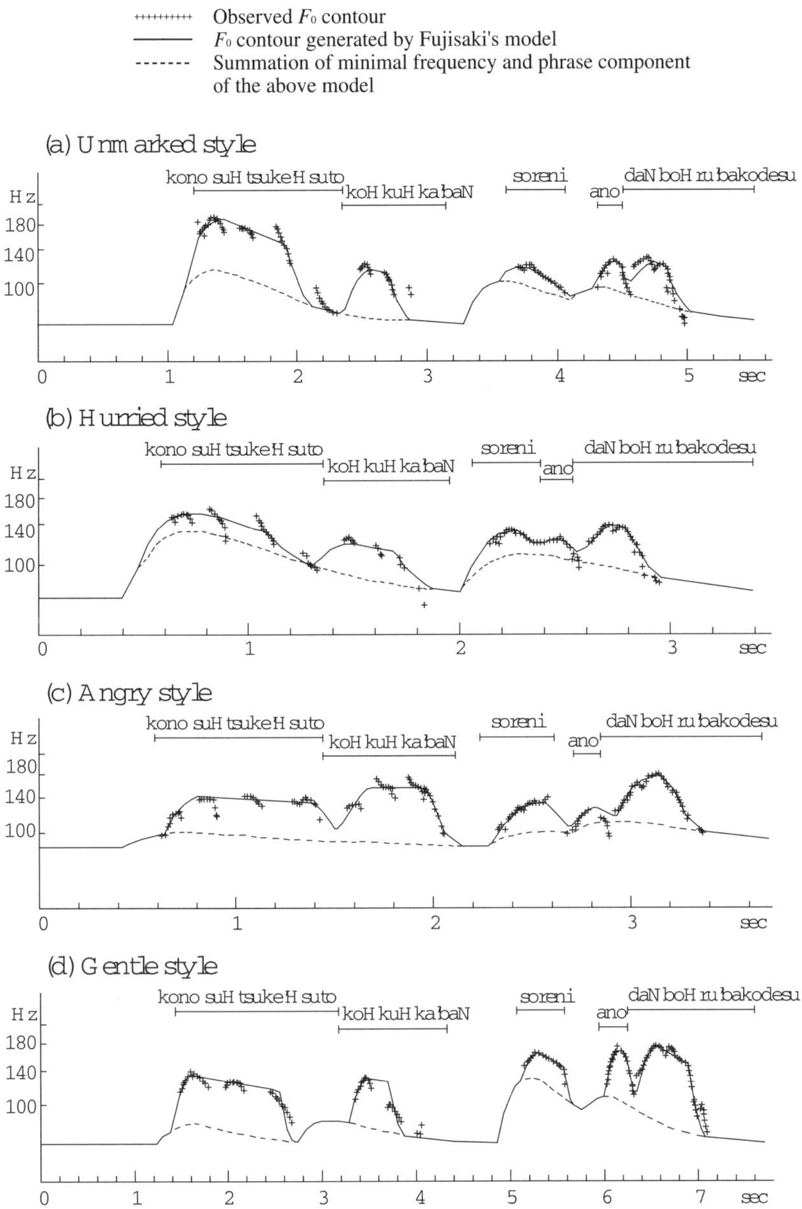

FIGURE 33.1. Example of approximation by Fujisaki's F_0 model. Each panel shows the F_0 contour and its best approximation using Fujisaki's F_0 model. The utterance is /kono suHtsukeHsuto koHkuHkabaN soreni ano daNboHrubakodesu/, meaning "(All I have are) this suitcase, a hand bag, and that cardboard box."

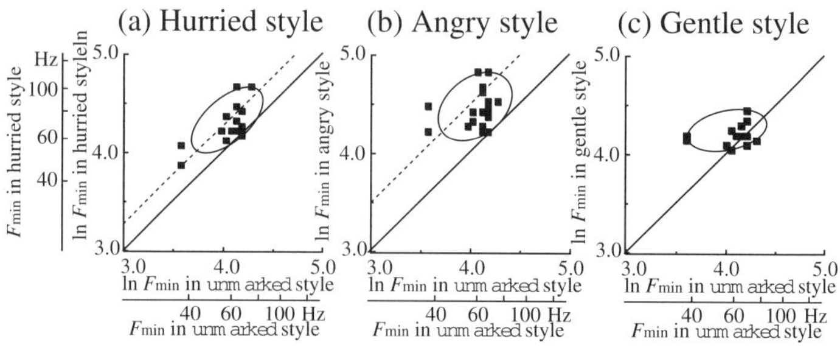

FIGURE 33.2. Extracted values of the logarithm of the minimal frequency. The dotted lines show the relation between the original parameter value extracted in unmarked style and the parameter value modified by conversion rule described in section 33.4.

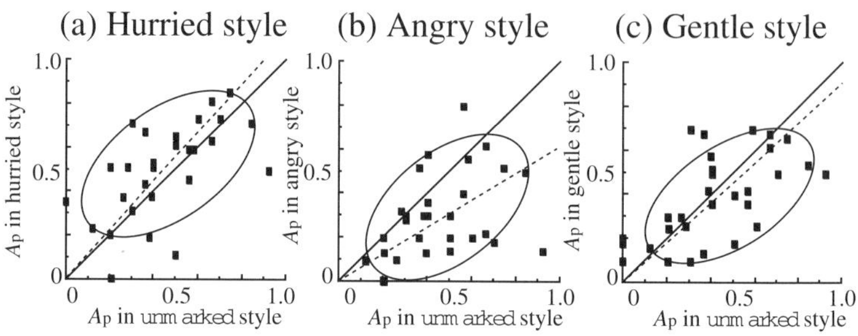

FIGURE 33.3. Extracted values of the amplitude of the phrase command. The dotted lines show the relation between the original parameter value extracted in unmarked style and the parameter value modified by the conversion rule described in section 33.4.

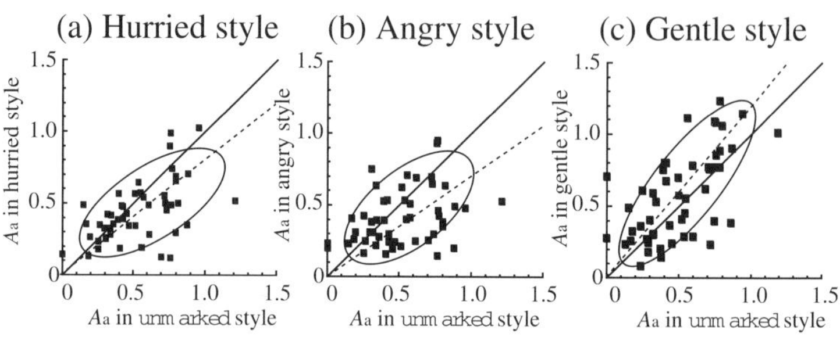

FIGURE 33.4. Extracted values of the amplitude of the accent command. The dotted lines show the relation between the original parameter value extracted in unmarked style and the parameter value modified by the conversion rule described in section 33.4.

on the horizontal and vertical axes in figure 33.3 indicate that phrase command was not found at the corresponding position in the utterance, whereas those in figure 33.4 indicate that the accent components were suppressed in the same word of the sentence. The position of the phrase command and the deletion of the accent component depend on the speech rate of the utterance. Although the utterance contents analyzed here are identical for the four different speaking styles, it was difficult to keep the same speaking style over all of the sentences.

The ranking of the mean of the amplitude of the phrase command (from highest to lowest) is hurried, unmarked, gentle, and angry, whereas the ranking of the mean of the amplitude of the accent command is gentle, unmarked, hurried, and angry.

In angry-style utterances, F_{min} is kept high, and the change due to both the phrase component and the accent component is very small. Consequently, the F_0 contours of angry-style utterances are flat. On the other hand, in gentle-style utterances, the dynamic range due to the accent component is greater than for the others, and in order to keep it high, the amplitude of the phrase component is accordingly suppressed. The F_0 contours of hurried-style utterances are similar except for the fact that the amplitude of the accent commands is slightly smaller than that of those spoken normally, perhaps because of the difference in speech rate.

33.4 Conversion of Speaking Style

33.4.1 Conversion Rules to Other Speaking Styles

In order to confirm the effectiveness of the above findings, simple F_0 conversion rules for speaking style were established by modification of the parameters of the fundamental frequency contour. The fundamental frequency contour was generated with the parameters derived from the conversion rules of F_0 parameters from unmarked style to the other three speaking styles. The rules are shown below.

Conversion rule for $\ln F_{min}$. The fixed conversion value was added to the extracted value of $\ln F_{min}$ of each sample sentence. $\Delta F_{min}^{hurried}$, ΔF_{min}^{angry} and ΔF_{min}^{gentle} indicate the fixed values for the conversions to hurried, angry, and gentle styles, which were 0.25 (4.33 semitones), 0.50 (8.66 semitones), and 0.00, respectively.

Conversion rule for A_p. The extracted A_p value of each sample sentence was multiplied by the fixed value for the conversion. $M_p^{hurried}$, M_p^{angry}, and M_p^{gentle} indicate the fixed values for the conversions to hurried, angry, and gentle styles, which were 1.1, 0.6, and 0.9, respectively.

Conversion rule for A_a. The extracted A_a value of each sample sentence was multiplied by the fixed value for the conversion. $M_a^{hurried}$, M_a^{angry}, and M_a^{gentle} indicate the fixed values for the conversions to hurried, angry, and gentle styles, which were 0.8, 0.7, and 1.2, respectively.

Segmental durations were modified based on labeled data from sample sentences spoken in different speaking styles. In this experiment, only the conversion of F_0 characteristics were tested, and original durational characteristics were simply copied.

33.4.2 Evaluation Tests of Converted Speech

The following utterances were used in subjective evaluation tests.

Original. The natural speech samples spoken in four speaking styles were used.

F_0 modification only. The speech samples resynthesized based on the F_0 parameters modified by the conversion rules described in section 33.4.1 were used with the analysis-synthesis speech sample spoken in an unmarked style.

F_0 *and duration modification.* The speech samples resynthesized based on the modified F_0 parameters and segmental duration measured in speech samples spoken in three different target speaking styles were used with the analysis-synthesis speech sample spoken in an unmarked style.

The procedure of the generation of speech samples used in the subjective evaluation tests of F_0 *modification only* and F_0 *and duration modifications* is shown in figure 33.5.

Three males judged which speaking style was represented by the samples. Each of the four styles were used five times. In order to decrease the differences in judgment due to the subject, the subjective evaluation test of *Original* was performed first. The listeners were given the following descriptions of the speaking styles.

Unmarked style. The standard speaking style, which is very easy to listen to.

Hurried style. The speaking style that gives the impression it was pronounced in a hurry.

Angry style. The speaking style that gives the impression the throat is always strained.

Gentle style. The speaking style that gives the impression it was pronounced in a relaxed manner.

33.4.3 Results of Evaluation Tests

Figure 33.6 shows the results of the subjective evaluation tests. The results indicate that it is possible to convert unmarked-style speech to hurried-style speech to some extent by modifying only the fundamental frequency contour without any change in segmental duration. However, in the case of the conversion from unmarked style to angry and gentle styles the rate of judgment of the expected speaking styles was not so high. Concerning *Sentence 2–1* spoken in unmarked and gentle styles, a large movement in the F_0 contour gave the subjects the impression of the angry style. *Sentence 2–1* was pronounced with hesitation, and accordingly the effect due to large movement in the F_0 contour was different from that in other cases.

By adding modification of segmental duration to that of the F_0 contour, most of the speech samples converted to hurried style and gentle style were judged as the expected styles. On the other hand, less than half of the samples converted

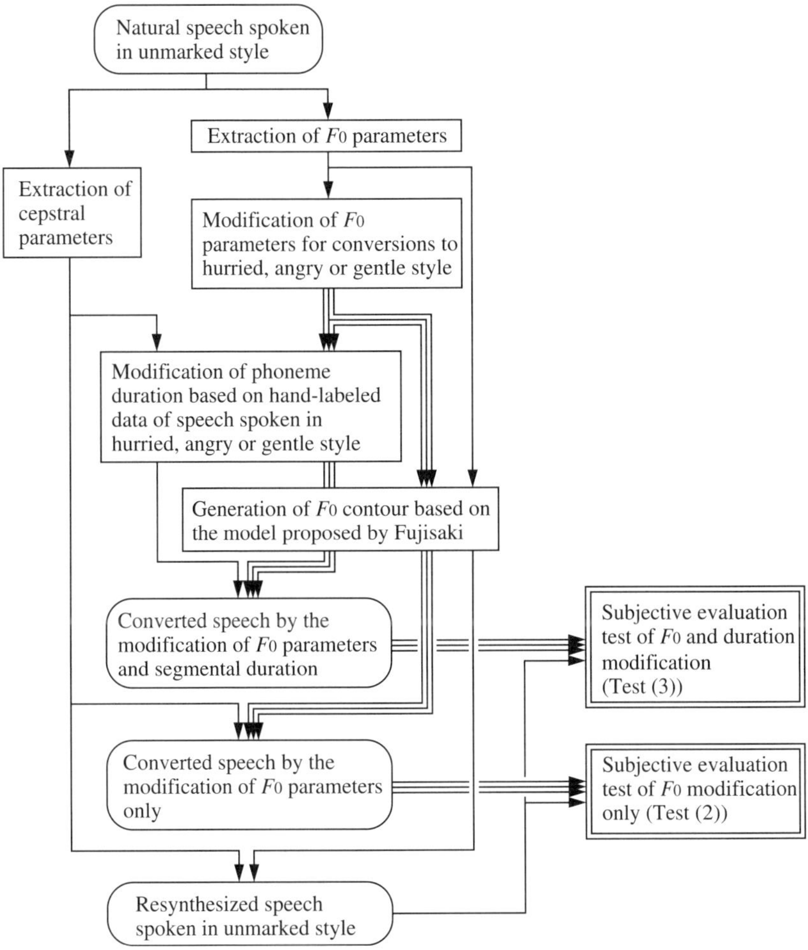

FIGURE 33.5. The procedure of the generation of speech samples used in the subjective evaluation tests of F_0 *modification only* and F_0 *and duration modifications*.

to angry style were judged as the angry style. The rest of the samples converted to angry style were judged as the hurried style because of the fast speech rate. It was found that conversion rules for F_0 parameters and segmental duration could express the difference among three speaking styles, unmarked, hurried, and gentle. But, it was also found that it was difficult to covert to the angry style using the proposed conversion rules. It was suggested that the difference in spectrum plays an important role, especially in the case of speech spoken in an angry style.

As shown in figure 33.6, the effectiveness of these conversion rules depends on the speech act, the sentence style, the length of a sentence, and so on. Therefore, further systematic analysis is needed.

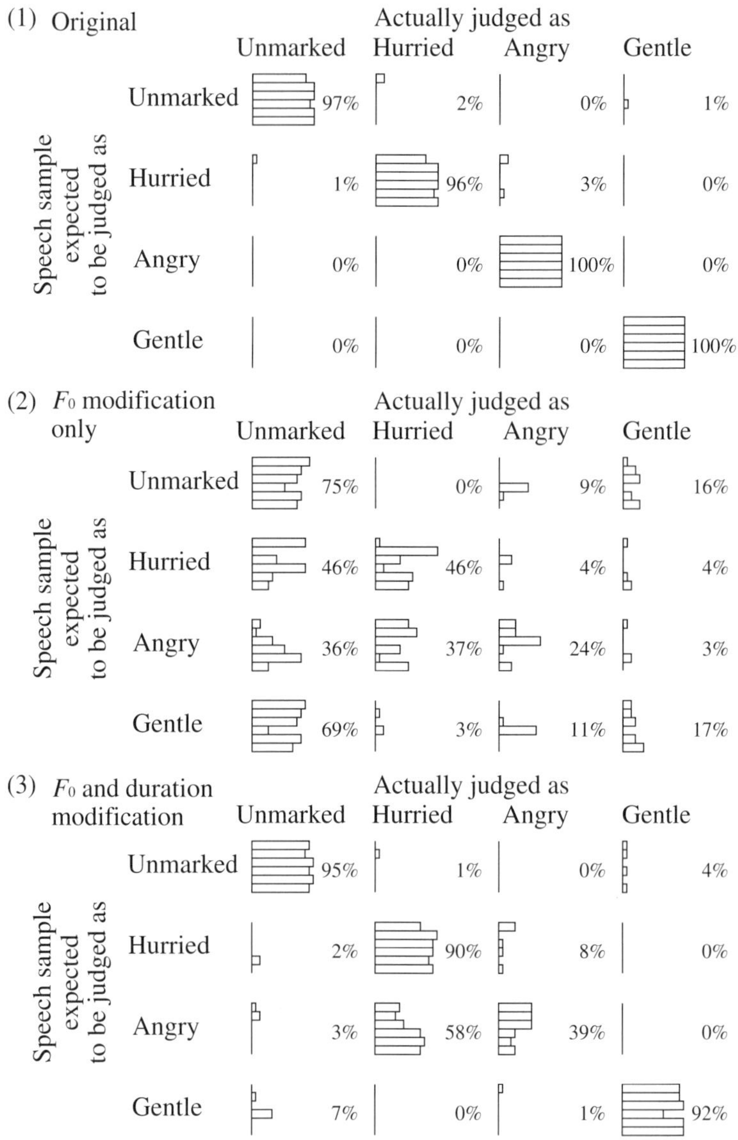

FIGURE 33.6. Results of the subjective evaluation tests for speaking styles. The upper, middle and lower panels indicate the rate of judgment in subjective evaluation tests using three sorts of manipulation, *Original*, F_0 *modification only* and F_0 *and duration modification*. Six bars included in the same row and the same column indicate the result of six sentences from *Sentence 1–1* to *Sentence 2–3*, respectively. Percentage scores after each set of six bars indicate the rate of judgment averaged across six utterances.

33.5 Summary

We have analyzed the F_0 contour of utterances spoken in four different speaking styles—unmarked, hurried, angry, and gentle—using the F_0 generation model proposed by Fujisaki, and established simple F_0 conversion rules from unmarked style to the other three speaking styles by modification of the F_0 contour. The results of subjective evaluation tests using converted speech samples indicate that it was possible to convert unmarked-style speech to hurried-style speech to some extent by modifying the fundamental contour. Most of the speech converted to hurried style and gentle style was judged as the expected style after modification of the fundamental frequency contour and segmental duration. Therefore, we conclude that modifying the F_0 parameters and segmental duration can express the difference due to speaking style, with the best results being for hurried and gentle speech.

However, the effectiveness of these conversion rules may depend on the content of utterance such as speech act, sentence style, and length of a sentence. Also, there may be several strategies to express the same speaking style, and the effectiveness also depends on the strategy selected by a speaker to express it. In order to realize more adequate control of speaking style, further research on the analysis of utterances spoken by different speakers is needed.

References

[FH84] H. Fujisaki and K. Hirose Analysis of voice fundamental frequency contours for declarative sentence of Japanese. *J. Acoust. Soc. Japan* 5(4):233–242, 1984

[KHSY94] H. Kawai, N. Higuchi, T. Shimizu, and S. Yamamoto. Development of a text-to-speech system for Japanese based on waveform splicing. In *Proc. ICASSP94*, Vol. I, 569–572, 1994.

[KS93] N. Kaiki and Y. Sagisaka. Prosodic characteristics of Japanese conversational speech. *IEICE Trans. on Fundamentals of Electronics, Communications and Computer Sciences* E76–A:1927–1933, 1993.

[LCRC94] D. Lambert, K. Cummings, J. Rutledge, and M. Clements. Synthesizing multistyle speech using the Klatt synthesizer. *J. Acoust. Soc. Amer.* 95(5):2979, 1994.

[OVWCR94] E. Ofuka, H. Valbret, M. Waterman, N. Campbell, and P. Roach. The role of F_0 and duration in signalling affect in Japanese: Anger, kindness and politeness. In Proc. ICSLP 94, Yokohama, Japan, 1447–1450, 1994.

[SKIM92] Y. Sagisaka, N. Kaiki, N. Iwahashi, and K. Mimura. ATR ν-Talk speech synthesis system. In *Proc. ICSLP92*, Vol. 2, 483–486, 1992.

[VCM93] J. Vroomen, R. Collier, and S. Mossiconacci. Duration and intonation in emotional speech. In *Proc. Eurospace'93*, 577–580, 1993.

Appendix: Audio Demos

There are twelve speech samples. The first four samples are natural spoken in each style. The following four are speech samples resynthesized based on the modification of F_0 parameters, and the last four are those resynthesized based on the modification of both F_0 parameters and segmental duration taken from natural speech.
Duration

1. Natural speech sample spoken in an unmarked style.
2. Natural speech sample spoken in a hurried style.
3. Natural speech sample spoken in an angry style.
4. Natural speech sample spoken in a gentle style.
   Duration
5. Analysis-synthesis speech sample spoken in an unmarked style.
6. Speech sample expected to be judged as a hurried style, resynthesized based on modified F_0 parameters.
7. Speech sample expected to be judged as an angry style, resynthesized based on modified F_0 parameters.
8. Speech sample expected to be judged as a gentle style, resynthesized based on modified F0 parameters.
   Duration
9. Analysis-synthesis speech sample spoken in an unmarked style.
10. Speech sample expected to be judged as a hurried style, resynthesized based on modified F0 parameters and segmental duration measured in speech sample spoken in a hurried style.
11. Speech sample expected to be judged as an angry style, resynthesized based on modified F0 parameters and segmental duration measured in speech sample spoken in an angry style.
12. Speech sample expected to be judged as a gentle style, resynthesized based on modified F0 parameters and segmental duration measured in speech sample spoken in a gentle style.
    Duration

Section VI

Synthesis of Prosody

34

Section Introduction. Text and Prosody

Sieb G. Nooteboom

34.1 Introduction

Work on text-to-speech systems is a challenging and interesting enterprise. For some time now, systems have been available that generate highly intelligible synthetic speech from unedited text. But with each new success, new and so far unanswered questions pop up. Such problems often require a closer look at the behavior of human speakers in controlled conditions in terms of the relations between underlying emotional, attitudinal, pragmatic and grammatical properties of the messages, on the one hand, and the characteristics of the output speech, on the other. This is particularly true for the domain of speech prosody. Prosody appears to be controlled by all kinds of higher-level information. Apparently, there is a technological desire to make machines speak like humans, with appropriate speech melodies and speech rhythms, in particular speaking styles, and expressing grammatical structures and referential relations in the text in the same ways that humans do. Because of this, there is a close and immediate link between text-to-speech technology and fundamental linguistic and phonetics research, each side profiting from the other. The chapters in this section on prosody in text-to-speech systems clearly combine technological goals with fundamental research interests.

34.2 Controlling Prosody in Text-to-Speech Systems

In phonetics the word "prosody" and its adjectival form "prosodic" are most often used to refer to those properties of speech that cannot be derived from the segmental sequence of phonemes underlying human utterances. Examples of such properties are the controlled modulation of the voice pitch; the stretching and shrinking of segment, syllable, and word durations; the intentional fluctuations of overall loudness; and articulatory precision and voice quality. On the perceptual level these properties lead, among other things, to perceived patterns of relative syllable prominences, perceived demarcation of coherent stretches of speech, attitudinal and emotional connotations, referential relations, and speaking style and speaker characteristics. A sophisticated development system for text to speech should have a host of abstract and less abstract representations for controlling speech prosody as well as facilities for trying out different rule sets and parameter settings. Published detailed descriptions of (parts of) such working systems are relatively rare because both writing them and reading them is not very exciting. Yet, it is a good thing to have them. Kohler is to be applauded for making his description of the prosodic part of the RULSYS/INFOVOX system for German available to us (see chapter 37). As far as someone who does not know the system intimately can judge, the description is neat and fair. Because it is a description and not a statement of scientific insights, there is little room for discussion or disagreement. This, apparently, has also been Kohler's intention, as there are hardly any explicit references to other models of prosody, either of German or of other languages. Chapter 37 will be particularly useful to those who are building such systems themselves and want to compare notes.

34.3 Controlling Speaking Styles in Text to Speech

Different kinds of messages are spoken or read aloud by human speakers in different speaking styles. Reading aloud from a novel requires a different way of speaking than reading an encyclopedia text, which is different again from sounding out advertisement texts. At the current state of the art in text to speech systems, it is reasonable that people involved in building text-to-speech systems are attempting to simulate such speaking styles convincingly. Chapter 39 by Abe describes a straightforward engineering approach to the problem of analyzing and synthesizing differences in speaking style, concentrating on the three speaking styles mentioned, as produced by a single professional narrator using the same language material. Acoustic analyses were made in terms of differences in three formant frequencies, in power, in spectral tilt, in the course of fundamental frequency, and in segment durations. Rules for generating different speaking styles were made on the basis of average speaking-style characteristics, and implemented in a text-to-speech system. The validity of the rules was perceptually tested in an ABX listening test, both with synthetic speech and with human speech artificially converted to another speaking style. A different sentence was used for A, B, and X. Given that the rules were based on rather rough, average measures of acoustic characteristics, the test turned out to be amazingly successful. It would be useful and interesting to see whether the same or a similar approach can successfully be applied to speaking styles that are wide apart than those mentioned. Can we, by applying this approach, find rules that make a machine speak like either a preacher or a newsreader or someone gossiping at the coffee table, or shouting or speaking very softly, etc.? Can we simulate not only differences between speaking styles from the same speaker but also differences between speakers?

34.4 Abstract Phonetic Structure and Phonetic Reality

In a particular branch of modern generative phonology, sometimes referred to as "prosodic phonology" (cf. [NV86]), the word "prosody" has taken on a meaning that differs from its meaning to phoneticians, as it refers to nonsegmental aspects of abstract linguistic structure, such as a particular type of constituent structure, often referred to as "prosodic structure" or "abstract prosodic structure," and the presence or absence of accents. Prosodic phonologists have argued convincingly for the need for an abstract prosodic structure, which can be derived from syntactic structure but is not identical to it. Prosodic structure is conceived of as a hierarchy of prosodic levels, from low to high: syllable, foot, prosodic word, clitic group, phonological phrase, intonational phrase, phonological utterance. The relations between phonological constituents are governed by the following four principles [NV86, p. 7]: (1) A given nonterminal unit of the prosodic hierarchy is composed of one or more units of the immediately lower category. (2) A unit of a given level of the hierarchy is exhaustively contained in the superordinate unit of which it is a part. (3) The hierarchical structures of prosodic phonology are n-ary branching. (4) The relative prominence relation defined for sister nodes is such that one node is assigned the value strong (s) and all the other nodes are assigned the value weak (w).

These principles imply, among other things, that prosodic structure contains no recursive constituents, an assumption sometimes referred to as the strict layer hypothesis. Furthermore, prosodic domains are subject to restructuring: Under certain conditions adjacent phonological phrases can be taken together to form a single new (and longer) phonological phrase, and an intonational phrase can be chopped up in maximally as many new (and shorter) intonation phrases as the number of phonological phrases contained within it. Because of the strict layer hypothesis, restructuring can never depend on syntactic categories.

These theoretical notions are of immediate interest to work on text-to-speech systems. If prosodic phonology is correct, a text-to-speech system should contain rules to derive from the input text those properties of syntactic structure that are needed to define an appropriate abstract prosodic structure. It should also contain rules for mapping syntactic structure into prosodic structure, and rules for restructuring, possibly as a function of such factors as accent position, constituent length, and required speech rate. Conversely, work on text-to-speech systems should be of immediate interest to prosodic phonology because they provide a testing ground for the relevance and correctness of the theoretical ideas involved. Potentially such questions come up as: Do we really need all the hypothesized levels to generate high-quality synthetic speech? Do we require other prosodic domains than those assumed in prosodic phonology? Is the strict layer hypothesis tenable, or do we need syntactic category information, for example, in prosodic restructuring rules? Two of the chapters on speech prosody—chapter 36 by Horne and Filipsson and chapter 38 by Marsi, Coppen, Gussenhoven, and Rietveld—are concerned with some aspects of the relation between abstract prosodic structure and speech data of some sort, in the context of working on text to speech.

Chapter 36 by Merle Horne and Marcus Filipsson, taking a lead from Nespor and Vogel [NV86], focuses on the enrichment of texts with abstract prosodic structure, needed for later generation of appropriate speech timing, speech pauses, and speech melodies. Their approach is common-sensical and modest in that they limit themselves to spoken texts of a particular type, Swedish stock market reports broadcast by radio. Also, the approach is data oriented, taking the relations between acoustic/phonetic patterns, on the one hand, and lexico-semantic and grammatical structure, on the other, as the raw data from which the properties of abstract prosodic structures may be inferred. In working with only three levels of prosodic structure—prosodic word, prosodic phrase, and prosodic utterance—these authors are unusually parsimonious. More levels will probably be needed in the end, when they continue modeling the prosodic aspects of real speech. The continuous back and forth between data and modeling in this chapter, gives, at least this reader, an interesting feeling of having a look in the kitchen where useful insights are being cooked up. A disadvantage of this style of presentation may be that it is not always immediately clear whether certain notions refer to the observable reality or to the model.

Chapter 38 by Marsi et al. focuses on a specific problem in the current approaches toward abstract prosodic structure. There frequently is an apparent mismatch between a so-called intonational phrase, as defined by Selkirk [Sel84] and by Nespor and Vogel [NV86], on the one hand, and actual intonational structure, as realized in F_0 patterning, on the other. More generally, "the stretch of speech covered by an intonation contour cannot consistently be identified with a particular prosodic constituent." The reason is that the intonational phrase is related to the pause structure of speech, and intonation contours often behave as if speech pauses have no relevance to them. The solution is to set up a new independent constituent over which a single intonation contour spreads. This new constituent is called the association domain. I wholeheartedly sympathize with this solution, first proposed by Gussenhoven [Gus91], although it seems somewhat unfortunate that the notion intonational phrase is reserved for something that has little to do with intonation, and that the notion association domain is reserved for what is covered by an intonation contour. The Marsi chapter further focuses on association domain restructuring, which, according to the results of a perceptual experiment, is related both to syntactic structure and to the length of the first of two association domains to be joined in a new one. Obviously, this restructuring cannot be expressed in more conventional prosodic constituency: the strict layer hypothesis is violated. This may be upsetting to some theorists, but examples, reasoning, and experimental results in this chapter are sufficiently convincing to be taken seriously.

References

[Gus91] C. Gussenhoven. Tone segments in the intonation of Dutch. In *The Berkeley Conference on Dutch Linguistics (1989)*, T. F. Shannon and J. P. Snapper, eds. University Press of America, Lanham, MD, 139–155, 1991.

[NV86] M. Nespor and I. Vogel. *Prosodic Phonology*, Foris, Dordrecht, 1986.

[Sel84] E. Selkirk. *Phonology and Syntax*, MIT Press, Cambridge, MA, 1984.

35

Section Introduction. Phonetic Representations for Intonation

Gérard Bailly
Véronique Aubergé

35.1 Introduction

Beside its intrinsic ability to help the listener in segmenting utterances into linguistically relevant parts of speech, intonation is an essential means for signaling *"how we feel about what we say, or how we feel when we say"* [Bol89, p.1]. How the cognitive representation of the utterance is encoded in the speech signal is still an open question. Two main approaches—bottom-up or top-down—may contribute to our understanding of how intonation contributes to both the capture of the overall meaning of the message and the speaker's position vis-à-vis its own discourse. A bottom-up approach aims to extract salient prosodic events with no linguistic a priori assumptions. Such a tentative approach, linking these events to phonological constructs, face the problems of automatic extraction [Mer93, HNE91], the perceptual relevance of these events [HCC90, HR94], and the coherence of labeling [BPOW92, Hd95]. A top-down approach aims to extract prosodic parameters that have sufficient statistical significance to carry given—contrastive—linguistic or paralinguistic tasks. This approach requires a careful design of large corpora. The resulting utterances should enable statistical analysis to extract significant prosodic correlates of underlying contrasts built in the corpus design. Careful design of the corpus and use of adequate statistical tools are the key problems of this top-down approach.

35.2 Building Phonological Representations with Bottom-Up Analysis

For the last twenty years, specific models have suggested phonetic constructs that could be the elementary contrastive elements on which phonological representations can be built. These phonetic models essentially concentrate on fundamental frequency (F_0). They incorporate global and local elements. Global elements are handled mostly at the utterance level. They include such constructs as declination lines [Pie79, Pie81, tC73] or tonal grids [Gar91]. These rigid templates could be replaced by a more flexible "down-stepping" and pitch reset (see chapter 37 by Kohler). Local elements code an F_0 curve into a sequence of pitch targets [Gar91, Hir83] connected by various interpolation functions, or pitch-pulses [FS71]. From a phonological perspective, these local pitch movements together with rhythmic patterns (pauses, syllabic lengthening) delimit phonological units such as accent, foot, prosodic phrase, or intonation phrase. Generative phonology thus offers a way to describe how these phonological units interact to produce acceptable language-specific intonation [PB88] and how it relates to linguistic descriptions [Sel84]. These constituents are assumed to also define the rhythmic structure of the utterance.[1] It has been shown that the rhythmic structure derived by psycholinguistic experiments [GG83, MG93] may carry fine details of the linguistic complexity [WSOP92], as emphasized by Hirst [Hir93, p.17]: *"Recent corpus-based studies of durational cues for prosodic constituency have suggested that more than two levels of intonation units can be systematically distinguished."* A solution consists of augmenting the abstract phonological representations with numerical values such as break indices [BPOW92] or stress levels (see chapter 37). In order to relate the variability of concrete phonetic representations describing pitch levels or movements—due to speech rate, speaking styles, or dialects—to these abstract phonological representations, different proposals have been made: Hirst [Hir91] has proposed linking the elementary tonal segments to high-order phonological constituents, whereas Gussenhoven [Gus91] has argued that the intonational domain of a pitch accent cannot be linked to any of the constituents, and introduces the *intonation domain* for each accent as "*an independent constituent over which a single intonation contour spreads*" (see chapter 38). These adaptations of surface phonological representations of the pitch curve and generation models are an indication that tonal segments are not independent and are often part of larger prosodic *contours* assuming different cognitive functions. Some stereotyped *melodic tunes* composed of a series of sustained pitches [Lad78] also called *cliché* tunes [FBF84] have been described in the literature. These tunes seem to function as a whole favoring a holistic analysis [HY91].

35.3 Building Phonetic Models with Top-Down Analysis

Another approach, introduced more recently in the field of speech synthesis, is to build and record large corpora based on strong linguistic or paralinguistic con-

[1] Note that so-called rhythmic structure mentioned above is often limited to the structural dominance within the utterance between accented syllables. Research [Cam92, BB94, Pas92] has shown that syllabic durations are organized in larger contrastive units within which gradual lengthening may occur.

straints. Statistical analysis can then be applied in order to capture prosodic regularities that arise from the implicit instructions (e.g., reading texts designed to cover the statistical range [Aub92, Tra92]) or by explicit instructions given to the speaker (e.g., different speaking styles for the same unmarked text). In this section, Abe presents in chapter 39 such a contrastive analysis. The same text is pronounced with three different speaking styles. Statistical analysis shows that styles differ in many dimensions including not only melodic contours and phonation/silence ratios, but also vocalic quality such as spectral tilt or formant trajectories.

Thus direct links between prosodic parameters and cognitive representation of the utterance controlled or biased by instructions can be established. More global phonetic models are implied by this approach: the kinematics of the F_0 curve is thus globally—and often directly—controlled by cognitive representations. Generation models with nonlinear dynamics [Sag90, Tra92, LF86] or superposition models ([Aub92], chapter 39 this volume). then convert cognitive representations into prosodic movements.

35.4 Phonetic Representations and Cognitive Functions

Both approaches may contribute to proposals and assessments of intermediate phonological representations that elucidate the real contrastive phonetic representations. We still lack significant psychoacoustic and perceptual data on these degrees-of-freedom, such as basic psycho-acoustic difference limens or perception thresholds [HR94], synchronization of prosodic trajectories [Koh90] (see for example the $\pm early$ and $\pm late$ distinctive features of the intonation component presented in chapter 37 by Kohler.

The question of how high-level phonological structures are encoded into prosodic trajectories is crucial for understanding how intonation is produced and identified in the communication process. In her introduction to the symposium Prosody in Situations of Communication at the Twelfth International Congress of Phonetic Sciences, Cutler noted that "*Two of the major uses of prosodic information in situations of communication are to encode salience and segmentation*" [CN91, p. 264]. In chapter 38 Marsi and colleagues study the influence of syntactic and phonotactic constraints on prosody as a cue for textual cohesion. In chapter 36, Horne and Filipsson present a computational framework for deriving prosodic structure using both syntactic and new/given information; complementary to other linguistic levels such as syntax, prosody contributes to "highlight" new information.

This suggests that listeners are able to clearly distinguish between prominence in accordance with "normal" linguistic structure and prominence that emphasizes speech segments. Consequently, listeners appear to have some expectations on what is a "normally" structured intonation for a given message and capture as salient something emerging from this basis or something that deviates from an expected pattern.

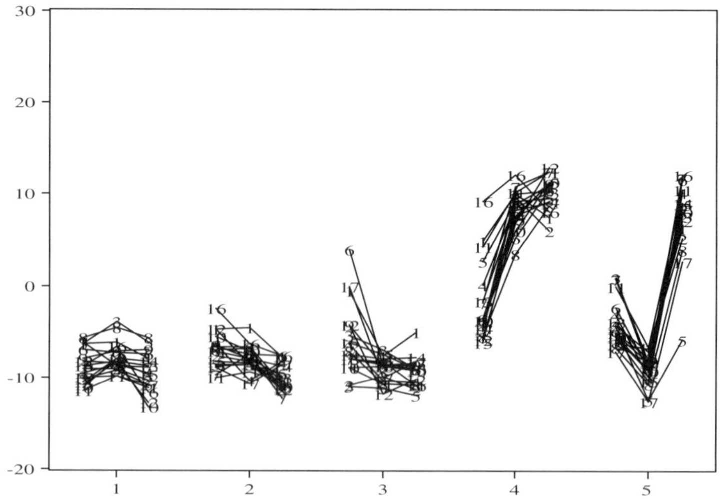

FIGURE 35.1. Superposition of melodic contours for utterances with five syllables typical of an incredulous interrogation [FBF84, p. 181]. The "accent" on the penultimate syllable is part of a global contour resisting variation of the syntactic and phonotactic substrate (from [MAB95]).

35.5 Prosodic Prototypes

Global approaches have been proposed previously in the literature. Fónagy states, for example, *"The opinion that any utterance is an instantaneous production . . . applies only to the domain of modern poetry. Except for this limited domain, our utterances are made of large prestored pieces"* (translated from [FBF84, p. 182]).

Books written by Fónagy [Fon82] and Bolinger [Bol81] are illustrated by multiple examples of such global intonational and rhythmic patterns as presented in figure 35.1. Does this suggest that such global patterns are instances of prosodic *prototypes*, which are the real basic elements of a prosodic *lexicon*? More experimental and psycholinguistic work is needed to assess such proposals. Some authors incorporate global contours in their phonetic model of melody. Eva Gårding models the F_0 as a superposition of global "slow" movements representing sentence or phrase intonation [Gar91] with local movements pertaining to an accent or tone. The first level is controlled by tonal grids connected by *pivots*. The second level is controlled by *turning points*. The tonal grid is thus controlled in a more complex way than two declination lines. This distinction between global and local models is also presented in [AS92], in which a simple superposition model sums two main contributions to the F_0 curve: (a) a global model assigns average an F_0 value for minor phrases and (b) a local model then assigns F_0 average values and slopes for each syllable within each phrase. Aubergé [Aub92] introduces a synthesis scheme by which the various linguistic levels (sentence, proposition, syntactic groups, and subgroups) are encoded into global prosodic multiparametric contours. Low-level contours modulate the higher-level contours. This latter approach differs from the preceding ones in the fact that no assumption is made a priori between levels in terms of rate of change (as, for example, the difference between cut-off frequencies of breath and word group commands in Fujisaki's model [FK88]): each syllable contributes to the encoding of each level and higher levels can use abrupt changes in intonation or rhythmic structure to encode cognitive representations.

35.6 Conclusions

Following Fónagy [Fon82] and Bolinger [Bol81], more attention should be paid in the future to prosodic means used by speakers to indicate how they "adhere" to their own discourse in order to emphasize global functions of prosody in the communication process and their interactions with salience and segmentation. The chapters in this section deal with different apsects of these interactions.

The links between generation and analysis models of intonation should also be reinforced in order to control efficient dialogs. Our assumption is that *holistic* analysis of prosodic representations could reinforce this link and explain why prosody can be memorized and identified precociously by the child, independently from any prelinguistic structuration [KM82, LT77].

References

[AS92] M. Abe and H. Sato. Two-stage f0 control model using syllable based f0 units. In *Proceedings, IEEE International Conference on Acoustics, Speech and Signal Processing*, vol. 2, 53–56, 1992.

[Aub92] V. Aubergé. Developing a structured lexicon for synthesis of prosody. In *Talking Machines: Theories, Models and Designs*, G. Bailly and C. Benoit, eds. Elsevier B.V., North-Holland, Amsterdam, 307–321, 1992.

[BB94] P. Barbosa and G. Bailly. Characterisation of rhythmic patterns for text-to-speech synthesis. *Speech Comm.* 15:127–137, 1994.

[Bol81] D. Bolinger. Some intonation stereotypes in English. *Studia Phonetica* 18:97–101, 1981.

[Bol89] D. Bolinger. *Intonation and its Uses*. Edward Arnold, London, 1989.

[Cam92] W. N. Campbell. *Multi-level Timing in Speech*. PhD thesis, University of Sussex, 1992.

[CN91] A. Cutler and D. Norris. Prosody in situations of communication: Salience and segmentation. In *Proceedings, International Congress of Phonetic Sciences*, vol. 1, Aix-en-Provence, France, 264–270, 1991.

[FBF84] I. Fónagy, E. Bérard, and J. Fónagy. Clichés mélodiques. *Folia Linguistica* 17:153–185, 1984.

[Fon82] I. Fónagy. *Situation et signification. Prolégomènes à un dictionnaire des énoncés en situation*. Benjamins, Amsterdam, 1982.

[FK88] H. Fujisaki and H. Kawai. Realization of linguistic information in the voice fundamental frequency contour of the spoken Japanese. In *Proceedings, IEEE International Conference on Acoustics, Speech and Signal Processing*, New York, 663–666, 1988.

[FS71] H. Fujisaki and H. Sudo. A generative model for the prosody of connected speech in Japanase. *Annual Report of Engineering Research Institute* 30:75–80, 1971.

[Gar91] E. Gårding. Intonation parameters in production and perception. In *International Congress of Phonetic Sciences*, vol. 1, Aix-en-Provence, France, 300–304, 1991.

[GG83] J. -P. Gee and F. Grosjean. Performance structures: A psycholinguistic and linguistic appraisal. *Cognitive Psychology* 15:418–458, 1983.

[Gus91] C. Gussenhoven. Tone segments in the intonation of Dutch. In *The Berkeley Conference on Dutch Linguistics*, T. F. Shannon and J. P. Snapper, eds. University Press of America, Lanham, MD, 139–155, 1991.

[Hd95] D. J. Hirst and A. di Cristo. *Intonation Systems: A Survey of Twenty Languages*. Cambridge University Press, Cambridge, in press.

[Hir83] D. Hirst. Structures and categories in prosodic representations. In *Prosody: Models and Measurements*, A. Cutler and A. Ladd, eds. Springer-Verlag, Berlin, 93–109, 1983.

[Hir91] D. Hirst. Intonation models: Towards a third generation. In *Proceedings, International Congress of Phonetic Sciences*, vol. 1, Aix-en-Provence, France, 305–310, 1991.

[Hir93] D. Hirst. Peak, boundary and cohesion characteristics of prosodic grouping. *Working Papers of Lund University* 41:32–37, 1993.

[HNE91] D. Hirst, P. Nicolas, and R. Espesser. Coding the f0 of a continuous text in French: An experimental approach. In *Proceedings, International Congress of Phonetic Sciences*, vol. 5, Aix-en-Provence, France, 234—237, 1991.

[HR94] D. J. Hermes and H. H. Rump. Perception of prominence in speech intonation induced by rising and falling pitch movements. *J. Acoust. Soc. Amer.* 96(1):83–92, 1994.

[HY91] J.-E. House and N. Youd. Stylised prosody in telephone information services: Implications for synthesis. In *Proceedings, International Congress of Phonetic Sciences*, vol. 5, Aix-en-Provence, France, 198–201, 1991.

[KM82] P. K. Kuhl and J. D. Miller. Discrimination of auditory target dimensions in the presence or absence of variation in a second dimension by infants. *Perception and Psychophysics* 31:279–292, 1982.

[Koh90] K. J. Kohler. Macro and micro f_0 in the synthesis of intonation. In *Papers in Laboratory Phonology*, J. Kingston and M. E. Beckman, eds. vol. I, Cambridge University Press, Cambridge, 115–138, 1990.

[Lad78] D. R. Ladd. Stylised intonation. *Language* 54:517–541, 1978.

[LF86] A. Ljolje and F. Fallside. Synthesis of natural sounding pitch contours in isolated utterances using hidden Markov models. *IEEE Transactions on Acoustics, Speech, and Signal Processing* 34:1074–1080, 1986.

[LT77] C. N. Li and S. A. Thompson. The acquisition of tone in Mandarin-speaking children. *J. Child Language* 4:185–200, 1977.

[MAB95] Y. Morlec, G. Bailly, and V. Aubergé Synthesis and evaluation of intonation with a superposition model. In *Proceedings, Eurospeech 95*, Madrid, 2043—2046, 1995.

[Mer93] P. Mertens. Intonational grouping, boundaries, and syntactic structure in French. *Working Papers of Lund University* 41:156–159, 1993.

[MG93] P. Monnin and F. Grosjean. Les structures de performance en français: caractérisation et prédiction. *L'Année Psychologique* 93:9–30, 1993.

[Pas92] V. Pasdeloup. Durée inter-syllabique dans le groupe accentuel en français. In *XIXe Journées d'Etudes sur la Parole*, 531–536, 1992.

[PB88] J. Pierrehumbert and M. Beckman. *Japanese Tone Structure*. MIT Press, Cambridge, MA, 1988.

[Pie79] J. Pierrehumbert. The perception of fundamental frequency declinaison. *J. Acoust. Soc. Amer.* 66:363–369, 1979.

[Pie81] J. Pierrehumbert. Synthesizing intonation. *J. Acoust. Soc. Amer.* 70(4):985–995, 1981.

[Sag90] Y. Sagisaka. On the prediction of global f0 shapes for Japanese text-to-speech. *Proceedings, IEEE International Conference on Acoustics, Speech and Signal Processing*, Albuquerque, NM, vol. 1, 325–328, 1990.

[SBPOW92] K. Silverman, M. Beckman, J. Pitrelli, M. Ostendorf, C. Wightman, P. Price, J. Pierrehumbert, and J. Hirschberg. Tobi: a standard for labeling English prosody. In *Proceedings of the International Conference on Spoken Language Processing*, 2:867–870, 1992.

[Sel84] E. Selkirk. *Phonology and Syntax*. MIT Press, Cambridge, MA, 1984.

[tC73] J. t' Hart and R. Collier. Intonation by rule: A perceptual quest. *J. Phonetics* 1:309–327, 1973.

[Tra92] C. Traber. F0 generation with a database of natural fo patterns and with a neural network. In *Talking Machines: Theories, Models and Designs*, G. Bailly and C. Benoit, Elsevier B.V., 287–304, 1992.

[WSOP92] C.W. Wightman, S. Shattuck-Hufnagel, M. Ostendorf, and P. J. Price. Segmental durations in the vicinity of prosodic boundaries. *J. Acoust. Soc. Amer.* 91(3):1707–1717, 1992.

36

Computational Extraction of Lexico-Grammatical Information for Generation of Swedish Intonation

Merle A. Horne
Marcus K. D. Filipsson

ABSTRACT By using a minimal amount of syntactic information in combination with information on lexical word class and morphological and lexico-semantic coreferential (identity of sense) relations, it is possible to generate an appropriate prosodic structure for Swedish texts. The structure of the algorithms involved in the prosodic preprocessing are presented. These include a referent tracker, a word-class tagger, a complex-word identifier, a clause boundary identifier, and a prosodic constituent parser.

36.1 Introduction

One of the goals of current research in text-to-speech systems is to improve the quality of intonation by developing algorithms for preprocessing texts in order to extract grammatical and discourse information necessary for the generation of appropriate prosodic patterns. This chapter describes the work we are currently carrying out aimed at using the information on coreferentiality obtained from the referent-tracking algorithm previously developed (see [HFLL93]) together with further information on lexico-syntactic category designation to group words together into a hierarchy of prosodic constituents. Whereas the referent-tracking process is important to the F_0-generating component in order to be able to predict the distribution of focal and nonfocal accents, information on prosodic structure is needed in order to be able to increase the textual cohesion by grouping words together into grammatically and semantically defined units whose boundaries are associated with specific prosodic correlates. One problem with the current synthesis of Swedish intonation, which has not included rules for generating prosodic structure, is that prosodic phenomena associated with the boundaries of clause-internal word groups cannot be modeled. For example, the transitions generated between word accents do not always correspond to those one finds in naturally occurring speech. As figure 36.1 shows, the end of the focused expression *för närvarande* "presently" coincides with a low F_0 point in the speech of the radio broadcaster we are modeling. This point, we claim below, corresponds to the end of the prosodic constituent, which we define as a [+focal] prosodic word. In figure 36.2, the synthetic F_0 curve generated using the current rule system cannot reproduce this pattern because no low point after the focal high is predicted in clause-internal position. The form of the F_0 transitions depends on the position of the word accents, which can be either focal (i.e., followed by a H) or nonfocal (i.e., without an additional following H). Thus, after the H*L word accent on the syllable *-när-*, there is a rise throughout the remainder of the word (due to an associated focal H) and the first syllable of the following word, *betecknas* is characterized, because the underlying accent pattern of an accent 1 word such as *betecknas* is HL*, with an H on the pre–main-stress syllable *be-* and an L* on the syllable *-teck-*. Thus, the low point at the end of *för närvarande*, such as in figure 36.1, cannot currently be generated.

Another problem with current synthesis is that one has not been able to predict the location of clause-internal prosodic phrase boundaries, that is, internal boundaries that are as strong as those that occur at the end of the majority of clauses/sentences. For example, figure 36.3 shows the F_0 contour associated with the sentence in (1), in which the internal boundary after *räntepunkt* (interest point) has the same strength as that after *procent* (percent). In order to be able to recognize such internal prosodic phrase boundaries, one must have access to more lexico-grammatical information than is currently available in text-to-speech systems.

(1) *Tolvmånaders statsskuldväxlar hade gått tillbaka 1 räntepunkt || till 10,58 procent || medan sexmånadersväxlar gått upp 5 punkter till 10,50 procent.*
"Twelve-month state-debt bonds had gone back 1 point to 10.58 percent while six-month bonds had gone up 5 points to 10.50 percent"
(where || represents a prosodic phrase boundary)

Following an approach similar to [BF90], [QK93], we have been conducting research to determine how one, using a minimal amount of syntactic parsing, can obtain enough information to construct a hierarchical prosodic structure for each sentence in a text. Unlike other researchers, however, we are also using contextual information such as coreference/identity relations in our approach to generating prosodic structure. This information is crucial, not only for accent assignment but also for the grouping of words into larger prosodically well-formed constituents.

36.2 Swedish Prosodic Structure

36.2.1 Prosodic Word

At least three levels of prosodic structure are required for Swedish in order to model all the prosodic information observed in our data [Hor94]. The smallest of these is the prosodic word, which we define as corresponding to a content word and any following function words up to the next content word within a given prosodic phrase. At the beginning of a prosodic phrase, the prosodic word can also begin with one or more function words. The prosodic word is characterized by a word accent

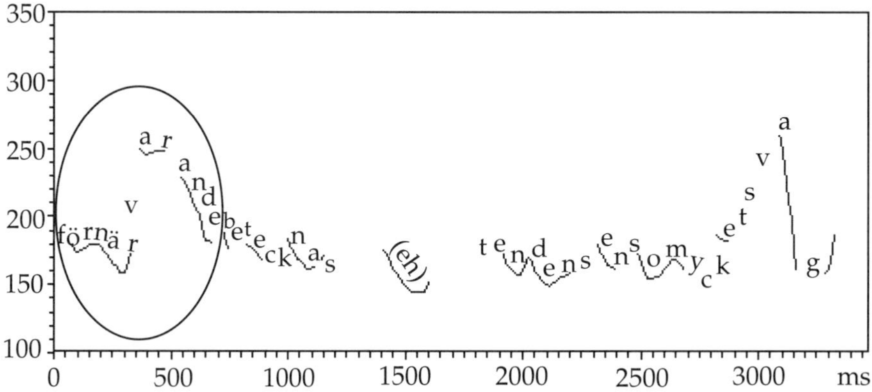

FIGURE 36.1. F_0 contour for the sentence *För närvarande betecknas tendensen som mycket svag* "At present the trend is characterized as very weak" produced by a professional radio reporter.

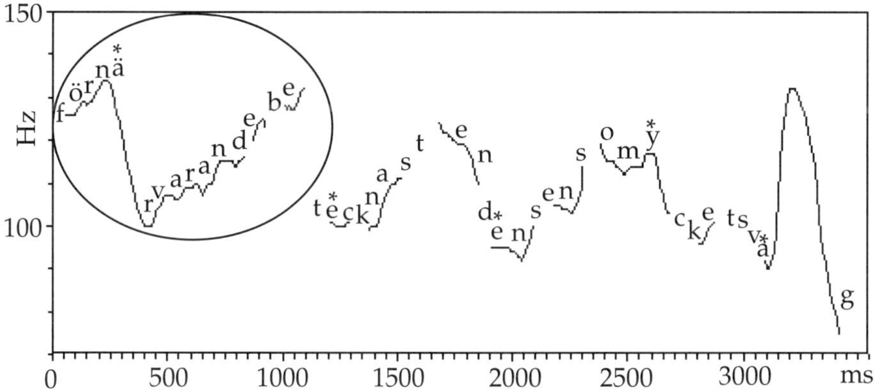

FIGURE 36.2. Synthesized F_0 contour for the same sentence as in figure 36.1.

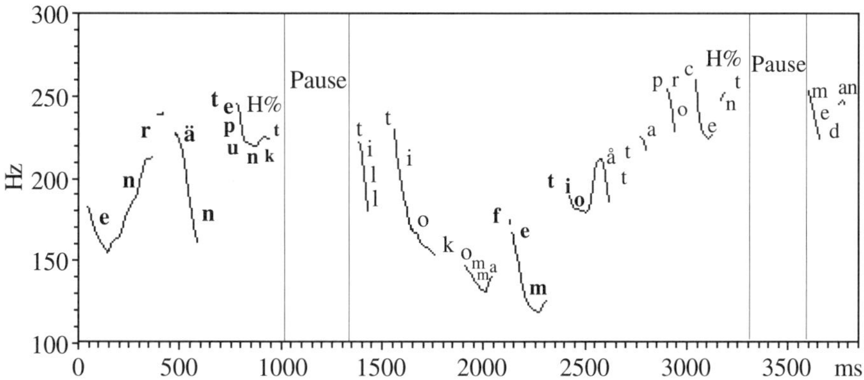

FIGURE 36.3. F_0 contour for a fragment of the sentence in (1) (*en räntepunkt* || *till 10,58 procent* || *medan*) with a clause-internal prosodic phrase boundary after en *räntepunkt.*

and potentially a focal accent (accent 1= HL*(H^-L^-); accent 2 = H*L(H^-L^-)). (We use H^- and L^- to represent, respectively, a focal high and the low tone accent following a focal high in order to distinguish them from the H and L associated with the word accents.) It is also marked by a boundary tone, which is realized by a final rise in the case when the content word is not focused (i.e., contextually given) (H#) or a fall when the content word is focused (L#). These boundary tones, we claim, play an important role in creating the transitions between consecutive prosodic words in a larger prosodic phrase. They are also points for potential pauses, for example, before focused content words (see [Gar67], [Str93]). The unit does not necessarily correspond to a syntactic constituent as the example in (2) illustrates ("—" represents the boundary between prosodic words). This type of nonsyntactic but rhythmic grouping is perhaps more characteristic of well-planned read texts or spontaneous speech than of non–well-planned texts read, for example, by a nonexpert/nonprofessional (see also [Cam93] for evidence of similar groupings in English).

(2) Kurserna på — Stockholmsbörsen — fortsätter att — falla.
Rates(det) on — Stockholm Stock Exchange(det) — continue to — fall
"Rates on Stockholm's Stock Exchange continue to fall"

Figure 36.4 illustrates the prosodic structure of (2) produced by the female speaker whose prosody we are modeling. She is an "expert" professional speaker who has detailed knowledge of the domain about which she is talking (stock market). The well-planned impression her speech gives probably results both from this fact and from her long experience as the principal reader of stock-market reports on Radio Sweden (she retired in 1992).

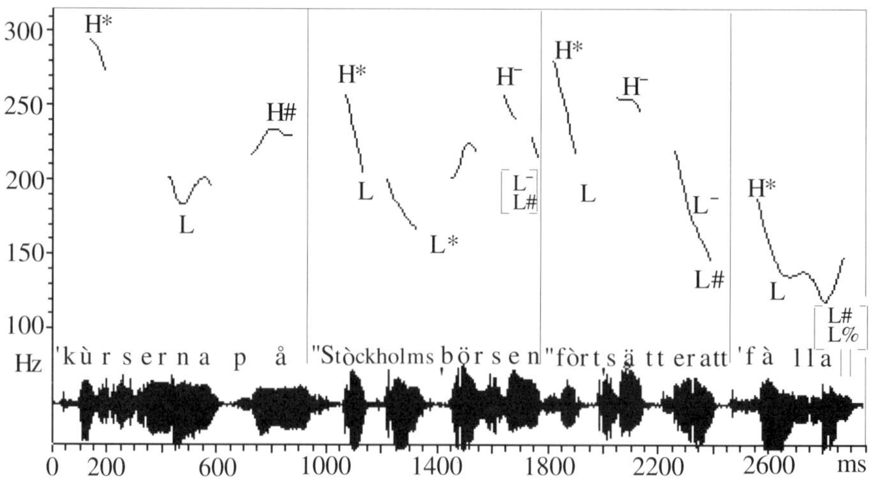

FIGURE 36.4. F_0 contour corresponding to the sentence in (2). Vertical lines correspond to prosodic word boundaries represented by L#; L% represents a prosodic phrase boundary.

36.2.2 Prosodic Phrase

One or more prosodic words make up a prosodic phrase, which is marked by a L% or H% boundary tone accent and a following pause. Factors that determine the location of prosodic phrase boundaries include the following:

Clause/sentence boundary: A clause boundary (see section 36.3 for identification of clause boundaries) corresponds in many cases to the end of a prosodic phrase. In an auditory analysis of a corpus of 36 radio broadcasts containing 724 clauses, where clauses also included elliptical clauses, 499, or 69%, of these were deemed to end in a boundary that was as strong or stronger than a prosodic phrase (404 were classified as prosodic phrase boundaries and 95 as prosodic utterance boundaries). Because we assume the strict layer hypothesis in the hierarchy of prosodic constituents, this means that the end of a prosodic utterance is also the end of a prosodic phrase; thus, 69% of the clauses ended in a boundary associated with a prosodic phrase at some level of analysis. Furthermore, 337 of these clause boundaries corresponded to sentence boundaries. In the whole corpus, there were 362 sentences. This means that 93% of the sentence boundaries were assigned a prosodic boundary equal to a prosodic phrase on some level of analysis.

Clause-internal prosodic phrase boundaries: Although prosodic phrase boundaries occur in the majority of cases in clause-final position, they may also occur optionally in clause-internal position. In our data of 724 clauses, we detected only 17 clause-internal cases (2%). Although the number was extremely small, we decided, nevertheless, to examine the lexico-syntactic structure of our data to determine whether one could make any generalizations concerning the environment for the insertion of these internal prosodic phrase boundaries. The following conclusions can be drawn from the investigation: In the domain-specific database dealing with the stock market, 12 clause-internal prosodic phrase boundaries occurred before focused complements (beginning with *till* (to) or *sedan* (since) to the verbs *gå upp* (go up), *gå ner* (go down), *falla* (fall), *stiga* (rise) if and only if these complements were preceded by another focused verb complement. Thus a prosodic phrase boundary (||) could occur before *till* in (3a) but not in (3b), where the first complement following the verb is not focused (relevant focused expressions are written in bold type):

(3) (a) *Fyra-åriga standardobligationen hade då fallit* **4 punkter** || *till en ränta på* **10,27 procent** ||
"The four-year standard bond had fallen **4 points** || to an interest rate of **10.27 percent** ||"
(b) *Tolvmånaders statsskuldväxlar hade också gått tillbaka 4 punkter till* **10,84 procent** ||
"Twelve-month national-debt bills had also gone back 4 points to **10.84 percent** ||"

The remaining five cases of clause-internal prosodic phrase boundaries occurred between a relatively long subject (on the average of 15 syllables) and a focused verb, as for example in (4):

(4) *Då hade den fyra-åriga standardobligationen* || **gått tillbaka 2 punkter sedan gårdagens slutnotering**
(word-for-word translation) "Then had the four-year standard bond || **gone back 2 points since yesterday's closing quotation**"

Length: A prosodic phrase probably will not exceed x syllables at a given rate of speech y, because prosodic phrases (at least in our data) correspond to what is often termed "breath groups" [Lie67]. In our material, where the average speech rate is about 5 syllables/second, prosodic phrases contained between 7 and 63 syllables, with the mean at 24 syllables (SD=10.3). Our data also suggest that a sentence-internal clause must be of a certain length in order for it to be associated with a prosodic phrase boundary; clauses containing less than 7 syllables in our database were assigned a weaker, that is, prosodic word boundary.

New/given distinction: A prosodic phrase must contain at least one focused prosodic word. This follows from the fact that a prosodic phrase is associated with a prominent phrase accent (L% or H%) and those are intimately tied to a preceding focal word accent. Thus, a string of words that constitutes a potential prosodic phrase but lacks a focused word will be attached to the end of a preceding prosodic phrase. This restriction is evident in our data only in cases involving clause-internal prosodic phrase boundary placement because all clauses contained at least one focused word. Thus in the first clause in (5) a clause-internal prosodic phrase boundary (||) is possible before *till* because 10,90 constitutes new information, but not in the second clause beginning with *medan* (while), where there is no new information after *till*:

(5) *Tolvmånaders statsskuldväxlar hade gått upp* **2** *punkter* || *till en ränta på* **10,90** *procent* || *medan sexmånadersväxlar stigit* **6** *räntepunkter till 10,90 procent* ||
"Twelve-month national-debt bills had gone up **2** points || to an interest rate of **10.90** percent || while six-month bills rose **6** interest points to 10.90 percent ||"

36.2.3 Prosodic Utterance

One or more prosodic phrases make up a prosodic utterance, which is bounded by extended pauses. These strong boundaries coincide with locations where a topic shift takes place (i.e., the end of a discourse segment [GH92]). In our data, 95 of the 727 clauses (13%) ended in a boundary that was classified as a prosodic utterance boundary. In the texts that were originally read on the radio, these correlate with the opening of a new paragraph immediately following the prosodic utterance boundary.

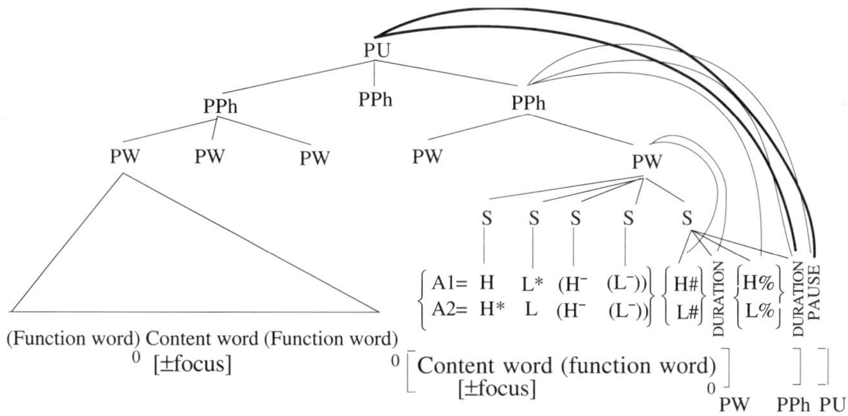

FIGURE 36.5. Schematic presentation of the prosodic hierarchy assumed for Swedish and the associated phonetic correlates. Accent 1 is represented as HL*(H⁻L⁻) and accent 2 as H*L(H⁻L⁻), where (H⁻L⁻) represents the focal high (H⁻) and potential low (L⁻) associated with the focal accent. H# and L# represent the prosodic word boundaries. H% and L% designate prosodic phrase boundaries. [PW=prosodic word, PPh=prosodic phrase and PU=prosodic utterance. (function word)$_0$ stands for zero or more function words.]

36.2.4 *Final Lengthening and Silent Intervals*

It is has been hypothesized that each prosodic constituent is characterized by a certain amount of final lengthening [GR92], [WSOP92]; in a recently completed study [HSH95] on Swedish, however, there was no evidence found for final lengthening below the level of the prosodic phrase in radio announcer–read speech. Final lengthening was observed to affect the final segment of the rhyme. A more or less gradual increase in the duration of the silent interval upon increase in the rank of the boundary was also observed. The negative correlation between silent interval duration and final lengthening reported in Fant and Kruckenberg's study of rhythmical prose reading [FK89] was found only at the lower-ranked boundaries (PW and PW-internal).

Figure 36.5 presents in schematic form the prosodic constituents assumed for Swedish and their phonetic correlates. The tone accents (H and L) are assumed to be associated with syllables (S) according to principles outlined in [Bru77]. It is also assumed that the realization of the tone accents is dependent to some extent on the number of syllables present in a particular word; that is, the number of syllables in a given word dictates to a great extent how many tones will be realized phonetically.

36.3 Design of the Prosodic Structure Component

In order to construct these prosodic constituents automatically, a number of different analyses are required. The present system (see figure 36.6) is based on a strictly modular approach, with each module having well-defined input/output formats. This will enable us to easily replace a module with a new one if a more efficient algorithm is developed at a later stage.

The first task is to tokenize the text into a list of words. This is performed by a module we call *explode*. At the same time, punctuation marks and paragraph boundaries (i.e., carriage returns) are recognized.

The next step is to look up the words in our domain-specific lexicon, which is an expanded subset of a larger computerized lexicon [HJL87]. The lexicon handles inflections, derivations, and compound words. In the look-up process, performed by the *stemfinder* module, inflections and derivations are decomposed into morphs, and the stem of the word is found. For compounds, each component word is handled separately and here also the stem is found for each word. The words are further labeled with one or more lexico-syntactic tags.

The *referent-tracker* module checks each word for its possible coreference with any previously mentioned word within an adjustable window (see [HFLL93]). In this case, the window has been chosen to be 60 words, but other domains are possible, such as, the paragraph (cf. [Hir90]). In the output from the *referent-tracker*, each word is marked as either new (N) or given (G). The given words are also marked with a letter indicating why they were considered given, and in most cases a number showing which earlier word triggered the given status. The referent-tracking algorithm consists of four parts: The first one tracks and marks words that are situationally/pragmatically given, such as *börs* (stock-market), and *krona* (crown) in the domain we have studied. These words are marked Gw. The second stage involves identifying cases of coreference due to reiteration of root morphs (or stems), as obtained from the earlier analysis using the lexicon. Given words detected in this manner are tagged with Gi and a number indicating with which earlier word the current word has a common stem. The third part uses domain-specific synonymy relations to identify cases of coreference. Examples from the stock-market domain are: *kurs–nivå* (rate) and *aktie–papper* (share). Synonyms are labeled with Gs and a number for identifying which word is the earlier-mentioned synonym. The fourth and final stage tracks words that are involved in hierarchical identity relations (hyponymy, part/whole relationships). In order to do this, hierarchical relations have been modeled using "is an example of" or "is a part of" pointers to establish the relation between pairs of lemmas, thus building up a forest of hierarchical, multibranch trees. These words are tagged with Gh and a number indicating the earlier word in relation to which the current word is a superordinate term or hyperonym.

The *stemfinder* module (see above) generates multiple tags for some words. For example, the word *som* (as), which is tagged both as a nonclausal conjunction and as a relative pronoun. The next step, performed by the module *tagger*, is therefore the disambiguation of these tags. In this endeavor, we are currently testing the performance of a stochastic parser based on lexical and sequential occurrence probabilities as well as overall tag probability [Eeg91]. The algorithm implements a first-order Markov chain and uses dynamic programming to estimate the best hypothesis for the whole sentence. A set of approximately 30 lexico-syntactic

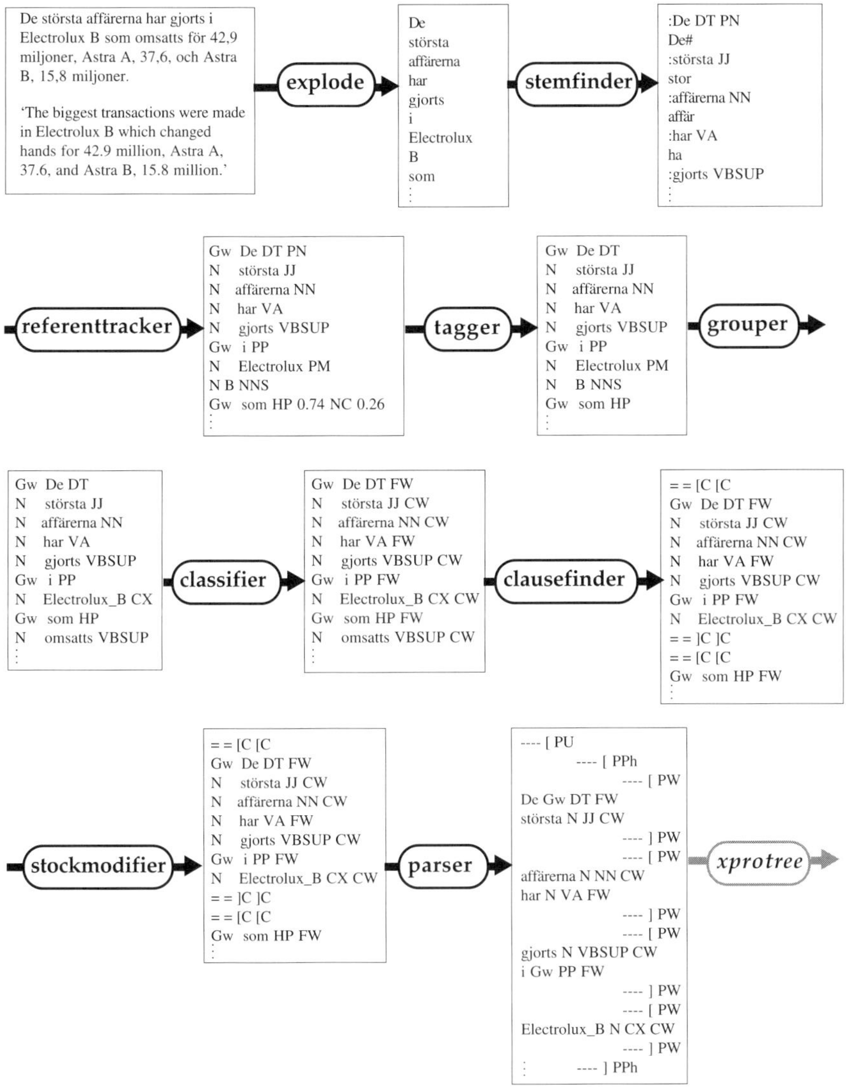

FIGURE 36.6. Schematic presentation of the present computer system for prosodic parsing. The modules in the system are represented within rounded corner rectangles. An excerpt of the output of each module (and consequently the input to the next module) is shown between each pair of modules. The first input is the stock-market report newspaper text; the final output is the prosodically parsed text. The very last module, *xprotree*, is not an actual part of the algorithm but a module for interpreting the output and displaying a graphic representation (see figure 36.7).

tags based the tag set of Ejerhed et al. [EKWA92] have been chosen to train the system. Domain-specific considerations also lead to the introduction of a number of unconventional tags, for example "specifer" nouns and adjectives that occur after the head noun in complex proper names such as *B fria* in the name *Electrolux B fria* (Electrolux B free (shares)). The tags have been further assigned to the words' lemma representations in the computerized lexicon, thus allowing recognition of all morphologically derived forms of a given head-word. Preliminary results indicate that the current algorithm works quite well, but we intend to compare it with other approaches. One involves a hidden Markov model such as in the Xerox part-of-speech tagger [CKPS92], which is currently being adapted to Swedish [LHSLF95].

After word classes are determined, the next stage (the *grouper*) is to recognize complex words, that is, strings of content words that function as a single prosodic unit. In the stock-market domain, these correspond to proper names (i.e., company/bank names and stock designations, such as *Avesta Sheffield*, *S-E Banken*, *Hennes & Mauritz*, and *Hasselfors Förvaltnings AB*). These strings are assigned a specific tag (CX–complex word), which, although it is not a lexical tag, is a member of the class of content word tags together with those associated with nouns, adjectives, verbs, adverbs, and so on.

The following stage, the *classifier* module, involves classifying each word as either a content word (CW) or a function word (FW). The assignment of words to one of these classes is not always straightforward, but in general, content words include the traditional categories of nouns, verbs, adjectives, adverbs, and numerals, whereas function words consist of prepositions, pronouns, determiners, auxiliary verbs, interrogative/relative adverbs, deictic adverbs, quantifiers, and so forth.

The next step is to recognize clause boundaries because the clause is the basic domain over which prosodic phrases are defined. This is performed by the *clausefinder* module. Clause boundaries occur at certain punctuation marks (e.g., full stop, colon, semicolon, some commas (those not occurring in lists of words having the same word class)), as well as before coordinate conjunctions (*och* (and), *men* (but)) and relative pronouns (e.g. *som* (that, who) and after subordinate conjunctions (*att* (that), *om* (if)).

In order to generate the clause-internal prosodic phrase boundaries, which, as we mentioned above, were optional and had a low frequency of occurrence, we decided to include a domain-specific module, *stockmodifier* in the prosody component. This was due to the fact that the locations of clause-internal prosodic phrase boundaries seemed to be quite domain-specific as regards their lexical specification. This is not the case with *clausefinder*, which is domain-independent. Thus, the *stockmodifier* module inserts clause-internal boundaries before the second focused prepositional complement to the verbs *gå upp* (go up), *gå ner* (go down), *falla* (fall), and *stiga* (rise). These internal boundaries are equated with clause boundaries in subsequent prosodic parsing. Notice also that the *stockmodifier* module is optional and the *parser* (see below) may thus apply directly to the output from *clausefinder* or to the output from *stockmodifier*, since the format is identical.

The next stage of the system is the actual prosodic *parser*, which parses the list of words into a hierarchical structure with three levels: prosodic word, prosodic phrase, and prosodic utterance. First, prosodic phrase boundaries are assigned to a clause boundary, if the clause contains a focused content word. If no such word is found, the clause is attached to the end of the preceeding prosodic phrase. Second, content words and function words are grouped together to form prosodic words. Finally, a prosodic utterance boundary is generated at each paragraph boundary in the present algorithm.

The final module in the system, *xprotree*, is not an actual part of the algorithm for generating prosodic structure; it is rather a program that interprets the prosodically parsed text and displays a graphic representation of the text as a tree structure.

Figure 36.7 illustrates two possible prosodic structures of a sentence, one without and one with the domain-specific modifications. Notice that the insertion of a clause-internal prosodic phrase boundary in the tree on the right also produces another grouping in terms of prosodic words so that the function words *till* (to) and *en* (a) are grouped together with the content word *ränta* (interest) at the beginning of the new prosodic phrase. This restructuring follows the constraint formulated in [Hir93] that the left end of a prosodic phrase corresponds to the left end of a syntactic constituent, in this case, a PP.

36.4 Performance

Our principal goal has been to construct a prosodic parser using mainly morphological and lexico-semantic information and a minimal amount of syntactic parsing. The programs and algorithms have therefore not been optimized in any way for speed. Nevertheless, we have done some preliminary tests on the performance of the system as a whole, and we find the results promising.

Twenty-one radio broadcasts were transcribed. The system was then applied to these texts and timed. The mean and standard deviation were calculated. The time consumed was related to the number of words in the text, and also to the number of "units" in the text (i.e., the number of words plus the number of periods, commas, and so forth). This is interesting because special characters, which include punctuation marks, are treated by the system much like ordinary words except that they are handled faster because they require no lexicon lookup. The results on a Sun SPARCStation LX were the following:

Words/second	mean 3.39	SD 0.23
"Units"/second	mean 4.02	SD 0.27

This means that a text of about 125 words, which corresponds to the length of the summaries at the beginning of the radio stock-market reports is preprocessed in roughly 37 seconds.

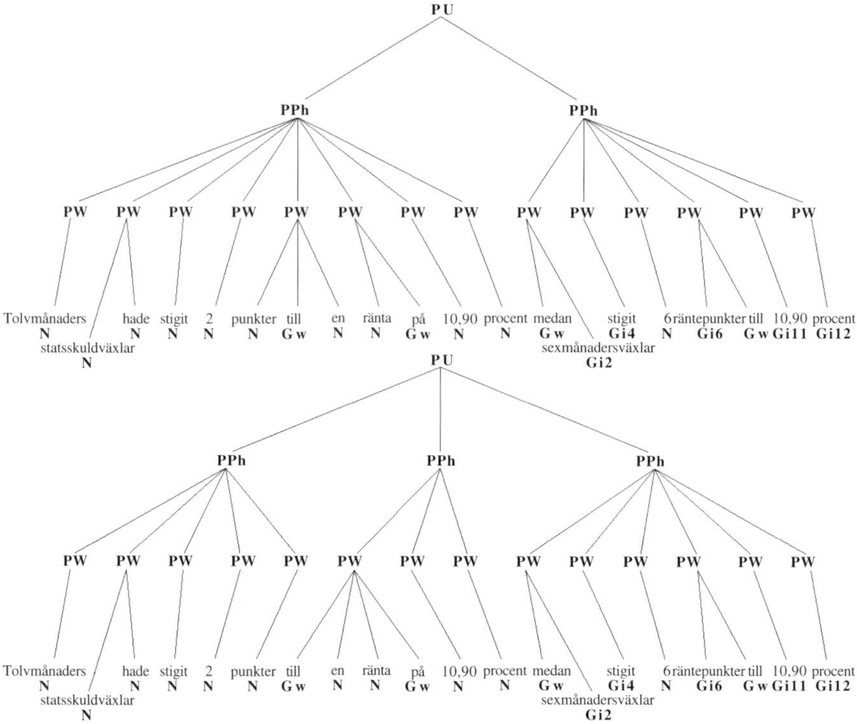

FIGURE 36.7. An example of the kind of variation in structure obtainable from the prosodic parsing system. It is also an example of the output produced by the *xprotree* graphic program. The left tree shows the sentence *Tolvmånaders statsskuldväxlar hade stigit 2 punkter till en ränta på 10,90 procent medan sexmånadersväxlar stigit 6 räntepunkter till 10,90 procent* "Twelve-month national-debt bills had risen 2 points to an interest rate of 10.90 percent while six-month bills rose 6 interest points to 10.90 percent" when it has been prosodically parsed by the domain-independent part of the system. The tree on the right shows another prosodic structure of the same sentence generated by using the additional rules in the domain-specific *stockmodifier* module. The letters N and G indicate the information status of each word as new or given, respectively.

36.5 Technical Data

This system was developed on a Sun SPARCStation, running SunOS UNIX. Ordinary programming was done in the C language, and compiled with the GNU C Compiler. Graphic programming was done with the XView Toolkit.

36.6 Conclusion

The results of the research reported on here indicate that it is possible to generate a suitable prosodic structure for Swedish text-to-speech by using different kinds of grammatical and discourse information extractable from texts. Domain-specific word-class information, along with information on the coreferential status of words, constitute crucial information used in the development of the prosodic parser. A minimum of syntactic information is required; clause and sentence boundaries are the most important syntactic information used by the algorithm. Generation of clause-internal Prosodic Phrase boundaries, however, requires the recognition of more syntactic information, such as verb complement structure. Recognition of domain-specific complex words also requires the definition of a limited number of derivational morphological patterns. Further research must be done, of course, in order to model the prosody of other speaking styles as well as to account for accentuation related to factors other than information focus, for example, contrastive accents. Our current efforts are being directed toward modeling contrastive accents arising in contexts of "parallel syntactic structure." Information on syllable count is also being built into the system in order to be able to prevent clauses that contain a very small number of syllables from being assigned a prosodic phrase boundary; rather, these short clauses will be grouped together with a neighboring clause into a larger prosodic phrase [HF95].

The algorithm for prosodic parsing presented here is still under development, so it is not currently able to automatically generate the acoustic prosodic correlates that we are assuming (see figure 36.5). We have therefore not yet tested the algorithm as to how well it generates appropriate prosodic patterns. This will be undertaken in the near future. On the accompanying CD-ROM, however, we have included three synthesized versions of a text (see Appendix), which allow one to informally compare the output without the prosodic parser (versions 1-2) with a hand-edited version synthesized with prosodic features that correlate with the prosodic constituent structure derived from the present algorithm.

Acknowledgments: This research has been supported by a grant from the HSFR/NUTEK Language Technology Programme.

References

[BF90] J. Bachenko and E. Fitzpatrick. A computational grammar of discourse-neutral prosodic phrasing in English. *Computational Linguistics* 16:155–170, 1990.

[Bru77] G. Bruce. *Swedish Accents in Sentence Perspective*. Gleerups, Lund, 1977.

[Cam93] W. Campbell. Automatic detection of prosodic boundaries in speech. *Speech Comm.* 13:343–354, 1993.

[CKPS92] D. Cutting, D Kupiec, J. Pedersen, and P. Sibun. A practical part-of-speech tagger. In *Proceedings of the Third Conference on Applied Natural Language Processing*, Trento, Italy, (Also available as Xerox PARC technical report SSL-92-01.)

[Eeg91] M. Eeg-Olofsson. *Word-class tagging. Some computational tools*. University of Göteborg, Department of Linguistics, 1991.

[EKWA92] E. Ejerhed, G. Källgren, O. Wennstedt, and A. Åström. *The linguistic annotation system of the Stockholm-Umeå corpus project*. Umeå, Department of Linguistics Report No. 33, 1992.

[FK89] G. Fant and A. Kruckenberg. Preliminaries to the study of Swedish prose reading and reading style. *Speech Transmission Laboratory, Quarterly Progress and Status Report* 2, Royal Institute of Technology, Stockholm, Sweden, 1989.

[Gar67] E. Gårding. Prosodiska drag i spontant och uppläst tal. In *Svenskt talspråk*, G. Holm, ed. Almqvist & Wiksell, Stockholm, 40–85, 1967.

[GH92] B. Grosz and J. Hirschberg. Some intonational characteristics of discourse structure. In *Proceedings of the International Conference of Spoken Language Processing*, Banff, 429–432, 1992.

[GR92] C. Gussenhoven and A. C. M. Rietveld. Intonation contours, prosodic structure and preboundary lengthening. *Journal of Phonetics* 20:283–303, 1992.

[HF95] M. Horne and M. Filipsson. Computational modelling and generation of prosodic structure in Swedish. In *Proceedings of the 13th International conference of Phonetic Sciences*, Stockholm, 1995.

[HFLL93] M. Horne, M. Filipsson, M. Ljungqvist, and A. Lindström. Referent tracking in restricted texts using a lemmatized lexicon: Implications for generation of prosody. *Proceedings Eurospeech '93*, vol. 3, Berlin, 2011–2014, 1993.

[Hir90] J. Hirschberg. Using discourse context to guide pitch accent decisions in synthetic speech. In *Proceedings, ESCA Workshop on Speech Synthesis*, Autrans, France, 181–184, 1990.

[Hir93] D. Hirst. Detaching intonational phrases from syntactic structure. *Linguistic Inquiry* 24:781–788, 1993.

[HJL87] P. Hedelin, A. Jonsson, and P. Lindblad. *Svenskt uttalslexikon:* 3 ed. Tech. Report, Chalmer's University of Technology, 1987.

[Hor94] M. Horne. Generating prosodic structure for synthesis of Swedish intonation. *Working Papers*, Department of Linguistics, University of Lund, 43:72–75, 1994.

[HSH95] M. Horne, E. Strangert, and M. Heldner. Prosodic boundary strength in Swedish: Final lengthening and silent interval duration. In *Proceedings of the 13th International Conference of Phonetic Sciences*, Stockholm, 1995.

[LHSLF95] A. Lindström, M. Horne, T. Svensson, M. Ljungqvist, and M. Filipsson. Generating prosodic structure for restricted and "unrestricted" texts. In *Proceedings of the 13th International Conference of Phonetic Sciences*, Stockholm, 1995.

[Lie67] P. Lieberman. *Intonation, Perception and Language*. MIT Press, Cambridge, MA, 1967.

[QK93] H. Quené and R. Kager. Prosodic sentence analysis without parsing. In *Analysis and Synthesis of Speech*, V. van Heuven and L. Pols, eds. Mouton de Gruyter, Berlin, 115–130, 1993.

[Str93] E. Strangert. Speaking style and pausing. *Phonum* 2:121–137, 1993.

[WSOP92] C. W. Wightman, S. Shattuck-Hufnagel, M. Ostendorf, and P. Price. Segmental durations in the vicinity of prosodic phrase boundaries. *J. Acoust. Soc. Amer.* 91:1707–1717, 1992.

Appendix: Audio Demos

This demonstration consists of three versions of the following three-sentence text in Swedish:

Kurserna på Stockholmsbörsen fortsätter att falla. För närvarande betecknas tendensen som mycket svag. Köpkurserna har stigit i 33 aktieslag, fallit i 121, medan 150 är oförändrade.

Translation: "Rates on Stockholm's Stock Exchange continue to fall. At present the trend is characterized as very weak. Buying rates have risen in 33 shares, fallen in 121, while 150 are unchanged."

Version 1: Infovox rule synthesis

Version 2: Illustrates how the same text would sound after *referent tracking* (see figure 36.7)

Version 3: Illustrates how the same text would sound after *prosodic parsing* (see figure 36.7)

37

Parametric Control of Prosodic Variables by Symbolic Input in TTS Synthesis

Klaus J. Kohler

ABSTRACT This chapter outlines the prosodic module of the RULSYS/INFOVOX TTS system for German, which represents the implementation of KIM (the Kiel intonation model). The first section sets out the categories of this model, and the second deals with their symbolization in input strings to the model and to the synthesis system for the control of parametric values.

37.1 KIM – The Kiel Intonation Model

37.1.1 Overview

The prosody model that we have been developing at Kiel for German and that has been implemented in the RULSYS/INFOVOX text-to-speech system incorporates the following domains (for further details see [Koh91]):

1. lexical stress—three levels: unstressed, secondary stress in compounds, and primary stress
2. sentence stress—four levels: reinforced, accented, partially deaccented, and completely deaccented
3. intonation:
    a. categories of pitch peaks and valleys as well as their combinations at each sentence stress position
    b. types of pitch category concatenation
4. synchronization of pitch peaks and valleys with stressed syllables—three steps: early, medial, late
5. prosodic boundaries (degrees of cohesion)—three variables: pause duration, phrase-final segmental lengthening, scaling of F_0 end points
6. overall speech rate between the utterance beginning and successive prosodic boundaries—four degrees: slow, medium, reduced, fast

7. downstep of successive pitch peaks/valleys and pitch reset

A system of prosodic distinctive features is used to specify the abstract symbolic, phonological categories in these domains. They are attributed to phonological units, which are either segmental (vowels and consonants) or nonsegmental (morphological and phrase boundaries). Attached to vowels is the fundamental distinction within the German prosodic system, namely, stress and intonation.

37.1.2 Stress

Lexical Stress

Within stress we have to differentiate between lexical and sentence stress. At the abstract level of phonological specifications in the lexicon, every German word has at least one vowel that has to be marked as potentially stressable, as being able to attract the feature specifications of sentence stress. Lexical stress is thus not a distinctive stress feature, it only marks a position that can, but need not, attract such a feature at the sentence level.

In noncompounded words one vowel has primary lexical stress. In compounds this also applies to at least one component word; the remaining compound word elements each have one vowel with secondary lexical stress. All vowels that are neither primary nor secondary-stressed are unstressed at the lexical level.

Sentence Stress

Sentence stress is attributed to a word as a whole; the lexical stress position in the word determines where the sentence stress manifests itself. By default a nonfunction word receives the category "accented." Deviations from this are either in the direction of emphatic reinforcement or of deaccentuation, which may be partial or complete. Function words are by default unaccented (= completely deaccented). Deviations are partially (de)accented, accented or reinforced.

37.1.3 Intonation

Pitch Categories at Sentence Stresses

All vowels of the reinforced, accented, and partially (de)accented sentence stress categories receive intonation features, which may be either valleys or peaks. Peaks may contain a unidirectional F_0 fall, classified as terminal, or rise again at the end, resulting in a (rise-) fall-rise, categorized as nonterminal. Rises in valleys and nonterminal peaks may be low, to indicate, for example, continuation, or high, for example, used in questions. All peaks and valleys may have their turning points (F_0 maximum in peaks or F_0 minimum in valleys) early or later with reference to the onset of the sentence stressed vowel, categorized as "(non-)early," and finally, for peaks, non-early may be around the stressed vowel center or toward its end, classified as "medial" or "late." This categorization of peaks captures the grouping of late and medial versus early peaks, as it showed up in perceptual experiments with stepwise peak shift from left to right [Koh90b].

Peaks are characterized by a quick F_0 rise confined to the vicinity of a sentence-stressed syllable. This rise precedes the onset of the latter, and is usually short and narrow in range for an early peak; it extends into the first half of the stressed nucleus in the case of a medial peak. In the late peak, it starts after the stressed vowel onset and continues into the second half of the nucleus or beyond. The exact timing of the maximum peak value depends on vowel type (duration according to quantity and quality), subsequent voiced/voiceless consonants, and number of immediately following unstressed syllables. There may even be a low stretch of F_0 in the stressed vowel before the rise. After the peak maximum is reached the F_0 descends immediately, especially on subsequent unstressed syllables. For chains of peaks, see subsection 37.1.3.2 below.

Valleys, on the other hand, have a continuous rise, starting before the stressed-syllable nucleus (early) or inside it (nonearly) and extending as far as the beginning of the following sentence-stressed syllable. If there are several unstressed syllables between two sentence stresses, a valley is thus realized as a more gradual F_0 ascent compared with the much quicker rise for a late peak. The less distance there is between stressed syllables, the more difficult it becomes to distinguish between a valley + peak and a late peak + peak sequence, especially if there is no F_0 dip between the first and second stress F0 maxima, as in a hat pattern (see 37.1.3.2).

Pitch Category Concatenation

In a concatenation of pitch peaks without prosodic boundaries between them (see subsection 37.1.4.3), F_0 may fall to a low or an intermediate level and then rise again for the next peak. This fall will be effected on intervening unstressed syllables between the two peaks, reaching the lowest point, to start the next rise, in the vicinity of the following stressed syllable, depending on peak position. If there are no unstressed syllables separating the two peaks, the dip can be accommodated between all peak combinations, except for late + early/medial, where a hat pattern is created; it combines the rise of the late peak and the fall of the early peak in a two-stress sequence.

This boundary case of the absence of an F_0 descent between peaks can also be extended to concatenations with intervening unstressed syllables. In such a hat pattern, an early peak is not possible initially, and a late one is excluded noninitially. If there are more than two stresses incorporated in a hat, the noninitial and nonfinal ones are unspecified as to peak position because they have neither a rise nor a fall but are simply integrated into the downstepped sequence of peak maxima (see 37.1.4.2). In the categorization of pitch patterns they are nevertheless grouped together with peaks. If in a two-stress rise-fall it is difficult to decide whether the rise represents a valley or a late peak in a hat pattern, the latter is chosen.

When prosodic boundaries intervene, any sequencing of peaks and/or valleys is possible, but the hat pattern is then excluded because it represents a very high degree of cohesion. On the other hand, a late peak with a full F_0 descent marks a dissociation from a following peak and will then normally be linked with a prosodic boundary, that is, final lengthening and F_0 reset afterwards.

Prehead

Unstressed syllables preceding the first sentence stress in a prosodic phrase may be either low or high: they represent different types of prehead.

37.1.4 Parametric Phonetic Control

The acoustic manifestations of the distinctive prosodic categories vary according to segmental and prosodic context, depending on the temporal alignment of peaks and valleys—defined by a small number of significant F_0 points—with different syllable types and sequences, on downstepping, speech rate, prosodic boundaries, and, finally, articulation-induced microprosody.

Temporal Alignment of Peaks and Valleys

Taking the default, medial peak as a reference, two significant F_0 points are defined. The first one, TF_0, is positioned at the beginning of the syllable containing the sentence-stressed vowel; the second, T2F_0, near the vowel center, the exact timing after voiced vowel onset depending on vowel quantity, vowel height, number of following unstressed syllables, and position in the utterance. The calculation of the time point T2F_0 after vowel onset is carried out on the basis of the segmental duration rules for German. They have adopted the principle proposed by Klatt [Kla79] for the rule synthesis of English (see also [Koh88]).

T2F_0 for medial peaks is derived from the basic vowel-type–related duration. It is essentially the intrinsic vowel duration that determines the point in time after stressed-vowel onset that it is positioned. But this has to be adjusted in the case of aspiration. On the one hand, aspiration lengthens the total vowel duration, compared with vowels in nonaspirated contexts, but this increase is not as large as the total aspiration phase; on the other hand, it shortens the stop closure duration compared with unaspirated cases, but again not by the total amount. So the larger part of the aspiration (three-quarters, according to comparative data analysis in the model construction [Koh91]) should be added to the vowel, but the rest attached to the plosive, and the F_0 peak placement has to take this ambivalence into account. That is, T2F_0 is shifted to the right by the period of aspiration time added to the vowel duration.

Sentence-final medial peaks receive a third F_0 point, T3F_0, which has been ascertained by data analysis and perceptual testing [Koh91] to occur 150 ms after the peak maximum in a medium speech rate (see 37.1.4.4, as well as item (b) in section 37.1.1 and item (b′) in section 37.2.1). In all nonfinal cases, the peak maximum of one sentence stress connects with the left-base point of the next sentence stress. As the absolute peak position is not affected by vowel duration modifications due to voiced/voiceless context, number of syllables in the word, sentence position, and so on, its relative position changes with vowel shortening or lengthening, moving closer toward or further away from the end. This way the microprosodic F_0 truncation before voiceless obstruents is automatically built into the rules.

Thus, in utterance-final short-vowel monosyllables ending in voiceless consonants, an F_0 peak fall is truncated, and has to be so to create the same perceptual peak pattern as in other, nontruncating contexts. The listener obviously takes the underlying constancy in absolute F_0 peak positioning within the same vowel type, and the difference across different vowel types, into account, disregarding contextual adjustments. He can always calculate the final F_0 point that should have been reached if the F_0 contour had not been curtailed, from the F_0 decline per unit of time up to the cut-off, and he can then compare this value with the likely low end of the speaker's speech voice range (about 60–80 Hz in a male voice, an octave higher in a female one). If the comparison is within a narrow margin the fall is terminal, otherwise it is not. Since the lower end of a speaker's F_0 range is a reliable reference value, truncation of falling F_0 patterns can be uniquely restored perceptually.

That no longer applies to rising F_0. Here the end point is not calculable because there is a large margin for the end point; anything up to or even above 1.5 octaves is possible. The position of the ceiling is not fixed, in contrast to the base line. That means that the intended high value of a valley always has to be physically reached, it cannot be deduced from what precedes, and, moreover, it does not change with different positions from early to late, which is different in peaks. There is thus a fundamental difference between peaks and valleys in the fixation of their offsets.

An early peak has its maximum value at the stressed-syllable onset, TF_0 100 ms before, and $T3F_0$—in sentence-final position—in an area where the medial peak has its maximum. A late peak has TF_0 at the same point as a medial peak, then an additional low F_0 point $T2F_0$ is inserted at vowel onset, and the late summit ($T3F_0$) occurs 100 ms after the point where a medial peak would have its center, or at the end of the last voiced segment in a nonfinal monosyllabic word if this distance is less than 100 ms. If there is an unstressed syllable following, the summit coincides with the unstressed vowel voice onset. In utterance-final position, a fourth F_0 point, $T4F_0$, occurs 100 ms after the summit; in monosyllables without final voiced consonants, $T3F_0$ has to occur at least 30 ms before vowel offset to signal the F_0 descent to $T4F_0$. The time positions of these significant F_0 points were obtained by interactive perceptual evaluation using the TTS development platform (see section 37.3).

Valleys have their left and center F_0 points at the same positions as TF_0 and $T2F_0$ in medial peaks; the right high point is located at the end of the last voiced segment.

Downstepping

Declination, the temporally fixed decline of F_0 prominent in, for example, the Dutch and Lund models [tCC90, Gar86], is not a feature of spontaneous speech production. It has therefore been replaced by downstepping in KIM, that is, a structurally determined pitch lowering from sentence stress to sentence stress, independent of the time that elapses between them.

Results of interactive testing make it clear that perception orient itself at structurally positioned and downstepped peaks, not at a time-based declination. For medium speech rate, the downstepping values used in the TTS implementation of KIM are 6% from peak to peak (starting from 130 Hz in a male voice), and 18% from a peak maximum to the subsequent base of the peak configuration. In valleys both the low and the high F_0 value are downstepped by 6%. Downstepping can be interrupted at any point by the reinforcement feature or by a resetting; it is stopped at a threshold value if the significant point(s) would go below it (set at 95 Hz for peak maxima in a male voice). Because downstepping is treated as automatic in the model, it is not given a phonological feature, nor is it symbolized.

Prosodic Boundaries

One of the functions of prosody is the sequential structuring of utterances and discourse, i.e. the signaling of prosodic boundaries and—at least partially—their hierarchical organization. To decode the syntagmatic chunking of messages in accordance with the speaker's intention, the listener requires signals that index degrees of cohesion or separation, respectively, between phrases, clauses, utterances, and turns. The parameters that achieve this are pause duration, phrase-final segmental lengthening, and scaling of F_0 end points at the respective boundaries. They can be controlled by parametric rules in the prosodic model upon appropriate symbolic input.

As at this stage the linguistically and phonetically relevant categorization of these boundaries is not well understood. The modeling cannot reduce the categories in this domain to the same small number as in the other areas of prosody discussed so far, but has to allow sufficient degrees of freedom for experimentation with data modeling. At each of the three parameters three degrees are therefore recognized, controlled by digit notation in the symbolic input to the model (see item (5′) in section 37.2.1). As our knowledge of prosodic boundary marking increases, the degrees of freedom can be reduced by establishing constraints between the three parameters in the signaling of the necessary and sufficient number of phonologically relevant distinctions.

Speech Rate

Speech rate changes within the same speaker not only alter the segmental durations (at varying degrees for different segment types, e.g., vowels versus consonants) but also the positioning of T2F_0 within a stressed vowel, moving it farther or less far into it in slower and faster speed, respectively (see subsection 37.1.4.1). This also implies slower or faster rises and falls, which lower or raise the perceived pitch level. The faster movements in turn mean that there comes a point when the complete F_0 excursions can no longer be performed. Since the peak and valley maxima are the essential target values controlled by the speaker, the leveling of F_0 movements in fast speech will particularly affect the low values. Finally, faster speech also means greater effort, which may produce a higher level of activation at the vocal folds right away. The correlation of F_0 level with speech rate perception has been shown in [Koh86b, Koh86c].

The control of speech rate is also coupled with articulatory reduction and elaboration [Koh90a]. The speech rate category in KIM and in its synthesis application therefore activates whole blocks of parametric rules that deal with F_0 timing, F_0 patterning, segment durations, and segmental adjustments (coarticulation, reduction, reinforcement, elision). To start with, the model distinguishes four degrees, one of them—reduced—taking reduction phenomena (at an otherwise medium speech rate timing) into account, as one way of changing speed of articulation.

Microprosody

In the generation of intonation, KIM separates two levels:

- the defining of phonology-controlled prosodic patterns by a small number of significant F_0 points
- the output of continuous F_0 contours influenced by articulation-related modifications [Koh90b]

This dichotomy implies the assumption that the underlying F_0 peak and valley patterns develop independently and in a very concrete physical and physiological sense in speech production, and are modified microprosodically by output constraints in the vocal apparatus. In particular, the model distinguishes five areas of microprosodic adjustments to the basic significant point patterns discussed so far.

1. In close vowels with medial peaks, the summit is raised by a factor of 1.08 compared with all other vowels.
2. An interpolation (cosine) is carried out between the significant F_0 points.
3. After voiceless obstruents, the F_0 value at vowel onset is raised by an additive constant of 15 Hz, the increment trailing off to 0 toward T2F_0.
4. In voiced plosives, all F_0 values are lowered by 10 Hz; in other voiced consonants by 5 Hz.
5. F_0 is masked in voiceless stretches.

37.1.5 Linguistic Environment of the Model

KIM is integrated into a pragmatic, semantic and syntactic environment. The input into the model consists of symbolic strings in phonetic notation with additional pragmatic, semantic, and syntactic markers. The pragmatic and semantic markers trigger, for example, the pragmatically or semantically conditioned use of peak and valley types or of sentence focus. Lexical stress position can largely be derived by rule; syntactic structure rules mark deaccentuation and emphasis in word, phrase, clause, and sentence construction. Phrasal accentuations are thus derived from the syntactic and semantic components that precede the prosodic model, and are given special symbolizations in the input strings to the model.

37.2 Symbolization of the Prosodic Categories

37.2.1 Symbolic Input to the Model

The following 7-bit ASCII characters are used to represent the prosodic categories of section 37.1.1 (using the corresponding numbering).

(1') Apostrophe or quotation mark [' "] are put in front of the primary or secondary stress vowel; vowels without a lexical stress marker are unstressed:

R'űck#s"icht (view to the rear) versus *R'űcksicht* (consideration)

(# marks the phonetically—especially prosodically—relevant word boundary in compounds.)

(2') Digits [3,2,1,0] are put in front of words that receive the reinforced, accented, partially, or completely deaccented sentence stress category, respectively, which in turn affects the manifestation of the respective lexically stressed vowel. Function words, marked by suffixed [+], have [0] as their default, nonfunction words [2]; in both cases the digit may be omitted from the symbolization:

2Max 0hat+ 0einen+ 2Brief 2geschrieben
(Max did write a letter)

2Max 0hat+ 0einen+ 2Brief 1geschrieben
(semantically unmarked rendering of Max wrote a letter)

2Max 0hat+ 0einen+ 2Brief 0geschrieben
(Max wrote a letter, not a card)

2Max 0hat+ 0einen+ 3Brief 0geschrieben .
(reinforcement of contrasted Brief in the previous example)

(3') Punctuation marks [. , ?] for pitch peaks, low and high rising valleys, and the character sequences [.,] and [.?] for fall-rises are put before prosodic boundaries (see (5')):

Ja .p: Ja ,p: Ja ?p: Ja .,p: Ja .?p:

A high prehead is symbolized by [=] after the prosodic boundary marker [p:] (see (5')).

p:= 0Wie+ 0sieht 0das+ 0bei+ 2Ihnen+ 0am+ 3Donnerstag 0aus .p:

(4') Parentheses [)(] for early and late peak positions, respectively, in sentence-stress syllables are put before the stressed word (after the sentence-stress digit); the medial peak position is regarded as the default case and remains unmarked:

Sie+ hat+ ja+ 2)gelogen
(She's been lying = summarizing, concluding statement)

Sie+ hat+ ja+ 2gelogen
(=start of a new argumentation)

Sie+ hat+ ja+ 2(gelogen
(as the preceding example but with a contradictory note)

In connection with valleys, there are only early and nonearly positions in the sentence-stressed syllable. Either one or the other category may be taken as the unmarked default case, depending on their frequency of occurrence: early for [,], nonearly for [?].

(5') The prosodic boundary (cohesion) marker [p:] is put after the word at which boundary indices occur. It is preceded by two digits, the second of which refers to pause length, the first to utterance-final lengthening. In the case of pitch peaks, there is a third boundary-related digit to the left of these two, referring to the scaling of the F_0 end point. Each of the digits may range from [0] (= absence of pause, of final lengthening or of F_0 descent) through [1] (= short pause - $\leq$ 200ms; default utterance-final lengthening; intermediate F_0 descent) to [2] (= long pause - $\geq$ 200ms; hesitation lengthening; full F0 descent):

zehn .212p: (110p:) minus+ zwei .0/100p: mal+ drei
(10 − 2 × 3)
zehn .0/100p: minus+ zwei .211p: (.110p:) mal+ drei
((10 − 2) × 3)

(6') The digit string associated with the phrase boundary marker [p:] is preceded by a further digit, ranging from 0 to 3 to mark four degrees of speech rate, which include degrees of reduction or elaboration: [2] refers to medium overall speed and default reduction (and may be omitted from the symbolization as an implicit default), [1] refers to the same speed but a higher degree of reduction; for [0], degrees of reduction and speed are increased, for [3] they are both decreased from [2]. The rate digit (or default) applies to the stretch of speech between the [p:] marker and its predecessor or the utterance beginning, respectively. In this modeling of speech rate, segment durations are not changed by uniform and proportionate up- or down-scaling across the whole sequence, but vowels and consonants are dealt with separately according to sets of rules including segmental reduction, assimilation, and elision.

mit+ roten gelben blauen schwarzen .0-3212p:
(with red, yellow, blue and black ones)

(7') Downstep is not indicated symbolically. Pitch reset is associated with a prosodic boundary. It is marked by [+] before the digit sequence at the preceding [p:].

mit+ roten gelben .+2110p: blauen schwarzen .2212p:

37.2.2 Symbolic Input in the German RULSYS/INFOVOX TTS System for Parametric Control

KIM has been implemented in the RULSYS/INFOVOX TTS for German. The Kiel development of this TTS system (for details see [CGH90, Koh91]) makes use of a very simple adaptation of 7-bit ASCII to the phonetic transcription of German:

- upper-case letters for segmental phonemes
- lower case with for allophones

1. [•] the characters listed in (1')–(7') of section 37.2.1, but with [%] replacing the compound boundary marker [#].

These phonetic symbols are either derived by rule from orthographic input or entered into the system directly, enclosed between the metacharacter #. In the latter case the input string can be either entirely phonetic or mixed orthographic/phonetic, as illustrated by the examples in the Appendix.

The TTS application of the prosodic system is based on default assumptions for certain orthographic characters that are identical with prosodically defined symbols, e.g., <?> - [?]:

- Orthographic <?> is rule-converted into phonetic [.] in question-word, but into phonetic [?] in yes-no questions.
- The deviations from default have to be specially indicated
  (a) by [.] in *Kommt sie*, for example
  (b) by [?] combined with the special marking of the otherwise default stress [2] category, in *Wie 2heißt du?* for example
  (c) by [?] combined with the special marking of the stress [2] category on the question word of a confirmatory question, which has a continuous F_0 rise throughout

In peaks the medial position is treated as default; in low valleys it is the early, in high valleys the late postion.

The period <.> has two functions in the TTS-system:

- a trigger symbol for the generation of terminal intonation patterns, that is, various degrees of falling F_0

- a command to the system (by the side of <?> and <!>) to start sentence processing

These functions cannot be dissociated from each other. This means that the input string must not contain a non–sentence-final <.> because the string following <.> will be processed separately, and, on the other hand, every sentence must be terminated by either <.> or <?> or <!> to initiate the sentence-level processing. So if a low rise, symbolized by <,>, is to occur in sentence-final position, the notation must be <,.>. This also implies that [p:] must precede punctuation marks in the TTS input strings (as against the KIM notation in section 37.2.1 (3')), and that [.] in the function of a sentence-medial falling intonation category (see (5') of section 37.2.1) cannot be input into the system as <.> but must be generated internally from [p:] if the latter stands on its own, that is, is not followed by <,>. A further necessary adaptation of the KIM symbolization system in its TTS implementation concerns the falling-rising patterns, which in KIM are mnemonically represented as [.,] and [.?]. The adjustment uses <„> and <,?> for falling + low/high rising, and sentence-final <„> is followed by <.> in the input string.

In the KIM symbolization system, intonation and prosodic boundary markings are kept strictly separate (see section 37.2.1 (3'), (5')), whereas in TTS, punctuation marks receive a (pause) duration: <,>=300ms, <. ? !>=600ms. This means that #(2)212p:.#, as default, may be reduced to <.> in the TTS input string, similarly #(2)12p:,# to <,>. And the pause length between TTS sentences can be graded in 600-ms steps by the addition of the respective number of <.> to the sentence-final punctuation mark (see Appendix, II. Texts).

The greater part of the prosodic notations in section 37.2.1 (2')–(7') have to be entered as such because the syntactic component of the system is not powerful enough to derive them by rule from orthographic input. Moreover, in many cases semantic and pragmatic rules would be required to generate the correct prosodic output. The symbolic prosody markers trigger hierarchical sets of symbolic distinctive feature rules, followed by sets of parametric F_0 and duration rules in the phonetic-to-acoustic output component. In the case of F_0, the rules define significant points for peak and valley configurations, synchronize them with lexically stress-marked vowels according to sentence-stress and intonation symbolizations, and modify them contextually as well as microprosodically. A cosine function then interpolates between the final sequence of significant F_0 values.

The speed-control digit at the [p:] marking attributes a parametric rate variable to every segmental symbol within its domain and sets it to a value representing the respective category. Blocks of duration, segment, and F_0 rules in the phonetic module are then activated by the particular rate-variable value and the appropriate calculations along the three phonetic scales are performed. This means that for a particular speed it is not only the segment durations that are adjusted across the whole chain to which the particular rate factor applies, but F_0 is also raised for speeding up or lowered for slowing down, and segmental reductions or elaborations are effected simultaneously, in accordance with natural speech production. The segment durations are scaled separately for vowels and consonants and also as a function of a number of other conditioning factors (vowel height, consonant category, stress, number of syllables in the word). The digit before [p:] controlling phrase-final lengthening triggers a more local increase or decrease of segment durations within the set global speech rate.

The TTS implementation of KIM allows the calculation of speech timing at a hierarchy of levels from segment to segment chain to phrase to utterance, according to a Klatt type model for segment timing with factors determined by stress, utterance position, number of syllables in the word, and overall speech rate (see [Kla79, Koh86a, Koh88]).

The Kiel prosodic model for German is comprehensive and detailed enough for its TTS realization to be capable of generating highly intelligible and natural-sounding synthetic output for very intricate phrasing structures in complicated continuous text.

37.3 A Development System for Prosodic Modeling, Prosodic Labeling and Synthesis

The TTS implementation of KIM constitutes a research tool for the further development of the prosodic model as well as for the improvement of prosodic synthesis. The categories of the prosodic phonology of German, based on extensive speech production and perception experiments, can be tested in quick interactive auditory evaluations and in formal, more costly listening experiments with carefully prepared synthetic speech output files. The development system allows rule-driven parameter control by symbolic input categories of the model as well as systematic changes of values in graphic parameter displays, in both cases with immediate acoustic output. It is thus possible to check (a) the validity of category differentiations, (b) the need for category extensions or reductions, and (c) the adequacy of defined parameter values for the categories. The results can be incorporated in a revised version of the prosodic TTS rules, and this loop of interactive or formal listening test evaluations and rule adjustments can be repeated until an optimization in the intelligibility and naturalness of the synthetic output is achieved.

By referring parametric variables to phonological categories of a prosodic model, the degrees of freedom of the TTS generation have been reduced considerably, without forgoing the flexibility and potentially exhaustive coverage of empirical speech data that a generative framework provides. However, the degrees of freedom in category combination and concatenation, such as in connection with phrase boundaries, are still quite large. The TTS development system offers a powerful device for the testing of constraints between prosodic categories. KIM will have to be preceded by a filter within the linguistic environment of the model (section 37.1.5) that excludes a great many combinations on syntactic, semantic, and pragmatic grounds and thus reduces their degrees of freedom.

The KIM prosodic symbolization system is also used for consistent, systematic, and efficient prosodic labeling of recorded speech data to enlarge the empirical basis for prosodic modeling. For an outline of the labeling framework, its conventions, and its application see [KPS94, Koh95].

The application of the TTS research platform to prosodic model and synthesis evaluation is thus also guided by natural speech data labeled within the same category and symbolization framework. So prosodic modeling, its TTS implementation and testing, as well as model-driven labeling of natural speech data form an interrelated and mutually conditioning set of procedures in prosodic research at IPDS Kiel.

Acknowledgments: The development of KIM was carried out with financial support from the German Research Council (DFG grants Ko 331/19-1-4) in the project "Form and function of intonation peaks in German" between 1985 and 1989. Some of the initial implementation in the RULSYS/INFOVOX TTS system was made possible by a contract with the company Infovox, Solna/Sweden in the 1987–1989. Furthermore, I particularly acknowledge, with great gratitude, the continuous and extremely fruitful cooperation with Rolf Carlson and Bjőrn Granstrőm at KTH, Stockholm and their generous supply of support software, especially the MIX program.

References

[CGH90] R. Carlson, B. Granstrőm, and S. Hunnicutt. Multi-lingual text-to-speech development and applications. In *Advances in Speech, Hearing, and Language Processing*, W. A. Ainsworth, ed. JAI Press, London, 269–296, 1990.

[Gar86] E. Gårding. Superposition as an invariant feature of intonation. In *Invariance and Variability in Speech Processes*, J. S. Perkell and D. H. Klatt, eds. Laurence Erlbaum, Hillsdale, NJ, 292–299, 1986

[Kla79] D. H. Klatt. Synthesis by rule of segmental durations in English sentences. In *Frontiers of Speech Communication Research*, B. Lindblom and S. Őhman), eds. Academic Press, London, New York, 287–299, 1979.

[Koh86a] K. J. Kohler. Invariance and variability in speech timing: from utterance to segment in German. In *Invariance and Variability in Speech Processes*, J. S. Perkell and D. H. Klatt, eds. Lawrence Erlbaum, Hillsdale, NJ, 268–289, 1986.

[Koh86b] K. J. Kohler. Parameters of speech rate perception in German words and sentences: Duration, F_0 movement, and F_0 level. *Language and Speech* 29:115–139, 1986.

[Koh86c] K. J. Kohler. F_0 in speech timing. *Arbeitsberichte des Instituts fűr Phonetik der Universitắt Kiel* (AIPUK) 20:55–97, 1986.

[Koh88] K. J. Kohler. Zeitstrukturierung in der Sprachsynthese. In *Digitale Sprachverarbeitung. ITG-Tagung, Bad Nauheim*, A. Lacroix, ed. vde-Verlag, Berlin, 165–170, 1988.

[Koh90a] K. J. Kohler. Segmental reduction in connected speech in German: Phonological facts and phonetic explanations. In *Speech Production and Speech Modelling*, W. J. Hardcastle and A. Marchal, eds. Kluwer Academic Publishers, Dordrecht, 69–92, 1990.

[Koh90b] K. J. Kohler. Macro and micro F_0 in the synthesis of intonation. In *Papers in Laboratory Phonology I*, J. Kingston and M. E. Beckman, eds. CUP, Cambridge, 115–138, 1990b.

[Koh91] K. J. Kohler. A model of German intonation. *Arbeitsberichte des Instituts für Phonetik der Universität Kiel* (AIPUK) 25:295–360, 1991.

[Koh95] K. J. Kohler. PROLAB—the Kiel system of prosodic labelling. In *Proceedings ICPhS Stockholm*, 1995.

[KPS94] K. J. Kohler, M. Pätzold, and A. Simpson. *Handbuch zur Segmentation und Etikettierung von Spontansprache*, 2.3 VERBMOBIL, Technisches Dokument Nr. 16, 1994.

[tCC90] J. 't Hart, R. Collier, and A. Cohen. *A Perceptual Study of Intonation.* CUP, Cambridge, 1990.

Appendix: Audio Demos

Illustrations of the prosodic categories of KIM in the input strings to the RULSYS/INFOVOX TTS system: words, sentences, and texts. The words and sentences follow the numbering in section 37.2.1. All segmental and prosodic transcription notations are enclosed in # #. The accompanying CD-ROM provides the acoustic TTS output of all the listed materials; each block of word and sentence examples is presented for auditory comparison and repeated.

I. Words and Sentences

(1') Lexical stress and compounding

(a) *Rűck-sicht.* (view to the rear)
#R'YKSICHT#. (consideration)

(b) *Frei-tag.* (free day) *Freitag.* (5th day of the week)

(c) *Feier-tag.* *Donnerstag.*

(d) (wrong) reversal of the compounding in (c)

(2') Sentence stress

Max hat einen Brief #2# geschrieben.
Max hat einen Brief #1# geschrieben.
Max hat einen Brief #0# geschrieben.
Max hat einen #3# Brief #0# geschrieben.

(3') Intonation

(a) *ja. ja,. ja? ja,,. ja,?*

(b) *Wie heißt du?*
Wie #2# heißt du?
Wie #2)# heißt du?

(c) *#2# Wie heißt du?* *#2)# Wie heißt du?*

(d) *Wie heißt du,.* *Wie #(# heißt du,.*

(4') Peak and valley positions

(a) peaks
Sie hat ja #)# gelogen.
Sie hat ja gelogen.
Sie hat ja #(# gelogen.

(b) high valleys in (3') (b), (c)

(c) low valleys in (3') (d)

(5') Prosodic phrase boundaries

(a) *zehn #211p:# minus #+# zwei #100p:# mal #+# drei.*
zehn #100p:# minus #+# zwei #211p:# mal #+# drei.

(b) *zehn #110p:# minus #+# zwei #000p:# mal #+# drei.*
zehn #000p:# minus #+# zwei #110p:# mal #+# drei.

(c) contrasting

i. absence of phrase boundary
ii. phrase boundary marked by lengthening
iii. phrase boundary marked by resetting of downstep
iv. phrase boundary marked by lengthening and resetting of downstep

rote gelbe blaue weiße graue schwarze.
rote gelbe #110p:# blaue weiße #110p:# graue schwarze.
rote gelbe #+100p:# blaue weiße #+100p:# graue schwarze.
rote gelbe #+110p:# blaue weiße #+110p:# graue schwarze.

(6') Speech rate: slow, medium, medium reduced, fast

mit roten gelben blauen braunen #3212p:#.
mit roten gelben blauen braunen #2212p:#.
mit roten gelben blauen braunen #1212p:#.
mit roten gelben blauen braunen #0212p:#.

(7') Downstep (default) and reset

mit roten gelben #110p:# blauen schwarzen #212p:#.
mit roten gelben #+110p:# blauen schwarzen #212p:#.

II. Texts

1. "*Die Buttergeschichte.*" Medium speed, not reduced.

Es war in Berlin zu einer Zeit, als Lebensmittel nicht genügend vorhanden waren. Vor einem Laden stand bereits um sieben Uhr eine beachtliche Menschenmenge; denn man hatte dort am Abend vorher auf einem Schild schon lesen können, daß frische Butter eingetroffen sei. Jeder wußte, daß die Butter schnell ausverkauft sein würde, und daß man ganz früh kommen müsse, um noch etwas zu erhalten. Da das Geschäft erst um acht geöffnet wurde, stellten sich die Leute vor der Laden-tür in einer Reihe an. Wer später kam, mußte sich hinten anschließen. Je näher der Zeiger auf acht kam, desto unruhiger wurden die #)# Leute. Da kam endlich ein kleiner Mann mit grauem Haar und drängte sich ziemlich rücksichtslos nach vorn. Die wartenden Menschen waren empört über solches Verhalten und forderten ihn auf, sich ebenfalls hinten anzustellen. Aber auch als mit der Polizei schon gedroht wurde, #0# ließ sich der Mann nicht beirren, sondern drängte sich weiter durch. Er bat, man solle ihn doch #3# durchlassen. Oder glaubte man, daß diese Drängelei für ihn vielleicht ein Vergnügen sei?..Das war für die Leute nun doch zu viel.. Alle kochten bereits vor Wut, und der Mann konnte jetzt von allen Seiten Schimpfwörter hören.. Er aber zuckte resigniert mit den Schultern und bemerkte : "Nun #3)# gut. Wie Sie #)# wollen. Wenn Sie mich nicht vorlassen, dann kann ich die Tür nicht aufschließen, und Sie können meinetwegen hier stehenbleiben, bis die Butter ranzig geworden ist."

2. Address to a tutorial at the Konvens meeting in Vienna, 27 September 1994. In this text <;;> is shorthand for #000p:# (the hat pattern).

Meine sehr ;; verehrten ;; Damen und Herren, #1# liebe ;; Teilnehmer am Tutorium #+211p:#, Aussprache-lexika in der signalnahen Sprachverarbeitung.. Es #2#begrüßt Sie die klare #+211p:#, etwas metallische #+210p:#, aber dennoch melodische #+211p:#, und vor allen ;; Dingen ;; rhythmische #+3210p:# synthetische Stimme des Nordens. Sie #2# basiert auf dem TTS-System #+2100p:# VOX #1# PC #+211p:#, der #0# Firma Infovox in Stockholm #+2212p:# und auf dem Softwehr Entwicklungswerkzeug Ruhlsys #+210p:#, der Technischen #3# Hochschule #0# Stockholm. Entscheidend für die Geburt dieser Stimme war aber die Entwicklung von Regeln #+210p:#, zur akustischen Wandlung orthographischer Symbolketten #+211p:#, im Institut für Phonetik #+210p:#, und digitale Sprachverarbeitung #+211p:#, der Christian ;; #'ALBR[CHTS%UNIVERZIT'[:T# zu Kiel...
An #2# sich sollte #0# Professer #3# Kohler diese #1# einleitenden #1# Worte #0# sprechen. Da er aber noch ein #0# bißchen unter Zeitverschiebung #+2210p:# nach einer USA und einer Japan-reise #0# leidet, #0# braucht er #1# heute morgen #+210p:#, noch einen etwas längeren Anlauf. #0# Deshalb ist er sehr froh #0DAR'Y:Br#,, daß er diese Aufgabe #2# mir #+2210p:# einer Sprechmaschine übertragen kann....
Ehe Herr Kohler #+2100p:# mit Ihnen die Struktur #2110# und die Generierung #2110p:# von Aussprache-lexika #+212p:#, sowie ihren Einsatz in Forschung und Anwendung #+2000p:# auf verschiedenen ;; Ebenen erläutert #+212p:#, möchte ich nicht versäumen, den Organisatoren der #1# Tagung #211p:#, auch in #1(# seinem #010p:#, Namen #+011p:#, für die Einladung zur Ausrichtung des Tutoriums #+2100p:# sehr herzlich zu ;; #)#

danken.. #2# Ihnen, den Teilnehmern, gebűhrt ebenfalls #0# Dank #210p:#, daß Sie sich #0# dafűr entschieden haben..
Wir haben die Einladung #0#natűrlich sehr ;; gern #1#aufgegriffen #+2210p:# #2# nicht nur weil sie Herrn Kohler die Mőglichkeit #1# gibt #200p:#, #3# Wien zu #1#besuchen #+211p:#, und den Heurigen zu #1# genießen #+211p:#, sondern um vor allem die Kieler ;; Forschung #1# vorzustellen #+2210p:# und #)# Interesse an ihr zu #1# wecken....
Jetzt darf ich aber Herrn Kohler nicht lǻnger von seiner Arbeit abhalten. Er ist #0# inzwischen aufgewacht #2111p:# und schon #3(# unruhig geworden. Und er findet vor allem die Text-eingabe sehr erműdend #+2211p:# da er nur mit zwei Fingern tippen kann. Ich ziehe mich also zurűck, und wűnsche Ihnen viel Vergnűgen ;; beim Tutorium.

38

Prosodic and Intonational Domains in Speech Synthesis

Erwin C. Marsi
Peter-Arno J. M. Coppen
Carlos H. M. Gussenhoven
Toni C. M. Rietveld

ABSTRACT An intonational domain corresponds to the part of an utterance spanned by one intonation contour. We lay out a theory of intonational domains that is rooted in intonational (autosegmental) phonology and prosodic phonology. We focus on restructuring—the process that joins two intonational domains together to form a single domain. We report on a perception experiment about restructuring involving synthetic speech. The results indicate that restructuring is constrained by: (1) syntactic structure, at least the distinction between a PP that is internal and a PP that is external to an NP; and (2) the length of the initial domain before restructuring. Finally, we discuss the consequences of our results for phonological theory and the intonational component in speech synthesis.

38.1 Introduction

Now that segmental synthesis has reached a certain quality, further improvement of synthetic speech is expected from improvements in the synthesis of prosody. Intonation is often considered the most salient aspect of prosodic structure, and consequently the synthesis of well-formed and contextually appropriate intonation contours has received much attention. Theoretical work on intonation, supported by both linguistic intuitions and experimental data, has accumulated over the years to constitute a field known as *intonational phonology* [Lad92]. Autosegmental descriptions of intonation, inspired by the work of [Pie80], have become available for a number of languages, including Dutch [Gus88b, Gus91]. Prosody has also been studied within the field of *prosodic phonology* [Sel84, NV86]. We feel that the synthesis of intonation can benefit from these developments in intonational and prosodic phonology. The experiment reported in this chapter is intended as a contribution to the application of autosegmental models of intonation to the synthesis of intonation.

We use *intonational domain* to refer to each part of an utterance spanned by a single intonation contour. It is intended to be a theory-neutral expression for what is called, for example, *intonational phrase* [NV86, Sel84], *intermediate phrase and intonation phrase* [BP86], *tone domain* [Lad86], or *association domain* [Gus88a]. In what follows we concentrate on intonational domains and abstract away from other aspects of intonation, such as the particular shape of the intonation contour (*tune*) or the prominence of pitch accents. In the first part of this chapter we lay out our theory of intonational domains, which is based on linguistic intuitions and practical experience with speech synthesis, and compare it to the conventional conception of intonational domains in prosodic phonology. Subsequently we focus on the problem of the *restructuring* of intonational domains—the phenomenon that under certain conditions adjacent intonational domains are joined together to form a single such domain. The problem of restructuring directly motivated the experimental work that is described in the second part of this chapter. This work concerned the effect of syntactic structure and length on the distribution of intonational domains.

38.2 A Theory of Intonational Domains

38.2.1 Intonational Domains and the Prosodic Hierarchy

The status of the intonation domain in phonological structure has been discussed in a number of recent publications. When the theory of prosodic phonology was first introduced, phonologists assumed that the intonation contour corresponded to a constituent called the *intonational phrase* (Ip), which took its place in a hierarchy of other prosodic constituents [Sel84, NV86]. Prosodic constituents are defined by means of mapping rules that derive the prosodic constituency of an utterance from its morphosyntactic structure. An utterance such as "A spoken message makes it clearer" might look like (1), where the Ip is slotted in between the *utterance* (Ut) and the *phonological phrase* (Pp), below which are the *phonological word* (Pw), the *foot* (Fo) and the *syllable* (Sy).[1]

The general motivation for prosodic phonology is that prosodic constituents appear to figure in the structural description of segmental phonological rules, as amply shown in [NV86]. It has also been assumed that these constituents define the rhythmic structure of the utterance, either directly or in more recent accounts, after translation into a metrical grid [NV89]. In the case of the Ip, there appears to be a problem when the criterion for the constituency provided by the intonation contour is confronted with the criteria provided by pausal and segmental phenomena. The difficulty is that the intonation contours that are appealed to in the definition of the Ip may have internal prosodic breaks that in other cases typically occur at the boundaries of the Ip. As an example, consider the utterances given in (2). Example (2a), where "John" is the indirect object of "tell," is a single intonation contour that

[1]The specific way in which function words have been attached in (1) is not relevant to the subsequent discussion.

(1)

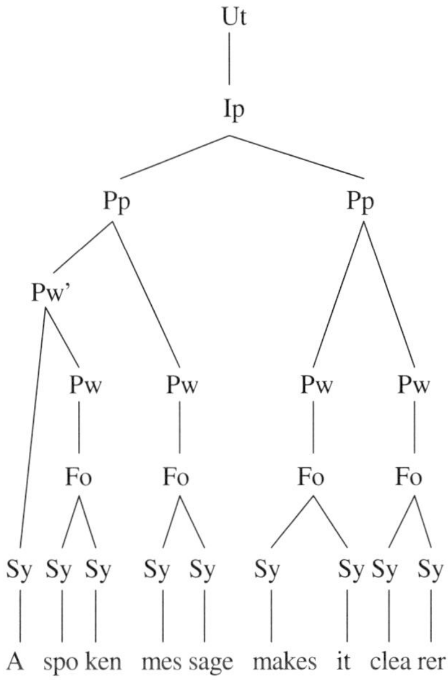

(2)

a.

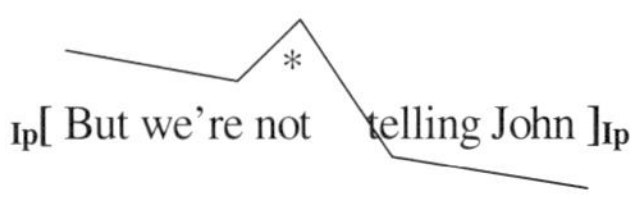

b.

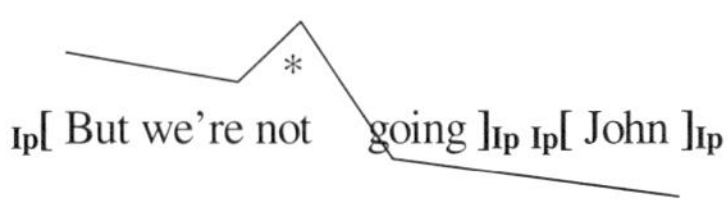

corresponds to a single Ip. Example (2b), where "John" is a vocative tag, has the same contour as (2a) but is analyzed as containing two Ip's. Primarily because of the pause that separates them from the preceding clause, vocatives are interpreted as separate Ip's. But because final vocatives are unaccented, the intonation contour chosen for the accent on "not" (the only accent in the Utterance) necessarily spans two Ip's (cf. [Tri59]). Thus, in (2b), a single intonation contour is mapped onto more than one Ip.

In accented Ip's, too, this conflict may arise. For example, in (3b), the preposed adverbial "Tomorrow" occurs in exactly the same intonational contour as does the verb "to borrow" in (3a). That is, on the basis of the intonational criterion, both (3a) and (3b) ought to be single Ip's. However, on the basis of the pausal and segmental

(3)

a. $*$ $*$

${}_{Ip}[$ To borrow the book may be cheaper $]_{Ip}$

b. $*$ $*$

${}_{Ip}[$ Tomorrow $]_{Ip}$ ${}_{Ip}[$ the book may be cheaper $]_{Ip}$

phenomena, they are distinct. There is a rhythmic break following "Tomorrow" in (3b) which is absent after "To borrow" in (3a)[2]

38.2.2 Association Domains

A response to this problem is given in [Gus88a]. Gussenhoven argues that the solution should be based on the recognition that the intonational domain of a pitch accent cannot consistently be identified with any one constituent in the prosodic hierarchy. The intonational domain of a pitch accent is determined primarily by the location of any following accent, and only secondarily by constituent structure. Therefore, he assumes that we need an independent constituent over which a single intonation contour spreads: the *association domain* (AD). The AD is determined by two factors: the pitch accent distribution and the prosodic constituency. The properties of ADs and their relation to accent distribution in combination with prosodic constituency can be stated as a sequence of constraints.

Constraints on Association Domains:

A Every pitch accent has its own unique AD and every AD belongs to a unique pitch accent.

B The AD of an accent ends with the *highest* prosodic constituent that dominates the accent but does *not* dominate a following accent.

C Every AD has a maximum size without overlapping other ADs.

As a consequence of constraint A, ADs and pitch accents are in a one-to-one relation. The situation of an AD without an accent or an accent without an AD does not occur[3] Constraint B expresses the important fact that ADs depend on the number and locations of accents in the utterance. Furthermore, this constraint guarantees that the end of every AD is aligned with the end of a prosodic constituent. Notice that divorcing intonational domains from prosodic constituency does not imply that intonational domains do not respect the boundaries of prosodic constituents. Rather, the consequence of our view is that instead of there being one particular

[2]Cf. [GR92b] for segmental phenomena that also indicate a distinction.

[3]There is one exception: ADs without a pitch accent may occur for reporting clauses. In such cases a rule called *tone copy* provides a pitch contour to the empty AD (cf. [Tri59].

(4)

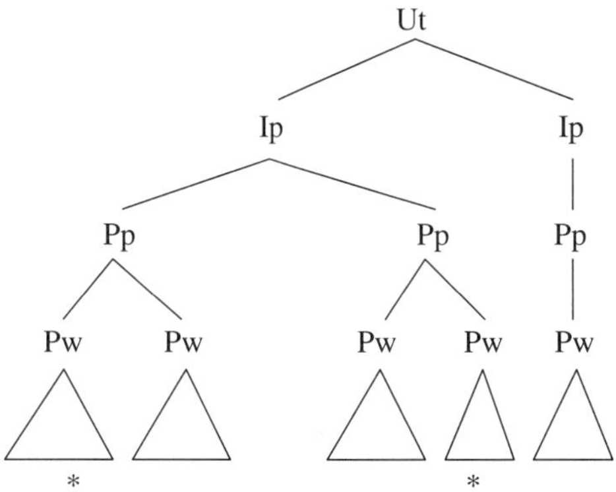

$_{AD}$[A spoken message]$_{AD}$ $_{AD}$[makes it clearer, he said]$_{AD}$

prosodic constituent that can be identified with the intonational domain of a pitch accent, any one of a number of constituents from the foot onward can define the rightmost boundary of a domain. Finally, constraint C implies that the left-hand boundary of an AD will be positioned to the left as far as possible, coinciding either with the end of a previous AD or the beginning of the utterance.

In (4), an example is given that shows how we arrive at a unique distribution of ADs by application of the constraints. Consider the pitch accent on "spoken." As a consequence of constraint A, it must have its own unique AD. Constraint B says that the end of this AD coincides with the highest prosodic constituent that dominates the accent, but does not dominate the next accent (on "clearer"). This can only be the first Pp. It cannot be the Pw that dominates "A spoken" because this is not the highest constituent. Neither can it be the first Ip or any higher constituent because these also dominate the next accent. The AD starts at the beginning of the utterance by virtue of constraint C. Next, consider the accent on "clearer." As the Ut is the highest constituent that dominates it and there are no accents following, the end of its AD coincides with the end of the Ut. To maximize the size of the AD without overlap, the AD starts where the previous one ends. Now, all constraints are satisfied, and every accent has its own AD. The shape of the actual intonation contour depends on the type of pitch accent with which the accents will be realized.

38.2.3 AD-Restructuring

It has so far been assumed that there is a one-to-one relationship between pitch accents and ADs. However, this situation is characteristic of slow and emphatic speech only. The intonation of ordinary natural speech shows that often several pitch accents actually share a common intonational domain. We assume there is an optional cyclic process called *AD-restructuring* that joins two adjacent ADs to form a single AD. The intonational consequence of restructuring can be illustrated with the help of the representations in (5). The "*" marks an accented syllable. The bottom line represents the tonal string that consists of pitch accents. The two strings

(5)

a.

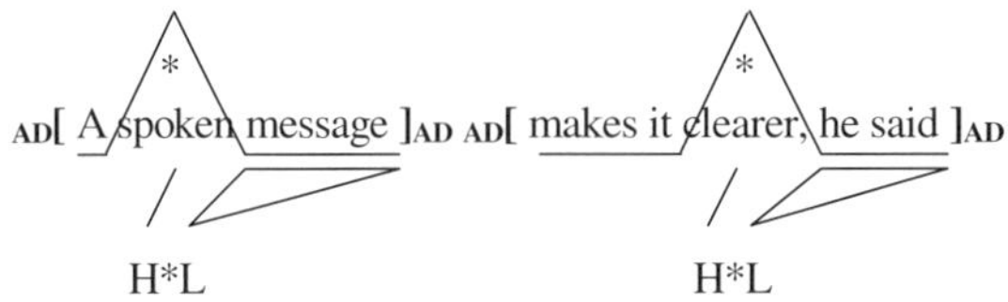

b.

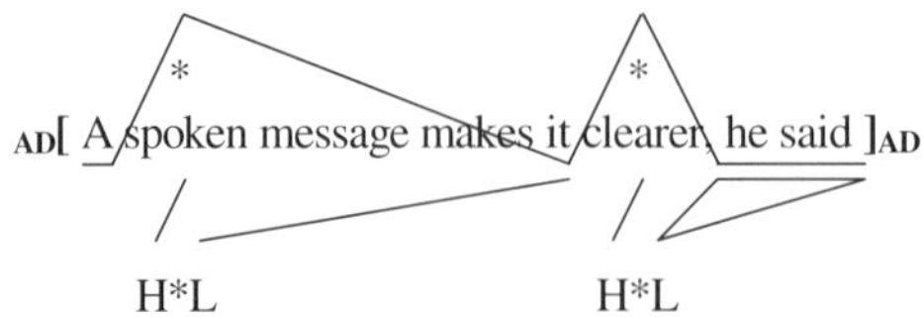

are aligned by means of association lines. The starred tone of the pitch accent, H* here, goes to the accented syllable, "spo" in this case. The second tone of the pitch accent, L here, *spreads*. This means that the pitch in the stretch of speech until the right-hand AD boundary is determined by the L. Spreading of a tone is indicated by a triangle.[4] In (5b), the two ADs have been restructured, causing the first AD to lose its right-hand boundary. As a result, the final tone of the first pitch accent moves to the syllable just before the second accented syllable.[5] After interpolation between the targets corresponding to the tones, the intonation contour has a different shape. In addition to the intonational consequences of restructuring, there are also durational consequences. [GR92b] provides experimental evidence for the claim that an AD boundary enhances *preboundary lengthening*, that is, a syllable before an AD boundary is longer than a syllable that is not AD-final. Furthermore, a right-hand AD boundary is often followed by a pause, which may also disappear after restructuring (see also [SC94]).

38.2.4 Constraints on AD-Restructuring

One of the problems concerning restructuring can be illustrated with the help of example (6). Part (a) shows the intonation contour before any restructuring took place. Restructuring of the first two ADs results in part (b), which is a perfectly normal intonation. Conversely, restructuring of the last two ADs gives rise to part

---

[4]The unaccented stretches preceding the first accent in an AD, i.e., "a" in the first and "makes" in the second, will typically have low or mid-pitch. We account for this by assuming an AD-initial *boundary tone* that spreads up to the first pitch accent. Also, we assume utterance-final boundary tones: L% or H%. We have omitted the boundary tones in the examples.

[5]In our phonological model the symbols on the tone string are associated with syllables on the segmental string. The phonetic model has the task of determining the exact alignment within syllables (cf. [GR92a]).

(6)

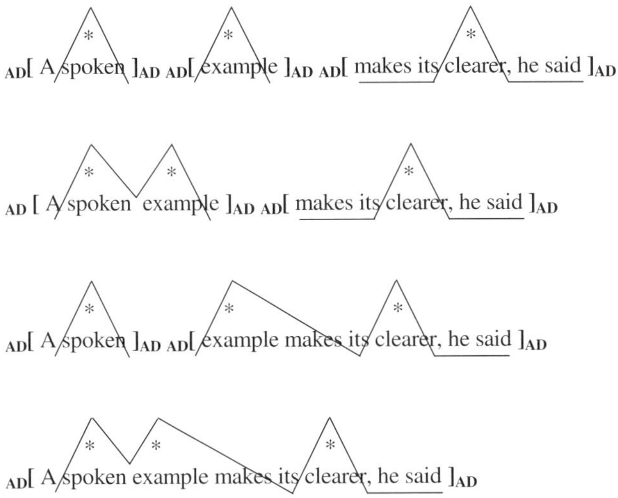

a. $_{AD}$[A spoken]$_{AD}$ $_{AD}$[example]$_{AD}$ $_{AD}$[makes its clearer, he said]$_{AD}$

b. $_{AD}$ [A spoken example]$_{AD}$ $_{AD}$[makes its clearer, he said]$_{AD}$

c. $_{AD}$[A spoken]$_{AD}$ $_{AD}$[example makes its clearer, he said]$_{AD}$

d. $_{AD}$[A spoken example makes its clearer, he said]$_{AD}$

(c), which is a very unnatural, if not ill-formed, intonation. However, restructuring of all three ADs results in a normal intonation again (part (d)). On the basis of examples such as this, [GR92b] proposed the following rule: *restructure two adjacent ADs, giving precedence to ADs that are separated by a lower-ranking prosodic boundary*. Notice that the first two ADs are separated by a Pw boundary whereas the last two ADs are separated by a PP boundary (cf. (4)). Thus, in the case of example (6), the proposed rule correctly excludes the restructuring in part (c), because the one in part (b) should have precedence.

Nevertheless, there are still many situations in which this rule offers no solution, simply because two or more pairs of adjacent ADs are separated by a prosodic boundary of the same strength. Examples like (7) suggest that the order of restructuring is nevertheless meaningful. This is an example of a well-known ambiguity caused by two possible interpretations of the PP (prepositional phrase) "with a telescope." The restructuring in (7a) favors the interpretation of the PP as a modifier of the verb "saw," that is, the policeman saw the criminal by means of a telescope. The restructuring in (7b), on the other hand, favors the interpretation of the PP as a modifier of the noun "criminal," that is, the policeman saw the criminal who possessed a telescope. Examples such as those in (7) suggest that restructuring is somehow related to the syntactic structure of an utterance. The experimental work described in the next section tries to establish this relation in the case of nonambiguous sentences.

(7)

a. * * *

${}_{AD}$[The policeman saw the criminal]$_{AD}$ ${}_{AD}$[with the telescope]$_{AD}$

b. * * *

${}_{AD}$[The policeman]$_{AD}$ ${}_{AD}$[saw the criminal with the telescope]$_{AD}$

38.3 Restructuring Intonational Domains: An Experiment

38.3.1 Hypotheses

In section 38.2, we pointed out a problem with AD-restructuring. We concluded that prosodic constituency does not sufficiently constrain the restructuring of ADs. We hypothesized, on the basis of ambiguous sentences, that syntactic structure might govern the restructuring of ADs. In addition, the length of the first AD to be restructured appears to influence the probability of restructuring. This led us to an experiment in which we investigated the dependence of restructuring on two factors: (1) the syntactic structure of an utterance; (2) the length of the first AD.

We selected two syntactic structures whose analysis is relatively uncontroversial, and moreover seem intuitively likely to influence restructuring in opposite ways. In fact, they are the syntactic structures that correspond to each of the two interpretations of the ambiguous example in (7). In the first structure the PP serves as a noun modifier. In terms of tree structures, this means that the PP is attached somewhere internal to an NP (noun phrase). In the second structure, by contrast, the PP serves as either a predicate modifier or a separate argument. This means, again in terms of tree structures, that the PP is not attached internal to an NP.

In view of the fact that intonation can contribute to linguistic processing, that is, the recovering of the structural aspects of the message, it seems reasonable to consider intonational domains as one of the cues to the attachment of a PP. For syntactically ambiguous utterances such as those shown in (7), the intonational domains might in fact be the only cue to the right interpretation. But even if the sentence in question is not ambiguous, intonational domains could still form an auxiliary cue. Moreover, incompatible intonational domains would clash with syntactic cues, thereby complicating the processing. In the case of an external PP, for example, we expect an intonational boundary (i.e., the start of a new AD in terms of our theory) to coincide with the beginning of the PP. Conversely, we expect no intonational boundary in the case of an external PP, that is, we expect AD-restructuring. Hypothesis 1 reflects these expectations concerning the relation between the attachment of the PP on the one hand and restructuring on the other.

Hypothesis 1: Restructuring based on syntactic structure

a If a PP is internal to an NP, restructuring will force them in the same AD.

b If a PP is external to an NP, no restructuring will occur.

Clearly, there are limits to the acceptable length of an intonational domain. In cases where restructuring would give rise to unacceptably long ADs, restructuring is not allowed. Conversely, restructuring of a relatively short AD could take place even if this would cause an NP and an external PP to be in the same AD. In other words, we expected that the factor of the length of the initial AD would sometimes override the factor of syntactic structure. This expectation is reflected in hypothesis 2: The conditions that length places on restructuring can overrule restructuring based on syntactic structure.

Hypothesis 2: Restructuring based on length

a If the length of an AD is "relatively long," restructuring will be blocked (overruling hypothesis 1a).

b If the length of an AD is "relatively short," restructuring will be forced (overruling hypothesis 1b).

The experiment was meant to obtain evidence for the two hypotheses as well as more specific indications of what "relatively long" and "relatively short" might mean.

We decided to use synthesized speech in this experiment. This allowed us to manipulate the intonational aspects only. Therefore, we could produce several versions of the same utterance that differed only with respect to their intonation contours and intonational domains. The decision to use synthesis instead of resynthesis was motivated by our aim to improve the intonational component of our speech synthesizer.

38.3.2 *Material*

We devised four sets of sentences with PPs internal to NPs, and four sets with PPs external to NPs. Each set contained four sentences that differed only in the length of the stretch from the start of the sentence to the beginning of the PP, measured in terms of the number of syllables. There were four corresponding ranges: [5–7] (mean value 5.9), [9–12] (mean value 10.4), [13–17] (mean value 15.1), and [17–20] (mean value 19.1) syllables. The stretch from the beginning of the PP to the end of the sentence was kept relatively short ([6–8] syllables, mean value 7). Examples of these sets are given in (8a) and (9a); (8b) and (9b) provide a gloss.

None of these sentences were ambiguous for humans. We accented all the words except verbs,[6] pronouns, and complementizers. Although in Dutch an optional rhythmic adjustment rule allows for the deletion of some accents, we decided to distribute the accents in a conservative but clearly acceptable way. For each sentence, we derived the corresponding strictly layered prosodic structure by means

[6]Verbs in Dutch can remain unaccented even when they are included in the focus of the sentence. The same is true for English and German [Sel95].

(8)

a.1 Tenzij hij zijn gedachten
over het schaakspel bedoelde.

a.2 Tenzij hij zijn filosofische gedachten
over het schaakspel bedoelde.

a.3 Tenzij hij zijn onorthodoxe filosofische gedachten
over het schaakspel bedoelde.

a.4 Tenzij hij zijn bijzonder onorthodoxe filosofische gedachten
over het schaakspel bedoelde.

b. Tenzij hij zijn bijzonder unorthodoxe filosofische gedachten
unless he his very unorthodox philosophical thoughts
over het schaakspel bedoelde.
about the chess-game meant
"Unless he meant his very peculiar philosophical thoughts
about the game of chess."

c. $_{S}$(tenzij hij zijn $_{NP}$($_{NP}$(zijn ... gedachten)$_{NP}$ $_{PP}$(over het schaakspel)$_{PP}$)$_{NP}$
bedoelde)$_{S}$

d. $_{Ut}$($_{Ip}$($_{Pp}$(tenzij hij zijn ... gedachten)$_{Pp}$ $_{Pp}$(over het schaakspel)$_{Pp}$
$_{Pp}$(bedoelde)$_{Pp}$)$_{Ip}$)$_{Ut}$

e.1 $_{AD}$[tenzij hij zijn ... gedachten]$_{AD}$ $_{AD}$[over het schaakspel bedoelde]$_{AD}$

L H*LH L H*L L%

e.2 $_{AD}$[tenzij hij zijn ... gedachten over het schaakspel bedoelde]$_{AD}$

L H*L H*L L%

of the mapping rules described in [NV86]. (8d) and (9d) exemplify the prosodic structures derived from the respective syntactic structures in (8c) and (9c). Notice that although their syntactic structures differ, their prosodic structures are identical; this is exactly the situation we wanted to investigate. Next, we determined for every accent its AD on the basis of accent distribution and prosodic structure. We assumed that all ADs separated by only a Pw boundary were obligatorily restructured. The net effect was that sentences consisted of two ADs, with the beginning of the second AD corresponding to the beginning of the PP. We produced two utterances out of each sentence by means of a rule-based allophone speech synthesizer for Dutch [KRGE94]. The first version contained two ADs, whereas the second restructured version contained only a single AD. The contrast in the shape of the intonation contours is exemplified in example (8e.1) versus (8e.2). The final pitch accent of the initial AD in the versions containing two ADs was realized as H*LH in order to obtain an H boundary tone, whereas all other pitch accents were realized as H*L. These choices of pitch accents are conservative in the sense that the resultant tunes are maximally neutral. In pilot experiments, it had become clear that our naive raters experienced difficulties when judging stylized intonation contours. We decided to increase the naturalness of the synthetic utterances by marking the nonfinal right-hand AD boundaries by a lengthening of preboundary syllable (10–30 ms) and by adding a small postboundary pause (35 ms).[7] We also added an accentual and phrasal *downstep* of 0.9 to each of the utterances, which caused the subsequent F_0 peaks to be lower than the preceding peak by a constant factor [vGR92].

38.3.3 Method

The restructured and nonrestructured versions of each sentence were compared by 25 naive raters. Raters could read the orthographic version of the sentence on a screen and could listen to both versions as many times as they wished. In the case of naive raters, it is, of course, impossible to ask directly which intonation contour is the more appropriate one, given the syntactic structure of the sentence. Instead, we asked them to judge which of the two versions was more adequate, given the meaning of the sentence. The total of 32 pairs of utterances were presented in random order. Raters recorded their ratings on a 7-point scale that ran from "first version much better" via "equally good" to "second version much better." In order to focus their attention on the phenomenon that we wanted them to judge, we started every session with four familiarization pairs that contained an idiomatic expression or proverb. In expressions such as these, the choice of intonational domains is fixed by convention. One version of every training pair had the conventional AD boundary, and the other version contained a very unlikely AD boundary inside the idiomatic expression. Raters received feedback about the correctness of their rating during these familiarization utterances.

[7] It is generally accepted that these amounts of lengthening and pausing are by themselves not sufficient to produce a subjective break.

(9)

a.1 Tenzij hij zijn gedachten
in een vakblad publiceert.

a.2 Tenzij hij zijn filosofische gedachten
in een vakblad publiceert.

a.3 Tenzij hij zijn onorthodoxe filosofische gedachten
in een vakblad publiceert.

a.4 Tenzij hij zijn bijzonder onorthodoxe filosofische gedachten
in een vakblad publiceert.

b. Tenzij hij zijn bijzonder unorthodoxe filosofische gedachten
unless he his very unorthodox philosophical thoughts
in een vakblad publiceert.
in a professional-journal publishes
"Unless he publishes his very peculiar philosophical thoughts
in a professional journal."

c. $_{S}($ tenzij hij zijn $_{NP}($ zijn ... gedachten $)_{NP}$ $_{PP}($ in een vakblad $)_{PP}$ publiceert $)_{S}$

d. $_{Ut}($ $_{Ip}($ $_{Pp}($ tenzij hij zijn ... gedachten $)_{Pp}$ $_{Pp}($ in een vakblad $)_{Pp}$
$_{Pp}($ publiceert $)_{Pp}$ $)_{Ip}$ $)_{Ut}$

38.3.4 Results

An analysis of variance was carried out on the data, with three within-subject factors: (1) syntactic structure, (2) length of the initial AD, and (3) sentence set. First of all, the analysis revealed no significant interaction among the factors syntactic structure and length of initial AD. Furthermore, it showed that the syntactic structure is a significant factor ($F_{1,24} = 13.79$, $p < 0.001$). This implies that restructuring is indeed sensitive to whether attachment of a PP is internal or external to an NP. The length of the initial AD is also a significant factor ($F_{3,72} = 7.25$, $p < 0.001$, Huynh-Feldt corrected). As expected, restructuring is sensitive to the length of the initial AD as well.

Figure 38.1 shows the combined effect of the factors syntactic structure and length of initial AD. Apparently, restructuring is less acceptable for utterances with an external PP than for utterances with an internal PP. Likewise, the acceptability of restructuring decreases when the length of the initial AD increases. z-ratios (one-tailed) were used to determine whether some combinations of the two factors made the preference scores significantly larger or smaller than zero ($p < 0.05$),

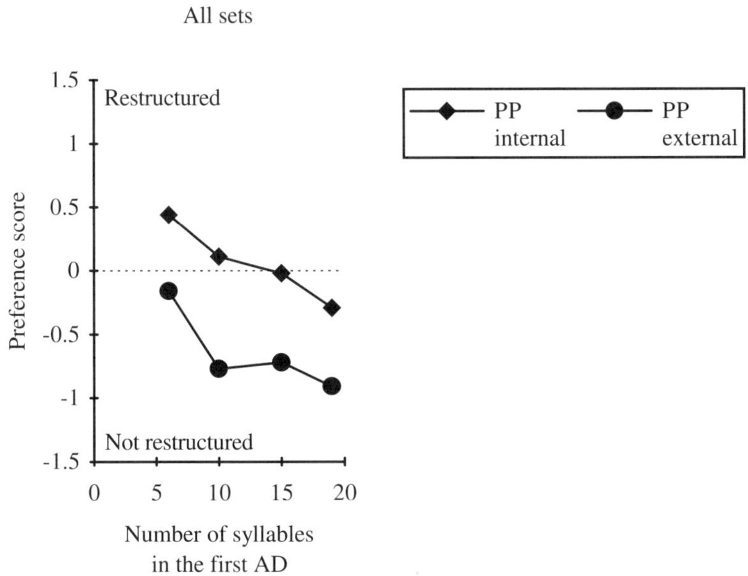

FIGURE 38.1. Mean preference scores over all four sets.

as this would imply a significant preference for either restructuring or a boundary. Restructuring is significantly preferred for those utterances with an internal PP and an initial AD of six or fewer syllables. On the other hand, a boundary instead of restructuring is significantly preferred in the case of an internal PP and an initial AD of 19 or more syllables. Finally, utterances with an external PP and a initial AD of 10 or more syllables are significantly preferred without restructuring. The only significant interaction, according to the analysis of variance, occurred among the factors length and sentence set.

These results support the claim that both the syntactic structure of an utterance and the length of its ADs are relevant to AD-restructuring. However, they are incompatible with hypothesis 1a, which states that if a PP is internal to an NP, restructuring will cause them to be in the same AD. Notice that it would be supported if we restricted ourselves to only those utterances with the shortest initial AD. Here, the preferred restructuring is more appropriately explained by referring to the length of the initial AD, as in hypothesis 2. The results do, however, support hypothesis 1b, which states that if a PP is external to an NP no restructuring will occur. The indecisive behavior of those utterances with the shortest initial AD is, again, more appropriately explained by the factor length.

For restructuring based on length we arrive at a similar conclusion. Hypothesis 2a states that restructuring will be blocked if the length of the initial AD is "relatively long." If we substitute the value "19 or more syllables" for "relatively long," this hypothesis is in fact compatible with the data (cf. the first point from the right for PP internal in figure 38.1). The expectation that restructuring will be forced if the length of an AD is "relatively short," as formulated in hypothesis 2b, can be neither rejected nor supported. The mean preference score for utterances with an external PP in combination with the shortest initial AD does not significantly differ from zero (cf. the first point from the left for PP external in figure 38.1). Although we cannot decide on hypothesis 2b in general, it is supported if we restrict ourselves to utterances with an internal PP and substitute "six or fewer syllables" for "relatively short" (cf. the first point from the left in the graph for PP internal in figure 38.1). In addition, we can remark that a boundary is no longer significantly preferred for utterances with an external PP and the smallest initial AD, as opposed to utterances with an external PP and a larger initial AD (cf. the first point from the left in the graph for PP external in figure 38.1).

38.4 Discussion

38.4.1 Effect Size

Although some of the results turned out to be statistically significant, they do not indicate a great sensitivity of the raters to any of the examined factors when compared to the full range of the scale. A possible explanation is that most raters considered the task to be rather difficult. An examination of the scores reveals that raters were somewhat erratic. For example, 16 out of 25 raters recorded at least once a difference of 4 points between pairs of utterances that differed in only one additional word. However, in support of our results, we can mention that they agree with other experimental results. Sanderman and Collier [SC94] report that both the length of the initial intonational domain and attachment of the PP influence the *perceptual boundary strength* (PBS).

Another point is the relation to speech rate. It is generally assumed that increasing the speech rate will reduce the number of intonational boundaries; see [Cas94] for experimental evidence. In our stimuli, we aimed to reproduce a normal speech rate such as that used in news broadcasting. We can only speculate about the effects on our results of an increased speech rate. Perhaps this would shift up all points in figure 38.1 and confirm hypothesis 1b.

Finally, we have seen that the critical AD lengths for restructuring are somewhere around 5 and 15 syllables. It would be interesting to see if these figures can be confirmed in a follow-up experiment.

38.4.2 Consequences for Our Theory of Intonational Domains

One of the reasons for assuming prosodic structure is that syntactic structure is too informative for the purposes of phonology. Prosodic phonology interfaces between syntax and phonology. In the process mapping the syntactic constituents to phonological constituents the amount of structural information is reduced. Recursive structure, for example, is one of the syntactic aspects that will be removed. The claim that prosodic structure contains no recursive constituents is known as the

(10)

Syntactic structure: Prosodic structure:

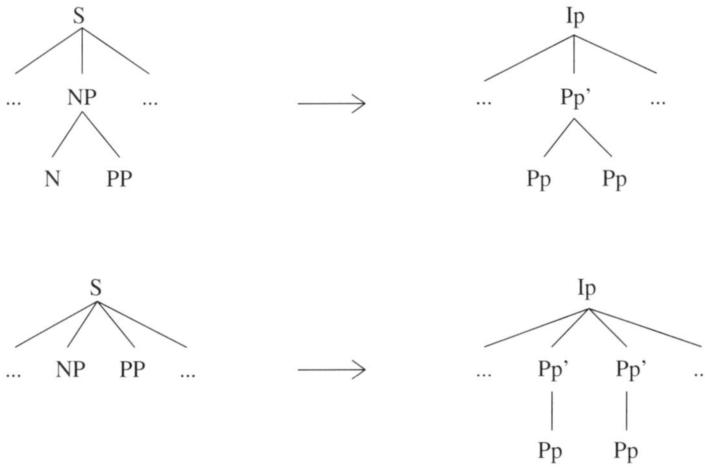

strict layer hypothesis. Exactly this claim is responsible for the problem that the results pose for our theory of intonational domains. Restructuring apparently depends on the way a PP is attached, internal or external, to a NP, but this information is lost during the translation from syntactic to prosodic structure. Both syntactic structures map onto the same prosodic structure (cf. example (8cd) and (9cd)). Consequently, the constraints on restructuring cannot be adequately expressed.

A possible solution is to augment the prosodic hierarchy with a recursive phonological phrase, let's say, a Pp'. The mapping rules would have to be adapted to give the result as in (10), which would enable us to express the difference in restructuring. Deleting the AD boundary between a noun and an internal PP, which coincides with a Pp boundary, has priority over the deletion of the AD boundary between an NP and external PP, which coincides with a Pp' boundary. Of course, this solution amounts to a rejection of the strict layer hypothesis. In fact, other research exists that suggests that the restrictions the strict layer hypothesis imposes on the prosodic structure are actually too strong [GG83, Lad86, dS94, p. 2046].

38.4.3 Consequences for Speech Synthesis

The present results allow us to improve our intonational domains in speech synthesis. In text generation, the information about the attachment of PPs as well as the length of the ADs is readily available. Real ambiguous sentences (to humans), in which the intonational domain is the only cue left to the right interpretation, are quite rare. On the other hand, PPs themselves are very frequent. In the latter case, adequate intonational domains will most likely contribute to processing ease and perceived naturalness.

38.5 Conclusion

Intonational domains cannot be consistently identified with the intonational phrase of the prosodic hierarchy. We need an independent domain, called the association domain, which is determined primarily by the pitch accent distribution and secondarily by the prosodic constituency. In principle, each accent has its own AD, but the process called AD-restructuring can join two ADs to form a single one. Restructuring seems to depend on the syntactic structure: It becomes more likely when an AD boundary corresponds to the beginning of a PP that modifies a noun. In addition, restructuring seems to depend on the length of the ADs involved: It is negatively related to the length of the first AD. The constraints on restructuring cannot be adequately expressed in terms of conventional prosodic constituency, which respects the strict layer hypothesis.

References

[BP86] M. Beckman and J. Pierrehumbert. Intonational structure in English and Japanese. *Phonology Yearbook* 3:255–309, 1986.

[Cas94] J. Caspers. *Pitch Movements Under Time Pressure: Effects of Speech Rate on the Melodic Marking of Accents and Boundaries in Dutch.* (HIL dissertations 10) Holland Academic Graphics, Den Haag, 1994.

[dS94] J. R. de Pijper and A. A. Sanderman. On the perceptual strength of prosodic boundaries and its relation to suprasegmental cues. *JASA* 96(4):2037–2047, 1994.

[GG83] J. P. Gee and F. Grosjean. Performance structures: A psycholinguistic and linguistic appraisal. *Cognitive Psychology* 15:411–458, 1983.

[GR92a] C. Gussenhoven and T. Rietveld. A target interpolation model for the intonation of Dutch. In *Proceedings of ICSLP 92*, vol. 2, 1235–1238, 1992.

[GR92b] C. Gussenhoven and T. Rietveld. Intonation contours, prosodic structure and preboundary lengthening. *J. Phonetics* 20:283–303, 1992.

[Gus88a] C. Gussenhoven. Intonational phrasing and the prosodic hierarchy. In *Phonologica 1988*, W. U. Dressler et al., eds. Cambridge University Press, Cambridge, 89–99, 1988.

[Gus88b] C. Gussenhoven. Adequacy in intonation analysis: The case of Dutch. In *Autosegmental Studies on Pitch Accent*, H. v. d. Hulst and N. Smith, eds. Foris Publications, Dordrechet, 95–121, 1988.

[Gus91] C. Gussenhoven. Tone segments in the intonation of Dutch. In *The Berkeley Conference on Dutch Linguistics 1989*, T. F. Shannon and J. . Snapper, eds. University Press of America, Lanham, MD, 139–155, 1991.

[KRGE94] J. Kerkhoff, T. Rietveld, C. Gussenhoven, and L. Elich. NIROS: The Nijmegen Interactive Rule Oriented Speech Synthesis System. Proceedings, Department of Language and Speech 18, University of Nijmegen, 107–119, 1994.

[Lad86] D. R. Ladd. Intonational phrasing: The case for recursive prosodic structure. *Phonology Yearbook* 3:311–340, 1986.

[Lad92] D. R. Ladd. An introduction to intonational phonology. In *Papers in Laboratory Phonology II*, G. J. Docherty and D. R. Ladd, eds. Cambridge University Press, Cambridge, 321–334, 1992.

[NV86] M. Nespor and I. Vogel. *Prosodic Phonology*. Foris, Dordrecht, 1986.

[NV89] M. Nespor and I. Vogel. On clashes and lapses. *Phonology* 6:69–116, 1989.

[Pie80] J. Pierrehumbert. *The Phonology and Phonetics of English Intonation*. Garland Press, New York, 1980.

[SC94] A. A. Sanderman and R. Collier. Prosodic phrasing at the sentence level. In *Festschrift for K. Harris of Physics, Modern Acoustics and Signal Processing Series*, American Institute of Physics, 1994.

[Sel84] E. Selkirk. *Phonology and Syntax.* MIT Press, Cambridge, MA, 1984.

[Sel95] E. Selkirk. Sentence prosody: Intonation, stress and phrasing. In *Handbook of Phonological Theory*, J. Goldsmith, ed. Blackwell, 550–569, 1995.

[Tri59] J. L. M. Trim. Major and minor tone groups in English, *Le Maître Phonétique* 112:26–29, 1959.

[vGR92] R. van den Berg, C. Gussenhoven, and T. Rietveld. Downstep in Dutch: Implications for a model. In *Papers in laboratory phonology II*, G. J. Docherty and D. R. Ladd, eds. Cambridge University Press, Cambridge, 255–310, 1992.

39

Speaking Styles: Statistical Analysis and Synthesis by a Text-to-Speech System

Masanobu Abe

ABSTRACT This chapter considers the enhancement of text-to-speech (TTS) systems by the synthesis of various speaking styles. The first part statistically analyzes speech uttered in three speaking styles. The speaking styles are indicated by text content: a paragraph of an literary novel, advertisement phrases, and a paragraph of an encyclopedia. A professional narrator uttered the three texts in speaking styles that he thought were appropriate. Characteristics of each speaking style are observed in F_0, duration, power, formant frequency, and spectral tilts. Based on the analysis results, we propose a strategy that permits a TTS system to synthesize speech in various speaking styles. Rules are integrated into a conventional TTS system, and listening tests show good performance of the proposed TTS system.

39.1 Introduction

To extend the application area of synthesized speech, the technology necessary for the synthesis of various speech styles is critical [Car92, Arg92, Esk92, Esk93]. For human/machine problem-solving dialog systems, the output speech should be capable of expressing information in a natural manner by reproducing natural emphasis and emotion [KT89]. Text-to-speech (TTS) systems will be required to synthesize not only the standard reading style, but also task-specific styles such as TV news, commercials, and warnings. Another requirement is to synthesize speech with unique characteristics [ANSK88, ML91]. As a first step in achieving these goals, we synthesize speech in three different speaking styles.

In the first part of this chapter, we examine the characteristics of distinct speaking styles. In some cases, speaking styles are determined by text content. Weather forecasts, for example, typically sound monotonous, whereas a slogan at a demonstration might be shouted with emotion. We carefully selected three different style for each text, and a listener could clearly distinguish them. All texts were Japanese. When a person uses a different speaking style, he changes prosodic characteristics such as F_0 range, speaking rate and speech power. However, prosodic parameters alone are insufficient for synthesizing realistic speaking styles because spectral parameters are also changed by moving articulators quickly or slowly, changing mouth opening and so on. Therefore, we analyzed spectral parameters as well as prosodic parameters. Because statistical methods were used to analyze data, we report the average and dominant characteristics of different speaking styles.

The second part of this chapter refers to the analysis results to propose a strategy that allows a TTS system to synthesize various speaking styles. According to the strategy, rules are integrated into a conventional TTS system, and its performance is evaluated by listening tests.

39.2 Speech Material

Three texts from different fields were selected as material to which style could be clearly assigned: a paragraph of a literary novel, advertisement phrases, and a paragraph from an encyclopedia. For convenience, we refer to the three speaking styles as the novel, advertisement, and encyclopedia speaking styles. These three fields were selected not only so that a narrator would consciously utter them differently, but also so that listeners would easily recognize them as different styles. A set of common texts (i.e., a set of 100 sentences and 216 phonetically balanced words) were also selected. For speech recording, a professional narrator first uttered the text of a field in a speaking style that he judged appropriate. This constituted a preparation session to fix the speaking style. Just after this session, he uttered the common text (100 sentences and 216 phonetically balanced words) in the same speaking style. All recorded speech data were manually assigned phonetic transcriptions. To extract the characteristics of speaking styles, we used the common text speech because it is free of text-dependent effects.

39.3 Spectral Characteristics of Different Speaking Styles

39.3.1 Analysis Method

To investigate the spectral characteristics of the speaking styles, we applied a voice conversion algorithm based on codebook mapping [ANSK88]. The algorithm generates a pair of codebooks using training data from two speaking styles, and the codevectors of one codebook have a one-to-one correspondence to the codevectors of the other codebook. Therefore, comparing pairs of codevectors makes it possible to reveal the spectral differences between the two speaking styles. Six mapping codebook pairs (all combinations of the three styles) were generated. Table 39.1 shows the analysis conditions.

Formant frequencies (F_1,F_2,F_3), formant bandwidths, and spectral tilts were extracted from each codevector as spectral parameters. Formant frequencies were determined by referring to the pole frequencies of an AR model, and formant corre-

TABLE 39.1. Analysis Conditions

sampling frequency	12kHz
window length	256 points(21.3msec)
window shift	36 points (3.0msec)
LPC analysis order	14
pre-emphasis	$1 - 0.97 z^{-1}$
clustering measure	cepstral Euclid distance
training samples for clustering	24000 frames
codebook size	256
training words for mapping codebook	200

spondences between codevectors were manually determined. Spectrum tilts were defined as the first-order regression coefficients of the LPC spectrum envelopes.

Phonetic attributes were also assigned to codevectors using phonetic transcriptions. We defined X as the total frequency of a codevector in the training data, and Y as the frequency of the codevector in phoneme Z. If Y/X was greater than 70%, the codevector was assigned to phonetic attribute Z.

39.3.2 Characteristics of Formant Frequency

Figure 39.1 shows the F_1-F_2 plane of the three speaking styles. In all vowels, the first formant frequency increases in the order of novel, encyclopedia, and advertisement speaking style. The difference in F_1 frequency is about 10% to 20% of the F_1 value of the encyclopedia speaking style. On the other hand, the F_2 differences among the three styles are about 5%. In terms of the third formant frequency (data not shown) the novel speaking style has the lowest frequency, and the amount of the difference is 5%–10%. A previous study [KO87] showed that voice personality is perceptually lost when an individual formant is shifted 15% toward either the high or low frequency region. Judging from the reference, in terms of F_1 and F_3, formant differences among speaking style are large enough to confuse voice personality. These results suggest that modifying formant frequency is important in synthesizing realistic speaking styles.

39.3.3 Characteristics in Spectral Tilt

The same trends in spectral tilt characteristics were observed in all vowels. We will explain the tendency using the phoneme /o/ as an example. Figure 39.2 shows the spectral tilts of the three speaking styles. Each circle represents the spectral tilts extracted from a pair of associated codevectors. The empty circles plot the relationship between the encyclopedia speaking style and the novel speaking style, and the filled circles show the relationship between the encyclopedia speaking style and the advertisement speaking style. Regression lines are drawn for both empty and filled circles.

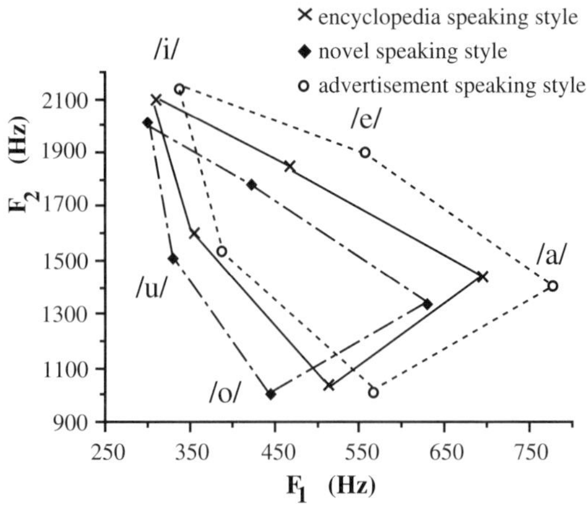

FIGURE 39.1. F_1-F_2 plane of the three speaking styles.

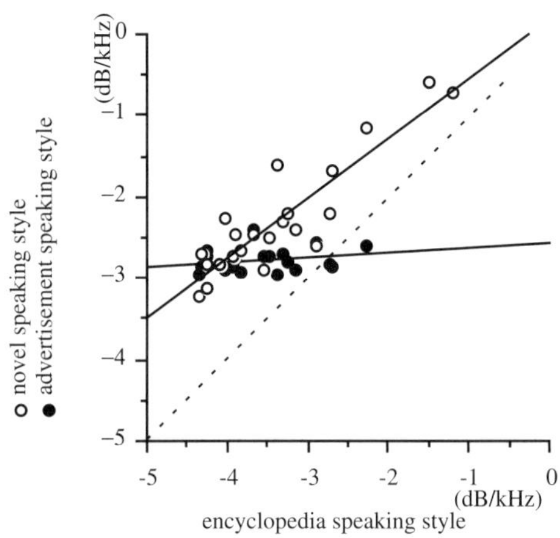

FIGURE 39.2. Spectral tilts in phoneme /o/.

The spectral tilt of the novel speaking style is always flatter than that of the encyclopedia speaking style. The spectral tilt values of the novel and encyclopedia speaking styles are distributed from -3dB/kHz to -1dB/kHz, and the distribution is considered to be caused by slight spectrum differences in the phoneme /o/. This may be because speech is produced in a free and easy manner in the novel and encyclopedia speaking styles. On the other hand, the spectral tilts of the advertisement style are almost constant around -3 dB/kHz. This could be a result of the way in which advertisements were articulated louder (see figure 39.7) and with a bigger jaw opening. This suggests hyperarticulation.

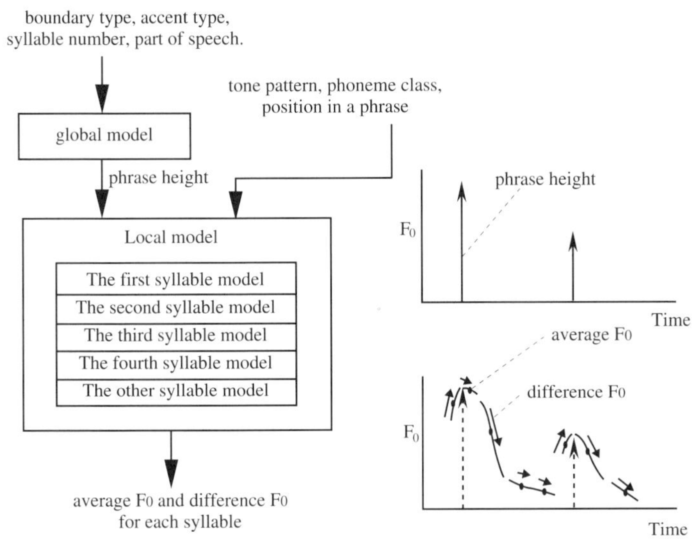

FIGURE 39.3. Two-stage F_0 control model.

39.4 Prosodic Characteristics of Different Speaking Styles

39.4.1 Characteristics in Fundamental Frequency

Analysis Method

To investigate the fundamental frequency (F_0) characteristics of the three speaking styles, a global-local model was applied [AS92]. Figure 39.3 shows a block diagram of the model. The global model estimates the maximum F_0 values for minor phrases that constitute the sentence, and the local model estimates detailed F_0 movements within each minor phrase. The reason for the global-local separation is that the control factors of each model are quite different. That is, in the global model they are speaker's intention, discourse structure and so on, and in the local model they are accent types, phonemes, and so on.

The local model consists of syllable-based F_0 units (SBUs). A SBU, which is generated for each syllable position, characterizes the F_0 pattern of the syllable. Currently, the F_0 pattern is specified by the average F_0 value and the slope of the F_0 movement within the syllable. For speech synthesis, the F_0 pattern is generated using these two parameters. To obtain smooth pitch change, linear interpolation is performed across syllable boundaries if necessary.

The following linear equation is used in both global and local models.

$$\hat{y}_i = \bar{y} + \sum_f \sum_c x_{fc}\delta_{fc}(i) \tag{39.1}$$

where $\hat{y}_i$ is the estimated value of the i-th sample, $\bar{y}$ is the mean of all samples, and $\delta_{fc}(i)$ is the characteristic function:

$$\delta_{fc}(i) = \begin{cases} 1 & \text{if } i\text{-th sample falls into category } c \text{ of factor } f \\ 0 & \text{otherwise} \end{cases} \quad (39.2)$$

x_{fc} is obtained by linear regression through the minimization of the following prediction error [Hay50].

$$(\hat{y}_i - y_i)^2 \quad (39.3)$$

One global model was generated for each speaking style because phrase height assignments differ among the speaking styles. Control factors are boundary type, accent type, syllable number, and part of speech. The local model is mainly controlled by accent types and phonemes. These factors seem to change little with speaking styles; for example, in Japanese, changing the accent type changes word meaning. Therefore, we have a hypothesis that a single local model is applicable for various speaking styles. To confirm this point, the local model was modified to output the difference F_0 value between the encyclopedia speaking style and the other speaking styles. The encyclopedia style was used as the reference. Because we also synthesized the encyclopedia style using the conventional TTS system, this comparison is useful. Control factors are tone pattern, phoneme class, position in a phrase, and difference of global model's outputs for each speaking style.

Analysis Result

Table 39.2 shows the parameters of the global model for the three speaking styles. The average F_0 indicates that the F_0 ranges are quite different for the three speaking styles. The multiple correlation coefficient means the model's preciseness; if it is 1.0, the model is perfect. Judging from the values of the encyclopedia and advertisement speaking styles (0.834 and 0.804, respectively), good global models were constructed for both styles. On the other hand, the global model for the novel speaking style is relatively poor (0.689). This result implies that other factors such as emotion, speaker's intention, and so on are also important in forming the novel speaking style.

The partial correlation coefficient indicates how important a factor is in a model. The experimental results show that the importance of the factors are the same in all speaking styles (i.e., boundary type is most important, syllable number is second, and so on). However, the components of the boundary type factor have quite different values for each speaking style. Figure 39.4 shows the values of boundary factors. In the figure, "tight connection" and "loose connection" are attributes of sentence structure. "Tight connection" means that a current minor phrase directly modifies the following phrase or is directly modified by the preceding phrase, and "loose connection" means there is no direct relation between two phrases. Judging from figure 39.4, the main differences are that the novel speaking style has a narrower dynamic range than the others, and that the advertisement speaking style has a different phrase height than the others when the phrase is preceded or followed by a pause.

TABLE 39.2. Partial and multiple correlation coefficients of a global model for the three speaking styles.

speaking styles	partial correlation coefficients							multiple correlation coefficient	RMS error (Hz)	average F_0 (Hz)
	boundary type		accent type of the phrase			syllable number in current phrase	part of speech			
	preceding	following	preceding	current	following					
encyclopedia	**0.417**	**0.528**	0.130	0.337	0.018	0.385	0.231	**0.834**	18.79	161.7
advertisement	**0.475**	**0.537**	0.110	0.221	0.039	0.323	0.150	**0.804**	25.55	229.6
novel	**0.311**	**0.438**	0.226	0.167	0.094	0.246	0.159	**0.689**	16.86	121.7

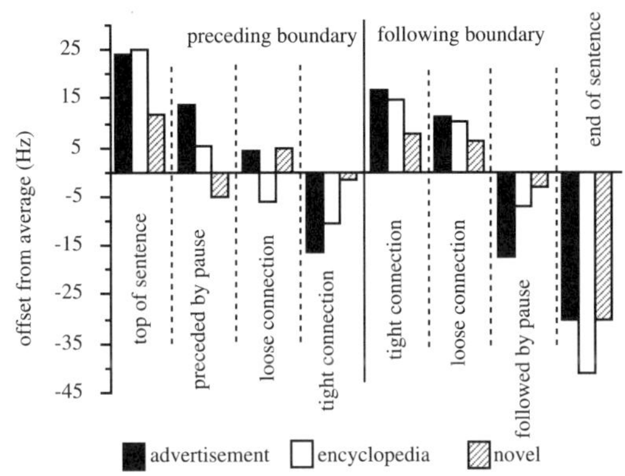

FIGURE 39.4. Effects of boundary for the three speaking styles.

Table 39.3 shows the local model parameters. In all models, the multiple correlation coefficients are quite high, and in terms of the partial correlation coefficient, the phrase height difference factor is dominant. This indicates that the influence of accent types and phonemes are quite similar for the three speaking styles. The results confirmed the hypothesis that a single local model is applicable for various speaking styles and only phrase height differences need be considered.

Judging from these results, we can conclude that one local model can be applied to the three speaking styles, but a unique global model should be constructed for each speaking style.

TABLE 39.3. Partial and multiple correlation coefficients of a local model (difference F_0 among the three speaking styles).

SBU models	partial correlation coefficients					multiple correlation coefficient	RMS error (Hz)/(Hz/3ms)
	tone pattern	consonant class of the syllable		max. difference F_0 in a phrase	position in a phrase		
		current	following				
1st syllable	0.25	0.13	0.09	0.97	–	**0.87**	18.4
2nd syllable	0.09	0.06	0.07	0.97	–	**0.97**	13.3
3rd syllable	0.21	0.12	0.10	0.98	–	**0.97**	13.7
4th syllable	0.15	0.06	0.17	0.95	–	**0.95**	16.7
other syllables	0.14	0.21	0.14	0.93	0.14	**0.93**	18.1

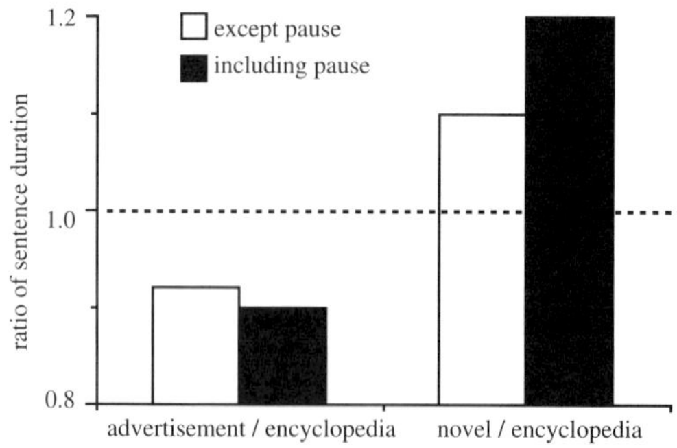

FIGURE 39.5. Sentence duration for the three speaking styles.

39.4.2 Characteristics of Segmental Duration

General Characteristics

To investigate the general characteristics of segmental duration, sentence duration was compared for the three speaking styles. The analysis considered only those sentences within which pauses were inserted at the same position in every speaking style (60 sentences). Figure 39.5 shows the ratio of sentence duration of the novel or advertisement speaking style to the sentence duration of the encyclopedia speaking style.

When sentence duration is calculated after eliminating pauses, the ratio shows the average lengthening or shortening of phoneme duration. Compared to the encyclopedia speaking style, on average, the novel speaking style has longer phoneme duration (1.1 times), whereas that of the advertisement style is shorter (0.9 times). The existence or absence of pause duration causes more difference in the novel speaking style than in the advertisement style. That is, the novel speaking style has much longer sentence duration (1.2 times) when sentence duration includes pauses. This indicates that the speaker emphasized pause length to decrease the speech rate.

Vowel Duration with Different Syllable Positions

For the 60 sentences analyzed in the previous section, vowel duration was classified for several different syllable positions, such as the beginning or end of the sentence, if preceded or followed by a pause and others. Figure 39.6 shows the average duration and standard deviation for the three speaking styles.

When a syllable is followed by a pause, vowel duration is lengthened [TSK89]. The increase is about 80 ms in both encyclopedia and advertisement speaking styles, but the increase is much longer (about 150 ms) in the novel speaking style. Moreover, in that style, vowel duration is lengthened (about 40 ms) when the syllable is located at the sentence end. Judging from the above results, we can conclude that syllable lengthening preceding a pause is an important phenomenon in recreating a novel speaking style.

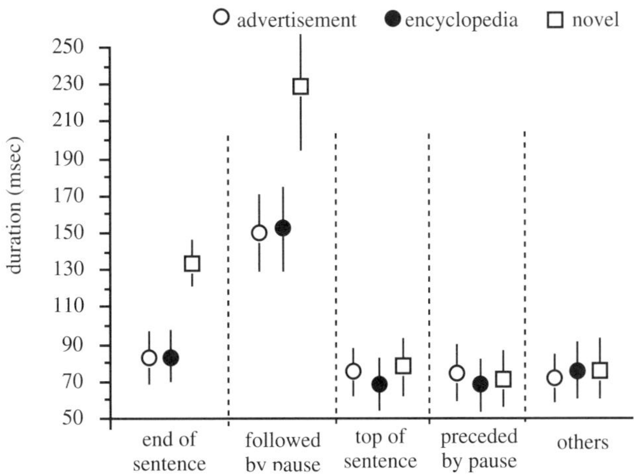

FIGURE 39.6. Average duration and standard deviation of vowels in different syllable positions.

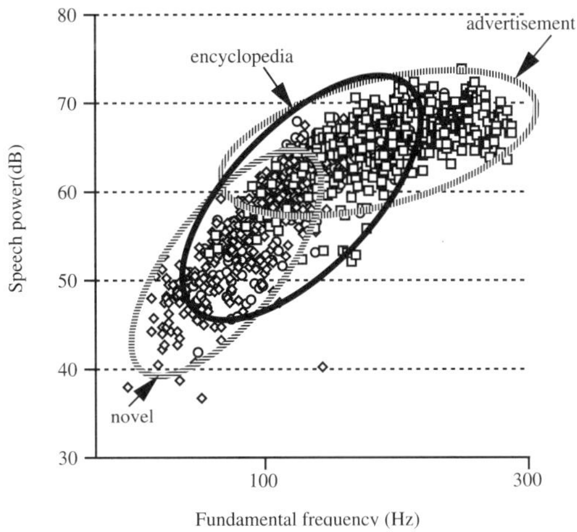

FIGURE 39.7. Relationship between fundamental frequency and power.

Characteristics of Speech Power

It has been proposed that speech power is controlled by the fundamental frequency [IHS93]. We examined speech power in the same way. Figure 39.7 shows the relationship between the average power of a vowel segment and the F_0 value in the center of the vowel. In general, the higher the F_0 is, the greater the speech power is. The relationship is almost linear in the range from 100 Hz to 200 Hz, but the speech power increase saturates above 200 Hz. It is also interesting that the speaker used the range above 200 Hz only in the advertisement speaking style. This confirms that the advertisement speaking style was uttered in an extreme mode. Judging from the results, it is reasonable to control speech power according to F_0.

Characteristics of Devocalization

In Japanese, vowel devocalization occurs very frequently in some phoneme environments, irrespective of speaking style. Examples are /i/ or /u/ in /ki/ or /su/. In this section, we do not analyze such environments, but only the phoneme environment wherein devocalization rarely occurs. We define a vowel as being devocalized when a vowel has clear formant structure and has no fundamental period in the waveforms. Decisions regarding devocalization were made manually.

Devocalization was noted 209 times, 41 times, and 1 time in the novel, encyclopedia, and advertisement speaking styles, respectively. This means that devocalization is an important phenomena to express the novel speaking style. Moreover, in the novel speaking style, 80% of the devocalization occurred in the final phrase of the sentences. This is because the F_0 range of the novel speaking style is low, and the F_0 value keeps decreasing toward the end of the sentence.

39.5 A Strategy for Changing Speaking Styles in Text-to-Speech Systems

We propose to separate rules into two groups: specific rules (Srules), which are specific to a particular speaking style, and general rules (Grules), which are applied to all speaking styles. This strategy makes it easy to adapt a TTS system to a particular speaking style because only Srules need be changed to match the speaking style required. In this section, we develop Grules and Srules according to the analysis results given in sections 39.3 and 39.4, integrate the rules into a conventional TTS system, and confirm the resulting performance by listening tests.

39.5.1 Rules for Different Speaking Styles

The baseline speaking style was the encyclopedia speaking style used to develop the conventional TTS system, and the rules of this style are Grules. We summarize the analysis results in sections 39.3 and 39.4 from the Grule-Srule point of view below.

Formant Frequencies

Speech unit segments are commonly used for all speaking styles except the following Srules: (1) In terms of the first formant frequency, the value is decreased 10% for the novel speaking style and increased 20% for the advertisement speaking style. (2) In terms of the third formant frequency, the value is decreased 20% for the novel style.

Fundamental Frequency (F_0)

In terms of the global model, boundary type factors work quite differently for each speaking style. Therefore, the global model is generated as a Srule for each style. In terms of the local model, there is no difference between the speaking styles. Thus, no Srule is necessary for the local model.

Duration

Phoneme duration rules are commonly used for all speaking styles except the following Srules: (1) Average phoneme duration of the novel speaking style is longer (1.1 times), while that of the advertisement style is shorter (0.9 times). (2) In the novel speaking style, duration lengthening in the syllable followed by a pause is emphasized by 1.8 times. (3) In the novel speaking style, vowel duration is lengthened (about 40 ms) when the syllable is located at the sentence end. (4) In the novel speaking style, pause duration is lengthened.

Power

Speech power is controlled according to F_0. In general, the relationship is almost linear, but the power-F_0 ratio depends on the speaking style. In our system, the ratio is approximated using the output from the F_0 global model.

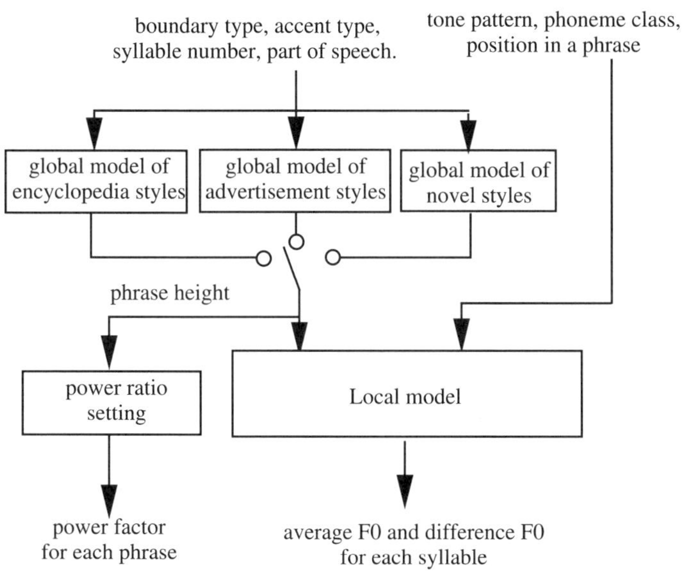

FIGURE 39.8. Prosody rule modules for three speaking styles.

39.5.2 *Proposed TTS System*

We have developed a text-to-speech system based on time-domain prosodic modification [HIS92]. In this system, phoneme segment units are selected from a large database (45,000 phoneme segments) containing word and sentence utterances, and optimum phoneme segments are selected according to various factors such as phoneme environment, phoneme duration, pitch frequency contour, and speech power. This is the baseline system that outputs the encyclopedia speaking style. The Srules explained in section 39.5.1 were integrated into the baseline system. A block diagram of F_0 generation is shown in figure 39.8. Two global models for the novel and advertisement speaking styles were added. Inputs of the global models are boundary type, accent type, syllable number, and part of speech. Inputs of the local models are tone pattern, phoneme class, and position in a phrase. Output of the global model is used to calculate the power factor.

Formant frequencies are modified by an algorithm [MA94]. In the algorithm, a formant is specified by its pole frequency, pole bandwidth, and spectral intensity at the pole frequency. After automatically extracting pole frequencies, target formant structure is set according to the Srules, and the specified formant structure is then generated by iteratively modifying pole bandwidth.

39.5.3 *Evaluation by Listening Test*

ABX listening test

An ABX listening test was carried out to check how effective the rules were. In addition to synthesizing speech in different speaking styles, using the same rules, human speech spoken in the encyclopedia speaking style was converted to other speaking styles.

Stimuli A and B were sentences uttered in the encyclopedia and the target speaking style, respectively, and stimulus X was either synthesized in the target speaking style or speech converted to the target speaking style from the encyclopedia speaking style or the speech uttered in the encyclopedia or the target speaking style. Different sentences were used for A, B and X. Two sets of three different sentences were used for human speech conversion and speech synthesis, and the total number of stimuli was 36. Eight listeners were asked to select either A or B as being closest to X.

Figure 39.9 shows the experimental results; figure 39.9(a) is for human speech conversion, and figure 39.9(b) is for synthetic speech. Even in the case of human speech, the speaking styles were not always judged to be the same as the speaker's intention (93.8%, 95.8%, and 83.3%). This indicates the difficulty of this task in some sense. Referring to human speech identification, the synthesized speech and converted human speech were effectively reproduced as the target speaking styles. These results indicate that the Srules are effective in generating different speaking styles.

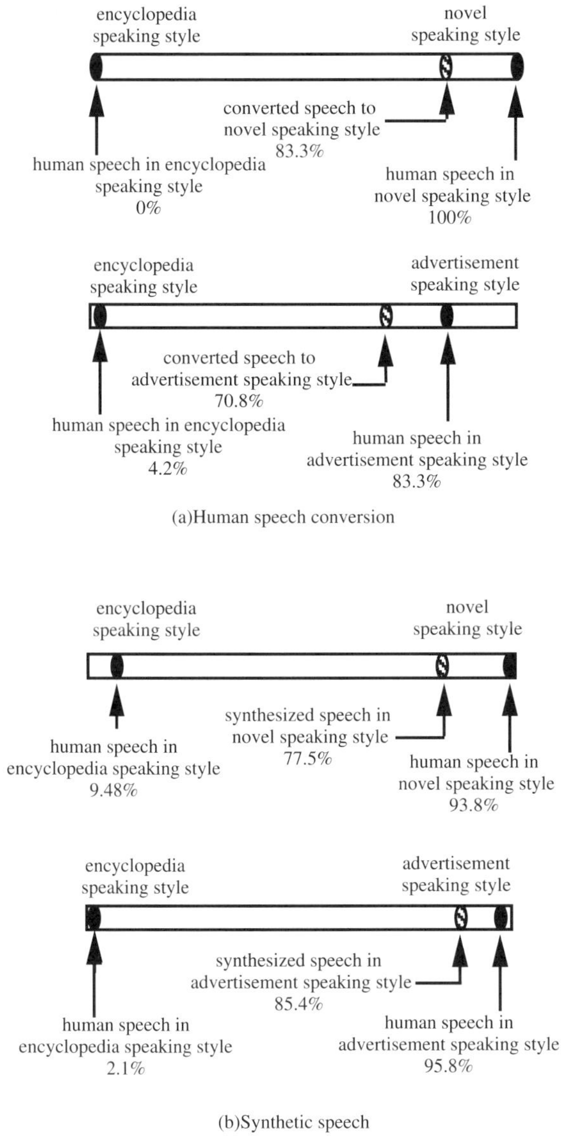

FIGURE 39.9. ABX experiment results.

Opinion Test

The proposed rules were evaluated by an opinion test. To obtain only effects of the rules, rules were applied to human speech instead of synthetic speech. Each speech pair consisted of two different sentences from five different groups. The groups were the encyclopedia style speech, the advertisement style speech, the novel style speech, and the speech converted into either advertisement or novel

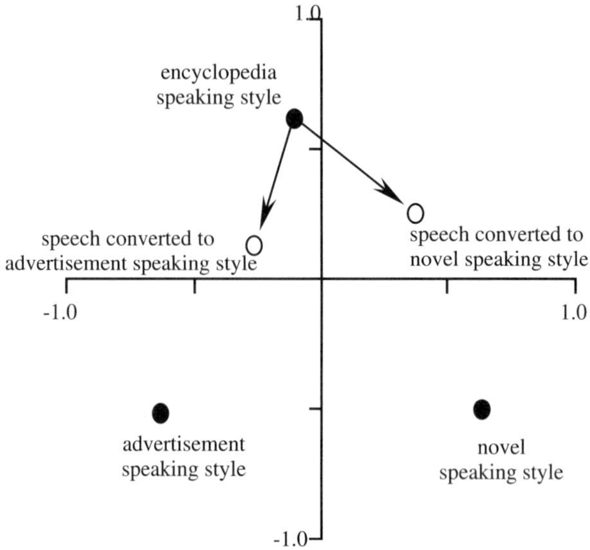

FIGURE 39.10. Distribution of psychological distance of three speaking styles and converted speech.

style from the encyclopedia speaking style. Two different sentences were used, and the total number of stimuli was 40. Eight listeners were asked to rate the similarity of each pair into five categories ranging from "similar" to "dissimilar." Hayashi's fourth method of quantification [Hay50] was applied to the experimental data obtained by the listening test. This method places a sample in a space according to the similarities between the samples.

The projection of the results onto a two-dimensional space is shown in figure 39.10. This figure shows the relative similarity-distance between stimuli.
In terms of speaking styles, the three speaking styles were clearly separated by the listeners. The converted speech is plotted midway between the encyclopedia speaking style and the target speaking style. Judging from the results, the converted speech is not close enough to the target speech. This is mainly because we changed just the average speaking style characteristics.

39.6 Conclusion

The spectral and prosodic characteristics of three speaking styles were statistically analyzed. Analysis results showed that the characteristics specifying speaking styles involve both spectral and prosodic parameters. To synthesize speech in various speaking styles, based on the analysis results, we proposed the strategy of isolating rules specific to a particular speaking style from general rules that are applied to all speaking styles. We applied the strategy to synthesize the three speaking styles, and listening tests confirmed the good performance of the strategy.

This initial research considered only three speaking styles from the same speaker. We expect that even for the same style, each speaker will have a slightly different style so that speaker-dependent characteristics as well as other speaking styles remain topics for future research. In terms of spectral parameters, we analyzed or modified only the average characteristics of the speaking styles. To obtain more realistic speaking styles, it is necessary to control more detail characteristics, such as formant transition rates, voice source characteristics, and so on. This is also work for the future.

Acknowledgments: We are grateful to the members of the Speech Processing Department for their helpful discussions. We also thank Dr. Sugamura, group leader, and Dr. Kitawaki, department head, for their continuous support of this work.

References

[ANSK88] M. Abe, S. Nakamura, K. Shikano, and H. Kuwabara, Voice conversion through vector quantization. In *Proceedings, ICASSP88*, 655–658, 1988.

[AS92] M. Abe and H. Sato. Two-stage F_0 control model using syllable based F_0 units. In *Proceedings, ICASSP92*, vol. 2, 53–56, 1992.

[Arg92] J. A. Argente. From speech to speaking styles. *Speech Comm.* 11:325–335, 1992.

[Car92] R. Carlson. Synthesis: Modeling variability and constraints. *Speech Comm.* 11:159–166, 1992.

[Esk92] M. Eskenazi. Changing speech styles: strategies in read speech and casual and careful spontaneous speech. In *Proceedings, ICSLP'92*, Banff, Alberta, Canada, 755–758, 1992.

[Esk93] M. Eskenazi. Trends in speaking styles research. In *Proceedings, Eurospeech'93*, Berlin, 501–509, 1993.

[Hay50] C. Hayashi. On the quantification of qualitative data from the mathematicostatistical point of view. *Ann. Inst. Statist. Math* 2, 1950.

[HIS92] T. Hirokawa, K. Itoh, and H. Sato. High quality speech synthesis based on wavelet compilation of phoneme segments. In *Proceedings, ICSLP'92*, Banff, Alberta, Canada, 567–570, 1992.

[IHS93] K. Itoh, T. Hirokawa, and H. Sato. Segmental power control for Japanese speech synthesis. In *Proceedings, ICSLP92*, 1143–1146, 1993.

[KT89] Y. Kitahara and Y. Tohkura. Prosodic components of speech in the expression of emotion. *JASA* 84, suppl. 1, 1989.

[KO87] H. Kuwabara and K. Ohgushi. Contributions of vocal tract resonant frequencies and bandwidth to the personal perception of speech. *Acoustica* 63:120–128, 1987.

[MA94] H. Mizuno and M. Abe. Voice conversion based on piecewise linear conversion rules of formant frequency and spectrum tilt. In *Proceedings, ICASSP'94*, Adelaide, Australia, vol.1, 469–472, 1994.

[ML91] A. I. C. Monaghan and D. R. Ladd. Manipulating synthetic intonation for speaker characterization. In *Proceedings, ICASSP'91*, Toronto, Canada, 453–456, 1991

[TSK89] K. Takeda, Y. Sagisaka and H. Kuwabara. On sentence-level factors governing segmental duration in Japanese. *JASA* 86(6):2081–2087, 1989.

Appendix: Audio Demos

Examples of the three speaking styles. Two sentences were uttered by the same speaker in the encyclopedia, novel, and advertisement speaking styles. Two sentences were synthesized both by encyclopedia speaking style and the novel or advertisement speaking style.

Section VII

Evaluation and Perception

40

Section Introduction. Evaluation Inside or Assessment Outside?

Christian Benoît

40.1 Speech Technology and Standards

Some years ago, at the time when *Talking Machines*[1] was under its editing process, it was anticipated that standardized methods would emerge from the work carried on by international and multilingual teams of researchers. European money was then at its peak to facilitate collaboration in this area so that reference techniques would be developed, encouraged, and disseminated for wide use in the speech synthesis community. After the SPIN project, the SAM project gathered several European speech laboratories in a joint effort over eight years. SAM finally delivered a battery of tests suitable for various kinds of synthetic speech evaluation. This was also when the need for worldwide collaboration was urged, and the so-called COCOSDA[2] international body was thus created to activate joint efforts among Europe, North America, Japan, and Australia. It was also at that time that a US initiative created and supported the LDC[3] association. All these activities thus let Benoît and Pols [BP92, p. 436] express the wish that "in the near future, authors will systematically use multilingual assessment methods, such as those developed within the European SAM consortium, rather than publishing their work with results, more often than not, informally tested by colleagues." How far are we from this expectation today?

Chapter 41 by Pols and Jekosch reports a situation somewhat paradoxical: Researchers in the area complain that (i) there are too many existing methods, and (ii) too many test domains lack suitable tests. Those two remarks are the reason developers and testers often prefer to design their own evaluation method (or set)

---

[1] *Talking Machines: Theories, Models, and Designs*, G. Bailly & C. Benoît, eds. Elsevier Science Publishers (1992), was based on a selection of updated papers presented at the First ETRW on Speech Synthesis, held in Autrans, France, September 1990.

[2] COCOSDA (Coordinating Committee on Speech Databases and Assessment), is a nonprofit, not-funded international group of experts in speech technology.

[3] LDC (*Linguistic DATA Consortium*), is a nonprofit organization mostly subsidized by the U.S. government, but also funded by subscription fees from public and private laboratories.

for the particular module in which they are interested, or for the particular application for which they intend their system to be used, to be evaluated adequately. Before investigating what can be standardized, and what has little chance of ever being standardized, let me quote Fourcin's [Fou92] definition of evaluation and assessment: "Whilst assessment gives an overall operational rating, evaluation is the process which, based on assessment, gives an insight into the essential dimensions and factors which are basic to the operation of a system, and basic to the nature of human communication itself." In short, evaluation looks at the inside of the system, whereas assessment looks at the system from the outside.

40.2 Why Evaluate the System Inside?

In *Talking Machines*, Fourcin [Fou92], as well as Benoît and Pols [BP92], pointed out the needs for an evaluation of intelligibility, quality, acceptability, friendliness, and so on, for whatever application the tested system would serve. There were thus many good reasons to claim that standard tests would help evaluate all those basic values that are expected from a speech synthesizer. However, things are changing with the increasing usage of TTS systems. Although their Chapter 41 here shows that researchers are largely unsatisfied with the existing tests, Pols and Jekosch propose a taxonomy to help researchers link their system and its intended application to a suite of tests. A more structured approach is necessary indeed. I wonder, however, if it is sufficient. Chapter 42 by Belhoula and his colleagues clearly illustrates how detailed and specific a test must be today for an application-oriented system to be evaluated. TTS systems are becoming less and less marketed to be used as reading machines of unlimited text in a given language. To date, two articles devoted to synthetic speech evaluation in *Talking Machines* [CGN92, Fal92] aimed at evaluating unlimited TTS synthesizers, independently of their application, in Swedish and in Italian. Conversely, chapter 42 here presents an evaluation of the performances of a speech synthesizer developed for the sole purpose of pronouncing proper names (most of them German but some having a foreign origin) embedded within a carrier sentence uttered by a human speaker. No reference corpus, no standardized test can be of any use in such a limited application! This is why Belhoula et al. developed their own test to assess the performances of their system. This is fair, but it doesn't help the reader, nor the user, to compare the reported performances with those of other systems.

At Cupertino, a good bunch of people investigate how the user of a TTS system could personalize his computer. People at Apple develop user-friendly interfaces to help the buyer record his own set of diphones so that his Mac can easily be turned into a programmable answering machine [NMS94]. Others have worked on the animated display of real faces synchronized with the audio output of the speech synthesizer, the prestored sequences of faces taking the shape of the computer owner—or even that of his cat [HL94]. We see from these examples that the entertainment industry might well create a renewed interest for TTS systems.

There is no doubt that this market will be much larger than that of reading machines dedicated to blind people or to those remaining few screenless people eager to access remotely stored electronic text. Moreover, it is likely that the incredible enthusiasm for multimedia interfaces is now pushing all the web-surfers who browse around the Internet to read cybertexts on their screen rather than to hear them through earphones—and through an unnatural voice! In that perspective, I must certainly be less ambitious than I was as far as standardization may go. Developers will probably go on running their own perceptual experiments with ad hoc tests to evaluate the performances of a given module, and there is little chance that one can give lessons—or even recommendations—to those specialists.

40.3 How to Assess the System Outside?

To sum it up, the performances of a grapheme-to-phoneme converter can hardly be evaluated through the ultimate standardized test that would give scores comparable across languages and across applications. Nevertheless, it is highly feasible to provide potential buyers, or users, with basic and simple tests that give them a better idea of what they have on the shelf. The price of the system will certainly remain the first criterion for long, and for most of the customers. Besides that, technical figures of merit, comparable across alternative speech synthesizers, should be systematically provided. They should be obtained from reference tests on the overall intelligibility of the system, its average error rate on phonemic conversions, and so forth. Then, it is the seller's role to explain why the system is good enough (or not) for the needs of the buyer. This approach is analogous to that of a car buyer. The driver's needs are expressed in terms of economy, usage (long distance versus commuting, highways versus fields), yearly mileage, comfort, speed, and so forth. The objective characteristics of the vehicle are expressed in terms of price, number of wheel drives, weight, length, number of seats, and so on. The tested performances of the vehicle are expressed in terms of gas mileage, horsepower, maximum speed limit, acceleration, among others. Those last figures are obtained from a series of standardized tests run under normalized conditions, on a motor circuit. They are thus comparable over different vehicles. From those figures, the seller and the buyer may estimate how well the vehicle would run in more natural conditions, adapted to the driver's needs. Once the given performances of the vehicle roughly match the needs of the buyer, the decision between close candidates is made on the availability of gadgets, on the image one has of the car, its color, and so forth. This is strictly subjective and cannot be evaluated by a standardized test. A car is not a speech synthesizer. Nonetheless, if almost everybody has a driving license at age twenty, everybody speaks at four. Even if almost no adults know how a car engine works, most pupils are taught the basic rules of grammar and spelling in their native language. In that sense, the comparison is not as silly as it may look. Nobody will ever buy a Jaguar to haul a caravan or to drive in mountain fields. Neither will anyone buy a TTS synthesizer from the Bell Labs or from the

CNET to synthesize proper names in Wolof or to develop a speaking clock! The characteristics of those systems don't match the Wolof needs. They also are technologically overcomplex for a speaking clock, since it would take far too many megabytes for such a simple purpose. Furthermore, a very simple speaking clock might easily be more intelligible than the best TTS synthesizer, if natural digits are adequately uttered, recorded, edited, and finally concatenated. The whole program would ultimately require few kilobytes of speech segments and few lines of code. It is easier to understand that than how an automatic gearbox works! So, why should all components of a TTS system be evaluated through standardized methods? No buyer would care, as long as the modules cannot be interchanged between synthesizers. Figures of merit would certainly please the author of a given module. But then, does the entire system perform better than another? This can be assessed only through a holistic test. This is also what the buyer wants to know. This is thus why we must work in priority on holistic tests and evaluate their robustness to serve as references in a small battery of standardized methods. I still hope that researchers will first look at existing tests when assessing their system or a module of it before developing their own tests. This would help the speech community make progress on how those tests should be better used. In that perspective, the EAGLES[4] consortium is currently publishing a handbook on speech technology. It includes a long chapter on speech synthesis assessment, which presents in great detail all the existing methods as well as recommendations for their use. A wide (and probably free) distribution of this handbook should allow researchers to use more systematically existing methods. Beside this need for academic references, COCOSDA should select, standardize, promote, and disseminate a limited set of tests, among those suggested in the chapter by Pols and Jekosch, so that developers can provide potential buyers with comparable figures of merit from holistic assessment.

40.4 And Finally, What About Humans?

Evaluation techniques allow researchers to improve the quality of a synthesizer under development. And this is rather good. Assessment methods allow a buyer to estimate the overall performances of a given system. This is not too bad, either. But what evaluation of speech synthesis brings to human knowledge goes far beyond these technological (or commercial) aspects. One not only evaluates a synthesizer when running a subjective test, whatever it is, but subject responses can—and must—be analyzed too. To me, this is the most attractive aspect of the game. Speech synthesis evaluation needs engineers and psychologists to collaborate on our most interesting challenge: To understand how speech communication works between humans, and how a system can be improved so that listeners can make

[4]EAGLES (*Expert Advisory Group on Language Standards*), is a European funded project.

the best use of it. Pisoni's chapter 43 emphasizes the role of human listeners in the evaluation process of the machines, and, in turn, how those machines allowed psycholinguists to investigate deeper how humans perceive speech. This chapter, with its historical flavor, is appealing because it compares the expectations of a prominent psychologist 15 years ago with the results obtained today. In short, it reminds us of something that should be obvious: The human is at the center of our investigation. Not only do we need humans to evaluate our systems, but all our efforts are directed toward the imitation of their voices, and to do so, we need to understand better how humans speak and how they perceive speech. In return, our machines to give some answers to the latter fundamental question. Thanks to Pisoni, we don't forget the human behind the machine. Thanks to the machine, we better understand the human mechanisms of speech perception. To conclude, speech synthesis hasn't taught us that much about the human processes of speech production so far. However, articulatory synthesis is today considered, by a large majority of the speech synthesis community, as the most promising technique in the future. There is no doubt that this area of inquiry will bring humans back to our everyday investigation. It will then be a benefit both to speech technology and to speech knowledge, that is, to speech sciences. We shall look forward to it!

References

[BP92] C. Benoît and L. C. W. Pols. On the assessment of synthetic speech. *Talking Machines: Theories, Models and Designs*, G. Bailly and C. Benoît, eds. Elsevier Science Publishers B.V., North-Holland, Amsterdam, 435–441, 1992.

[CGN92] R. Carlson, B. Granstrom and L. Nord. Segmental evaluation using the ESPRIT-SAM test procedures and mono-syllabic words. *Talking Machines: Theories, Models and Designs*, G. Bailly and C. Benoît, eds. Elsevier Science Publishers B.V., North-Holland, Amsterdam, 443–453, 1992.

[Fal92] A. Falaschi. Segmental quality assessment by pseudo-words. *Talking Machines: Theories, Models and Designs*, G. Bailly and C. Benoît, eds. Elsevier Science Publishers B.V., North-Holland, Amsterdam, 455–472, 1992.

[Fou92] A. Fourcin. Assessment of synthetic speech. *Talking Machines: Theories, Models and Designs*, G. Bailly and C. Benoît, eds. Elsevier Science Publishers B.V., North-Holland, Amsterdam, 431–434, 1992.

[HL94] C. Henton and P. Litwinowicz. Saying and seeing it with feeling: Techniques for synthesizing visible, emotional speech. In *Proceedings of the Second ESCA/IEEE Workshop on Speech Synthesis*, New Paltz, NY, 73–76, 1994.

[NMS94] S. Narayan, S. Meredith and K. Silverman. Automated speech analysis for text-to-speech systems. Handout distributed at the Second ESCA/IEEE Workshop on Speech Synthesis, New Paltz, NY, 1994.

41

A Structured Way of Looking at the Performance of Text-to-Speech Systems

Louis C. W. Pols
Ute Jekosch

ABSTRACT Via the COCOSDA Bulletin Board an extensive questionnaire was distributed in 1993, upon which 16 reactions have been received. Almost all these reactions were very interesting and detailed. They contain a wealth of ideas and suggestions. In this chapter we combine suggestions from the questionnaire respondents with the knowledge and experience collected in various projects (SPIN, SAM(-A), Eagles, and (Euro)COCOSDA), into a new approach for a structured way of evaluating the performance of text-to-speech systems. The basic idea is to define a set of keywords/descriptors that specify the *system* under study, with special emphasis on its application. In a similar way the available and to-be-developed tests should be characterized. System and application can then be linked, in a matrix way, to the suite of tests, and a proper selection can then be made, or it might become apparent that additional specific tests are still required.

41.1 Questionnaire

As the coordinator of the Synthesis Group within the International Coordinating Committee on Speech Databases and Assessment (COCOSDA), as well as within the European interface to it (EuroCOCOSDA; LRE-project 62-057), the first author took the initiative to distribute a questionnaire on synthesis development and assessment in June 1993 via the COCOSDA bulletin #9. It was distributed again in October 1993 via bulletin #13. This initiative was discussed at the COCOSDA meetings in Banff (ICSLP'92) and Berlin (Eurospeech'93). A somewhat similar initiative was intended within SAM-A (Esprit project 6819), but never was effectuated because of the premature end of that project. Also within the Spoken Language Working Group 5 of Eagles (Expert Advisory Group on Language Standards; LRE project 61-100), this questionnaire evaluation was supported.

TABLE 41.1. List of 16 respondents to the questionnaire that was distributed twice via the COCOSDA bulletin board.

Europe	TNO, Soesterberg, NL	H. Steeneken
	CSELT, Torino, IT	S. Sandri
	Phonetics, Aix-en-Provence, FR	C. Pavlovic
	Ruhr University, Bochum, GE	U. Jekosch
	KTH, Stockholm, SW	R. Carlsson
	IKP, Bonn, GE	W. Hess
	UCL, London, UK	J. House
	CNET, Lannion, FR	C. Sorin
	LIMSI, Orsay, FR	C. d'Alessandro
USA	Bellcore, Morristown, NJ	M. Spiegel
	SRI Int., Menlo Park, CA	J. Bernstein, later P. Price
	Bell Laboratories, Holmdel, NJ	A. Syrdal
Japan	Utsunomiya Univ.	H. Kasuya
	ATR, Kyoto	N. Campbell
	JEIDA, Univ. Tsukubu	S. Itahashi
Australia	Univ. Sydney	J. Vonwiller

41.1.1 Responses

The response was not overwhelming, but the 16 (often very detailed) reactions do represent worldwide coverage (Europe, USA, Australia, and Japan) from most of the major speech laboratories, representing academic, governmental, and industrial research (see table 41.1).

Unfortunately, the application side and the user organizations are underrepresented. Apart from the unavoidable questions related to name, affiliation, function, and coordinates, most of the remaining questions were related to the awareness and the actual use of various types of tests. We also asked whether specific tests were missing for the language concerned. These tests ranged from text preprocessing and segmental tests, to prosodic tests and field evaluation. For lack of a better scheme, the subdivision as presented in table 41.2 was chosen. Further details about these tests can be found in various overview papers, such as [Pol91, PSAM92, Eag95].

The respondents indicated some missing entries, such as calibration and adverse conditions. Also multilingual aspects, or rather equilibration over languages, were considered to be underrepresented. Some concern was raised about non-European languages. Various aspects of linguistic processing, such as concept-to-text, word class assignment, pausing, and sentence accent, should also require more attention. From the numbers in table 41.2 one can see that most of the segmental tests were well known and had been used somehow in the past, but probably more for in-house system improvement than for final product evaluation. Also, some global overall quality tests are in use, including telecommunication applications. Finally, there was some positive response on certain application-specific tests, such as

TABLE 41.2. The suite of tests about which reactions were asked from the respondents. The figures represent the number of respondents (max. 16) being aware of these tests / actually using them.

	aware of / using
- text preprocessing	7 / 5
- grapheme-to-phoneme conversion	10 / 6
- segmental evaluation	
Diagnostic (DRT) or Modified Rhyme Test (MRT)	14 / 6
Phonetically Balanced word lists (PB)	13 / 4
audiometric lists	11 / 3
lists with CV, VC, VCV, or other (nonsense) word types	11 / 4
spondaic or polysyllabic words	8 / 1
- intelligibility of (proper) names	10 / 5
- sentence level	11 / 3
Harvard sentences	11 / 3
Haskins sentences	13 / 3
Semantically Unpredictable Sentences	12 / 3
- paragraph level (comprehension)	12 / 6
- prosody	8 / 2
form	
function	
- overall quality	
Diagnostic Acceptability Measure (DAM)	9 / 0
Mean Opinion Score (MOS)	11 / 5
Qualitas	2 / 1
Magnitude Estimation (ME)	10 / 3
Categorical Estimation (CE)	11 / 4
Pair Comparison (PC)	12 / 9
- different voices and speaking styles	6 / 3
- linguistic and psychological evaluation	
word/sentence verification, lexical decision	8 / 2
naming, word recall	5 / 1
target detection	3 / 0
priming	0
- field tests	
secondary tasks	3 / 0
newspaper	4 / 1
telephone directory	4 / 1
dialogue	4 / 1
- audio-visual assessment	3 / 1
- objective evaluation	
overall speech quality	7 / 2
segmental and prosodic details	7 / 4
intelligibility prediction	0
objective measurements	0

those related to telephone-directory [Spi93] or train-table information services. Almost all the other tests had few supporters or were unknown. Most respondents admitted that several test domains, such as text preprocessing, paragraph-level comprehension, and prosody, would require good tests but that such tests were not available to the best of their knowledge. Some of the more conventional tests are incomparable over languages, or are available only for American English. In terms of needs, good prosodic training material and test methods were frequently mentioned. Needs for improved voice characteristics and various speaking styles were also mentioned regularly [Pol94b]. One respondent said: "There are already too many *methods*. What we need is an algorithm that would help people, given a particular problem, to know what existing method(s) can be used or what minimal adaptations to existing methods are needed."

41.1.2 Databases

We also asked about *databases* (speech and text) presently available, actually being used, and badly needed now or in the near future for synthesis development and assessment. There has been a proposal in COCOSDA, analogous to the Polyphone initiative for word recognition [DBVSB94, BTG94], to produce a phonetically dense database in various languages, ranging from the 1,000 most frequent words, whether or not in carrier phrases, to some 30,000 or more words from a dictionary, again isolated or embedded, from a few trained male and female speakers up to a representative selection of speakers. Reactions to this were rather diverse. Some felt it to be a good idea in principle, but that no money was available for it; others felt that anyone's needs would always be different. One said that he would prefer a prosodically rich database. Another indicated the trend in concatenative synthesis toward larger and more varied inventories for various units, prosodic implementation, and different speakers and speaking styles [Cam94]. This may require large (annotated) databases projected for future requirements. Actually, there is also the commercial value of specific annotated databases, such as lists of most common names in various languages, including pronunciation variants, and probably also information about frequency of occurrence, which makes them proprietary and not available to others. Not just in Europe, there is a growing interest in modular multilingual synthesis systems, which makes the need for multilingual training data and evaluation methods even greater. COCOSDA would be the ideal international organization to stimulate international cooperation in the precompetitive areas of database collection and analysis for research.

41.1.3 Bibliography

Finally we asked the respondents to include a *bibliography* of their publications in the area of speech synthesis development and assessment. This list has been added to an existing bibliography, resulting in a list of more than 700 references. This list plus a summary of the questionnaire is accessible via the COCOSDA WWW-site (http://www.itl.atr.co.jp/cocosda).

41.2 A Structured Way of Evaluating the Performance of Speech-Generating Systems

As stated above, there may be too many tests already, and we had better give them some structure. The second half of this chapter will give some constructive suggestions in this direction. These proposals mainly grew from discussions within SAM-A and Eagles; see also [Jek93] and [JP94].

41.2.1 System Specifications

The basic idea is to define a set of keywords or descriptors that specify the *system* under study, with special emphasis on its *applications* (see table 41.3).

This implies that the evaluation is not limited to the synthesis system as such, but rather refers to a speech-generating element in a multimodal, dialogue, or man/machine communication system. It is not difficult to apply this scheme to certain characteristic and realistic applications, such as multilingual name and address pronunciation for a reverse directory service (e.g., Onomastica, LRE project 61-004), or a standard form weather forecast for aviation purposes extracted from meteorological information, or a spoken newspaper for the visually impaired.

41.2.2 Test Specifications

In a similar way the available and to-be-developed tests, as well as the test conditions should be characterized. In the overview in table 41.4 we have emphasized the secondary, practically very important, aspects of evaluation tests next to the functionality of the test itself. The suite of tests as originally presented in the questionnaire (table 41.2) is mainly summarized under point 6 as test functionality in table 41.4. In [JP94] some further details are given.

In the final step, system and application descriptors are then linked, in a matrix-type way, to the qualifiers of the suite of tests. A proper selection then can be made, or it might become apparent that additional specific tests are still required.

41.2.3 Examples

For instance, in the situation where the quality of a speech output device that is meant to be used as a traffic information system in a driving car (variable speed, noise) should be assessed, no single test might be readily available for that language. On the other hand, given the characteristics of the input data (probably linguistically and prosodically correct text, which might well be generated as canned natural speech in which synthesized keywords are included), the emphasis should probably be on keyword intelligibility, overall acceptability, and ergonomical aspects. Initial intelligibility testing could probably be done in the laboratory under simulated listening conditions with an appropriate word set to which dummy words are added. The field test then could concentrate on acceptability.

TABLE 41.3. Set of keywords and descriptors to describe various aspects of the speech-generating part in a man-machine application *system*.

1. text coverage and required text processing
- language(s) (mono- or multilingual)
- unlimited text input
- text from concept
- database interpretation (e.g., tables, punctuation)
- fixed subtext plus keywords
- highly structured and annotated text
  (style specifications, spell options, etc.)

2. source
- coded, synthetic, or canned speech
- speaker and voice type
- dialect, style, and emotion
- adaptive to channel and user

3. communication channel
- high quality
- telephone (handset, mobile, earphone)
- interfering noise, reverberation, and/or competing speech

4. user characterization
- experience, training, age
- nativeness, second language use
- impairment
- cooperativeness

5. application characterization
- reading-machine type *or* information-retrieval type
- field *or* laboratory environment
- application specific *or* independent

6. functional system characterization
- comprehension, intelligibility, naturalness, or otherwise
- prosody
- secondary tasks
- benchmarking
- dialogue aspects

7. practical restrictions
- time, money, system availability, single modules

8. alternatives and/or combined modes
- multimedia, mouse, screen, visual, braille, tactile

9. technical details
- price, size
- interface, plug-in options, DSP board, modularity
- units for synthesis
- options: voices, style or tempo control, hand-tuning

TABLE 41.4. Set of keywords and descriptors to describe various aspects of *tests* for evaluating the performance of speech-generating systems.

1. text coverage
- language(s)
- frequency of occurrence
  - (phonemes, phoneme sequences (phonotactics), words, grammatical structures)

2. source
- text-to-speech synthesis-by-rule or otherwise
- speaking style, emotion, rate

3. communication channel
- loudspeakers or headphones
- single or multiple listeners
- interactive, feedback

4. user characterization
- homogeneity among subjects
  - (e.g., in terms of experience and age)
- potential response difficulties
  - (use of nonsense words, second language users, children, or elderly or hearing-impaired people)
- amount of training of subjects

5. test type
- field *or* laboratory test
- application specific *or* independent
- black box *or* glass box evaluation

6. functional test characterization
- text level
- segmental intelligibility
- comprehension, naturalness, or otherwise
- prosody, voice characteristics
- secondary tasks
- benchmarking
- dialogue aspects
- objective evaluation

7. practical restrictions
- availability of test in a specific language
- time constraints

8. alternatives and/or combined modes
- (beyond a listening test alone)

9. experimental details

Another relevant application could be the collection of public transportation information over the telephone via spoken requests and an interactive dialogue. Here the spoken output is probably not the most critical part, although proper operation of the system relies on it. Again, one could imagine that much attention is given to the intelligibility of the keywords (city or station names), whereas the prosody of the dialogue and of the informative sentences should certainly also be optimized and tested.

On the other hand, in an application in which the whole digitally stored newspaper can daily be "listened to" by visually impaired "readers," the ergonomy of the application appears to be much more important than the actual speech quality [vJ93]. Right now it looks as if such users are willing to accept any speech quality as long as they can get access to any self-selected part of the newspaper, preferably including the ads. With further progress in the development of text-to-speech systems, and with extended use, one should guess that also a higher quality of both the linguistic and the acoustic components becomes more of a necessity.

Just as linking the system descriptors of an application (table 41.3) with the qualifiers of the evaluation methods (table 41.4) will generally allow us to choose the appropriate test(s), or will tell us that additional speicific tests are required, so will such a matrix-type link, in principle, also inform us about other necessities. We are thinking of specific databases in specific languages (e.g., prosodic labeling, concatenative units), as well as standardized hardware and software tools, handbooks, and benchmark test results.

Acknowledgments: We thank all respondents for the effort they took in producing their extensive replies to this questionnaire. We also thank our colleagues in various projects for their contributions. This work was partly supported by LRE-project EuroCOCOSDA.

References

[BTG94] J. Bernstein, K. Taussig, and J. Godfrey. Macrophone: An American English telephone speech corpus for the Polyphone project. In *Proceedings, ICASSP-94*, Adelaide, I-81–I-83, 1994.

[Cam94] N. Campbell. Prosody and the selection of units for concatenative synthesis. In *Proceedings, ESCA/IEEE Workshop on Speech Synthesis*, New Paltz, NY, 61–64, 1993.

[DBVSB94] M. Damhuis, T. Boogaart, C. Veld, C. M. Versteylen, W. Schelvis, L. Bos, and L. Boves. Creation and analysis of the Dutch Polyphone corpus. In *Proceedings, ICSLP-94*, vol. 4, Yokohama, 1803–1806, 1994.

[Eag95] Eagles. *Spoken Language Systems*, Chapter 4 on Assessment of Speech Output Systems, in press.

[FHBH89] A. Fourcin, G. Harland, W. Barry and v. Hazands, eds. *Speech Input and Output Assessment. Multilingual Methods and Standards.* Ellis Horwood Ltd., Chichester, 1989.

[Jek93] U. Jekosch. Speech quality assessment and evaluation. In *Proceedings, Eurospeech'93*, vol. 2, Berlin, 13870-1394, 1993.

[JP94] U. Jekosch and L. C. W. Pols. A structured approach towards a framework for application-specific. speech quality assessment. In *Proceedings, ICSLP-94*, vol. 3, Yokohama, 1319–1322, 1994.

[Pol90a] L. C. W. Pols. How useful are speech databases for rule synthesis development and assessment? In *Proceedings, ICSLP-90*, vol. 2, Kobe, 1289–1292, 1990.

[Pol90b] L. C. W. Pols, ed. Speech input/output assessment and speech databases. Special issue of *Speech Comm.* 9(4):261–388, 1990.

[Pol91] L. C. W. Pols. Quality assessment of text-to-speech synthesis-by-rule. In *Advances in Speech Signal Processing*, S. Furui and M. M. Sondhi, eds. Marcel Dekker Inc., New York, chapter 13, 387–416, 1991.

[Pol94a] L. C. W. Pols. Speech technology systems: Performance and evaluation. In *The Encyclopedia of Language and Linguistics*, R. E. Asher, ed. Pergamon Press, Oxford, vol. 8, 4289–4296, 1994.

[Pol94b] L. C. W. Pols. Voice quality of synthetic speech: Representation and evaluation. In *Proceedings, ICSLP-94*, vol. 3, Yokohama, 1443–1446, 1994.

[PSAM92] L. C. W. Pols and SAM-partners. Multi-lingual synthesis evaluation methods. In *Proceedings, ICSLP-92*, vol. 2, Banff, 181–184, 1992.

[Spi93] M. E. Spiegel. Using the ORATOR synthesizer for a public reverse-directory service: Design, lessons, and recommendations. In *Proceedings, Eurospeech'93*, vol. 3, Berlin, 1897–1900, 1993.

[vJ93] R. van Bezooijen and W. Jongenburger. Evaluation of an electronic newspaper for the blind in the Netherlands. In *Proceedings, ESCA Workshop on Speech and Language Technology for Disabled Persons*, B. Granström, S. Hunnicutt, and K.-E. Spens, eds. Stockholm, 195–198, 1993.

[vP93] V. J. van Heuven and L. C. W. Pols, eds. *Analysis and Synthesis of Speech. Strategic Research Towards High-Quality Text-to-Speech Generation.* Speech Research 11, Mouton de Gruyter, Berlin, 1993.

42

Evaluation of a TTS-System Intended for the Synthesis of Names

Karim Belhoula
Marianne Kugler
Regina Krüger
Hans-Wilhelm Rühl

ABSTRACT A German speech synthesis system is presented for telephone-based inquiry systems supplying address-oriented information. Providing speech synthesizers for public inquiry systems is a very demanding task, as most users of such systems expect a speech quality close to natural and perfect intelligibility. To achieve these goals, a synthesis-by-concept scheme was developed for speech output allowing one to combine synthetically spoken names with naturally spoken carrier sentences, enhanced by special rule sets for grapheme-phoneme conversion of first names, and family names, business names, street names, and location names.
To test the performance of address-oriented synthesis by concept, an evaluation was done to determine the influence of the remaining pronunciation errors on intelligibility. Most pronunciation errors could be classified into either (1) wrong word accents, (2) a wrong vowel quality or quantity, or (3) a class covering errors caused by wrong morphological decomposition. For these three classes, the influence of errors on intelligibility was evaluated. It turned out that this influence is dependent both on the word class and on the familiarity of the presented names.

42.1 Introduction

With the advent of automatic telephone-based inquiry systems, there is an increasing demand for speech servers supplying address-oriented information. Due to the large amount of data in address databases, it is impossible to store this information for output in spoken form. Instead of stored speech, synthetic speech has to be used. But it is impossible to use an off-the-shelf text-to-speech converter for this task, due to the fact that address-related information often follows special pronunciation rules.

In Germany, as in most other countries, first names tend to change with time and fashion, so that names with German, French, English, Scandinavian, and Slavonic origin are in use. Several millions of immigrant workers (called *Gastarbeiter*) from southern and eastern European countries, and immigrants from the former Soviet Union and several other Near East and Far East countries further complicate the problem, both for first names and family names. Concerning business names, only a part may be classified as completely German in a linguistic sense. A large amount of international companies have German subsidiaries using the international name, and it is fashionable for German high-tech companies to have a fully or partly English business name.

On the other hand, location names, of course, are all of German origin. But their writing has been fixed in times preceding the standardization of the high German writing conventions, resulting in various local dialectal name notations that make pronunciation of unknown names difficult even for native speakers.

Compared to general text-to-speech (TTS), one of the advantages of address-oriented inquiry systems is that the syntax classes and name classes for the address parts to be synthesized are explicitly known. This means that the grapheme-to-phoneme conversion can be explicitly told to activate special conversion rules for a specific name class. Doing so helps to reduce pronunciation error rates significantly, to a rate that is likely to be accepted by the general public.

But standard TTS synthesizers, even without pronunciation errors, are not accepted by the public due to the fact that the naturalness of standard synthetic speech is inferior to human speech in a longer discourse. For this reason, it was decided to use synthetic speech only for pronunciation of proper names, and embed the names into naturally spoken stored utterances. An example of this concatenation scheme, which we call synthesis by concept, is shown in figure 42.1.

Synthesis by concept has been implemented as an adaptation of the general TTS synthesizer PHRITTS, described in more detail in [MR93].

To avoid creating the impression of a speaker change when switching between natural and synthetic speech, both the natural utterances and the nonsense words used to create the synthesizer's diphones were spoken by the same speaker. Doing so allows managing the dialogue using stored utterances, with only the few messages supplying address-oriented information created in mixed mode. As more than 90% of the output is natural speech, the influence of synthetic speech on the naturalness of an announcement is minimal. In fact, most users do not perceive that parts of the messages are synthetic. The majority of people sense a difference in speech quality for synthetic or natural speech, but usually relate it to distur-

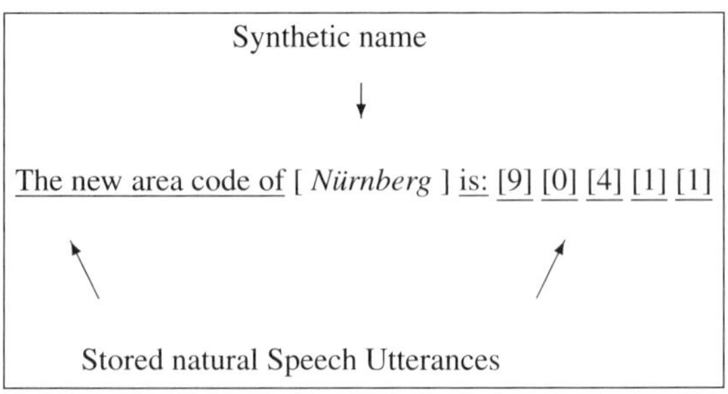

FIGURE 42.1. Message generation for synthesis by concept.

bances on the telephone line, as long as the speech intelligibility is not inflected by pronunciation errors.

The speech synthesizer was first applied in an area code information system [DM92, DMR93]. For this application, informal listening tests, done by the system developers and by unexperienced casual users of the system (e.g., at presentations or at trade shows) indicated that the achieved quality was accepted by most listeners and that the achieved speech intelligibility was sufficient for the implemented demonstration. But the informal tests did not provide sufficient information on the influence of the infrequent pronunciation errors. For this reason, a special intelligibility test was set up to evaluate the influence of pronunciation errors for names on name intelligibility. The results of this test are presented in this chapter.

The phoneme-to-sound part of the speech synthesizer has been described in detail in [MR93]. It uses the PSOLA technique [CM89] to synthesize a female voice with a frequency range adapted to the telephone and 8 kHz sampling rate. As speech units, 1,573 diphones are stored in A-law compressed form.

Due to synthesis-by-concept, only a limited set of intonation contours are needed to determine pitch values for the remaining open slots. The intonation contours are restricted to phrases with one up to at most three accents, and have been taken from [Adr91]. Duration control is based on a set of few preliminary rules that are not adequate for continuous text but work sufficiently for short phrases.

In the next section, the name-specific letter-to-sound conversion for names is presented. Section 42.3 presents the evaluation setup and its results, and in the final summary, conclusions concerning applications are drawn.

42.2 Grapheme-to-Phoneme Conversion of Names

In the last few years telecommunication applications of speech synthesis have increased in importance. Because most of the systems involve the names of locations (cities, streets, etc.) and proper names, and because no syntactic or semantic knowledge is available to support name understanding when we hear names in the context of a message, very good name pronunciation is required. Most commercially available TTS systems that are designed for a standard German vocabulary are unable to pronounce names correctly. These systems generally perform well for segmenting compound words into morphemes and for letter-to-sound rules, but these algorithms are not necessarily appropriate for the pronunciation of names. A preliminary test carried out with our moderate TTS system on two databases (21,000 city names and 22,000 surnames) has shown that more than 37% of the city names and 43% of the surnames were mispronounced. Furthermore, names deriving from other languages cannot be converted using the general rules of the German language.

An efficient grapheme-to-phoneme conversion of names therefore requires the appropriate algorithms for morphological decomposition and the appropriate rules for phonetic transcription [SMG93, CG93]. Additionally, much effort has to be put into the assignment of lexical stress. The main problems when pronouncing foreign names can be solved to a large extent if we can predict their origin. In this case, language-dependent letter-to-sound rules can be applied to the corresponding category of names. Focusing on the demand for such applications, we have extended our TTS system to the pronunciation of names. Sets of rules have been extracted from a telephone directory database, which is described below. Some of the key features of the integrated approaches are briefly described in this chapter.

42.2.1 System Overview

Database: Our investigation into creating appropriate rules was supported by a database that contains close to 1.5 million subscriber records. A large part of the data was taken from the telephone directory of Munich. The total number of different types comprises about 303,000 names. We constructed six databases according to the different name groups: city names, surnames, first names, street names, professions, and business names. All the records were sorted according to their decreasing frequency of occurrence. Using these databases, we extracted subsets from each group. Table 42.1 shows a survey of the number of names we selected and the percentage of data we covered.

The databases were automatically transcribed using our standard TTS system [Boh92]. When necessary, the entries were then corrected manually in a second step. It should be noted that the percentage of foreign surnames in the database was estimated to be 7%. The percentage of foreign first names is much larger because we worked with the entire database (see table 42.1).

Classification of names: One critical problem in the pronunciation of first names was their diverse etymological origins [SF93]. First names in our database are derived from more than 20 languages. We limited our approach to the most frequent categories: German, French, English, Greek, Italian, Slavonic, Arabic, Turkish, and Asian. The pronunciation of such names often deviates considerably from the standard German letter-to-sound rules. To ensure correct segmental pronunciation and stress, we classified them into the corresponding categories by a position-dependent and language-specific cluster analysis [Bel93a].

Morphological and graphotactical analysis: About 70% of the German location names and surnames result from the composition of morphemes. These morphemes

TABLE 42.1. Reference databases.

Name type	Selected	Data covered
city name	21,406	100%
surnames	21,634	75%
first names	18,667	100%
street names	11,446	50%
profession	5,760	100%
business names	4,358	none

could be classified into name-specific suffixes, premorphs, stems, and endings. According to this morphological structure, we developed appropriate sets of rules to support the segmentation of composed names [Bel93b]. First names do not follow the morphological structures described above. Our investigation indicated, however, that a wide number of suffix-like graphemes can be split off. In this manner, a context-sensitive graphotactical decomposition was developed to convert first names.

Letter-to-sound conversion: The standard letter-to-sound rules do not lead to satisfactory results when applied to location names or surnames. By developing appropriate sets of rules, we reached a good level of pronunciation. Letter-to-sound rules are morph-oriented, they were mostly extracted from the reference corpora. To each first name language category we applied appropriate language-specific letter-to-sound rules.

Lexical stress assignment: In German, lexical stress is usually located on the first syllable, if no stressable affix is available. This rule cannot be applied to locations or to proper names: an examination of the reference corpora led to an insight of variability in name accentuation. The module developed for this aim therefore deals with different cases of stress mark assignment. Endings holding or modifying the lexical stress are enclosed in special rule modules.

42.2.2 *Objective Evaluation of the Grapheme-to-Phoneme Conversion*

The error-statistics of the grapheme-to-phoneme conversion are achieved by automatically comparing the transcription of the reference databases with the transcription produced by our rules (symbol for symbol). Any difference between them was considered to be a phonetic error. It should be noted at this point that the reference corpora do not include alternate forms of pronunciation [SF93]. This method is nevertheless efficient, since it provides an overview of the current error rate and requires little effort. Furthermore, it automatically creates a list of frequently made mistakes. The results of the performance test (correct pronunciation) are shown in table 42.2.

This objective evaluation cannot supply any qualitative information, because no distinction is made between slight deviations and severe mistakes. Thus, even a nondistinctive phonetic variation is treated as an error equivalent to a serious mistake. A survey of the error-statistics from a qualitative point of view showed that there is, in fact, a large number of errors that seem tolerable in that they probably do not affect the intelligibility of the names.

The statistics on erroneous phonetic transcriptions revealed some regularities that allowed us to establish three classes to which a large number of mistakes could be assigned:

TABLE 42.2. System performance compared to the standard TTS-system.

Name type	Entries	Standard rules	Name rules
Locations	21,406	63%	92%
Surnames	21,634	57%	90.7%
First names (German)	18,647	43%	89.6%
Streets	11,435	53%	86.7%
Business names	4,358	38%	84%

1. *Misplaced stress mark:* The primary word stress is assigned to a vowel or diphthong that has only secondary stress or no stress at all. Example (reference transcription followed by TTS-transcription, and SAMPA-standard):

   Alexander [alEks'and=6] ['alEksand=6]

2. *Vowel quantity and quality:* This class can be subdivided into:

   a. Alteration in vowel quantity: The TTS-transcription contains a short vowel [V] where there should be a long vowel [V:] (or vice versa). Example:

      Rotraud [R'o:tRaUt] [R'otRaUt]

   b. Alteration in vowel quality: The TTS-transcription contains a lax vowel where there should be a tense vowel (or vice versa). Example:

      Dorothea [doRot'e:a] [doROt'e:a]

   c. Both vowel quantity and quality are different. Example:

      Philipp [f'IlIp] [f'i:lIp]

3. *Errors in morphological decomposition:* This is a special class of mistakes, due to the fact that an erroneous morphological decomposition may result in a variety of phonetic mistakes. For example, if the diminutive suffix "-*chen*" in the female first name *Dorchen* (derived from Dore) is ignored by the TTS-system, the result will be an alteration in vowel quantity and quality that prevents the assimilation of the following /R/-phoneme. Example:

   Dorchen [d'o:6C@n] [d'ORC@n]

42.3 Perceptual Evaluation of the Grapheme-to-Phoneme Conversion

In order to find out to what extent transcription inaccuracy affects the intelligibility of synthesized names, we carried out a listening test. Our experiment was based on the assumption that listeners tolerate a certain degree of variation that may even go beyond the range of phonetic variations (allophones). There may even be some phonemic variation that does not have a negative impact on intelligibility, since in interhuman communication intelligibility does not merely depend on the correct (i.e., standard) pronunciation, but also on many other perceptual factors. In our specific context—the intelligibility of names—it seemed reasonable to assume that the degree to which a name is familiar to the listener may be very important for the correct perception of the name.

42.3.1 Test Design

The basic corpus for our test comprised first names and surnames that were mistranscribed by the TTS system. On this basis we compiled a list of 40 randomly selected examples for each class of mistakes. Each list contained 20 first names and 20 surnames. This selection of list entries was used as the input for our TTS system. In the first part of the experiment, the synthesized entries (mispronounced by the TTS system) were acoustically presented word by word to a group of 67 subjects, who were instructed to type in the name they had just heard. The experiment was carried out in an office-like room at a computer terminal. Each stimulus was presented only once to the subjects via headphones. To evaluate the degree of intelligibility, the subjects' responses were then compared to the orthographic form of the TTS input. The examples in table 42.3 reflect a survey about the comparison.

In case of perfect correspondence between TTS input and the name typed in by the subject (example (a)), we could conclude there was a high degree of intelligibility. If the subject's orthographic response was different from the TTS input, but would be pronounced the same way (example (b)), we could not put this difference down to a lack of intelligibility, but to the fact that one phoneme may have multiple orthographic representations. Example (c) shows a similar situation. Furthermore, it reveals a peculiarity of names: Names often have homophonic heterographs, e.g., the surname Meier can be written in various ways (Meyer, Mayer, Mayr, etc.) without changing its pronunciation. Thus the difference in example (c) does not permit any statements about intelligibility either. Only in example (d) has the transcription error actually led to a wrong perception of the name (the name should be pronounced [z'a:kmaIst=6]).

In the second part of the experiment, the stimuli were presented visually in their orthographic form to the subjects. The subjects' task was to decide whether the name was familiar to them or not. This additional information allowed us to test our hypothesis that there may be a correlation between intelligibility and familiarity of a name.

TABLE 42.3. Examples of subjects' responses.

TTS-input	*TTS-output*	*subject's responses*	*correct transcription*
a) Tobias	t'o:bias	Tobias	tob'i:as
b) Brechenmacher	bR'ECEnmax=6	Brächenmacher	bR'EC nmax=6
c) Meier	m'aIER	Meyer	m'aI=6
d) Sagmeister	z'akmaIst6	Sackmeister	z'a:kmaIst=6

42.3.2 *Results*

Before we discuss the test results, it should be noted that the test was not carried out to investigate the intelligibility of the synthesizer. The main intention was to evaluate the grapheme-to-phoneme conversion and in particular the influence of multiple transcription errors on the intelligibility.

Assuming that the overall quality of the synthesizer did not vary for the selected names, the first point to be concluded from this study is that the segmental errors (labeled "segmentation" in figures 42.2 and 42.3) do have a much more severe impact on intelligibility than the mispronunciations caused by wrong vowel quality or quantity. This aspect could be seen in both name types: On average only 20% of the wrongly segmented surnames and 50% of the first names were identified by the subjects. Errors due to misplaced stress marks seem to be secondary to good intelligibility. More than 40% of surnames and 65% of first names were identified. The results collected from this test are shown in figures 42.2 and 42.3.

When examining the data we observed another detail relevant to the results achieved by first names: All subjects recognized more first names than surnames. By comparing the mean of the overall identification, we found that the identification score of first names was 28% higher than the score of surnames. This interesting performance can be related to the following facts: After doing the listening test, the majority of subjects confirmed that first names were easier to understand than surnames.

A first reason can be connected to the repertoire of names. The number of surnames is generally larger than the number of first names. In our database, for example, there are about 130,000 different surnames but only 18,660 first names. A second reason can be explained by the perception of the subjects: Words will be generally associated to their meaning, but names will be associated either to persons whom we know personally or to persons about whom we have read or heard something. In order to examine this aspect, we computed the extent of association between the following sets of attributes:

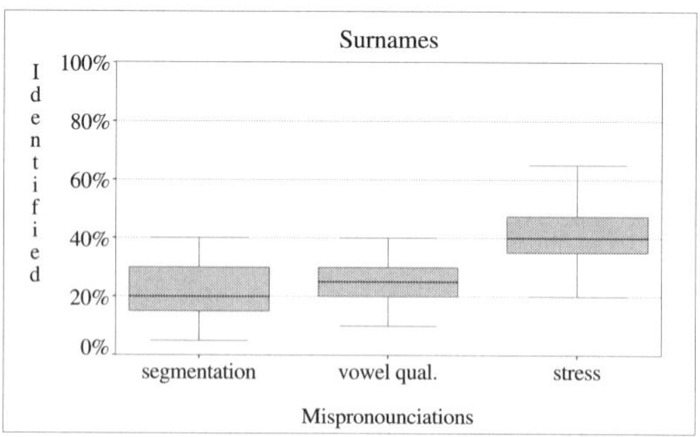

FIGURE 42.2. Identification of surnames.

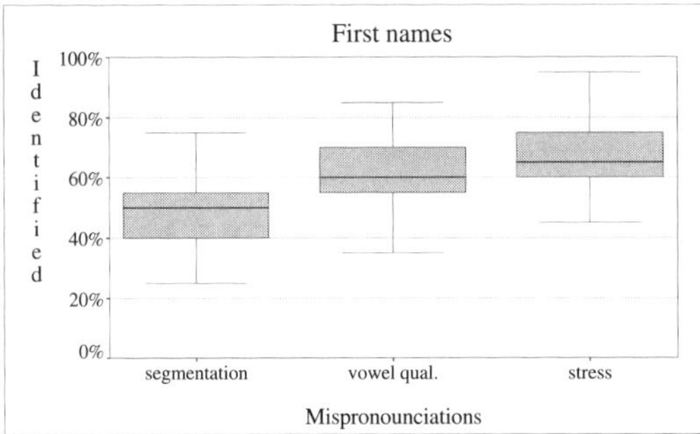

FIGURE 42.3. Identification of first names.

First attribute: name is identified / unidentified by the subject
Second attribute: name is familiar / not familiar to the subject

According to the nominal-scaled frequencies in the 2×2 contingency tables (see tables 42.4 and 42.5), we first made a χ^2-test to see if there is any dependence between the attributes. We obtained $\chi^2 = 476$ for the surname attributes and $\chi^2 = 504$ for the first name attributes. The extent of association or relation between the sets of attributes can be determined by calculating the contingency coefficient C:

$$C = \sqrt{\frac{\chi^2}{N + \chi^2}} \tag{42.1}$$

where N is the total number of cases.

The value of C is equal zero when there is no association, but it cannot attain unity. When the variables are perfectly correlated, the upper limit of C for the 2×2 table is $\sqrt{1(1+1)} = 0.707$. According to the values of χ^2, the computed value of the contingency coefficients are $C = 0.32$ for surnames and $C = 0.33$ for first names. By testing the null hypothesis (H_0: there is no correlation between the sets of attributes), we found that we could reject H_0 at the level of significance $p \leq 0.0001$. We can thus conclude that the observed association in our samples is not a result of chance but rather represents a genuine relation in the data.

On the basis of these results, we could first conclude that the morphological decomposition of names plays an important part in the pronunciation accuracy. Segmental errors lead frequently to such erroneous transcriptions, which degraded the

TABLE 42.4. Contingency table (surnames).

	unfamiliar	*familiar*	row total
identified	396	847	1243
not identified	720	842	2777
column total	1935	1689	4020

TABLE 42.5. Contingency table (first names).

	unfamiliar	*familiar*	row total
identified	435	1905	2340
not identified	860	820	1680
column total	1295	2725	4020

intelligibility. Deviations in the quality and/or quantity of vowels or the misplacement of lexical stress leads often to an alternative pronunciation only. The second conclusion drawn from this test is the correlation between the degree of familiarity of names and their intelligibility. The results show that names that were familiar to the subjects were better identified than unfamiliar ones. Because most of the surnames presented in the listening test were not common ones, their identification quota was globally lower than the identification of first names.

To be connected with our results, the intelligibility of names depends, apart from the pronunciation accuracy, on the individual experience and recollection of the listener. Subjects listen to names in a framework of anticipation and associations, and they process their auditory event cognitively. For instance, the surname Schimanski, which is very popular in the Ruhr area, was correctly identified and judged to be familiar by nearly all the participants. Since synthesis systems cannot make assumptions on the user's knowledge it is impossible for them to produce hyperspeech (i.e., pronouncing names slower or overarticulating them) like a human speaker does. In this context it would be interesting to carry out the same test with natural speech and to compare both results.

42.4 Summary

A system designed for speech synthesis-by-concept for address-oriented information systems was presented. It employs a special set of rules for pronunciation of German proper names, a diphone-based speech generation using the time domain PSOLA technique, and intonation and duration control modules adapted to the synthesis-by-concept task.

An evaluation was done to determine the influence of pronunciation errors on the intelligibility. For this purpose, two sets of first names and surnames containing three types of frequent pronunciation errors were presented to a group of test persons. It turned out that accent placement on a wrong syllable decreased intelligibility least. Determining vowel quantity or quality wrong reduced intelligibility significantly. Errors in analysis of the names' morpheme structure, usually resulting in clusters of wrong phonemes, had the most impact on intelligibility.

Absolute rate of misunderstandings is strongly dependent on the familiarity of the names to be recognized. More than half of the surnames were unfamiliar, and 30% of the surnames were recognized, whereas 2/3 of the first names were familiar, and 60% were recognized.

From the fact that intelligibility for familiar names is high, some guidelines for the potential acceptance of speech synthesis in its actual state may be derived.

- There is little use to develop applications for which users do not have an expectation horizon for the information to come, such as reverse directory assistance, as they may have difficulties in understanding totally unfamiliar names and addresses.

- In applications for which there is already some information available, chances for successful applications are higher. This is, for example, the case with standard telephone directory assistance, when synthetic speech is used to announce addresses and telephone numbers of few persons with the same name in the same town.

References

[Adr91] L. M. H. Adriaens. *Ein Modell Deutscher Intonation.* PhD thesis, Eindhoven, 1991.

[Bel93a] K. Belhoula. A concept for the synthesis of names. In *Proceedings, ESCA-Workshop on Applications of Speech Technology*, Lautrach, Bavaria, 167–170, 1993.

[Bel93b] K. Belhoula. Rule-based grapheme-to-phoneme conversion of names. In *Proceedings, Eurospeech-93*, Berlin, 881–884, 1993.

[Bel94] K. Belhoula. Evaluation of a TTS-system intended for the synthesis of names. In *Proceedings, Second ESCA-Workshop on Speech Synthesis*, New York, 211–214, 1994.

[Boe92] A. Böhm. *Maschinelle Sprachausgabe Deutschen und Englischen Textes.* Dissertation, Ruhr-Universität Bochum 1992, Verlag Shaker, Aachen 1993.

[CG93] R. Carlson and B. Granström, et al. Predicting name pronunciation for a reverse directory service. In *Proceedings, Eurospeech-93*, Berlin, 113–116, 1989.

[CM89] F. Charpentier and E. Moulines. Pitch synchronous waveform processing techniques for text-to-speech synthesis using diphones. In *Proceedings, Eurospeech-89*, vol. 2, Paris, 13–19, 1989.

[DM92] S. Dobler and P. Meyer, Peter, et al. A server for area code information based on speech recognition and synthesis by concept. In *Proceedings, Konvens 92*, G. Goerts, ed. Nüernberg, 353–357, 1992.

[DMR93] S. Dobler, P. Meyer, and H.-W. Rühl. Voice controlled assistance for the new German mail area codes. In *Proceedings, ESCA-Workshop on Applications of Speech Technology*, Lautrach, Bavaria, 23–26, 1993.

[MR93] P. Meyer and H.-W. Rühl, et al. PHRITTS—a text-to-speech synthesizer for the German language. In *Proceedings, Eurospeech-93*, Berlin, 877–880, 1993..

[SF93] M. Schmidt and S. Fitt, et al. Phonetic transcription standards for European names (Onomastica). In *Proceedings, Eurospeech-93*, Berlin, 279–282, 1993.

[SMG93] M. F. Spiegel, M. J. Macci and K. D. Gollhardt. Synthesis of names by a demisyllable-based speech synthesizer. In *Proceedings, Eurospeech-93*, Berlin, 117–120, 1993.

Appendix: Audio Demos

The sound example demonstrates our system.

43

Perception of Synthetic Speech

David B. Pisoni

ABSTRACT This chapter summarizes the results we obtained over the last 15 years at Indiana University on the perception of synthetic speech produced by rule. A wide variety of behavioral studies have been carried out on phoneme intelligibility, word recognition, and comprehension to learn more about how human listeners perceive and understand synthetic speech. Some of this research, particularly the earlier studies on segmental intelligibility, was directed toward applied issues dealing with perceptual evaluation and assessment of different synthesis systems. Other aspects of the research program have been more theoretically motivated and were designed to learn more about speech perception and spoken language comprehension. Our findings have shown that the perception of synthetic speech depends on several general factors including the acoustic-phonetic properties of the speech signal, the specific cognitive demands of the information-processing task the listener is asked to perform, and the previous background and experience of the listener. Suggestions for future research on improving naturalness, intelligibility, and comprehension are offered in light of several recent findings on the role of stimulus variability and the contribution of indexical factors to speech perception and spoken word recognition. Our perceptual findings have shown the importance of behavioral testing with human listeners as an integral component of evaluation and assessment techniques in synthesis research and development.

43.1 Introduction

My interest in the perception of synthetic speech dates back to 1979 when the MITalk text-to-speech system was nearing completion [AHK87]. At that time, a number of people, including Dennis Klatt, Sheri Hunnicutt, Rolf Carlson, and Bjorn Granstrom, were working in the Speech Group at MIT on various aspects of this system. Given my earlier research on speech perception, it seemed quite appropriate to carry out a series of perceptual studies with human listeners to assess how good the MITalk synthetic speech actually was and what differences, if any, would be found in perception between natural speech and synthetic speech produced by rule. Since that time, my research group at Indiana has conducted a large number of experiments to learn more about the perception and comprehension of synthetic speech produced by rule. This chapter provides a summary and interpretation of the major findings obtained over the last 15 years and some suggestions for future research directions.

TABLE 43.1. Needed research on the perception of synthetic speech [Pis81a].

1.	Processing time experiments
2.	Listening to synthetic speech in noise
3.	Perception under differing attentional demands
4.	Effects of short- and long-term practice
5.	Comprehension of fluent synthetic speech
6.	Interaction of segmental and prosodic cues
7.	Comparisons of different rule systems and synthesizers
8.	Effects of naturalness on intelligibility
9.	Generalization to novel utterances
10.	Effects of message set size

In placing our earlier work in context, it is important at the outset to draw a distinction between basic research on the perception of synthetic speech and more applied or practical work dealing with questions concerning assessment and the development of evaluation techniques. Although we have been involved with both kinds of activities, most of our research has been oriented toward basic research issues that deal with the perceptual analysis of speech. In particular, we have been concerned with trying to understand some of the important differences in perception between natural speech and several kinds of synthetic speech. By studying the perception of synthetic speech produced by rule, we hoped to learn more about the mechanisms and processes used to perceive and understand spoken language more generally [Kla87, PNG85].

A few years after this research program began, I generated a list of about a dozen basic issues that seemed at the time to be important topics for future research [Pis81a]. Table 43.1 lists these research issues. Many of these questions have been studied over the years since I constructed this list, but some of the topics still remain to be explored in future research.

A number of researchers have taken our initial set of findings and developed much more detailed assessment and evaluation techniques to test various types of voice output devices [Pol89a, Pol92, van94]. The goal of this recent work has been to develop reliable methods of evaluation and assessment so that standards can be formulated for use in a variety of languages across a range of applications [BB92, FHBH89]. One approach to assessment was proposed recently by Pols [Pol89b]. He suggests that assessment techniques be categorized into four broad classes: (1) *global*, including acceptability, preference, naturalness, and usefulness; (2) *diagnostic*, including segmentals, intelligibility, and prosody; (3) *objective*, including the speech transmission index (STI) and articulation index (AI); and (4) *application-specific*, including newspapers, reading machines, telephone information services, and weather briefings. Much of our early research on the perception of synthetic speech was concerned with global and diagnostic issues, topics that continue to occupy most researchers even today.

Although some aspects of our research have been concerned with evaluation and assessment, such as the intelligibility studies using the modified rhyme test (MRT), many other studies over the years have focused on why some types of synthetic speech are difficult to perceive and understand and how listeners compensate for the generally poor-quality acoustic-phonetic information in the signal. In the sections below, I provide a brief summary of the major findings and conclusions from our research program. Both the perceptual evaluation studies and the experimental work have suggested a number of general conclusions about the factors that affect the perception of synthetic speech. Finally, I offer several suggestions for future research.

43.2 Intelligibility of Synthetic Speech

A text-to-speech system can generate three different kinds of errors that may affect the overall intelligibility of the speech. These errors include incorrect spelling-to-sound rules, the computation and production of incorrect or inappropriate suprasegmental information, and the use of error-prone phonetic implementation rules that are used to convert the internal representation of allophones into a speech waveform [AHK87, PNG85]. In the studies described below, my collaborators and I have focused much of our attention on measures of segmental intelligibility, assuming that the letter-to-sound rules used by a particular text-to-speech system were applied correctly. For most of our research, we simply ignored the suprasegmentals because at the time this work was initially carried out in the late 1970s, there were no behavioral techniques available to assess these attributes of synthetic speech.

Phoneme Intelligibility. The task that has been used most often in previous studies evaluating synthetic speech and the one we adopted to measure the segmental intelligibility was the modified rhyme test. In the MRT, subjects are required to identify a single English word by choosing one of six alternative responses that differ by a single phoneme in either initial or final position [HWHK65, NG74]. All the stimuli in the MRT are consonant-vowel-consonant (CVC) monosyllabic English words; on half the trials, the response alternatives share the vowel-consonant portion of the stimulus, and on the other half the response alternatives share the consonant-vowel portion. Thus, the MRT provides a measure of how well listeners can identify either the initial or the final phoneme from a set of spoken words. In recent years, new tests specifically for the assessment of synthetic speech have been developed using this approach [BP89, CG89, SAMW89]. Some examples of data obtained using in the MRT from [LGP89] for 10 text-to-speech systems are shown in figure 43.1. The intelligibility data shown here reveal a wide range of performance levels across different synthesis systems.

In addition to the standard forced-choice closed-response MRT, we have also explored the use of an open-response format. In this procedure, listeners are instructed simply to write down the word that they heard on each trial. The open-response test provides a measure of performance that minimizes the constraints on the response set; that is, all CVC words known to the listener are possible responses compared

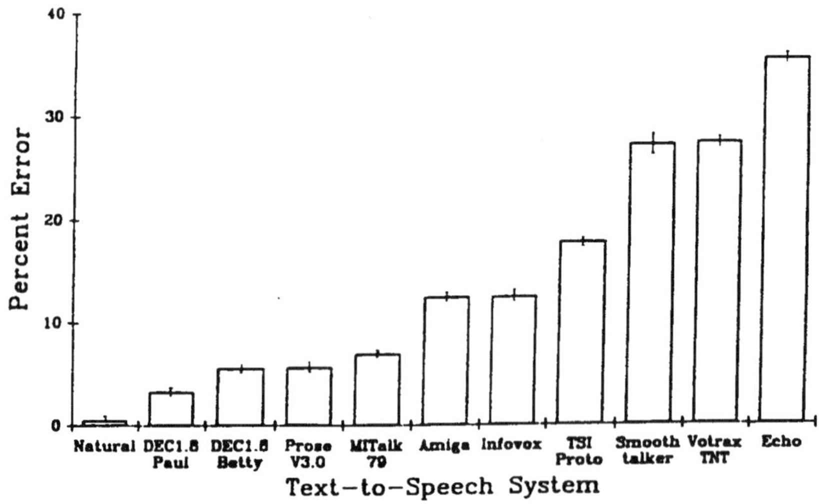

FIGURE 43.1. Overall error rates (in percent) for each of the 10 text-to-speech systems tested using the MRT. Natural speech is included here as a benchmark (from [LGP89]).

to the six alternative responses in the standard closed-response MRT. This open-set procedure also provides information about the intelligibility of vowels that is not available in the closed-response set version. Open-set tests require more cognitive effort and attention because the listener must first encode the auditory stimulus, search through his/her lexicon for one or more appropriate words, and then finally select one word as the match for the auditory stimulus [SKPO94]. By comparing the results obtained in the closed- and open-response versions of the MRT, we were able to obtain a great deal of useful information about the sources of error in a particular system [LGP89]. Indeed, detailed analyses of the stimulus-response confusions provided knowledge about the specific rules used to generate segmental contrasts in particular phonetic environments and how to modify them to improve intelligibility [Kla87].

Our results have shown large differences in segmental intelligibility between the open- and closed-response formats. Although the rank-ordering of intelligibility remains the same across the two forms of the MRT, it is clear that as speech becomes less intelligible, listeners rely more heavily on the response alternatives provided by the closed-set format to help their performance on this task [MHL51]. Comparisons between open- and closed-set performance are shown in figure 43.2 for 10 synthesis systems.

Nonnative Speakers of English. We have carried out several studies in which nonnative speakers of English listened to both natural and synthetic speech materials [Gre86]. Nonnative speakers reveal essentially the same pattern of results found for native speakers: Their performance is better when listening to natural speech than synthetic speech. Results were obtained for intelligibility of isolated words in the MRT task and for word recognition in sentences using a transcription

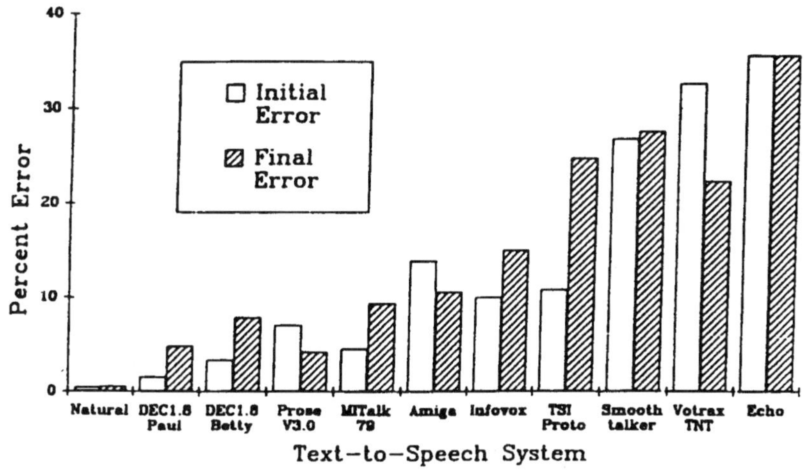

FIGURE 43.2. Error rates (in percent) for 10 systems tested using both the closed- and open-response format MRT. Open bars designate error rates for the closed-response format, and striped bars designate error rates for the open-response format (from [LGP89]).

task. However, the absolute levels of performance were substantially lower for these listeners than for native English speakers using the same materials.

Word Recognition in Sentences: Transcription Performance. To examine the contribution of sentence context and linguistic constraints on performance, we studied word recognition in two types of sentences: syntactically correct and meaningful sentences—"Add salt before you fry the egg"—and syntactically correct but semantically anomalous sentences—"The old farm cost the blood." Subjects listen to each sentence and simply write down what they hear on each trial.

In comparing word recognition performance for these two types of sentences, we found large and consistent effects of meaning and linguistic constraints on word recognition. For both natural and synthetic speech, word recognition was always better in meaningful sentences than in the semantically anomalous sentences. Not surprisingly, meaningful sentences narrow down response alternatives and help listeners understand words in context. Both top-down and bottom-up processes are required to carry out this transcription task. Furthermore, a comparison of correct word identification in these sentences revealed an interaction in performance, suggesting that semantic constraints are relied on much more by listeners as the speech becomes progressively less intelligible [Pis87, PH80]. Subjects also have great difficulty inhibiting the use of semantic constraints in word recognition even when it is not helpful to them. Some interesting examples of these response

TABLE 43.2. Consonant class labels used for American English.

Harvard and Haskins Sentence Targets and Examples of Responses
Harvard Sentences:
TARGET:
The juice of lemons makes fine punch.
RESPONSES:
The juice of lemons makes Hawaiian punch.
The goose lemon makes fine punch.
The juice of lemons makes a high punch.
Haskins Sentences:
TARGET:
The far man tried the wood.
RESPONSES:
The fireman dried the wood.
The fireman tried the wool.
The farm hand dried the wood.

strategies are shown in table 43.2. A recent modification of this task, called the semantically unpredictable sentences (SUS) task, uses five different syntactic structures for the anomalous sentences [Gri89, HG89]. More details of the methods and results of this test using other languages can be found in papers by Benoît [BvGHJ89, Ben90].

43.3 Comprehension of Synthetic Speech

In addition to our studies on segmental intelligibility and word recognition in sentences, we have had a long-standing interest in the process of comprehension. We carried out a series of experiments on the verification of isolated sentences as well as several studies on how listeners understand and answer questions about long passages of continuous synthetic speech produced by rule.

Sentence-Verification Studies. In the sentence-verification experiments, we used three- and six-word sentences that were either true or false: "Cotton is soft," "Snakes can sing." The sentences were pretested to determine whether the final word in each sentence was predictable and to insure that listeners could transcribe all the sentences correctly with no errors. In the sentence verification test, subjects were required to respond "true" or "false" after hearing each sentence. Results shown in figure 43.3 indicated that subjects were consistently faster in responding to natural speech than to synthetic speech. For both natural and synthetic speech, responses were faster for high-predictability sentences than for low-predictability sentences [PMD87]. The results showed that although the sentences were highly intelligible, even high-quality synthetic speech is not perceived in the same way

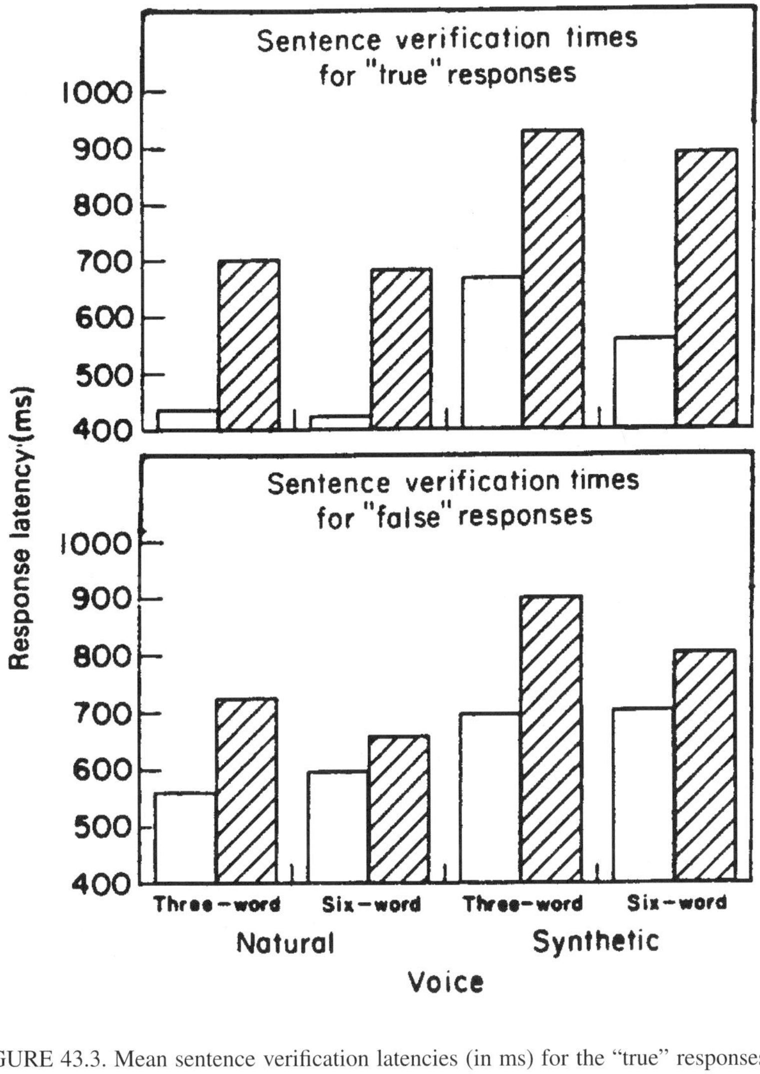

FIGURE 43.3. Mean sentence verification latencies (in ms) for the "true" responses (top panel) and "false" responses (bottom panel) for natural and synthetic speech for each of two sentence lengths. The high-predictability sentences are displayed with open bars; the low-predictability sentences are displayed with striped bars. The latencies displayed in this figure are based on only those trials in which subjects responded correctly and also transcribed the sentence correctly (from [Pis87]).

as natural speech. All of these sentences were easy to recognize and encode in memory as indexed by transcription scores, but there was additional cognitive effort required to understand the sentence and carry out the verification task, as shown by longer response times.

Comprehension of Connected Text. Spoken language understanding is a very complex cognitive process that involves the encoding of sensory information, retrieval of previously stored knowledge from long-term memory, and the subsequent interpretation and integration of various sources of knowledge available to a listener. Language comprehension therefore depends on a relatively large number of diverse and complex factors, many of which are still only poorly understood by cognitive psychologists. One of the factors that plays an important role in listening comprehension is the quality of the initial input signal—that is, the intelligibility of the speech itself, which is assumed to affect the earliest stage of processing, the encoding stage. But the acoustic-phonetic properties of the signal are only one source of information used by listeners in speech perception and spoken-language understanding. Additional consideration must also be given to the contribution of higher levels of linguistic knowledge to perception and comprehension.

In our first comprehension study, subjects either listened to synthetic or natural versions of narrative passages or they read the passages silently. All three groups then answered the same set of multiple-choice test questions immediately after each passage. Although there was a small advantage for the natural-speech group over the synthetic-speech group, the differences in performance appeared to be localized primarily in the first half of the test. The somewhat higher performance on natural speech was eliminated by the second half of the test. Performance by the subjects listening to synthetic speech improved substantially, whereas performance by the natural-speech group and the control group that simply read the texts and then answered questions remained about the same.

The finding of improved performance in the second half of the test for subjects listening to synthetic speech is consistent with our earlier results on word recognition in sentences. We found that recognition performance always improved for synthetic speech after only a short period of exposure and familiarization with the synthesis system [CGL76, NG74, PH80]. These results suggest that the overall differences in performance among the three comprehension groups are probably due to familiarity with the output of the synthesizer and not to any inherent differences in the basic strategies used in comprehending the linguistic content of these passages. One serious problem with this initial comprehension study was that we used multiple-choice questions presented immediately after listeners heard each passage. This test procedure therefore confounds the early stages of perceptual analysis and encoding with later stages of comprehension involving memory, inferencing and reconstruction, which are known to play an important role in studies of text processing, independent of which modality is used for input.

On-line Measures of Comprehension. Recently, we examined the comprehension process in greater detail using several on-line measurement techniques. In one study, Ralston and colleagues ([RPLGM91]) used a word-monitoring task to investigate comprehension of natural and synthetic speech. Subjects were required to monitor a spoken passage for a set of target words. Specifically, the listeners had to memorize a set of target stimuli, rehearse the items, and then press a response button whenever they heard one of the target words in a spoken passage. To make

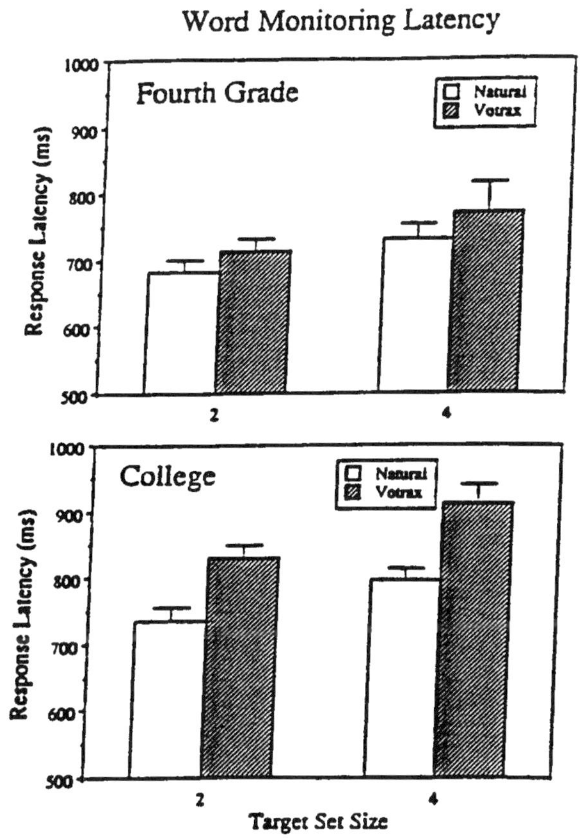

FIGURE 43.4. Word-monitoring latencies (in ms) as a function of target set size. The upper panel shows data for fourth-grade passages; the lower panel shows data for college-level passages. Open bars represent natural speech; striped bars represent latencies for passages of Votrax synthetic speech (from [RPLGM91]).

sure that subjects understood the content of the passage, we also had them answer a set of comprehension questions after each passage. As shown in figure 43.4, word-monitoring performance was better for subjects listening to natural speech compared to synthetic speech. Detection accuracy decreased and response latencies increased as the number of words in the target set became larger. Listeners were more accurate in answering questions following presentation of naturally produced passages than synthetic passages. Thus, both speech quality and memory load affected word-monitoring performance in this task.

In another comprehension study, we used a self-paced listening task to measure the amount of processing time subjects need to understand individual sentences in a passage of connected speech [RPLGM91]. As expected, we found that listeners required more time to understand synthetic speech than natural speech. When the sentences in the passages were scrambled, listeners required even more processing time for *both* natural and synthetic speech [RLP90]. Moreover, the listening times were much larger for the passages of synthetic speech that required greater cognitive effort and processing resources than the natural passages that were less demanding.

43.4 Mechanisms of Perceptual Encoding

The results of the MRT and word-recognition studies revealed that synthetic speech is less intelligible than natural speech. In addition, these studies demonstrated that as synthetic speech becomes less intelligible, listeners rely increasingly on linguistic knowledge and response-set constraints to facilitate word identification. The findings from the comprehension studies are consistent with this conclusion as well [DP92]. However, the results of these studies are descriptive and do not provide an explanation for the differences in perception between natural and synthetic speech. Several different experiments were carried out over the years to pursue this problem.

Lexical Decision and Naming Latencies. In order to investigate differences in the perceptual processing of natural and synthetic speech, we carried out a series of experiments that measured the time needed to recognize and pronounce natural and synthetically produced words [Pis81b]. To measure the time course of the recognition process, we used a lexical decision task. As shown in figure 43.5, subjects responded significantly faster to *natural* words and nonwords than to *synthetic* words and nonwords. Because the differences in response latency were observed for both words *and* nonwords alike, and did not appear to depend on the lexical status of the test item, the extra processing effort appears to be related to the initial analysis and perceptual encoding of the acoustic-phonetic information in the signal and not to the process of accessing words from the lexicon. In short, the pattern of results suggested that the perceptual processes used to encode synthetic speech require more cognitive "effort" or resources than the processes used to encode natural speech. Thus, synthetic speech appears to be encoded less efficiently than natural speech, presumably because there is less redundancy in the acoustic signal.

Similar results were obtained in a naming task using natural and synthetic words and nonwords [SP82]. The naming results demonstrated that the extra processing time needed for synthetic speech does not depend on the type of response made by the listener; the pattern of latencies were comparable for both manual and vocal responses. Taken together, these two sets of findings suggest that early stages of perceptual encoding for synthetic speech are carried out more slowly and therefore require more processing time than natural speech.

Consonant-vowel (CV) Confusions. To account for the greater difficulty of encoding synthetic speech, some researchers have suggested that synthetic speech

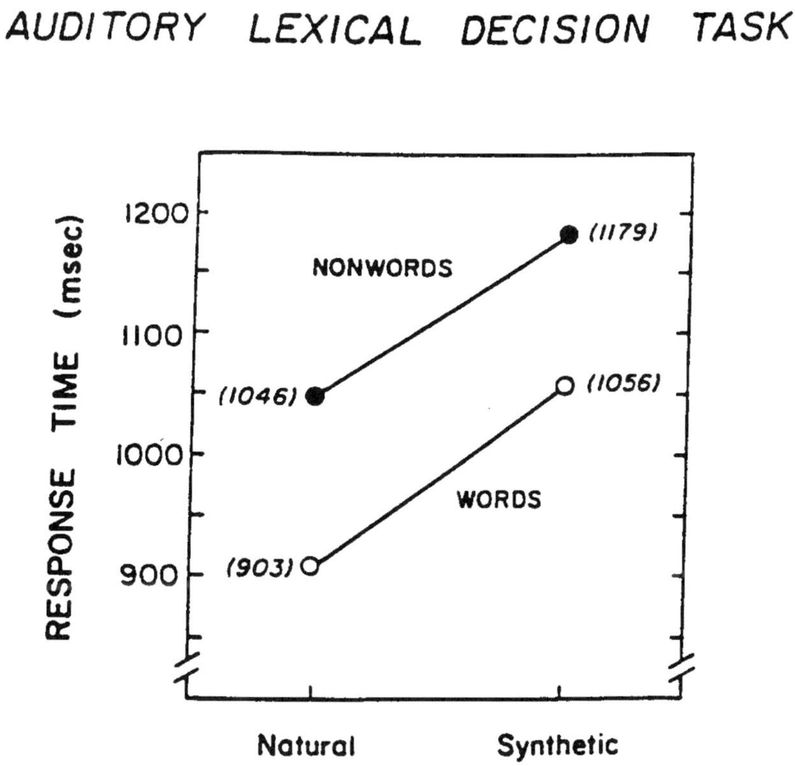

FIGURE 43.5. Response times (in ms) obtained in an auditory lexical decision task for words (open circles) and nonwords (filled circles) for natural and synthetic speech (from [Pis81b]).

should be viewed as natural speech that is degraded by noise. An alternative hypothesis proposes that synthetic speech is not like "noisy" or degraded natural speech at all, but instead may be thought of as a "perceptually impoverished" signal relative to natural speech because it lacks the additional redundancy and acoustic-phonetic variability found in natural speech. Thus, synthetic speech is fundamentally different from natural speech in both degree and kind because it contains only a minimal number of acoustic cues for each phonetic contrast. Recent findings have suggested that this redundancy is important for perceptual learning and retention of novel voices and novel words.

To test this proposal, we examined the perceptual confusions for a set of natural and synthetic consonant-vowel (CV) syllables [NDP84]. By comparing the confusion matrices for a particular text-to-speech system with the confusion matrices for natural speech masked by noise, we found that the predictions made by the "noise-degradation" hypothesis were incorrect. Some consonant identification errors were based on the acoustic-phonetic similarity of the confused segments, but others followed a different pattern that could be explained only as "phonetic miscues." These were confusions in which the acoustic cues used in synthesis simply specified the wrong segment in a particular phonetic environment.

Gating and Signal Duration. The results of the consonant-vowel confusion experiment support the conclusion that the differences in perception between natural and synthetic speech are largely the result of differences in the acoustic-phonetic properties of the signals and the initial process of encoding. In another study, we obtained further support for this proposal using the gating paradigm to investigate the perception of natural and synthetic words [Gro80]. This technique is used to manipulate the amount of stimulus information presented to a listener. We found that, on the average, natural words could be identified after only 67% of a word was heard, whereas for synthetic words it was necessary for listeners to hear more than 75% of a word for correct word identification [MP84]. These results demonstrate more directly that the acoustic-phonetic structure of synthetic speech conveys less information, per unit of time, than the acoustic-phonetic structure of natural speech, and thus the uptake of linguistic information from the signal appears to be less efficient [PNG85].

These results provide some converging evidence that encoding of the acoustic-phonetic structure of synthetic speech is more difficult and requires more cognitive effort and capacity than encoding of natural speech. Recognition of words *and* nonwords requires more processing time for synthetic speech compared to natural speech. The CV confusion study demonstrated that synthetic speech may be viewed as a phonetically impoverished signal. Finally, the gating results showed that synthetic speech requires *more* acoustic-phonetic information to correctly identify isolated monosyllabic words than natural speech.

Capacity Demands in Speech Perception. We also carried out a series of experiments to determine the effects of encoding synthetic speech on working memory and rehearsal processes [LFP83]. Subjects were given two different lists of items to remember: The first list consisted of a set of digits *visually* presented on a CRT screen; the second list consisted of a set of ten natural words or ten synthetic words. After the list of words was presented, the subjects were instructed to first write down all the visually presented digits in the order of presentation and then all the words they could remember from the list of items they heard. Recall for the natural words was significantly better than for the synthetic words. In addition, recall of both synthetic and natural words became worse as the size of the preload digit list increased. However, the most interesting finding was the presence of an interaction between the *type* of speech presented (synthetic versus natural) and the *number* of digits rehearsed (three versus six). As the size of the memory load increased, significantly fewer subjects were able to recall *all* the digits for the synthetic word lists compared to the digits from the natural word lists. Thus, processing of the synthetic speech impaired recall of the visually presented digits more than the processing of natural speech. These results demonstrate that synthetic speech requires more processing capacity and resources in short-term working memory than natural speech. The findings also suggest that synthetic speech should interfere much more with other concurrent cognitive processes because the perceptual encoding of synthetic speech imposes greater capacity demands on the human information processing system than the encoding of natural speech [Wic91].

Training and Experience with Synthetic Speech. We also carried out several experiments to study the effects of training on the perception of synthetic speech [SNP85, GNP88]. In the Schwab et al. study, three groups of subjects followed different procedures for eight days. When pre- and post-test scores were compared, the results showed that performance improved dramatically for only one group—the subjects who were specifically trained with the synthetic speech. Neither the first control group trained with natural speech nor the second control group, who received no training of any kind, showed any significant improvement in recognition of synthetic speech. These findings suggest that human listeners can easily modify their perceptual strategies and that substantial increases in performance can be realized in relatively short periods of time, even with poor-quality synthetic speech, if naive listeners become familiar with the synthesis system.

43.5 Some Cognitive Factors in Speech Perception

The literature in cognitive psychology over the last 40 years has identified several major factors that affect an observer's performance in behavioral tasks. These factors include: (1) the specific demands imposed by a particular task, (2) the inherent limitations of the human-information processing system, (3) the experience and training of the human listener, (4) the linguistic structure of the message set, and (5) the structure and quality of the speech signal. We consider here each of these in the context of our findings on the perception of synthetic speech.

Task Complexity. In some tasks, the response demands are relatively simple, such as deciding which of two words was presented. Other tasks are extremely complex, such as trying to recognize an unknown utterance from a virtually unlimited number of response alternatives, while simultaneously engaging in another activity that already requires attention. In carrying out any perceptual experiment, it is necessary to understand the requirements and demands of a particular task before drawing any strong inferences about an observer's performance. In our studies on the perception of synthetic speech, we have found that task complexity influences performance and affects the strategies that listeners use to complete their task. To take one example, large differences were observed in intelligibility of isolated words when the response set was changed from a closed-set to an open-set format.

Limitations on the Observer. Limitations exist on the human information-processing system's ability to perceive, encode, store, and retrieve information. The amount of information that can be processed in and out of short-term working memory is severely limited by the listener's attentional state, past experience, and the quality of the original sensory input that affects ease of encoding. These general principles apply to the study of speech perception as well as other domains of human information processing.

Experience and Training. Human observers can quickly learn new strategies to improve performance in almost any psychological task. When given appropriate feedback and training, subjects can learn to classify novel stimuli, remember complex visual patterns, and respond to rapidly changing stimuli presented in different sensory modalities. It comes as no surprise then that listeners can benefit from short-term training and exposure to synthetic speech.

Message Set. The structure of the message set—that is, the constraints on the number of possible messages and the linguistic properties of the message set—play an important role in speech perception and language comprehension [MHL51]. The arrangement of speech sounds into words is constrained by the phonological rules of language; the choice and patterning of words in sentences is constrained by syntax; and finally, the meaning of individual words and the overall meaning of sentences in a text is constrained by the semantics and pragmatics of language. The contribution of these levels varies substantially in perceiving isolated words, sentences, and passages of continuous speech. In each case, listeners exploit constraints to recover the intended message.

Signal Characteristics. The segmental and prosodic structure of a synthetic utterance also constrains the choice of response. Synthetic speech is an impoverished signal that represents phonetic distinctions with only the minimum number of acoustic cues used to convey contrasts in natural speech. Under adverse conditions, synthetic speech may show serious degradation because of the lack in redundancy in the signal, which is the hallmark of natural speech.

43.6 Some New Directions for Research

Much of the research carried out on the perception of synthetic speech over the last 15 years has been concerned with the quality of the acoustic-phonetic output. Researchers have focused most of their attention on improving the segmental intelligibility of synthetic speech. At the present time, the available perceptual data suggest that segmental intelligibility is quite good for some commercially available systems. Synthetic speech is not at the same level of intelligibility as natural speech, and it may take a great deal of additional research effort to achieve relatively small gains in improvement in intelligibility. Thus, it seems entirely appropriate at this time to move the focus of research efforts to a number of other issues and to pursue several new directions in the future. In this section, I review several topics that have not been studied very much in the past. Research in these areas may yield important new knowledge that will help improve performance to levels approaching natural speech.

Naturalness. Improving the naturalness of synthetic speech has been a longstanding problem in synthesis research. Everyone working in the field today believes this is the next goal. Why do listeners want to listen to "natural"-sounding speech? Why do listeners "prefer" to listen to natural speech? We believe these are important research questions for the future because they suggest a close association between the linguistic properties of the signal and the so-called "indexical" attributes of a talker's voice, which are carried in parallel in the speech signal. Recent studies have shown that knowledge of and familiarity with the talker's voice affects intelligibility of speech and may contribute to more efficient encoding of the message in memory [NSP94]. In the past, researchers have treated these two sets of acoustic attributes as independent sources of information in the signal. Now we are beginning to realize the importance of naturalness and, in particular, the role of talker-familiarity in spoken language processing. Familiar voices are easier to process and understand because the listener can access talker-specific attributes from long-term memory, which facilitates encoding the signal into a linguistic representation [Pis93]. In the next few years, we will see increased research on synthesizing specific voices, dialects, and even different speaking styles in an effort to improve the naturalness of the synthesizer's voice and therefore increase intelligibility.

Sources of Variability. In order to achieve naturalness, a great deal of new research will be needed on different sources of variability in speech and how to model and reproduce this information in the next generation of synthesis systems. Much of the early research on speech perception and acoustic-phonetics assumed the existence of abstract idealized representations for phonemes, words, and sentences. The research methodologies used over the years assumed that variability was a source of noise that had to be reduced or eliminated from the signal in order to recover the talker's intended message. Several researchers have argued recently that variability in speech is not a source of noise but rather is informative and useful to the listener [EM86, Pis93]. If we are ever going to have synthesis systems that can model different speaking styles and adjust to the demands of the listener and the environment, it will be necessary to learn more about the different sources of variability that occur in natural speech and how these factors can be incorporated into synthesis routines [MA93].

Audio-Visual Integration. Along with improving naturalness, a number of researchers have suggested developing multimodal synthesis systems that are able to produce visual displays of a synthetic talking face along with synthesis of the speech signal [Des92, BLMA92]. The case for multimodal synthesis is convincing on several theoretical grounds, the most important of which is that listeners are able to use the additional information contained in the visual display of a talker's face to improve intelligibility and recognition of the intended message [SP54]. Until recently, most of the research on speech synthesis has been concerned exclusively with the auditory modality despite the evidence that deaf and hearing-impaired listeners gain substantial information about speech from the optical display of a talker's face [Sum87]. It appears very likely that synthetic faces will become an integral part of speech synthesis systems in the next few years.

New Assessment Methods. As we look back over the past 15 years, it is obvious from the research findings that much more behavioral research with human listeners will be needed to study the complex relations between traditional measures of segmental intelligibility, comprehension performance, and listener preferences. For example, whereas more basic research on prosody and speech timing will no doubt help to eventually improve synthesis in the long term, new behavioral tests will need to be developed to measure and assess these gains in performance. We believe that prosodic characteristics are not perceived directly by naive listeners; rather, they exert their influence indirectly on the processes used to recognize words and understand the meaning of sentences and discourse. Because of this, we believe new perceptual tests will have to be developed to study and measure the effects of prosody as they affect other aspects of speech perception and spoken language processing. These tests might include measures of processing load, attention, memory, or real-time comprehension.

A good example of the problems in measuring prosody can be seen in the recent experiments of van Santen [van94] on the development of duration rules in the Bell Laboratories synthesis-by-rule system. Using very sophisticated methodologies, van Santen showed that a group of naive listeners consistently preferred a "new" set of duration rules over an "old" set of rules and were able to make explicit quality judgments about sentences containing various kinds of problems in pronunciation, stress, voice quality, and timing. The question of interest about these new duration rules is whether they produce speech that is more intelligible than the old rules. Is the synthetic speech easier to process, that is, recognize or recall, and are these new rules less susceptible to degradation from noise, other competing voices, or tasks requiring higher cognitive load? If a naive listener prefers one durational rule system over another, is that system therefore "better" on a variety of behavioral performance measures or are the preferences and quality judgments simply domain-specific? Assuming that we could develop a sensitive on-line measure of comprehension, would there be any difference in comprehension performance between the "old" and "new" rule systems?

I believe these are the kinds of questions we will have to address in the future in designing the next generation of synthesis-by-rule systems. Because spoken language processing in human listeners is extremely robust under a wide variety of conditions, it has been and will continue to be very difficult to identify precisely which component or subcomponent in the system is responsible for a particular problem or what aspects of the system control a listener's preference in one direction or another. Humans are able to adjust and adapt their perceptual processing and response criteria rapidly on the fly to changing conditions in their listening and speaking environments. In the years to come, we will need to continue several broad-based programs of basic research on human speech perception and spoken language processing in parallel with research and development efforts on speech synthesis by rule.

These are a few of the questions and new research directions that should be pursued over the next few years. Research on naturalness, variability, audio-visual integration, prosody, and comprehension are not only topics of practical concern with regard to the design and implementation of synthesis-by-rule systems, but these particular research issues are also at the center of current theoretical work in cognitive science and psycholinguistics. Answers to these questions will provide us with new insights into the reasons why synthetic speech is difficult to understand and why it requires more attention and effort for the listener to recover the intended meaning.

Acknowledgments: This research was supported, in part, by NIH Research Grant DC-00111 and NIH T32 Training Grant DC-00012 to Indiana University. I thank Beth Greene for her help over the years on the perceptual evaluation project and Steve Chin for his editorial comments.

References

[AHK87] J. Allen, S. Hunnicut, and D. H. Klatt. *From Text to Speech: The MITalk System.* Cambridge University Press, Cambridge, 1987.

[BB92] G. Bailly and C. Benoît. *Talking Machines, Theories, Models, and Designs.* Elsevier, North-Holland, Amsterdam, 1992.

[Ben90] C. Benoît. An intelligibility test using semantically unpredictable sentences: Towards the quantification of linguistic complexity. *Speech Comm.* 9:293–304, 1990.

[BLMA92] C. Benoît, T. Lallouache, T. Mohamadi, and C. Abry. A set of French visemes for visual speech synthesis. In *Talking Machines: Theories, Models and Designs*, G. Bailly, C. Benoît, and T. R. Sawallis, eds. Elsevier Science Publishers, North-Holland, 485–504, 1992.

[BP89] R. V. Bezooijen and L. Pols. Evaluation of text-to-speech conversion for Dutch: From segment to text. In *Proceedings of the ESCA Workshop on Speech Input/Output Assessment and Speech Databases*, Noordwijkerhout, The Netherlands, 1989.

[BvGHJ89] C. Benoît, A. van Erp, M. Grice, V. Hazan, and U. Jekosch. Multilingual synthesizer assessment using unpredictable sentences. In *Proceedings of the First Eurospeech Conference*, Paris, France, 633–636, 1989.

[CG89] R. Carlson amd B. Granstrom. Evaluation and development of the KTH text-to-speech system at the segmental level. In *Proceedings of the ESCA Workshop on Speech Input/Output Assessment and Speech Databases*, Noordwijkerhout, The Netherlands, 1989.

[CGL76] R. Carlson, B. Granstrom, and K. Larssen. Evaluation of a text-to-speech system as a reading machine for the blind. *Quarterly Progress and Status Report*, STL-QPSR 2-3. Stockholm: Royal Institute of Technology, Department of Speech Communication, 1976.

[Des92] R. Descout. Visual speech synthesis. In *Talking Machines: Theories, Models and Designs*, G. Bailly, C. Benoît, and T. R. Sawallis, eds. Elsevier Science Publishers, North-Holland, 475–477, 1992.

[DP92] S. A. Duffy and D. B. Pisoni. Comprehension of synthetic speech produced by rule: A review and theoretical interpretation. *Language and Speech* 35:351–389, 1992.

[EM86] J. L. Elman and J. L. McClelland. Exploiting lawful variability in the speech wave. In *Invariance and Variability in Speech Processes*, J. S. Perkell and D. J. Klatt, eds. Erlbaum, Hillsdale, NJ, 360–385, 1986.

[FHBH89] A. Fourcin, G. Harland, W. Barry, and V. Hazan. *Speech Input and Output Assessment*. Ellis Horwood, Chichester, England, 1989.

[GNP88] S. L. Greenspan, H. C. Nusbaum, and D. B. Pisoni. Perceptual learning of synthetic speech produced by rule. *Journal of Experimental Psychology: Human Learning, Memory and Cognition* 14:421–433, 1988.

[Gre86] B. G. Greene. Perception of synthetic speech by nonnative speakers of English. In *Proceedings of the Human Factors Society*, Santa Monica, CA, 1340–1343, 1986.

[Gri89] M. Grice. Syntactic structures and lexicon requirements for semantically unpredictable sentences in a number of languages. In *Proceedings of the ESCA Workshop on Speech Input/Output Assessment and Speech Databases*, Noordwijkerhout, The Netherlands, 1989.

[Gro80] F. Grosjean. Spoken word recognition processes and the gating paradigm. *Perception & Psychophysics* 28:267–283, 1980.

[HG89] V. Hazan and M. Grice. The assessment of synthetic speech intelligibility using semantically unpredictable sentences. In *Proceedings of the ESCA Workshop on Speech Input/Output Assessment and Speech Databases*, Noordwijkerhout, The Netherlands, 1989.

[HWHK65] A. S. House, C. E. Williams, M. H. L. Hecker, and K. Kryter. Articulation-testing methods: Consonantal differentiation with a closed-response set. *J. Acoust. Soc. Amer.* 37:158–166, 1965.

[Kla87] D. H. Klatt. Review of text-to-speech conversion for English. *J. Acoust. Soc. Amer.* 82:737–793, 1987.

[LFP83] P. A. Luce, T. C. Feustel, and D. B. Pisoni. Capacity demands in short-term memory for synthetic and natural word lists. *Human Factors* 25:17–32, 1983.

[LGP89] J. S. Logan, B. G. Greene, and D. B. Pisoni. Segmental intelligibility of synthetic speech produced by rule. *J. Acoust. Soc. Amer.* 86:566–581, 1989.

[MA93] I. R. Murray and J. L. Arnott. Toward the simulation of emotion in synthetic speech: A review of the literature on human vocal emotion. *J. Acoust. Soc. Amer.* 93:1097–1108, 1993.

[MHL51] G. A. Miller, G. A. Heise, and W. Lichten. The intelligibility of speech as a function of the context of the test materials. *J. Experimental Psychology* 41:329–335, 1951.

[Mil56] G. A. Miller. The perception of speech. In *For Roman Jakobson*, M. Halle, ed. Mouton, The Hague, 353–359, 1956.

[MP84] L. M. Manous and D. B. Pisoni. Effects of signal duration on the perception of natural and synthetic speech. *Research on Speech Perception Progress Report No. 10*, Indiana University, Bloomington, IN, 1984.

[NDP84] H. C. Nusbaum, M. J. Dedina, and D. B. Pisoni. Perceptual confusions of consonants in natural and synthetic CV syllables. *Speech Research Laboratory Technical Note 84-02*. Indiana University, Speech Research Laboratory, Bloomington, IN, 1984.

[NG74] P. W. Nye and J. Gaitenby. The intelligibility of synthetic monosyllabic words in short, syntactically normal sentences. *Haskins Laboratories Status Report on Speech Research* 38:169–190, 1974.

[NSP94] L. C. Nygaard, M. S. Sommers, and D. B. Pisoni. Speech perception as a talker-contingent process. *Psychological Science* 5:42–46, 1994.

[PH80] D. B. Pisoni and S. Hunnicutt. Perceptual evaluation of MITalk: The MIT unrestricted text-to-speech system. In *1980 IEEE Conference Record on Acoustics, Speech and Signal Processing* IEEE Press, New York, 572–575, 1980.

[Pis81a] D. B. Pisoni. Perceptual processing of synthetic speech: Implications for voice response systems in military applications. Paper presented at the *Conference on Voice-Interactive Avionics, Naval Air Development Center*, Warminster, PA, 1981.

[Pis81b] D. B. Pisoni. Speeded classification of natural and synthetic speech in a lexical decision task. *J. Acoust. Soc. Amer.* 70:S98, 1981.

[Pis87] D. B. Pisoni. Some measures of intelligibility and comprehension. In *From Text to Speech: The MITalk System*, J. Allen, S. Hunnicutt, and D. H. Klatt, Cambridge, Cambridge University Press, Cambridge, 1987.

[Pis93] D. B. Pisoni. Long-term memory in speech perception: Some new findings on talker variability, speaking rate and perceptual learning. *Speech Comm.* 13:109–125, 1993.

[PMD87] D. B. Pisoni, L. M. Manous, and M. J. Dedina. Comprehension of natural and synthetic speech: Effects of predictability on the verification of sentences controlled for intelligibility. *Computer Speech and Language* 2:303–320, 1987.

[PNG85] D. B. Pisoni, H. C. Nusbaum, and B. G. Greene. Perception of synthetic speech generated by rule. *Proceedings of the IEEE* 73(11):1665–1676, 1985.

[Pol89a] L. C. W. Pols. Improving synthetic speech quality by systematic evaluation. In *Proceedings of the ESCA Tutorial Day on Speech Input/Output Assessment and Speech Databases*, Noordwijkerhout, The Netherlands, 1989.

[Pol89b] L. C. W. Pols. Assessment of text-to-speech synthesis systems. In *Speech Input and Output Assessment*, A. J. Fourcin, G. Harland, W. Barry, and V. Hazan, eds. Ellis Horwood, Chichester, England, 55–81, 1989.

[Pol92] L. C. W. Pols. Quality assessment of text-to-speech synthesis by rule. In *Advances in Speech Signal Processing*, S. Furui and M. M. Sondhi, eds. Marcel Dekker, New York, 387–416, 1992.

[RLP90] J. V. Ralston, S. E. Lively, and D. B. Pisoni. Comprehension of normal and scrambled passages using a sentence-by-sentence listening task. *Research on Speech Perception Progress Report No. # 16*. Indiana University, Speech Research Laboratory, Bloomington, IN, 1990.

[RPLGM91] J. V. Ralston, D. B. Pisoni, S. E. Lively, B. G. Greene, and J. W. Mullennix. Comprehension of synthetic speech produced by rule: Word monitoring and sentence-by-sentence listening times. *Human Factors* 33:471–191, 1991.

[SAMW89] M. Spiegel, M. J. Altom, M. Macchi, and K. Wallace. A monosyllabic test corpus to evaluate the intelligibility of synthesized and natural speech. In *Proceedings of the ESCA Workshop on Speech Input/Output Assessment and Speech Databases*, Noordwijkerhout, The Netherlands, 1989.

[SKPO94] M. S. Sommers, K. I. Kirk, D. B. Pisoni, and M. J. Osberger. Some new direction in evaluating the speech perception abilities of cochlear implant patients: A preliminary report. Poster presented at *ARO*, St. Petersburg Beach, FL, 1994.

[SNP85] E. C. Schwab, H. C. Nusbaum, and D. B. Pisoni. Effects of training on the perception of synthetic speech. *Human Factors* 27:395–408, 1985.

[SP54] W. H. Sumby and I. Pollack. Visual contribution to speech intelligibility in noise. *J. Acoust. Soc. Amer.* 26:212–215, 1954.

[SP82] L. M. Slowiaczek and D. B. Pisoni. Effects of practice on speeded classification of natural and synthetic speech. *J. Acoust. Soc. Amer.* 71 Sup. 1, S95–S96, 1982.

[Sum87] Q. Summerfield. Some preliminaries to a comprehensive account of audio-visual speech perception. In *Hearing by Eye: The Psychology of Lip-Reading*, B. Dodd and R. Campbell, eds. Erlbaum, Hillsdale, NJ, 3–51, 1987.

[van94] J. P. H. van Santen. Assignment of segmental duration in text-to-speech synthesis. *Computer Speech and Language* 8:95–128, 1994.

[Wic91] C. D. Wickens. Processing resources and attention. In *Multiple Task Performance*, D. Damos, ed. Taylor & Francis, Washington, DC, 334, 1991.

Section VIII

Systems and Applications

44

Section Introduction. A Brief History of Applications

Wolfgang J. Hess

Two chapters have been grouped into this section: chapter 45 by Sproat and Olive entitled "A modular architecture for multilingual text-to-speech," and chapter 46 by J. R. de Pijper, "High-quality message-to-speech generation in a practical application." These chapters address rather different topics.

An extensive discussion about the aspects of system architectures or the choice of the right system for a particular application could easily exceed the lengths of the chapters to be introduced here. The introductory remarks shall thus be confined to such aspects discussed in these chapters.

A text-to-speech system that converts orthographic input into synthetic speech principally consists of three major building blocks [Hes92]: (1) conversion of the orthographic input into a string of phonetic and prosodic symbols including information about phrasing, intonation, and duration; (2) concatenation, that is, conversion of this string of symbols into a continuum of speech parameters; and (3) acoustic synthesis, the step from the parametric representation to the speech signal. In systems with time-domain concatenation of natural speech segments, the acoustic synthesis may be confined to manipulating the three parameters, amplitude, duration, and fundamental frequency, within these segments. The three building blocks can be split up into a larger number of modules that act independently from each other. Therefore a modular architecture is appropriate for such a TTS system, and the chapter by Sproat and Olive is a good example of how this can be done.

The first TTS systems date back to the late 1960s [Kla87]. Since the early 1980s, systems have been developed that go multilingual; that is, they allow synthesizing several languages. Especially for these systems, it is important that language-independent and language-dependent modules and knowledge sources be separated. A good architecture makes it easy to add a new language to a TTS system that is already operational for other languages. The so-called NewTTS system described in chapter 45 by Sproat and Olive has such an architecture. It consists of 13 pipelined modules ranging from text processing to acoustic synthesis. At present, the system works for nine languages including French, German, Russian, and Japanese. The speech signal is synthesized using natural speech data in a parametric representation (LPC) with a sophisticated glottal source and dyads as phonetic units. With respect to multilinguality, the ultimate goal is "to provide a single set of programs that can load language-specific tables and therefore function as a synthesizer for multiple languages."

TTS systems can have many applications. However, not all possible applications really require a TTS system with its ability to synthesize an unlimited vocabulary. A selection of such applications is discussed in De Pijper's chapter on high-quality "message-to-speech" generation. We must remember that, in speech synthesis, no module whatsoever is as good as the original (i.e., human speech), and that any module involved in the process of synthesizing speech contributes to the overall degradation of the synthetic speech signal. If the vocabulary to be synthesized is limited, it may be advantageous to store natural speech samples in larger units (i.e., words or even sentences) and synthesize the output messages by concatenating these units. Usually this involves manipulating the three parameters amplitude, duration, and fundamental frequency, which is readily done by the well-known PSOLA (pitch synchronous overlap add) method or comparable techniques.

Chapter 46 in particular discusses such techniques under the aspect of multimedia applications. The system "Appeal" (which stands for "a pleasant personal environment for adaptive learning") provides an interactive tool that guides students through individual interactive computer sessions, such as in language learning. The synthesis technique applied is word concatenation. The chapter also includes a number of acoustic examples, which can be heard on the accompanying CD-ROM.

An emerging application for speech synthesis that has not been addressed in these chapters is speech synthesis as a component of a man-machine dialogue system. Such an application, for instance a database query via telephone with voice input and output, may be covered by a full text-to-speech system because the module that generates a natural-language answer from the formal representation of the data in the database usually yields orthographic text as output. However, if the output of such a system is to be a synthetic speech signal, the answer generator may directly convert the formal representation of the database information into the phonetic/prosodic string of symbols required for the concatenation module of the TTS system. This means a shortcut to the (up to now still) weakest and most error-prone modules of TTS systems: syntactic analysis and prosody generation. The necessary syntactic and prosodic information, even delicate information such as focus, can be derived from the formal representation in the database. It is obvious that the more difficult problems of such dialogue systems (speech understanding and recognition) have to be solved first, and it is likely that a future workshop on speech synthesis will describe such a concept-to-speech synthesis module.

References

[Hes92] W. J. Hess. Speech synthesis—a solved problem? In *Signal Processing VI - theories and applications. Proceedings of EUSIPCO-92, Sixth European Signal Processing Conference*, Brussels, Elsevier B.V., Amsterdam, 287–304, 1992.

[Kla87] D. H. Klatt. Review of text-to-speech conversion for English. *J. Acoust. Soc. Amer.* 82:737–793, 1987.

45

A Modular Architecture for Multilingual Text-to-Speech

Richard W. Sproat
Joseph P. Olive

ABSTRACT We describe the architecture of the Bell Laboratories New text-to-speech system. The modular pipeline design of NewTTS enhances capabilities for research on TTS, because it facilitates revision and replacement of modules of the system and it makes it easy to insert into the pipeline tools that modify various TTS parameters. While we illustrate most of these points with reference to the English version of NewTTS, we will also sketch how systems for other languages are being developed within the NewTTS framework.

45.1 Introduction

This chapter gives a brief synopsis of the architecture of the new Bell Laboratories TTS system (which we currently call NewTTS), a completely modular system for TTS conversion. We start with a description of the current architecture for our English synthesizer, describe the general structure of individual modules, and argue for the advantages of such a design over more monolithic approaches. We then describe how this architecture is being used to build synthesizers for languages other than English.

45.2 Architecture

45.2.1 English NewTTS

The English version of NewTTS consists of 13 modules, each of which is responsible for one piece of the problem of converting text into speech.[1] These modules include:

- *Text preprocessing*, including end-of-sentence detection, "text normalization" (expansion of numerals and abbreviations), and limited grammatical

[1]The programs can also be compiled into a single executable.

analysis, such as grammatical part-of-speech assignment (on the latter see [Chu88])

- *Lemmatization* of word forms
- *Accent assignment*—the assignment of levels of prominence to various words in the sentence [Hir93, Spr94]
- *Word pronunciation*, including the pronunciation of names and the disambiguation of homographs [CCL90]
- *Intonational phrasing*—the breaking of (usually long) stretches of text into one or more intonational units [WH92]
- *Phrasal accent assignment* [APL84]
- *Segmental durations*—the determination, on the basis of linguistic information computed thus far, of appropriate durations for phonemes in the input [van94]
- F_0 *contour computation* [APL84]
- *Amplitude assignment*
- *Source parameter computation*—parameters of the glottal source (glottal open quotient, spectral tilt, and aspiration noise) are computed for each sentence — this information to be used later in the synthesis module
- The selection of appropriate concatenative units given the phoneme string to be synthesized [Oli90]
- The concatenation of those units
- The synthesis of a speech waveform given the units, plus a model of the glottal source [Oli93]

The particular ordering of the modules is to some degree determined by the logical structure of the task, and to some degree arbitrary. For example, segmental duration assignment must, of course, come after the computation of word pronunciation, but it seems at the same time to logically precede intonation assignment (because the computation of appropriate contours depends upon timing information). Thus the three modules in question must be ordered as shown. Similarly, accent assignment precedes intonational phrasing because the particular model of phrasing used depends upon prior computation of pitch accent placement [WH92]. On the other hand, there is no particular reason why accenting must precede word pronunciation, and we could easily have ordered these two modules in the opposite order. In a similar fashion, each module obviously corresponds to a fairly well-defined subproblem, although the division is necessarily somewhat arbitrary. For example, the module that selects the units to be concatenated and the module that actually performs the concatenation could clearly be combined into a single module, and indeed were so combined in previous versions of our system. Again, the division of text-preprocessing from word-pronunciation reflects the fairly standard assumption that these are separate problems. However, as argued in [Spr95], there are some reasons to view them as really being facets of the same problem, and so at least these two modules will most likely be collapsed into one in future versions of the system. In any event, although the choices are in some cases arbitrary, the chosen architecture has certainly proven quite adequate for the task of synthesizing English.

45.2.2 Communication Between Modules

With the exception of the dyad concatenation module, which outputs a stream of LPC parameters and source state information for the waveform synthesizer, and the synthesis module itself, which reads the LPC and source-state stream, each module in the pipeline communicates with the next via a uniform set of data structures. Information is passed along on a sentence-by-sentence basis,[2] and consists of a set of arrays of linguistic structures. Each array is preceded by the following information:

1. The structure's type T: e.g., word, syllable, phrase, phoneme, polyphone, and so forth
2. N: the number of structures of type T for this sentence
3. S: the size of an individual structure of type T

After each of these headers is an $N \cdot S$ byte array consisting of N structures of type T.

Each module in the pipeline reads in the structures a sentence at a time using a library function, performs some processing on the input, and then writes out the structures (again using a library function) for the next module. The flow of information from module to module is basically monotonic, in that a module typically just adds information. For instance, the accent module would add accent structures to the data stream. Of course, structures of different types need to be aligned in some fashion with one another. For example, in English, accent structures need to be associated with words. (In the Mandarin synthesizer, accents—called tones—are aligned with syllables rather than words.) These associations are represented with pointers, a property inherited from earlier versions of the Bell Labs synthesizer [OL85].[3]

[2] What counts as a sentence is determined by the text-processing module.

[3] Because pointers denote addresses in the memory of a given process, pointers cannot be passed as such between processes. We get around this by passing to the next module not an actual pointer to a structure of type T, but rather an offset into the array of type T. This number is then converted into a pointer upon read-in of the structures.

The monotonicity of NewTTS has some commonality with the Delta system [Her88] and its descendants [vt93]. In Delta, data are represented as a series of streams, each stream representing one type of linguistic information, such as orthographic representation, phonemes, or accent status; associations between levels of representation are handled by "sync" marks that align information at the various levels rather than by pointers, as in NewTTS.

45.2.3 Advantages of Modular Structure

This architecture confers a number of advantages over a more monolithic approach. The first advantage is a standard observation about systems with modular architectures: if the modules of a system share a common interface then it is relatively easy for different people to work on different modules of the system independently of one another, as long as they agree on the input-output behavior of each module and the order in which the modules should go. See [vt93] for similar arguments concerning the modular *Speech Maker* system. In any large TTS effort it will necessarily be the case that different researchers will be responsible for different modules, so a well-designed modular architecture is crucial. The NewTTS architecture has already proven fruitful in this regard and has allowed for the easy addition of various new modules to our English system, including work on accentuation [Hir93, Spr94] and segmental duration [van94], as well as the addition or modification of some modules of wider applicability—for example, Oliveira's [Oli93] work on source modeling.

Second, the pipeline design makes it easy to cut off the processing at any point in the pipeline. During the design of a segmental duration module, for example, one might want to test the system by running it over a large corpus of text and sampling the segmental durations that are computed. For that task one would not normally care about what happens in modules subsequent to duration, such as intonation, amplitude, or dyad selection. In the NewTTS architecture, one can simply run all modules up to and including the duration module, and then terminate the pipeline with a program that prints out information about segments and their durations. (A general purpose pipefitting called **print structures** is provided, but since this prints out more information than one generally wants, it may be desirable to write a more special-purpose pipefitting, something that is very easy to do.) Naturally, one can also *initiate* the pipeline at any point.

As noted by a reviewer of this chapter, in a message-to-speech system, one can presume that the system has some notion of the linguistic structure of the message it is producing, and it should therefore be unnecessary to do any text analysis. Rather, one would like to have the message generation module generate appropriate linguistic structures, annotated with information about phrasing, accentuation, and possibly even word pronunciation. In the NewTTS architecture this is perfectly straightforward: one would merely have to replace the text processing module with a text-generation module that produces an appropriate set of NewTTS structures. Finally, it is easy to insert anywhere one desires in the pipeline programs that modify TTS parameters in various ways. One particularly useful program is a

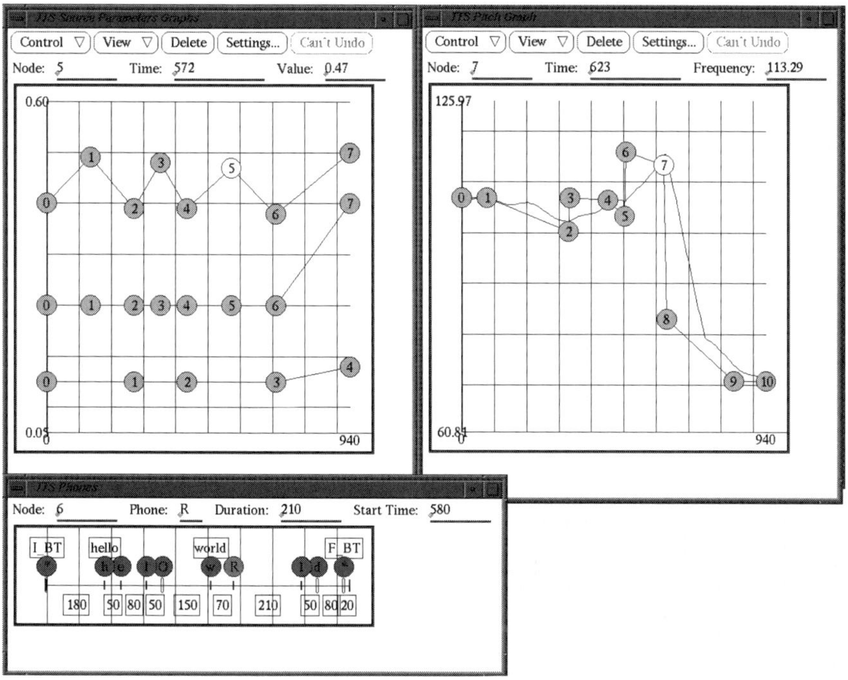

FIGURE 45.1. Partial IPEX display for the input sentence *Hello world*. Depicted are the *glottal source* parameters (top left panel) consisting of (from top to bottom), *glottal open quotient*, *spectral tilt*, *aspiration noise*; F_0 (top right panel); *phones and durations* (bottom panel). A panel displaying the amplitude parameters is also available, but is not shown here. The circles in the top two panels represent computed target values for the contours in question, and these can be interactively manipulated; for example, the highlighted node labeled "7" in the F_0 panel represents one of the targets associated with the H* accent assigned to *world*, and this target can be raised or lowered in pitch, or moved earlier or later in time as desired. Target nodes can also be added or deleted, resulting in modifications of the curves. The circles in the *phones and durations* panel represent individual phones; these can be selected and changed, or moved to decrease or increase their segmental duration.

graphical interface tool that we call IPEX, an interactive prosody editor that runs under the X window system. IPEX fits into the pipeline after the glottal source module, and allows for interactive manipulation of segments, segmental durations, pitch contours, three glottal source parameters, and amplitude contours. This tool has proved to be particularly useful in the development of intonation models for various languages (see also section 45.3), because it allows one to immediately see the consequences of a particular set of higher-level prosodic specifications; one can then experiment interactively with various ways of correcting any problems before rewriting one's higher-level specification. An example showing some of the display panels in IPEX is given in figure 45.1.

More jocular applications are also easy to incorporate into the NewTTS architecture. For example, one of the authors wrote a simple singing module that replaces the intonation module and causes the system to sing a text according to a hand-specified tune: The musical notes along with their durations are represented in the input text via a special escape sequence that instructs the escape processor to put an arbitrary string into the *comment* field of the subsequent word structure; this comment field is later read by the singing module.

45.3 Application to Other Languages

The modular pipeline architecture of NewTTS has also, we believe, been an important aid in our development of synthesizers for new languages. We are currently in the process of building synthesizers for Mexican Spanish, French, Italian, German, Russian, Japanese, Mandarin Chinese, and Taiwanese:[4] Some aspects of some of these systems are described elsewhere ([PvH95, HP94, Tzo94], and chapter 31, this volume).

Some of the modules for the English NewTTS can simply be taken and used as is—with appropriately different parameter settings or language-specific tables—in a TTS pipeline for a new language. Modules that are already language-independent in this sense are the synthesis module; the dyad selection and concatenation modules, which must, of course, be provided with a language-specific set of units; the glottal source module; and the amplitude module. An intonational phrase module can be trained and compiled for a new language using appropriately tagged text data for that language [HP94]. We are currently working on table-driven, language-independent duration and intonation modules, as well as a language-independent text-analysis module, incorporating text-processing, morphological analysis, and word pronunciation (on the latter, see [Spr95]). The ultimate goal is to provide a *single* set of programs that can load language-specific tables and therefore function as a synthesizer for *multiple* languages.

Needless to say, in the early phases of the development of a system for a new language, one may well lack much of the information that one would need to construct the necessary tables to drive all of these modules. For example, initially one might typically have only a phonetic transcription scheme for the new language, along with a dyad inventory; data for constructing duration, intonation, phrasing, accenting, or text-analysis modules would be collected later on. Clearly what is desirable is a "skeletal" TTS system that can at least allow one to get reasonable speech output, given a phonetic transcription, along with a tool such as IPEX for manipulating TTS parameters. Toward this end we have developed a set of "vanilla" (i.e., minimal, default) modules that do minimal analysis corresponding to various components that we have discussed in connection with the

[4] To complete the set, a partial system for Navajo using the NewTTS architecture was built in 1993 as a summer project for a graduate student, Robert Whitman; and a fairly complete TTS system using this architecture was built for Romanian in 1995 by an undergraduate summer student, Karen Livescu.

English NewTTS. For example, the vanilla text analysis module does little more than read a phonetic transcription of a sentence in a predefined phonetic alphabet for the language in question; the vanilla duration module computes its best guess of segmental durations based upon its knowledge of English durations; and the vanilla phrasing module will make intonational phrase breaks only at punctuation marks. Other modules, such as the dyad selection and concatenation modules, are, of course reasonably vanilla already and merely need to be provided with a table for the language in question. The vanilla version of NewTTS is being used in all of the new language systems currently under development.

45.4 Audio Files

The audio files accompanying this chapter contain an introduction read by our American English synthesizer (male and female voices), plus examples of our current state of synthesis for the following languages:

- German
- Mexican Spanish
- French
- Romanian
- Russian
- Japanese
- Mandarin
- Taiwanese
- Navajo

Acknowledgments: Many people have aided in the development of NewTTS. In addition to those who have contributed modules to the English version and who have been cited in the text, various people have helped with the implementation, and of these we would particularly like to thank Mark Beutnagel, James Rowley, and Michael Tanenblatt. Tanenblatt also wrote IPEX; this work also benefited from advice from Doug Blewett upon whose *xtent* system IPEX is based. Michael Riley provided much useful discussion about software design issues. We also thank two anonymous reviewers for useful comments on a previous version of this chapter for this volume. Finally we thank Evelyne Tzoukermann, Pilar Prieto, Bernd Möbius, Yuriy Pavlov, Elena Pavlova, and Chilin Shih for providing us with recent examples of multilingual synthesis for the accompanying audio files.

REFERENCES

[APL84] M. Anderson, J. Pierrehumbert, and M. Liberman. Synthesis by rule of English intonation patterns. In *Proceedings of the International Conference on Acoustics, Speech, and Signal Processing*, vol. 1, San Diego, 2.8.1–2.8.4, 1984.

[CCL90] C. Coker, K. Church, and M. Liberman. Morphology and rhyming: Two powerful alternatives to letter-to-sound rules for speech synthesis. In *Proceedings of the ESCA Workshop on Speech Synthesis*, G. Baily and C. Benoît, eds. Autrans, France, 83–86, 1990.

[Chu88] K. Church. A stochastic parts program and noun phrase parser for unrestricted text. In *Proceedings of the Second Conference on Applied Natural Language Processing*, Morristown, NJ, 1988. Association for Computational Linguistics, 136–143, 1988.

[Her88] S. Hertz. Delta: Flexible solutions to tough problems in speech synthesis by rule. In *The Official Proceedings of Speech Tech 88*, Media Dimensions, Inc., New York, 1988.

[Hir93] J. Hirschberg. Pitch accent in context: Predicting intonational prominence from text. *Artificial Intelligence* 63:305–340, 1993.

[HP94] J. Hirschberg and P. Prieto. Training intonational phrasing rules automatically for English and Spanish TTS. In *Proceedings of the Second ESCA/IEEE Workshop on Speech Synthesis*, 159–162, 1994.

[OL85] J. P. Olive and M. Y. Liberman. Text to speech—An overview. *J. Acoust. Soc. Amer.* Suppl. 1, 78(Fall):s6, 1985.

[Oli90] J. Olive. A new algorithm for a concatenative speech synthesis system using an augmented acoustic inventory of speech sounds. In *Proceedings of the ESCA Workshop on Speech Synthesis*, 83–86, 1990.

[Oli93] L. Oliveira. Estimation of source parameters by frequency analysis. In *Eurospeech-93*, ESCA, 99–102, 1993.

[PvH95] P. Prieto, J. van Santen, and J. Hirschberg. Patterns of F_0 peak placement in Mexican Spanish. In *Proceedings of the Second ESCA/IEEE Workshop on Speech Synthesis*, Mohonk, USA, 1994.

[Spr94] R. Sproat. English noun-phrase accent prediction for text-to-speech. *Computer Speech and Language* 8:79–94, 1994.

[Spr95] R. Sproat. A finite-state architecture for tokenization and grapheme-to-phoneme conversion for multilingual text analysis. From *Texts to Tags: Issues in Multilingual Language Analysis, Proceedings of the ACL SIGDAT Workshop*, Dublin, Ireland, 65–72, 1995.

[Tzo94] E. Tzoukermann. Text-to-speech for French. In *Proceedings of the Second ESCA/IEEE Workshop on Speech Synthesis*, Mohonk, NY, 1994.

[van94] J. van Santen. Assignment of segmental duration in text-to-speech synthesis. *Computer Speech and Language* 8:95–128, 1994.

[vt93] H. van Leeuwen and E. te Lindert. Speech Maker: A flexible framework for constructing text-to-speech systems. In *Analysis and Synthesis of Speech*, L. Pols and V. van Heuven, eds. Mouton de Gruyter, Berlin, 317–338, 1993.

[WH92] M. Wang and J. Hirschberg. Automatic classification of intonational phrase boundaries. *Computer Speech and Language* 6:175–196, 1992.

46

High-Quality Message-to-Speech Generation in a Practical Application

Jan Roelof de Pijper

ABSTRACT This chapter describes the use of concatenation of prerecorded words to generate spoken messages in a practical application. This simple technique can lead to quite acceptable results if the utterances are provided with appropriate prosody using a waveform manipulation technique.

46.1 Introduction

Today's buzzword in the world of computers is *multimedia*. Although nobody can give an exact definition of what multimedia is, it is clear that it involves the use of sound and graphics in the interaction between human and computer.

One of the things achieved by the use of multimedia is that computer applications are made more easily accessible to more people. This is the user-interface aspect of multimedia. More and more, the computer application adapts itself to the user, rather than the other way around.

Modern multimedia applications employ still pictures, full-motion video and audio, and both speech and nonspeech, all of potentially high quality. Computers are still becoming more powerful every day and (personal) computer systems with full support for sound and graphics are becoming commonplace, bringing these high-powered applications within the reach of many.

Speech, of course, is just another manifestation of sound, and there is thus no technical impediment any longer to stand in the way of the use of speech in man-machine interaction. Also, speech is still man's most natural and preferred means of communication. On these grounds, we might expect to frequently encounter speech in the context of input to and output from computers. But we do not.

The reason for this is that we simply do not yet know enough about everything involved in human-computer interaction to do a good job. Even if we ignore such important issues as how to handle and maintain a dialogue between man and machine, and concentrate only on speech recognition (so that *we* can talk to *machines*) and speech synthesis (so that *machines* can talk to *us*), our present knowledge is insufficient for full-scale integration of speech in man-machine interfaces. Speech recognizers do not recognize well enough, and synthetic speech still sounds too unnatural.

This is not to say that speech cannot be used at all in man-machine interaction. It can be and it is. If the context within which the interaction is to take place can be made restricted enough, a speech recognizer will be able to perform better. The same applies if it is possible to train the speech recognizer for a specific voice. A visually handicapped person will gladly listen to a machine that gives him or her access to books and newspapers, even if the synthetic speech is of poor quality and perhaps even difficult to understand. This same synthetic speech, however, is completely unacceptable in the context of an application with high-quality graphics and high-fidelity sound.

There are various techniques that can be used to provide applications with speech output. One can simply play back prerecorded speech, one can use a text-to-speech system, or one can concatenate prerecorded phrases. These techniques will be briefly reviewed in the next sections, after which another technique, which involves concatenation of prerecorded words or phrases followed by prosodic postprocessing will be proposed.

46.2 Speech Output Techniques

46.2.1 Playback of Prerecorded Speech

The easiest and most straightforward way to incorporate speech output in a computer application is to simply make voice recordings of all sentences that need to be pronounced and then play them back at the appropriate moments. This is easy to do and will provide high-quality speech output, but it is impracticable in all but the simplest of applications. For one thing, memory limitations will usually severely restrict the number of possible sentences. A simple sentence such as *Press the spacebar to move ahead after you read a screen. Try it now.* [Demo 1 on the CD-ROM][1] lasts only 4.79 s, but in Microsoft Windows Wave format (mono, unsigned 8 bit, sampled at 22,050 Hz), it takes up 105,633 bytes, or roughly 100 Kb of disk space. For another, this scheme will work only if it is exactly known beforehand what sentences have to be produced.

46.2.2 Text-to-Speech

The opposite extreme is to use an unrestricted text-to-speech program. Ideally, an application builder in need of speech output would simply obtain a speech synthesis program for the language in question, and build it into the application. Input to such a program is unrestricted text, and the output consists of speech samples, to be saved in a file or sent directly to a digital-analog converter to be made audible.

This scheme offers great flexibility. It is not necessary to restrict the number of possible sentences to be generated by the application on the grounds of limited

[1]All examples discussed in this paper are provided on the CD-ROM included in the Appendix.

memory resources, or for any other reason: Anything goes. This, however, is not so easy and the resulting speech is clearly less acceptable.

Text-to-speech synthesis is being worked on all over the world, as testified in this book, and the quality is constantly improving. But it will be some time yet before speech synthesis on the basis of unrestricted text is a feasible proposition in general-purpose applications.

To give the reader an impression of the quality level that can be expected with text-to-speech synthesizers today, I have included samples of synthetic speech on the CD-ROM as produced by four different systems. These are by no means the only systems available, but I have chosen them because they are all capable of generating American English speech, because I had access to samples of their speech, because they are commercially available, and because I do believe that the quality of the speech generated by these systems is among the best that can be expected today.[2]

For an index to the CD-ROM demonstrations, see the Appendix. These synthetic speech examples should not only serve to demonstrate the quality level that can be expected today, but should also highlight its shortcomings.

46.2.3 The Prosodic Problem

The speech quality in the cited demonstrations clearly varies from one system to the next, but all systems demonstrated generate speech that can be easily understood. In particular the segmental quality, i.e., the quality of the speech segments themselves and the transitions between the speech sounds, has greatly improved over the past years. It is also interesting to note the kind of improvement made between one version of a system and the next.

The most noticeable weakness for all systems would appear to be in the prosody, especially the intonation: As soon as longer sentences or whole paragraphs have to be generated, the intonation quickly becomes repetitive and often clearly incorrect. In this context, note that some of the above demonstrations contain only short sentences. This is because prosody is a complex phenomenon, with multiple functionality, involving a variety of phonetic cues such as tempo, rhythm, and speech melody or intonation. Two important functions of prosody are:

- highlighting *informational* structure, for example, by accenting important and new information while deaccenting less important or old information

- highlighting *linguistic* structure, for example, by marking certain boundaries and the use of pauses at appropriate places

---

[2]There is a World Wide Web (WWW) page containing synthetic speech examples. The URL (internet address) of this page is
http://www.cs.bham.ac.uk/ jpi/synth/museum.html and it is managed by Jon Iles (e-mail: j.p.iles@cs.bham.ac.uk). The systems mentioned here, and others, are also referred to there.

Providing a sentence or paragraph with correct (acceptable, natural) prosody is a two-stage process. First, position and type of sentence accents and boundaries to be realized prosodically have to be determined. This implies both a discourse analysis and a linguistic analysis of the input text. Second, the abstract representation for accentuation and phrasing resulting from these analyses is translated into a phonetic realization.

To do all this properly on the basis of plain text, more sophisticated linguistic and discourse analysis techniques than are currently available are needed, and we need to know more about how to phonetically realize this. For this reason, text-to-speech-synthesis systems tend to use a fairly low-profile prosody scheme. The result is usually a speech melody, tempo, and rhythm that is neutral, unobtrusive, and consequently often tedious to listen to, but in which accentuation and phrasing errors are not too conspicuous.

46.2.4 Concatenation of Prerecorded Phrases

Above, I have argued that straightforward playback of prerecorded utterances is usually not flexible enough for speech output in applications, whereas the quality of unrestricted text-to-speech synthesis is insufficient. In some applications, we find a different approach being used: Entire phrases are prerecorded, and these phrases are played back in different orders to form complete utterances. For instance, in a system that gives information about the weather over the telephone, the utterance *In Eindhoven, the sky is heavily overcast* could be a concatenation of the prerecorded phrases *In Eindhoven*, *the sky*, and *is heavily overcast*. Similar utterances could be easily concatenated to indicate different locations (*In Amsterdam, the sky is heavily overcast*) or conditions of the sky (*In Eindhoven, the sky is clear*). In this way, a large number of utterances can be pronounced on the basis of a limited number of prerecorded phrases, saving memory space and increasing flexibility.

In this technique, the prosodic realization of the prerecorded phrases is critical. For instance, the phrase *In Eindhoven* that was used for the utterance *In Eindhoven, the sky is heavily overcast* cannot be used to pronounce the sentence *The sky is heavily overcast in Eindhoven*. The intonation would be all wrong, probably starting too high and containing some sort of continuation rise on the final syllable which is inappropriate in sentence-final position. Here, too, it is the prosodic realization that sets limits to the usability of the technique. Whenever a certain phrase is to be used in different contexts, it has to be recorded multiple times, with different and carefully pronounced prosodic realizations.

A program that I happen to know and that I think makes particularly good use of this technique is the chess program *The Chessmaster 3000 Multimedia*. At any point in a chess game, one can ask the program for an analysis of the current position and advice on what to do next. This advice can be given in both written and spoken form. An example of such an advice is:

[Demo 2 on the CD-ROM]

You move your queen to e3, which moves it out of take, pins Black's knight at e4, and attacks Black's knight at e4. Black responds with the pawn to d5, which removes the threat on Black's knight at e4 and attacks your bishop. You move your knight to c3, which enables the long castle. Black replies by moving the bishop to e6, which releases the pin on Black's knight at e4. Your knight captures knight. Black's pawn takes bishop.

As a result of this sequence of moves, you win a knight for a bishop. In addition, Black's pawn structure is somewhat weakened.

On careful listening, it can be heard that this is actually a concatenation (very well done) of separate phrases as indicated by slashes in the following:

You move your / queen / to / e / 3 / which / moves it out of take / pins / Black's / knight / at / e /4 / and / attacks / Black's / knight / at / e / 4 / Black / responds with the / pawn / to / d / 5 / which / removes the threat / on / Black's / knight / at / e / 4 / and / attacks / your / bishop / You move your / knight / to / c / 3 / which / enables / the long castle / Black replies by moving the / bishop / to / e / 6 / which / releases the pin on / Black's / knight / at / e / 4 / Your / knight / captures / knight / Black's / pawn / takes / bishop /

As a result of this sequence of moves / you / win / a / knight / for / a / bishop / In addition / Black's / pawn structure / is somewhat weakened

Although the number of possible different positions in a chess game is virtually infinite, a limited list of phrases can be drawn up to describe them all, and thus this concatenation technique is suitable here. Moreover, the program is distributed on CD-ROM, so enough memory is available to hold the prerecorded speech samples needed. In fact, there are 1,120 sound files (Microsoft Windows Wave format: mono, unsigned 8 bit, sampled at 22,050 Hz) on the CD-ROM, totaling no less than 411,516,286 bytes of disk space. The shortest of these is 7,760 bytes (the single word *check*), the longest is the description of a complete game by Karpov: 7,709,006 bytes.

46.2.5 Prosodic Postprocessing

Also this technique of concatenating prerecorded phrases will not always be a feasible solution for the speech output problem. Disk space may not be so readily available as in the above example, so that the number of phrases has to be limited. If a greater variety of sentences is to be generated it pays to have smaller concatenation units, but then getting the prosody right becomes more difficult. If a wider range of syntactic structures is used, then also a larger number of prosodic realizations has to be recorded for each phrase.

In the next section, a technique is described that combines concatenation of prerecorded words or phrases with prosodic postprocessing. It is argued that for some applications this may be a suitable technique to provide the application with speech output. This method has been used in *Appeal*, a multimedia educational application developed at IPO.

46.3 Word Concatenation and PSOLA

46.3.1 Appeal

At the Institute for Perception Research, a multidisciplinary project has been under way for a little over a year whose goal is to develop a multimedia-based educational system that helps students learn through individual interactive computer sessions. The system is designed as a computer environment that educators can make use of to write courses. To make the system more interesting to use, and therefore more motivating for the student, full use is made of modern multimedia techniques: The student is presented with a window-like environment in which still pictures, full-motion video, and both speech and nonspeech audio are integrated. This system has been dubbed *Appeal*, which is an acronym for "a pleasant personal environment for adaptive learning."

As a feasibility study, we have implemented a small number of lessons, supposedly part of a language course. A first prototype of Appeal, featuring Dutch as a foreign language, was recently completed.

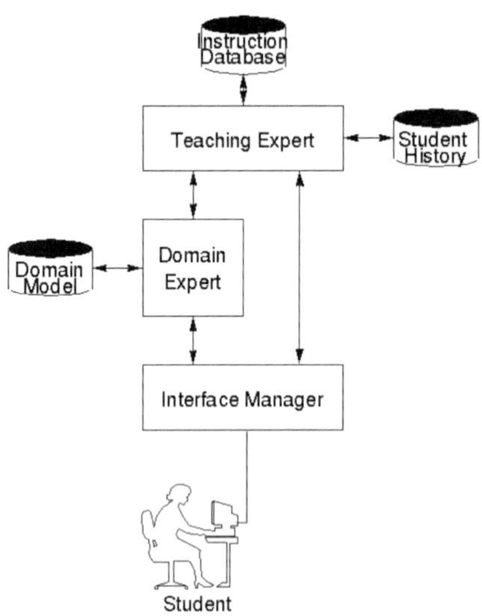

FIGURE 46.1. Appeal system architecture

Figure 46.1 gives a schematic representation of the Appeal system. For a complete description of the Appeal system, the reader is referred to [vMAHd94]. Also, a video demonstrating the system is available on the CD-ROM in the Appendix.

What distinguishes Appeal from other computer-assisted language learning (CALL) programs (apart from the multimedia aspects) is its adaptivity: The system adjusts itself to the student rather than the other way around. The path that the student follows through the course material is not predetermined, as in the traditional approach, but depends on the ability, performance, and wishes of the student.

This adaptivity is achieved by the use of so-called *software agents* (See [Mv94]). The agent approach makes it possible for the system to take action on the basis of the student's performance in a flexible way. Also, the use of agents introduces an element of unpredictability: It is impossible to tell exactly what the system will do at any particular point in a study session, only that it will satisfy the student's needs as much as possible.

An important aspect of the adaptivity built into Appeal is the fact that exercises are not selected from a database, but are created on the fly, according to predefined exercise models. This allows a great deal of variation in training material, thus reducing repetitiveness and enhancing adaptivity.

Exercise creation in Appeal is possible because the system contains a language-generation component. For instance, Appeal recognizes so-called transformation exercises, in which the student is asked to transform a declarative sentence into an interrogative, negative, or passive one, or to change its tense from present to past or future. A stimulus sentence is generated by the system on the basis of its lexicon, grammar rules, and awareness of the student's current level of competence. In order to evaluate the student's response, the system also generates the correct answer itself; there is no need to prestore any sentence material at all.

46.3.2 Speech in Appeal

Sentences generated by Appeal are not only printed on screen, but also have to be made audible as speech. In the context of Appeal, we were faced with the following two points concerning speech output:

- The use of language generation and on-the-fly exercise creation in Appeal implies that the exact contents of the speech to be produced are not known in advance. Concatenation of prerecorded phrases appeared to be too limited a device to do justice to the flexibility of the language-generation component.

- The quality of the output speech should be sufficiently high to be acceptable in the context of a multimedia-oriented system such as Appeal. At the Institute for Perception Research, both allophone and diphone-based speech synthesis algorithms are available, but we felt that in both cases the quality of the synthetic speech was too poor for use in Appeal.

In view of these considerations, it was decided to try a slightly different approach than concatenation of prerecorded phrases and see if a *word* concatenation technique could be used instead. The word as the basic concatenation unit for speech synthesis in Appeal seemed a feasible proposition because, although the number of *phrases* that can be produced by the language-generation component in Appeal is very large, the number of different *words* to be used can be kept within manageable limits.

The idea is to store on disk naturally spoken samples of all words to be used and then retrieve and concatenate them as needed to pronounce sentences generated by the system. As intimated earlier, the catch is that it is not possible to concatenate prerecorded units as small as words and still get natural-sounding prosody. The solution that we tried in Appeal is a combination of concatenation of prerecorded words and postprocessing of the resulting waveform to get the prosody right.

Until recently, manipulation of pitch and duration of a speech wave was not possible directly. The speech signal first had to be analyzed, using, for example, the linear predictive coding technique. After manipulation, the modified speech signal can then be resynthesized and made audible. In this process, some loss of quality occurs.

Now, we can make use of PSOLA, a technique to manipulate a waveform directly, so that it is possible to change pitch and rhythm of a spoken utterance without ever subjecting it to an analysis/resynthesis process. In theory, this should result in hardly any loss of quality at all. It was the existence of PSOLA that prompted us to revisit the concept of word concatenation, which is by no means new. For details on PSOLA, the interested reader is referred to [CM89] and [Ham89].

46.3.3 Dedicated Prosody

In an earlier section, we pointed out that prosody is presently the weakest point in general-purpose text-to-speech systems. In a system that has to produce synthetic speech from plain text that may be written in any style, varying from e-mail messages to literary novels, it is difficult to come up with rules to provide the utterances generated with adequate prosody. The prosodic realization has to sound acceptable in a wide variety of syntactic, semantic, and pragmatic contexts.

In Appeal, the situation is quite different. Here, we are dealing with message-to-speech generation in a known and restricted setting, rather than plain-text-to-speech synthesis. Quite a lot is known about the messages to be spoken, and this knowledge can be used when calculating prosodic parameters.

For instance, if it is known that a certain message is in fact the confirmation of a correct answer given by a student in response to a task set in a language-learning context, then it seems reasonable to adopt a rather emphatic speech style, using rich accentuation and keeping the speech tempo low.

When a sentence is pronounced in Appeal, the discourse context is known, as is the informational structure. Moreover, because in Appeal the message is actually generated by the system itself, complete and reliable syntactic and lexical information is available. This information can be used to determine where sentence accents should go and where there are boundaries important enough to be realized prosodically.

This introduces the notion of *dedicated prosody*: The prosodic rules used in the system do not have to satisfy a criterion of general applicability. The purpose here is to generate speech that is acceptable in one particular application.

46.3.4 An Example

The Appeal system is still in the early stages of development, and so far only a simple set of prosodic rules for a few types of exercises has been implemented. The point is that, in spite of its simplicity, the set serves its purpose well, because the context and syntactic structure of the sentences to be generated are known.

As an example, consider one type of exercise (the *transformation exercise* mentioned earlier), in which the student is asked to transform simple declarative sentences of the type [$NP_{subject}$ V NP_{object} PP] into the passive form. In this exercise, the system may generate the sentence *Het kind legt het boek op de tafel (The child puts the book on the table).*

To calculate an adequate prosodic realization for this type of fairly simple sentence, we found that we could do without any rules for boundary marking and lengthening of words or syllables in preboundary or accented position. It proved enough to apply a fairly straightforward set of accentuation rules to determine the position of sentence accents and use those to calculate a pitch pattern. For the pitch patterns themselves we can capitalize on the long tradition in intonation research that we have at the Institute for Perception Research. The principles developed over the past years are outlined in [tCC90].

When the time comes for this sentence to be pronounced, it is first determined which words are to be accented. Because the speech style is emphatic, most content words are accented. An abstract pitch contour is then calculated as shown in stylized form in figure 46.2. Slightly simplified, the algorithm used stipulates that the last two accented words are realized with a flat hat (i.e., an accent-lending rise (1) on the stressed vowel of the penultimate accented word and an accent-lending fall (A) on the last one), whereas all preceding accented words (only one in this case) are assigned an accent-lending rise (1) on the stressed vowel and a non–accent-lending fall (B) whose onset is located just before the end of the word.

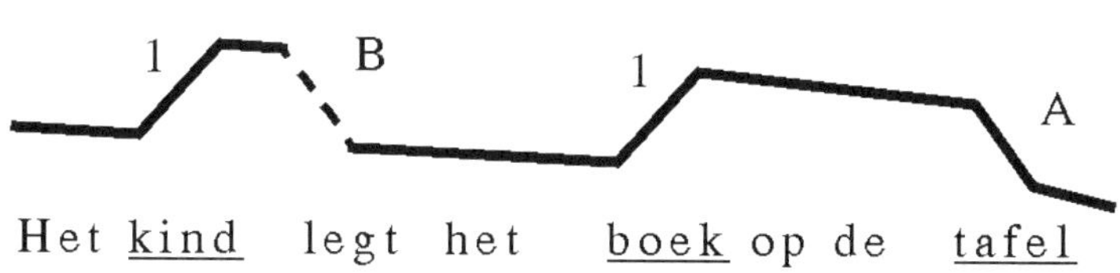

FIGURE 46.2. Stylized representation of the pitch contour calculated for the sentence *Het kind legt het boek op de tafel.* The syllables associated with accent-lending pitch movements are underlined.

The sampled data files corresponding to the words in the sentence are now retrieved from disk and concatenated. The abstract pitch contour calculated above is used to derive an array of F0 values (using rules given in [T93]) and the PSOLA module in the system then calculates the waveform that is played to the student.

This process is illustrated on the CD-ROM [Demo 3]. The demonstration contains the separate words as they are stored on disk, then the concatenated utterance before PSOLA manipulation, and finally the same utterance after implementation of the prosodic rules.

In the next release of Appeal, we intend to introduce more prosodic variation by implementing results from recent research concerning prosodic boundaries, reported in [dS94, SC94, S94].

46.3.5 Some Remarks

The Appeal system is very much in the early stages of development. As a consequence, formal evaluations of the various components of the system, including the speech output, have not yet been performed. Reactions of people who have been experimenting with the system in a more informal setting are generally positive, also concerning the speech quality. This is not to deny the necessity of formal testing, however.

Currently, the Appeal prototype is implemented on a Sparc X workstation, not the kind of machine that you can expect prospective Appeal users to have at home. At the moment, the system is being implemented on a PC. No fundamental obstacles are expected in this respect.

The system requirements of the speech component in Appeal are not very high. One module determines which speech files are to be concatenated; another module calculates the pitch contour and duration parameters; a third module performs waveform manipulations using the PSOLA technique; and finally, the speech is made audible, for which a sound card is obviously needed. The disk space required by a fully functional test program for the text-to-speech component of Appeal is currently only about 150 Kb.

The disk space needed for the speech data is variable, and depends on two factors: the size of the vocabulary to be made available and the sampling frequency with which the material is recorded and stored. A larger vocabulary makes for more variety in the exercises, but takes up more disk space. A higher sampling frequency (typically, 16 kHz) gives a clearly higher speech quality than a lower one (typically, 8 kHz), but is also more costly in terms of disk space. The current Appeal prototype has a vocabulary of 161 words. They are in SUN audio format (mono, 8 bit, mu-law encoded, sampled at 8 kHz) and take up about 700 Kb of disk space.

The recording of the words has proven to be of critical importance for the overall quality of the speech generated by the system. Unfortunately, this has turned out to be to some extent a laborious trial-and-error affair. The speech of some speakers stands up better to PSOLA manipulation than that of other speakers, but it is impossible to predict who the "better" speakers are. The same goes for different realizations of the same word by the same speaker. Nevertheless, some general guidelines can be given:

- The quality of the recording must be high; professional studio recordings are preferred.
- It is best to let the speaker pronounce the target words in carrier phrases rather than in isolation.
- At the cost of a larger database, the speech quality can be improved by recording both accented and unaccented versions of content words, and selecting the appropriate one during concatenation.
- Function words are especially critical: they are very frequent and should be recorded in appropriate contexts, doing justice to the fact that they are always strongly reduced, in both the spectral and the temporal domains. Here it especially pays to try out various instances of each.

46.4 Discussion and Conclusion

There is little doubt that speech will continue to gain importance as an output medium in all sorts of applications. In this contribution, we have tried to demonstrate that concatenation of prerecorded words followed by prosodic postprocessing is a feasible method of generating speech messages in at least one working practical application.

To make the speech sound coherent, prosodic rules are implemented using the PSOLA waveform manipulation technique. In this way, the quality of the concatenated speech is affected as little as possible. We feel safe in claiming that, if the recording of the words is carefully done, a better speech quality can be obtained through this technique than by using allophone or diphone synthesis.

There are, however, disadvantages to the approach. It has to be known exactly what words will be used, and preparing the words for the database is a rather laborious trial-and-error affair. Using diphones as the basic building block for concatenation rather than words is a serious alternative. With diphone synthesis, any message may be synthesized, and it is not necessary to go back to the recording studio if an application is extended to produce messages containing new words. The demonstrations of text-to-speech systems on the CD-ROM show that the segmental quality of top-of-the-line diphone systems can be quite high. This quality can be further improved if enough is known about the message to provide it with natural-sounding prosody.

The problem here is that not only should a good diphone database be available for the language in question (Dutch, in the case of Appeal), but the system designer should also have access to the concatenated diphone data in order to apply a dedicated prosody scheme. Most commercially available text-to-speech systems do not allow this kind of access.

Another alternative is working with prerecorded carrier phrases with slots into which prerecorded words are inserted. This can give the best results, if good care is taken to record the words and phrases with the right prosody, but is a great deal less flexible and a lot more trial-and-error in nature than word concatenation.

In this chapter, I have shown that an application builder in need of speech output has a number of alternatives, ranging from simple playback of prerecorded speech to full- scale text-to-speech synthesis. These methods each have their own advantages and drawbacks. Which technique is most appropriate depends on the requirements and limitations set by each individual application. For certain types of application, word concatenation in combination with prosodic postprocessing using PSOLA is a practical possibility.

References

[CM89] F. Charpentier and E. Moulines. Pitch-synchronous waveform processing techniques for text-to-speech synthesis using diphones. In *Proceedings Eurospeech* 89, Paris, France, 2, 13–19, 1989.

[Ham89] C. Hamon et al. A diphone synthesis system based on time-domain prosodic modifications of speech, *ICASSP* 89:238–241, 1989.

[tCC90] J. 't Hart, R. Collier, and A. Cohen. *A Perceptual Study of Intonation*. Cambridge University Press, Cambridge, 1990.

[dS94] J. R. de Pijper and A. A. Sanderman. On the perceptual strength of prosodic boundaries and its relation to suprasegmental cues. *J. Acoust. Soc. Amer.* 96(4):2037–2047, 1994.

[Mv94] J. F. M. Masthoff and R. R. G. van Hoe. A multi-agent approach to interactive learning environments. In *Proceedings of the Modeling Autonomous Agents in a Multi-Agent World (MAAMAW) Workshop*, Odense, 1994.

[San94] A. A. Sanderman. How can prosody segment the flow of (synthetic) speech? In: *Proceedings of the Second ESCA/IEEE Workshop on Speech Synthesis*, New Paltz, NY, 1994.

[SC94] A. A. Sanderman and R. Collier. Prosodic phrasing at the sentence level. In *Festschrift for K. Harris, Physics, Modern Acoustics and Signal Processing Series*. American Institute of Physics, 1994.

[T93] J. M. B. Terken. Synthesizing natural sounding intonation for Dutch: rules and perceptual evaluation. *Computer Speech and Language* 7:27–48, 1993.

[vMAHd94] R. R. G. van Hoe, J. F. M. Masthoff, E. Appelo, H. Harkema, J. R. de Pijper, and J. H. G. Rous. Appeal: A multimedia environment for adaptive learning. *IPO Annual Progress Report* 29, 1994.

Appendix: Audio Demos

This appendix contains the text of demonstrations of four representative text-to-speech synthesis systems. The order of presentation is alphabetic. To make it easier to compare the output of the different systems, all examples have been converted to the same SUN/NEXT format: mono, mu-law coded, with 8,000 Hz sampling frequency. This means that for some systems the general sound quality may in reality be better.

DecTalk

DecTalk is probably the best-known commercially available speech synthesis system. It is the only one of the systems demonstrated here that is allophone-based. It is descended from MITalk and Dennis Klatt's later work with Klattalk.

> [Demo 4 on the CD-ROM]
>
> *DecTalk PC is a new and improved version of the well- known DecTalk speech synthesizer. It represents the state of the art in text-to-speech synthesis. This demonstration will provide some examples of DecTalk's capabilities.*

Eurovocs

Eurovocs is a stand-alone text-to-speech synthesizer. The synthesizer was developed by Technologie & Revalidatie (T&R) and is now commercially available. The synthetic speech is diphone-based. Today, Eurovocs is available in Dutch, French, German, and American English, and other languages such as Spanish are expected to be released very soon.

> [Demo 5 on the CD-ROM]
>
> *Today, a new form of wireless services is generating a lot of excitement. Whereas cellular digital packet data, packet radio, and eventually digital cellular will play the dominant roles in long-distance wireless communication, a new set of services called personal communication services could change the face of short-distance wireless.*

Eurovocs has announced a new and improved version of their system for the first quarter of 1995 and provide the following demonstration:

> [Demo 6 on the CD-ROM]
>
> *Hi! Welcome to our demonstration of the American English TTS system. This program converts any written text into a phonetic representation. From this representation, a spoken version is then synthesized. Thus, the artificial voice which you have just been listening to. Thank you for your attention. Bye.*

Bell Laboratories

The multilingual text-to-speech system developed at Bell Laboratories is concatenative, using mostly diphones. The American English version is now available as a commercial product under the name TrueTalk.

[Demo 7 on the CD-ROM]

The history of speech synthesis research at Bell Laboratories goes back several decades, actually predating the digital computer. Dudley's "Voder," demonstrated at the 1939 World's Fair, was an early analog system that could be played by an expert and made to produce fairly intelligible speechlike sounds. A description of this device can be found in Dennis Klatt's 1987 review article. Early work on articulatory synthesis was done by Cecil Coker in the mid to late 60s. Our work on concatenative synthesis was started by Joseph Olive in the mid 70s. All of our text-to-speech systems are concatenative systems, though there is continuing interest in other approaches, including articulatory synthesis.

Naturally, most of our work over the past couple of decades has been aimed towards developing full text-to-speech systems for American English. Our American English systems can produce very high quality speech in both a male and a female voice. The system also has good text-analysis capabilities, as well as good word pronunciation and proper name pronunciation, prosodic phrasing, accenting, segmental durations, and intonation. We have been particularly active in developing statistical methods for handling these various problems.

Recently, we have begun expanding our set of languages to include Peninsular Spanish (in collaboration with Telefonica Research and Development), Mexican Spanish, French, Italian, German, Russian, Japanese, Mandarin, and Southern Min. We have also done a summer project on Navajo. As of Autumn 1994, all of these systems, except for Peninsular Spanish, are still under development at Bell Labs.

Orator

ORATOR is a commercially available text-to-speech synthesis system developed by Bell Communications Research (Bellcore). The synthesis is based on demisyllable concatenation.

The examples below compare the ORATOR synthesizer (Release 10.1) to the laboratory ORATOR II system. Bellcore claims that the ORATOR II system produces speech that is more natural sounding than the current ORATOR system.

[Demo 8 on the CD-ROM]

A cat lives nine lives.
Close the book after a close look.
Be content with the content of the course.

[Demo 9 on the CD-ROM]

Hello. My name is orator 2. I am a speech synthesizer. I convert English text to speech. Address e-mail to: orator@bellcore.com

Infovox

[Demo 10 on the CD-ROM]

Text-to-speech conversion is performed in two stages. First, text input is converted into a phonetic transcription, using a combination of lexical look-up and general pronunciation rules. Special rules care for the pronunciation of digits, dates, measures, and so on. Phonetically transcribed text can also be entered directly by the user. Secondly, phonetic transcription is converted into speech sounds using a model of human speech production.

Index